VOLUME FIFTY ONE

ADVANCES IN
INSECT PHYSIOLOGY

VOLUME FIFTY ONE

ADVANCES IN INSECT PHYSIOLOGY

Progress in Mosquito Research

Edited by

ALEXANDER S. RAIKHEL
Department of Entomology and Institute of Integrative Genome Biology, University of California Riverside, Riverside, CA, United States

ELSEVIER

AMSTERDAM • BOSTON • HEIDELBERG • LONDON
NEW YORK • OXFORD • PARIS • SAN DIEGO
SAN FRANCISCO • SINGAPORE • SYDNEY • TOKYO
Academic Press is an imprint of Elsevier

Academic Press is an imprint of Elsevier
125 London Wall, London, EC2Y 5AS, United Kingdom
The Boulevard, Langford Lane, Kidlington, Oxford OX5 1GB, United Kingdom
525 B Street, Suite 1800, San Diego, CA 92101-4495, United States
50 Hampshire Street, 5th Floor, Cambridge, MA 02139, United States

First edition 2016

ISBN: 978-0-12-802457-7
ISSN: 0065-2806

Publisher: Zoe Kruze
Acquisition Editor: Kirsten Shankland
Editorial Project Manager: Hannah Colford
Production Project Manager: Vignesh Tamil
Cover Designer: Miles Hitchen

Typeset by SPi Global, India

CONTENTS

CONTRIBUTORS

Z.N. Adelman
Fralin Life Science Institute, Virginia Tech, Blacksburg, VA, United States

P.A. Armbruster
Georgetown University, Washington, DC, United States

S. Asgari
Australian Infectious Disease Research Centre, School of Biological Sciences, The University of Queensland, Brisbane, QLD, Australia

J.K. Biedler
Department of Biochemistry, The Fralin Life Science Institute, Virginia Tech, Blacksburg, VA, United States

M.R. Brown
University of Georgia, Athens, GA, United States

F. Catteruccia
Harvard T.H. Chan School of Public Health, Boston, MA, United States

G.K. Christophides
Imperial College London, London, United Kingdom

D.L. Denlinger
The Ohio State University, Columbus, OH, United States

G. Dimopoulos
Bloomberg School of Public Health, Johns Hopkins University, Baltimore, MD, United States

K. Etebari
Australian Infectious Disease Research Centre, School of Biological Sciences, The University of Queensland, Brisbane, QLD, Australia

M. Hussain
Australian Infectious Disease Research Centre, School of Biological Sciences, The University of Queensland, Brisbane, QLD, Australia

L. Johnson
Graduate Program in Molecular, Cell and Developmental Biology, University of California Riverside, Riverside, CA, United States

C. Montell
Neuroscience Research Institute, University of California, Santa Barbara, CA, United States

F.G. Noriega
Florida International University, Miami, FL, United States

M.A. Osta
American University of Beirut, Beirut, Lebanon

J. Overcash
Fralin Life Science Institute, Virginia Tech, Blacksburg, VA, United States

N. Petchampai
Vector-Borne Infectious Disease Research Center, School of Public Health and Tropical Medicine, Tulane University, New Orleans, LA, United States

P.M. Piermarini
Ohio Agricultural Research and Development Center, The Ohio State University, Wooster, OH, United States

M. Povelones
School of Veterinary Medicine, University of Pennsylvania, Philadelphia, PA, United States

A.S. Raikhel
Department of Entomology and Institute of Integrative Genome Biology, University of California Riverside, Riverside, CA, United States

S. Roy
Department of Entomology and Institute of Integrative Genome Biology, University of California Riverside, Riverside, CA, United States

T.T. Saha
Department of Entomology and Institute of Integrative Genome Biology, University of California Riverside, Riverside, CA, United States

P.Y. Scaraffia
Vector-Borne Infectious Disease Research Center, School of Public Health and Tropical Medicine, Tulane University, New Orleans, LA, United States

V. Smykal
Department of Entomology and Institute of Integrative Genome Biology, University of California Riverside, Riverside, CA, United States

A. South
Harvard T.H. Chan School of Public Health, Boston, MA, United States

M.R. Strand
University of Georgia, Athens, GA, United States

Z. Tu
Department of Biochemistry, The Fralin Life Science Institute, Virginia Tech, Blacksburg, VA, United States

S. van Tol
Bloomberg School of Public Health, Johns Hopkins University, Baltimore, MD, United States

K.J. Vogel
University of Georgia, Athens, GA, United States

J. Zhu
Virginia Tech, Blacksburg, VA, United States

Z. Zou
State Key Laboratory of Integrated Management of Pest Insects and Rodents, Institute of Zoology, Chinese Academy of Sciences, Beijing, China

L.J. Zwiebel
Vanderbilt University; Vanderbilt Brain Institute, Program in Developmental Biology, and Institutes of Chemical Biology and Global Health, Vanderbilt University Medical Center, Nashville, TN, United States

PREFACE

The medical significance of mosquitoes as the most dangerous vectors of numerous devastating human diseases cannot be overstated. Diseases transmitted by mosquitoes cause over a million deaths and hundreds of millions of new infections every year. Malaria incidences have modestly declined in recent years, mainly because of the use of insecticide-treated bed nets. In contrast, Dengue fever and Chikungunya are on the rise due to an unprecedented spread of their vectors, *Aedes aegypti* and *Aedes albopictus*, respectively. The recent stunningly rapid invasion of Zika virus, which causes microcephaly and is transmitted by *Ae. aegypti*, is the latest chapter in human suffering from mosquito-borne diseases. Current mosquito-control approaches are insufficient in the face of the numerous negative factors aiding the spread of such diseases; these factors include growing insecticide and drug resistance, uncontrolled urbanization with poor sanitation, and climate change. Unavailability of vaccines for malaria and viral vector-borne diseases is worsening this already tragic situation. A deeper understanding of mosquito biology could provide a foundation for developing novel mosquito-control methodologies.

Recent years have witnessed impressive progress in the field of mosquito biology, heralded by genome sequencing of several mosquito species and adaptation of genetic manipulation technologies, such as CRISPR/Cas9 for mosquitoes. Simultaneously, major advances have been accomplished in investigating molecular aspects of the physiological processes specific to mosquitoes as blood-feeding insects. This volume "Progress in Mosquito Research" is aimed at highlighting the recent achievements in elucidating mosquito adaptations that have contributed to the astonishing evolutionary success of these insects.

Chapter 1 explores in great detail recent advances in the application of novel tools of genetic modifications, signified by introduction of CRISPR/Cas9, that permit us to become immersed in mosquito molecular biology and physiology at an unprecedented level, comparable to that done in model organisms such as *Drosophila melanogaster*. These tools are invaluable for devising innovative strategies for mosquito control.

Chapter 2 examines the significance of molecular characterization of the sex-determination pathway in mosquitoes and the importance of the Male factor. With the advent of genetic modification tools, knowledge of this pathway should permit manipulations of sex ratio and a reduction in blood-sucking female mosquito populations. Chapter 3 reviews novel data regarding the evolution of mating systems in mosquitoes. It discusses unique adaptations of Anopheles mosquitoes, including the transfer of a steroid hormone, 20-hydroxyecdysone (20E), as a male input in the regulation of female reproduction.

The sesquiterpenoid juvenile hormone (JH) has long been recognized as an essential factor controlling maturation of female mosquitoes for subsequent blood feeding and a response to 20E. Characterization of JH biosynthesis in mosquitoes is dealt with in Chapter 4, which further reviews recent discoveries regarding the action of JH at the molecular level involving intracellular and membrane receptors. In female vector mosquitoes, major reproductive events are intimately linked to the intake of blood. Regulatory pathways that are affected by blood feeding and control egg maturation in female mosquitoes, which include 20E, amino acid/target of rapamycin and insulin-like peptides, are discussed in Chapter 5. This chapter also reviews genomic studies of temporal coordination of differential gene expression in mosquitoes. A plethora of peptide hormones is attuned to diverse physiological processes. Availability of genome sequences for several mosquito species has provided a remarkable opportunity for identification of peptide hormones and their putative receptors in these insects. Chapter 6 summarizes the state of current knowledge of these molecules and their functions in mosquitoes.

Small non-coding RNAs play essential roles in regulating gene expression, anti-pathogen defence and genome stability. In Chapter 7, a detailed account of the latest discoveries of small RNA functions in mosquitoes is presented.

Immune defence is essential for species survival. Mosquitoes possess an efficient immune system, providing resistance against microorganisms and modulating infection levels by human pathogens such as *Plasmodium* parasites that cause malaria. Chapter 8 describes new discoveries concerning immune mechanisms affecting *Plasmodium* infection and parasite evasion tactics that permit the persistence of the parasite in the mosquito host. The importance of microbiota for various physiological functions has been recognized in recent years. Chapter 9 examines current progress in our understanding of mosquito microbiota with respect to its contribution to basal immunity and its effect on vector competence for human pathogens. This chapter further discusses potential exploration of the mosquito gut microbiome's functions for devising disease-control methodologies based on blocking pathogen infection in the mosquito and shortening the mosquito's life span.

Mosquitoes have evolved sophisticated sensory systems, permitting them to efficiently find their vertebrate hosts. The structure and functions of these systems as well as the cellular and molecular basis underlying multiple sensory pathways are reviewed in great detail in Chapter 10.

Survival of adverse environmental conditions is an essential adaptation for any species. A dormant state, the diapause, permits many mosquitoes to endure cold winters in temperate zones or tropical dry seasons. The considerable progress in the understanding of physiological and molecular processes underlying mosquito diapause is discussed in Chapter 11.

Physiological adaptations in nitrogen metabolism are essential for mosquitoes to deal with a massive blood meal. Chapter 12 provides fascinating insight into our understanding of complex pathways underlying nitrogen metabolism in these insects.

During their life cycle, mosquitoes shift from aquatic to terrestrial environments, which pose various osmoregulatory challenges. Moreover, hematophagous females face the additional task of rapid solute excretion after utilization of a huge blood meal. Chapter 13 highlights recent accomplishments in the comprehension of physiology and neuroendocrine regulation of the mosquito renal excretory system. It also discusses possible vector control approaches based on targeting the renal system.

Overall, this volume, which is a collective effort of the leaders in various aspects of mosquito biology, should be regarded as an extensive repository of existing knowledge and a significant contribution to this field. The focus audience of this work are scientists and students working in and studying mosquito biology. I also hope that this volume will create interest among researchers and students, planning on transitioning from other fields, to enter this important area with new ideas that would help mosquito research to expand and flourish even further. I would like to thank the authors for taking time from their busy schedules and contributing to this volume, making it an excellent resource of current knowledge of mosquito molecular biology and physiology.

Alexander S. Raikhel
Department of Entomology and
Institute of Integrative Genome Biology,
University of California Riverside, Riverside,
CA, United States

CHAPTER ONE

Progress in Gene Editing Transgenesis Genome Manipulation in Mosquitoes

J. Overcash, Z.N. Adelman
Fralin Life Science Institute, Virginia Tech, Blacksburg, VA, United States

Contents

Abstract

Recent advances in site-specific gene editing technologies, split genetic systems, recombinases and transposon remobilization have been applied to disease vector mosquitoes with great success. These in turn now enable a suite of hypothesis-driven questions concerning the physiology and behaviour of important mosquito species, guided by previous advances in genome sequencing and transcriptomics. In this chapter, we will review progress in CRISPR/Cas9-based gene disruption, including target selection, mutation identification, tracking homology-dependent gene insertion, CRISPR-mediated gene drive mechanisms and CRISPR/Cas9 variants. We will also review progress in using split genetic systems such as Gal4/UAS, recombinase-mediated cassette exchange using ΦC31, as well as *piggyBac*-mediated forward genetic screens using transgene remobilization.

Advances in Insect Physiology, Volume 51
ISSN 0065-2806
http://dx.doi.org/10.1016/bs.aiip.2016.05.001

ABBREVIATIONS

C-NHEJ classical nonhomologous end joining
CRISPR clustered regularly interspaced palindromic repeats
DSB double-stranded break
GD gene drive
HDR homology-directed repair
RMCE recombinase-mediated cassette exchange
TE transposable element

1. PROGRESS IN MOSQUITO GENE EDITING

Mosquitoes act as vectors of some of the world's most burdensome disease-causing pathogens including malaria and dengue (The Centers for Disease Control and Prevention, 2014; World Health Organization, 2014a,b). Advancements in genomics have given molecular biologists and geneticists a wealth of information from which to further understand the basic biology of these vectors with the hope of generating novel solutions to reduce or beneficially modify vector populations. While the technology to generate transgenic mosquitoes has been established for almost two decades, recent advancements stand to completely revolutionize the specificity and efficacy with respect to gene editing, transgenesis and genome manipulation. The most recent, and perhaps the most exciting of these advancements come as part of the CRISPR revolution.

1.1 Clustered Regularly Interspaced Palindromic Repeats/Cas9

Clustered regularly interspaced palindromic repeats (CRISPR) in conjunction with CRISPR associated (Cas) proteins are found in a wide range of bacteria and play a role in antiviral immunity (Marraffini and Sontheimer, 2010). CRISPR sequences are variable in nature, 30–40 nucleotides in length and are interspersed by direct repeats. Transcription of CRISPR loci followed by processing generates short RNA sequences which then form a complex with the Cas protein(s) to target viral DNA previously integrated into the bacterial genome (Jinek et al., 2012). A key requirement for the Cas nuclease to generate double-stranded breaks (DSBs) at its target site is the protospacer adjacent motif (PAM). While present on the viral genome, the PAM is absent in the corresponding bacterial sequence, allowing for specific cutting of foreign viral DNA. The most common variant of the Cas proteins used to date is Cas9 (derived from *Streptococcus pyogenes*), which has a PAM site of NGG (Cong et al., 2013). While there are several classes of Cas proteins, the Cas9 variant is particularly useful because it only requires a single protein to function. Furthermore, it has been shown that the Cas9 PAM site can be modified to accept alternative nucleotide sequences (Kleinstiver et al., 2015). While initial gene editing efforts utilized two small RNA molecules (crRNA and tracrRNA), as in bacteria, they are now virtually always fused together for synthetic use (synthetic guide RNA, sgRNA). The CRISPR system has been adapted for use in a range of insects including *Tribolium castaneum* (Gilles et al., 2015), *Gryllus bimaculatus* (Awata et al., 2015), *Bombyx mori* (Wei et al., 2014; Zeng et al., 2016), *Danaus plexippus* (Markert et al., 2016), *Drosophila suzukii* (Li and Scott, 2016) and *Drosophila melanogaster* (Bassett et al., 2013), and of particular interest for the purposes of this review, the mosquitoes *Aedes aegypti* (Basu et al., 2015), *Anopheles stephensi* (Gantz et al., 2015) and *Anopheles gambiae* (Hammond et al., 2016). The CRISPR system utilizes two key components, the sgRNA and the Cas9 protein, to search out a specific sequence within the genome of interest and generate a blunt end DSB (Fig. 1).

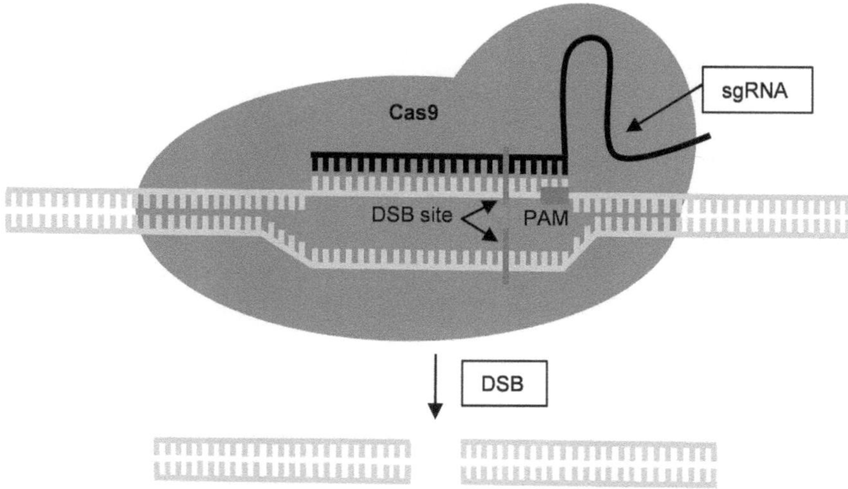

Fig. 1 Model of sgRNA-directed Cas9 binding to DNA target. DSB (*red vertical lines*) are 3 bp upstream of the PAM site. The Cas9 protein cuts both strands of the targeted DNA generating a blunt ended double-stranded break, which will be repaired by one of several DSB repair pathways. (See the colour plate.)

Following DNA cleavage, DSBs are repaired by one of several endogenous cellular repair pathways including the classical nonhomologous end-joining (C-NHEJ) pathway and the homology-directed repair (HDR) pathway. Any resulting mutation, or lack thereof, is dependent upon which pathway is utilized. Mistakes made during C-NHEJ repair may result in small insertions or deletions (Williams et al., 2014). These indels may generate frameshift mutations within the open-reading frame of a gene of interest, effectively generating knockout mutants or truncated proteins (desired location within the ORF). It is also plausible that portions of a gene may be removed by utilizing several guide RNAs in tandem. In contrast, if the DSB is repaired utilizing the HDR pathway then perfect repair of the break may be accomplished using the homologous chromosome or an exogenously introduced template (Lamarche et al., 2010). Providing a templated sequence for use in HDR may allow the experimenter to remove undesired alleles from a population or insert foreign DNA in the form a transgenic construct (Fig. 2). In addition to its power as a highly specific endonuclease, alternative uses and varieties of the CRISPR system will be discussed later in the chapter.

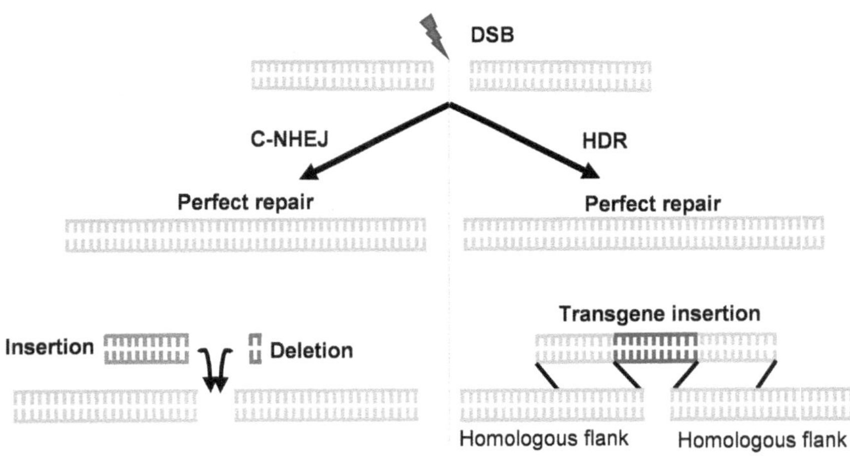

Fig. 2 DNA repair following DSB induction is a competitive process. Two primary repair pathways are the classical nonhomologous end-joining pathway (*C-NHEJ, left*) and the homology-directed repair pathway (*HDR, right*). The C-NHEJ pathway does not require a homologous template to repair a double-stranded break, but may result in the generation of indels (insertions or deletions). The homology-directed repair pathway requires a homologous template to complete the repair of a double-stranded break. If a synthetic template with homologous flanks is present, the HDR pathway can be used to insert transgenes into a highly specific location.

1.2 Progress in CRISPR in *Aedes*

The ability to utilize CRISPR was first demonstrated by Kistler et al. (2015) in *Ae. aegypti* with the goal of investigating the genetic basis of innate chemosensory behaviours (Kistler et al., 2015). Kistler et al. (2015) provide a step-by-step bench manual for the utilization of CRISPR/Cas9 in *Ae. aegypti*, as well as providing data on the optimization of injection mixes (suggesting that 400 ng/μL protein displays the highest cutting efficiency), and testing the efficiency of C-NHEJ- and HDR-based repair mechanisms with respect to six different genes. These researchers noted a correlation between Cas9 concentration and embryo survival, with lower injection concentrations (333 ng/μL) yielding survival rates between 46% and 63% and higher injection concentrations (500 ng/μL) yielding an average survival of 18.6%. Other reports (Basu et al., 2015; Dong et al., 2015) indicate typical survival rates lower than 30%, again using higher concentrations of Cas9 (summarized in Table 1).

After optimization of Cas9 concentration, Kistler et al. (2015) injected embryos with sgRNAs and single-stranded oligodeoxynucleotides (ssODN) 200 bp in length to quantify mutation rates attributed to indel creation by the C-NHEJ pathway and insertion of the ssODN via the HDR pathway. In G_0 survivors, an indel generation rate of 24% and an ssODN insertion rate of 0.71% were observed, suggesting a dominating preference for C-NHEJ repair. In G_1 progeny, an indel inheritance of 18.9% and ssODN inheritance of 0.6% were also observed, suggesting that somatic mutation rates may be a viable way to guide future germline experiments. Basu et al. (2015) also successfully utilized CRISPR/Cas9 in *Ae. aegypti* targeting a range of genes associated with DNA repair, RNAi and sex determination. Basu et al. (2015) observed mutation rates as high as 90% following prevalidation of sgRNAs in embryo-based assays (Basu et al., 2015). This may have been due to a higher concentration of Cas9 (600 vs 400 ng/μL), the larger number of sgRNAs screened in a prevalidation process, or might simply reflect variation between independent genomic target sites. Hall et al. (2015) further investigated the targeting of the sex determining gene *nix* which is believed to play a role as the master regulator of *doublesex* and *fruitless* splicing, and hence sexual dimorphism in *Ae. aegypti* (Hall et al., 2015). Using the same injection specifications as Basu et al. (2015), two-thirds of transgenic male survivors that had been injected with CRISPR/Cas9 as embryos showed some level of feminization and deformities in sexually dimorphic organs, indicating that useful gene disruptions can be obtained in a single generation (albeit from a haploid target in this case). Lastly, Dong et al. (2015) also

Table 1 Mosquito Gene Editing Experiments Using CRISPR/Cas9

References	Cas9 Source	Mosquito	Cas9 (ng/μL)	sgRNA (ng/μL)	NHEJ Mutation Rate (%)	DNA Repair Manipulation	HDR Insertion Size (kb)	Flanks (kb)	Insertion Rate[a] (%)
Kistler et al. (2015)	Protein	*Ae. aegypti*	300	40	24	N/A	2.15–2.32	0.8/1.5	0.2–8.7
Basu et al. (2015)	mRNA	*Ae. aegypti*	600	100	24–90	Ku70 dsRNA	2.4	2.2/1.7, 1/1.7	1.7–2.7
Dong et al. (2015)	mRNA	*Ae. aegypti*	1000	50	5.5	N/A	N/A	N/A	N/A
Hall et al. (2015)	mRNA	*Ae. aegypti*	600	100	~70	N/A	N/A	N/A	N/A
Gantz et al. (2015)	mRNA	*An. stephensi*	100	100	N/A	Ku70 dsRNA	~17	~1	1.6
Hammond et al. (2016)	Plasmid	*An. gambiae*	300	300 (on same plasmid)	N/A	N/A	~1	2/2	11.1–18.8

[a]Insertion rates were estimated as a percentage of founders of G_0 survivors. When G_0 survivors were pooled 50% fertility was assumed.

validated CRISPR/Cas9 in *Ae. aegypti* by targeting the enhanced cyan fluorescent protein gene which had been previously introduced as a transgene (Dong et al., 2015). In contrast to previous studies, Dong et al. (2015) only observed a 5.5% knockout efficiency of the targeted fluorescent gene. While this may have been due to the low number of sgRNAs utilized and screened, the vast variation in CRISPR/Cas9 mosquito mutagenesis within a single species suggests that target selection, gRNA design and optimization can be further improved.

1.3 Progress in CRISPR Editing in *Anopheles*: Gene Drive

In contrast to the basic research described for *Ae. aegypti*, in *An. stephensi* and *An. gambiae*, CRISPR has been utilized to generate an active gene drive (GD) system as part of applied strategies to combat malaria (Gantz et al., 2015; Hammond et al., 2016). These papers quickly followed the initial utilization of a CRISPR GD first reported in *D. melanogaster*, where CRISPR/Cas9 invaded the *yellow* gene at hyper-Mendelian rates (Gantz and Bier, 2015). In each case, the CRISPR/Cas9 GD was constructed encoding the gene for Cas9 driven by a gonad specific promoter and the sgRNAs utilized for targeting the homologous sequence driven by a pol III promoter into the target site of the sgRNA. In this fashion, the Cas9–sgRNA complex targets the homologous chromosome; repair via the HDR pathway ensures the GD construct is copied to the homologous chromosome, essentially doubling its copy number in each generation. This technique can be used to interrupt genes of interest, or by including transgenes into the GD construct, drive a transgene into a population at hyper-Mendelian rates. For further information on CRISPR GDs, the reader is directed to reviews by Champer et al. (2016), Esvelt et al. (2014) and Gantz and Bier (2016).

Building on previous work in *An. stephensi*, Gantz et al. (2015) utilized CRISPR Cas9 along with a DNA template to introduce via the HDR pathway a ~17-kb multigene unit (Gantz et al., 2015). Once integrated, the GD cassette was able to copy itself to its homologous chromosome with greater than 98% efficiency in initial generations. However, G_4 males and females derived from transgenic G_2 females showed inheritance rates more in line with expected Mendelian inheritance rates (1.33:1). In contrast, the offspring from G_2 males retained hyper-Mendelian rates of inheritance (96.9–98.5%). Gantz et al. (2015) suggested that this may have to do with C-NHEJ-based repair occurring during embryogenesis prior to the availability of the homologous chromosome, and recommends further restricting the expression of Cas9 from the GD construct to the parental germline. These findings provide insight into the use of GDs in mosquitoes suggesting that the misexpression of Cas9 expression can undermine the efficiency of GD constructs, and reinforces the necessity to survey promoters capable of controlling Cas9 to maximize HDR within a GD system. Similarly, Hammond et al. (2016) utilized CRISPR and TALENS (transcription activator-like effector nucleases) to insert docking constructs into three haplosufficient female sterility genes (Hammond et al., 2016). In contrast to insertion of the multigene GD into a known phenotypic gene, Hammond et al. (2016) used a flexible two-step process to build their multigene unit. In the first step, they generated EGFP-marked HDR constructs loaded with φC31 *att*B sites (This docking system is discussed in more detail later in this chapter). The EGFP-marked HDR constructs were then inserted into their respective target genes and the impact on female fertility assessed. Once reduced fertility was confirmed (suggesting the target site was useful), the φC31 docking mechanism was used to insert the GD portion (Cas9 and sgRNA expression cassettes). Of the three genes targeted, one provided to be a viable candidate for a sustainable release program (AGAP007280), with the average homing rate per generation being 98% for transgenic males and 98% transgenic females. One very interesting finding that Hammond et al. observed was a distinct drop in fertility after the GD construct was exchanged into the targeted genes; they suspect that this is due to somatic expression of Cas9. Both Gantz et al. (2015) and Hammond et al. (2016) prove that CRISPR-based GDs can function efficiently in mosquitoes. Each study raises questions which require further investigation, specifically which promoters will provide the highest GD efficiency and how does the presence of Cas9 within a transgenic line effect the overall fitness of a given transgenic line.

While there have been no reports of a successful GD in *Ae. aegypti*, the construct(s) in principle should be the same with respect to its utilization in *An. stephensi* and *An. gambiae*. Perhaps one of the more interesting suggestions for its use is to generate a system to drive maleness through a population utilizing male determining factors such as the *Nix* gene (Adelman and Tu, 2016; Hall et al., 2015). This sex changing GD system could lead to rapid population and collapse.

1.4 Future Prospects of Mosquito GD Using CRISPR

The validation of highly efficient CRISPR-based GD constructs, while extremely promising for novel malaria and dengue control strategies, poses special problems with how such transgenic mosquito strains can be safely evaluated in a field or semifield environment where the release (intentional or otherwise) may result in the permanent fixation of the transgene in nature. Thus, there has been substantial theorizing as to how to further refine or redesign GD constructs to make them more controllable, or to inactivate them from wild populations. Two recently proposed concepts are those of the ERASR (elements for reversing the autocatalytic chain reaction) and the CHASR (construct hitchhiking on the autocatalytic chain reaction), described by Gantz and Bier (2016). ERASRs are constructs which can be introduced into a transgenic line harbouring a GD mechanism in an attempt to effectively "erase" its GD capabilities. By introducing an sgRNA expression cassette targeting the Cas9 open-reading frame itself at the same locus as the GD construct, Cas9 is predicted to disrupt its own ORF and allow for the ERASR sequence to be copied to the chromosome which the CRISPR GD previously inhabited. In this fashion, the GD system may drive itself out of a population through the broken Cas9, though any associated transgenic sequences would remain. In contrast, CHASRs are proposed to piggyback on the presence of an already present CRISPR GD system to drive themselves into a population, while also possibly targeting other genes. While both ERASRs and CHASRs need to be validated experimentally, a similar construct has already been validated in *Drosophila* (Wu et al., 2016). Termed the CATCHA construct, it works by inserting a guide RNA targeting the Cas9 protein flanked by Cas9 homologous sequences on either side. If Cas9 is present then it will target itself and drive the CATCHA construct into the population. Using a system involving the *ebony* gene, Wu et al. (2016) showed that the CATCHA construct was 93.4% efficient at disrupting Cas9 activity and had a conversion rate of 85%. The ERASR, CHASR and CATCHA constructs provide just a few of what promises to be many creative ways to utilize GD systems in the future. Another factor that may play a critical role in the future of CRISPR GDs is alternative endonucleases. All studies to date have used Cas9 endonuclease, but one issue that may arise is the proximity of the induced DSB site to the critical PAM site. As NHEJ competes with HDR, small indels at the DSB site may destroy the PAM and prevent the possibility of the endonuclease to generate another DSB, generating a drive-resistant allele. To circumvent this issue, one might use an endonuclease which induces DSB formation further downstream of the PAM site (DSB induction at 18 and 23 bp from PAM), such as Cpf1 (Zetsche et al., 2015). Cpf1 differs from Cas9 in that it lacks the requirement for a transactivating RNA, it has an alternative PAM site

(TTN vs NGG), and the cut site is staggered and further upstream of the PAM site; traits that may help to facilitate HDR (Zetsche et al., 2015). A recent assessment of Cpf1 in *Drosophila* suggested that it is less efficient than Cas9, with phenotypic mutant offspring only occurring at rates between 4% and 6% (Port and Bullock, 2016), suggesting further improvements may be necessary before the technology can be applied to mosquito GD strategies. Other possibilities could include but are not limited to systems which transiently modify DNA repair pathways to boost HDR and limit error prone fixes and additional endonucleases which may prove to be better candidates for GD (Shmakov et al., 2015). Current studies, as well as hypothetical constructs and untested endonucleases, suggest that there is much to be tested before we can fully maximize the potential of CRISPR GD systems, yet they appear to have a bright future in the advancement of basic and applied mosquito research.

1.5 Progress in CRISPR Expression sgRNAs and Multiplexing

The wide range of efficiencies seen by the CRISPR/Cas9 in mosquitoes leaves room for much to be learned. One interesting question is what source of Cas9 should be used to deliver the highest DSB efficiency while keeping costs down. Providing Cas9 to embryos via a DNA plasmid would provide cheap stable constructs for immediate use in injections. However, Kistler et al. (2015), Basu et al. (2015) and Dong et al. (2015) all tested the use of Cas9 expression plasmids using strong constitutive promoters but in all cases failed to observe mutations. In contrast, Hammond et al. (2016) was able to utilize a plasmid construct to generate DSBs utilizing the gonad specific promoter Vasa2. Another question that should be addressed are the advantages and disadvantages associated with introducing multiple guide RNAs targeting a single gene (multiplexing). Utilizing several sgRNAs targeting a single gene can lead to large deletions (Ren et al., 2013), this was confirmed in mosquitoes by Kistler et al. (2015). Large deletions may or may not be advantageous depending upon the targeted gene. It may also be that the ability to multiplex increases efficiency so much that G_0 survivors display sufficient biallelic disruptions become viable experimental subjects. Port et al. (2016) showed that the utilization of four sgRNAs targeting the sepia eye gene in *Drosophila* lead to 100% removal of sepia in almost all cases (Port and Bullock, 2016). While undoubtedly not all sgRNAs can achieve this efficiency, if such high efficiency is seen with most multiplexing attempts, the screening process is likely to be much less laborious in the future. Not only may multiplexing lead to high-efficiency single target mutant generation but may also be used to target several genes simultaneously (Cong et al., 2013).

1.6 Progress in CRISPR Target Selection

CRISPR/Cas9 target selection is a multidimensional process requiring the experimenter to appropriately target the gene of interest (keeping in mind functional domains, alternative start codons/splice variants), while also considering the possibility of off-target effects. An initial consideration to take into account is the so-called seed sequence found directly upstream of the PAM site (between 8 and 12 bp) (Doudna and Charpentier, 2014). The seed sequence typically shows reduced tolerance for mismatches with respect to the target of interest; others factors which may be worth including are the GC content of the particular sgRNA and the underlying chromatin state (Wu et al., 2014b) of the targeted region, as there is a high correlation between the GC content of the first five bases upstream of the PAM site and mutagenesis efficiency (Ren et al., 2014). Kistler et al. (2015) designed their sgRNAs manually by searching sequences of interest and locating PAM sites. While this group did not empirically examine regions for off-target effects, they utilized two web tools to avoid sgRNAs with predicted off-target binding sites: http://zifit.partners.org/ZiFiT and http://crispr.mit.edu. Both Dong et al. (2015) and Hammond et al. (2016) also utilized ZiFiT to predict and avoid sgRNAs with strong off-target potential. Similarly, Basu et al. (2015) manually designed sgRNAs and searched for off-target effects by utilizing the software flyCRISPR (Gratz et al., 2014). In this latter study, the top four predicted off targets for each gene were screened for mutations via high-resolution melt curve analysis (HRMA); no evidence of off-target effects was observed. Basu et al. (2015) also showed that in *Ae. aegypti* highly active sgRNAs did not conform to previously described optimal design criteria of avoiding high and low GC content and targeting the transcribed vs nontranscribed strand (Wang et al., 2014). These differences could be due to low numbers (dozens, not thousands) of sgRNAs screened in mosquitoes compared to those screened in mammalian cell lines, or to elements of sgRNA design not yet fully understood. As further studies continue to emerge on how sgRNA design (as well as other factors) impacts efficiency, a more extensive set of rules guiding sgRNA optimization is likely to form. Similarly, Dong et al. (2015) selected their sgRNAs using ZiFiT, as opposed to flyCRISPR, and searched selected sgRNAs against the *Ae. aegypti* genome to minimize off-target effects.

At a certain level, the importance of predicting and avoiding all off-target effects may very well depend upon the goals of the experimenter. For basic research investigating gene function, off-target effects on experimental conclusions may be minimized with standard genetic practices such as outcrossing for multiple generations and restricting analysis to transheterozygotes derived from independent mutation events/sgRNAs. However, for GD constructs built for field application, stability and off-target analysis may be crucial to the success or failure of the technology. Hammond et al. (2016) used both ZiFiT and Chop Chop (https://chopchop.rc.fas.harvard.edu) to search for predicted off-target sites with respect to their genes of interest. An alternative to software suites which search for predicted off-target effects to examine individually, techniques like GUIDE-seq and Digenome-seq allow for unbiased assessment of DSBs occurring within the genome following CRISPR treatment (Kim et al., 2015; Tsai et al., 2015). GUIDE-seq utilizes double-stranded oligodeoxynucleotide tags to "capture" a DSB; genome fragmentation and amplification with a tag-specific primer followed by high-throughput sequencing can reveal an extensive catalogue of off-target sites that is distinct from computational prediction. Digenome-seq utilizes a comparison of sequence reads after digestion of the genome. If no mutation is present, a clear delineation between 5 and 3′ reads is apparent at digestion locations, if sequences overlap the digestion location an indel is present. Neither of these techniques has been utilized in mosquitoes and may provide unbiased insight into the impact off-target effects may generate. For further review of CRISPR prediction software suites, the reader is directed to several thorough reviews (Brazelton et al., 2015; Peng et al., 2016; Wu et al., 2014a).

In conjunction with sgRNA design and off-target prediction, significant advancements have been made with respect to the fidelity of the Cas9 protein itself. Recently, a highly specific version of the Cas9 protein has been developed by altering four amino acids which directly interact with the target DNA's phosphate backbone (Kleinstiver et al., 2016) which in combination with improved sgRNA selection may remove any off-target activity. This high-fidelity Cas9 variant was tested against its wild-type Cas9 variant using the nonbiased GUIDE-seq to test eight sgRNAs for genome-wide off-target effects. While the wild-type Cas9 variant induced multiple off-target cuts in seven of the eight sgRNAs tested, the high-fidelity Cas9 variant only induced a single off-target effect in one of the eight sgRNA tested. This high-fidelity Cas9 may be extremely beneficial for basic and applied researches in mosquitoes. The precision it provides may allow for stable

Cas9 driver lines to be generated and boost the confidence that can be placed on knockout experiments. Similarly, it may reduce any fitness costs associated with the introduction of Cas9 in GD systems while ensuring that any inserted sgRNAs are only able to cleave the desired targeted sequence.

1.7 Progress in Mutation Identification

Following introduction of the CRISPR reagents, the desired mutation must be identified and selected for. In the absence of a physical marker, common methods for doing this including PCR amplification of the target region followed by Surveyor or T7 endonuclease treatment (Reyon et al., 2012), Sanger sequencing (Brinkman et al., 2014), HRMA (Dahlem et al., 2012) and deep sequencing (Gagnon et al., 2014). Kistler et al. (2015) suggest deep sequencing as the most cost effective for the high-resolution data which is provided. To identify if CRISPR/Cas9 was indeed generating indels in each target gene, these researchers sequenced amplicons covering the site of interest derived from G_0 pupae or adults; barcoding amplicons allowed for substantial multiplexing and cost savings. Once the deep sequencing reads were aligned to the reference sequence, deletions, insertions (including insertions of ssODN) and inversions could be readily identified. However, not every laboratory will be equipped to generate and analyse large volumes of sequence data, and this method does not lend itself to situations where genotype information is required on shorter time-scales. Basu et al. (2015) utilized HRMA to identify mutations in *Ae. aegypti*. This technique uses intercalating dyes which fluoresce strongly when bound to dsDNA; when the strands melt the dye dissociates and fluoresces at a much lower level. Using the proper equipment, the dissociation of the intercalating dye can be observed under "high resolution" (ie, 10–20 data points per °C) and a melt curve can be generated for a specific amplicon. Differences in melt curves between similar amplicons can detect single nucleotide polymorphisms as well as indels as small as 1 bp. The strength of HRMA is that it allows relatively high-throughput genotyping of mutant individuals with a delay of only a few hours (Pasay et al., 2008). Amplicons displaying differences in melt curves can be compared to known deletions (genotyping a previously identified mutation in the absence of a physical marker) or grouped and sequenced to identify novel out of frame mutants (Fig. 3). One drawback of HRMA (and other amplicon-based genotyping methods such as T7 and Surveyor) is that they can give false positive results in the presence of preexisting genetic polymorphisms. Thus, assays should be

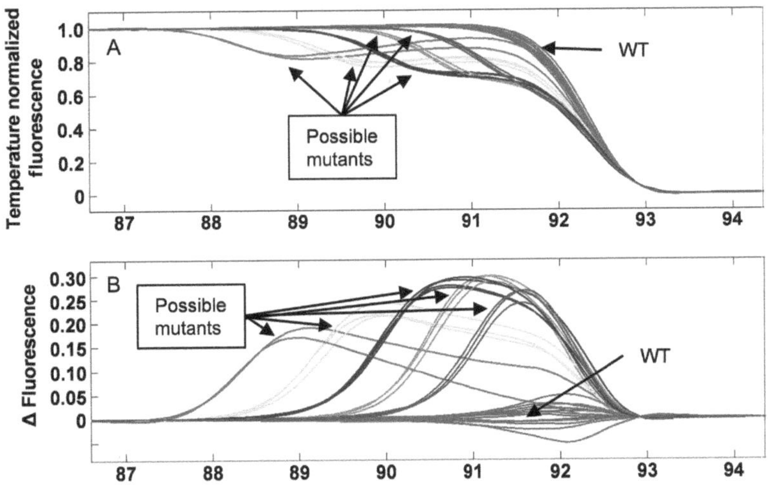

Fig. 3 High-resolution melt curve analysis identification of mutations. PCR amplification of mosquito tissue from single individual (typically from a clipping of a leg) followed by melting in the presence of a fluorescent dye. As the dsDNA melts, the fluorescent dye dissociates from the dsDNA generating a melt curve. Melt curves can be viewed directly with different curves segregating from WT (A). Alternatively, WT curves can be subtracted from unknowns to generate difference curves (B). After putative mutant groups are identified, sequencing is required to confirm any insertions or deletions, as differences from WT may also be caused by preexisting polymorphisms. (See the colour plate.)

designed around regions devoid of such preexisting polymorphisms whenever possible. A comparison of the sensitivity, cost and time associated with each technique is listed in Fig. 4.

In addition to deep sequencing and HRMA, Kistler et al. (2015), Basu et al. (2015), Dong et al. (2015), Gantz et al. (2015) and Hammond et al. (2016), all utilized phenotypic markers incorporated through HDR to tag disrupted regions. Given the capacity of a laboratory to screen for fluorescent markers, screening mutations generated by HDR insertion of a fluorescent marker may be the easiest way to screen and maintain mutant groups. Given the examples provided, not only can fluorescent markers allow for easy maintenance of lab colonies for basic research but also they may be required for proper tracking of GD strains. Given the variety of methods that can be utilized for mutation tracking, there are numerous ways in which experimenters can begin to apply CRISPR to their research. The continued advancement of HDR techniques (which will be discussed in more depth in the next section) suggests this technology may be the most appealing way to identify and monitor mutations within a mosquito population.

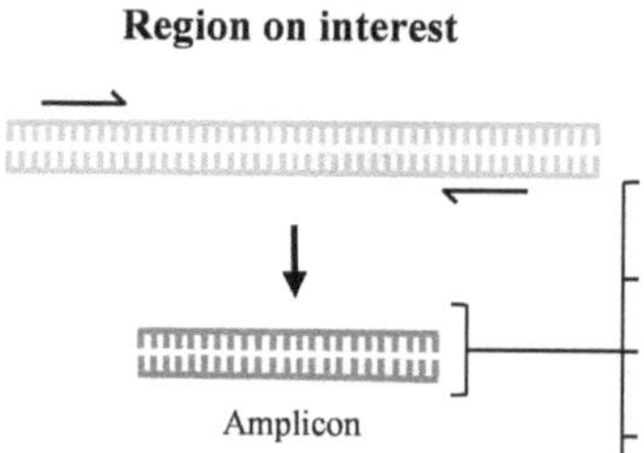

Technique	Sensitivity	Cost	Time
Deep sequencing	+++	+++	+++
HRMA	++	++	+
Surveyor	+	+	+
T7	+	+	+
Sanger	+	+	+++

Least = + Most = +++

Fig. 4 Comparison of amplicon-based mutation detection techniques. Following amplification of a target region, the resulting product can be queried for the presence mutations through digestion with mismatch-specific nucleases (T7 or Surveyor), analysis of differences in melting properties (HRMA), or through traditional (Sanger) or next-generation sequencing technologies. Each techniques is associated with its own strengths and weaknesses, which should be considered based on the goals and resources of the experimenter.

1.8 Progress in Gene Insertion via Homology-Directed Repair

The ease of use of inserted fluorescent markers suggests an attractive future for CRISPR-based basic and applied research. As mentioned earlier, the HDR pathway utilizes a homologous template to repair a DSB. If these homologous sequences are placed on either end of a marker gene (along with other cargo), then the cargo can be inserted within the genome. HDR has been utilized with alternative endonucleases in mosquitoes within the past few years (Bernardini et al., 2014; Liesch et al., 2013; McMeniman et al., 2014) with rates of insertion less than 0.1%. Kistler et al. (2015) was able insert ssODN at a rate of 0.71% in G_0 somatic tissue and 0.6% in germline tissue but the ssODN was only 200 bp in length with the majority of the construct (excluding the restriction enzyme site) consisting of the homologous flanks. When testing the ability to insert larger constructs (such as those that harbour a fluorescent marker) Kistler et al. (2015) saw highly variable frequencies of insertion, while making the observation that lower G_0 survival rates correlated with higher levels of HDR. Specific characteristics of the construct design and injection mix concentration have been shown to play a role in insertion rate via HDR. First, studies have shown that the larger the construct to be inserted, the larger the homologous sequence must be (Kung et al., 2013). Manipulation of DNA DSB repair pathways via inhibition of key C-NHEJ components in the G_0 embryos may also increase HDR outcomes. For example, in *Drosophila* the knockout of Lig4 was utilized to enhance HDR (Beumer et al., 2008), while the knockout of Ku70 was used to enhance HDR in *B. mori* (Ma et al., 2014). Basu et al. (2015) utilized dsRNA transiently targeting Ku70 which increased insertion rates to 1.7–2.7%. Basu et al. (2015) also utilized dsRNA targeting Ligase IV but found that its inhibition in *Ae. aegypti*, while leading to an increase in HDR, also lead to an increase in indel generation (by what is suspected to be an alternative nonhomologous end-joining pathway). Whether this observation is paralleled in other mosquitoes is still unknown, but it is an interesting question which may provide insight into the diversity of DSB DNA repair pathways in mosquitoes. Gantz et al. (2015) also utilized dsRNA transiently targeting Ku70 when they introduced their GD construct into the kh^{w} locus. Of the 680 embryos injected, 251 survivors were collected which yielded two transgenic progeny. Assuming 50% fertility, an insertion rate of 0.8–1.6% was observed, comparable to that reported by others (Basu et al., 2015; Hammond et al., 2016; Kistler et al., 2015). Interestingly, Hammond et al. (2016) utilized HDR without the use of a NHEJ inhibitor to insert their EGFP docking constructs and saw rates similar to transposable element (TE) insertion. This may be due to other variables not yet identified within the injection mix, variation in microinjection technique or differences in the species used (*An. gambiae* vs *Ae. aegypti*). Regardless, it suggests that continued collection and analysis of injection mix data may yield insight regarding rates of insertion via HDR. Recently, it has been suggested that the nature of the homologous sequence also plays a role in how often HDR will occur. The 3′ end of the DSB is released prior to the 5′ end, suggesting that asymmetric templates which favour the 3′ section of DNA lead to higher rates of HDR (Richardson et al., 2016).

As suggested earlier, a primary difficulty associated with the utilization of HDR is competing DSB repair pathways. While in some organisms, such as yeast, the utilization of the homology-directed repair pathway is highly prevalent, in other organisms, such as mammals, the C-NHEJ pathway may be the most prevalent pathway (Mansour et al., 2008). DNA repair pathway distinctions in cell cycle phase and tissue type will also play a role. The ability for the experimenter to manipulate the DNA DSB repair hierarchy is key for the utilization of the HDR pathway in organisms in which other pathways may be utilized prior to the HDR pathway. One possible way to achieve this is the successful identification and utilization of small molecules to inhibit a desired DNA repair pathway. One molecule that has been shown to be effective at inhibiting C-NHEJ in vertebrates is

SCR-7 (Chu et al., 2015; Maruyama et al., 2015). An inhibitor of the critical C-NHEJ protein Ligase IV, SCR-7 has been shown to increase the rate of HDR in human and mouse cell lines. It remains to be determined whether SCR-7 will be useful in invertebrates such as mosquitoes, but given that many molecules that inhibit DNA repair are researched for their anticancer therapeutic properties, it is possible that new molecules will surface which may play a role in aiding the genetic engineering of mosquitoes including inhibitors of other key repair proteins and chromosomal relaxers (Murr et al., 2006; Pinder et al., 2015; Yu et al., 2015).

Chemical inhibitors may prove to be advantageous additives to the genetic toolbox in mosquitoes but traditional techniques which involve the use of knockout lines and RNA interference are still highly relevant to DNA repair manipulation. For these techniques to be implemented a firm understanding of how DNA repair pathways work and which genes/proteins are the most appropriate to target. Another thought to be considered is whether a specific NHEJ component should be removed transiently or permanently. While transient knockdown of Ku70 resulted in an increase in HDR in *Ae. aegypti*, such mutations may or may not be sustainable in mosquito populations and may even heavily depend upon the species of mosquito. For instance, Basu et al. (2015) found that Ligase IV mutants were unstable as heterozygotes, with mutations disappearing from the population by the third generation even after selective breeding. While Ligase IV and Ku70 are good starting points, there is a long list of genes which may be targeted to the same effect, or to greater effect (Overcash et al., 2015; Zhu et al., 2015). In summary, HDR allows for the insertion of easily detectable and tractable marker genes which can be highly beneficial to both basic and applied mosquito research. Some basic tenets, such as large homologous flanks on transgenic constructs and DNA repair manipulation, increase the rate of insertion via HDR. Lastly, there are several more steps which can be taken to further increase insertion rate including identification of possible chemical modulators, identification of alternative DNA repair genes for targeting and the overall continued optimization of injection mixes.

1.9 Future Prospects of Gene Editing: CRISPR Variants

The CRISPR revolution has taken hold and is opening up several new avenues for genetic engineering. While on a basic level the CRISPR systems act as a highly specific nuclease, there are several modifications and variants which promise to further enhance genetic engineering in mosquitoes. Since the discovery of the Cas9 system much effort has been put into finding alternative versions of Cas9 which may be more specific (such as the high-fidelity Cas9 mentioned earlier), or allow for different target sites (different PAM site) (Leenay et al., 2016). There have been a variety of the proteins found with some having alternative PAM sites and others being smaller in size (Garneau et al., 2010; Kleinstiver et al., 2015; Zhang et al., 2013). These proteins could be used to generate smaller GD constructs or expand the targetable regions within mosquito genomes. The current Cas9 protein has also been modified to provide enzymatic action as a nickase, generating only a single-stranded cut (instead of a DSB) (Mali et al., 2013a). This allows for two sgRNAs targeting opposite strands in close proximity to generate a DSB, generating higher specificity and removing some of the risk of off-target effects. A variety of catalytically inactive Cas9 fusion proteins have been developed to take advantage of sgRNA-directed DNA binding (Mali et al., 2013b). For instance, Cas9 fusion proteins containing an activator have been shown to upregulate the expression of human genes (Maeder et al., 2013). The CRISPR system has also been retooled to interfere with mRNA (Qi et al., 2013). It should also be noted that creativity within the CRISPR system is not limited to Cas9 protein modifications. An interesting example is the generation of synthetic constructs that provide tissue-specific expression of sgRNAs of interest. Ubiquitous expression of the sgRNA is typically achieved via a strong pol III-directed promoter so that the sgRNA is not given a 5′ cap, a poly A tail and maintains its nuclear localization. A disadvantage with such pol III-based transcription is that they are not tissue specific. Port and Bullock (2016) recently described a method that allows for experimenters to link a tissue-specific pol II-based promoter to a synthetic construct whereby sgRNAs of interest are flanked by tRNAs *in cis* (Port and Bullock, 2016). Following initial transcription, RNases P and Z cleave the tRNAs releasing the intact sgRNAs within the nucleus. Port and Bullock (2016) observed that pol III-based sgRNA production resulted in leaky expression and incomplete reduction of the targeted protein in imaginal wing discs of developing *Drosophila* (only 60% imaginal wing discs were deficient in protein of target gene and 60% of imaginal wing discs targeted showed some level of protein reduction of the targeted gene protein in adjacent tissue). In contrast, utilizing tRNA-flanked sgRNAs, no detectable protein of the targeted gene was observed in targeted tissue and only 10% imaginal wing discs showed some protein removal in adjacent tissue. These observations suggest that there is still room for improvement with respect to tissue-specific expression of transgenes and that the tRNA flanking technique is worth exploring in mosquitoes for generating sgRNAs. The creativity with which the CRISPR system can be used is almost limitless. Now that the utilization of CRISPR in mosquitoes is becoming more well established, the Cas9 variants which provide the most promise with respect genetic manipulation of disease vectors should be assessed and characterized with respect to mosquitoes. This burgeoning field should provide ample research opportunities which could drastically enhance mosquito research.

2. PROGRESS IN TE-BASED MOSQUITO TRANSFORMATION

Routine transformation with TEs in mosquitoes was first established in *Ae. aegypti* more than a decade after the successful use in *Drosophila*, with the use of Class II (DNA elements) *MosI*, *Hermes* and *piggyBac* TEs (Coates et al., 1998; Jasinskiene et al., 1998; Kokoza et al., 2001). Shortly afterwards, TEs were adapted for use in *An. stephensi* using *minos* and *An. gambiae* using *piggyBac* (Catteruccia et al., 2000; Grossman et al., 2001). TEs have to this point served as the work horse of transgenic experiments in mosquitoes allowing vector biologists to confer phenotypes such as pathogen resistance (Corby-Harris et al., 2010; Kokoza et al., 2010; Meredith et al., 2011), sex bias (Galizi et al., 2014), flightlessness (Fu et al., 2010) and conditional sterility (Phuc et al., 2007) on laboratory mosquito populations. While routine and reliable, TEs are limited by the fact that their integration site in the genome cannot be predetermined. For many applications such as the development of antiviral/malarial or GD, it is critical that transgenes are integrated in stable locations within the genome which do not impact fitness (Amenya et al., 2010) and avoid transcriptional gene silencing (Vaucheret and Fagard, 2001). What TEs do allow for, is for the experimenter to conduct forward genetic screens where an optimal location is not known a priori. By inserting genetic elements which may have antiviral or antiparasitic properties at random within the genome, the investigator can sample locations to identify those that might be most favourable. For example, Labbé et al. (2012) generated 18 transgenic lines (nine for *Ae. aegypti* and nine for *Ae. albopictus*) carrying the same construct, but reported varying results with respect to female-specific flightlessness in each (Labbé et al., 2012). Ultimately, screening a number of unique insertions was critical to finding an optimal position, emphasizing the impact of TE integration site on construct efficiency. In addition to embryonic injection, another way to sample multiple locations is remobilization of the TE housing the transgenic construct. O'Brochta et al. (2012) provide an excellent example of forward genetic screens by remobilizing the *piggyBac* TE for the generation of Gal4–UAS lines (O'Brochta et al., 2012). Utilizing TEs to generate multiple genome-wide insertions, the researchers observed various expression profiles which were then selected from for future Gal4–UAS experiments; these experiments will be discussed in more detail in Section 3.

Sufficient mosquito transformation experiments have been performed to allow meta-analysis of various parameters and success rates. Gregory et al. (2016) analysed metadata on *piggyBac* usage in a range of insects with the goal of assessing appropriate transformation efficiency (Gregory et al., 2016). To this end, Gregory et al. (2016) developed a decision-making model which they termed as "Goldilocks" (https://mammykins.shinyapps.io/App-gold), allowing researchers to predict the number of embryos they need to inject to achieve a desired number of transgenics. These researchers found that experimenters can utilize analysis of survival rates and transformation efficiency to minimize wasted efforts and resources resulting from too few or too many injections. Their method could also be extrapolated to other germline transformation methods given that the proper data is available for analysis.

Recently, TEs have been used in combination with highly specific DNA binding domains to increase transposition efficiency in *B. mori* (Ye et al., 2015). Ye et al. (2015) found that when a transcription activator-like effector was fused to the *piggyBac* transposase a stable transposition efficiency as high as 63.9% was seen, as well as a nearly sixfold increase in transgene positive individuals. While no site-specific integration of the TE was observed, the significant increase in overall TE integration warrants further investigation. If applied in mosquitoes, such a large increase in transformation efficiency could greatly reduce the labour and cost associated with forward genetic screens. While no reports have been published to this date, it may very well be the case that CRISPR and TEs can be utilized in a similar fashion with the advantage of being able to multiplex sgRNAs. In summary, TEs still have an important role to play in mosquito transgenesis. They are still a viable option when it comes to inserting transgenes within the genome at a reliable rate, and allow for the systematic testing of stability, fitness and transgene expression in an unbiased manner. Assessment of current and future TE studies with respect to integration rates should be conducted to optimize injection parameters and potentially unnecessary labour. Lastly, ways that TE can be used in conjunction with the expanding genetic toolbox (especially CRISPR) should be explored.

3. PROGRESS IN CONTROLLING TRANSGENE EXPRESSION USING BIPARTITE SYSTEMS

Bipartite systems involve the use of two independent transgenes in conjunction to achieve an activity that neither alone possesses. The separate nature of the transgenes allows for spatiotemporal control of a gene of interest via proper crossing of transgenic lines. Typically, one portion of the bipartite system acts as a "driver" which produces a specific transcription factor and the other portion acts as a "responder" which is the target of that particular transcription factor; those mosquitoes with both the driver and the responder will express the particular gene that the responder promotes. For the purposes of this chapter, we will discuss the Gal4–UAS system, the Q system and the Tet-On/Off systems.

The Gal4–UAS system derives from yeast, with the Gal4 transactivator promoting the transcription of the target gene by binding to an upstream activation sequence (UAS). The workhouse of *Drosophila* genetics (reviewed in Griffin et al., 2014), the Gal4–UAS system has been successfully deployed in several mosquito species including *Ae. aegypti*, *An. gambiae* and *An. stephensi* (Kokoza and Raikhel, 2011; Lynd and Lycett, 2012; O'Brochta et al., 2012; Zhao et al., 2014). Kokoza and Raikhel (2011) utilized the fat body-specific, bloodmeal-induced Vg promoter to drive Gal4 expression. When crossed to a UAS–EGFP responder line, expression of EGFP was observed within the fat bodies of bloodfed females. The Vg-Gal4 driver subsequently allowed for the interrogation of 20E activation of the Vg promoter. In a follow-up study, Lucas et al. (2015) developed *Ae. aegypti* transgenic strains with the UAS element controlling expressing of a "miRNA-sponge" in order to deplete miR-8 levels specifically in the fat body when crossed with the same Vg-Gal4 driver. These experiments demonstrated that miR-8's critical role in regulating yolk protein precursor secretion (and thus nutrient acquisition by the developing ovary) occurred specifically in the fat body, data that could not be obtained by systemic depletion experiments. To date this remains the only report utilizing the GAL4/UAS system beyond proof of principle to interrogate the function of a novel gene, but emphasizes the power and potential of bipartite systems to control gene expression. Lynd and Lycett (2012) utilized the midgut-specific bloodmeal-induced carboxypeptidase promoter to control Gal4; these driver lines were crossed with UAS-luciferase and UAS-YFP responders recapitulating the sex, tissue and stage specific expression of carboxypeptidase in *An. gambiae*. Similarly, Zhao et al. (2014) also established an *Ae. aegypti* Gal4 driver line based on the homologous midgut-specific carboxypeptidase A gene promoter; this was also validated by crossing to a UAS–EGFP responder line. Despite the power of the Gal4–UAS system to control transgene expression and the number of mosquito gene promoters that have already been characterized (reviewed in Adelman et al., 2015), relatively few have been utilized to generate reusable Gal4 drivers. Increasing the number of driver lines available to the mosquito genetics and physiology community would likely accelerate basic research and should be pursued when possible.

In addition to reverse genetic approaches described earlier, bipartite systems such as Gal4–UAS represent a powerful tool to perform forward genetic screens in mosquitoes. O'Brochta et al. (2012) took advantage of the fact that *piggyBac* can be efficiently remobilized in *An. stephensi* (O'Brochta et al., 2011) to build a *piggyBac*-based enhancer trap using the Gal4–UAS system. Remobilization of the *piggyBac* element containing the Gal4 enhancer trap in the presence of UAS-tdTomato (red fluorescent protein) allowed the identification of novel insertion sites yielding tissue-specific expression patterns in the midgut, fat body and salivary glands. The identification of unique enhancer elements provides a catalogue of valuable insertion sites; combined with the power of CRISPR and site-specific gene insertion, such sites can be reused even if the initial strains cannot be propagated indefinitely. The unique ability to perform such screens is so far restricted to *An. stephensi*, as such transposon remobilization has not yet been possible in *Ae. aegypti* or *An. gambiae* (Palavesam et al., 2013; Sethuraman et al., 2007; Wilson et al., 2003). Thus, this work is extremely valuable in identifying enhancer elements which may be utilized for basic or applied research approaches, as the successful introduction of antipathogen genes in mosquitoes is dependent upon utilizing an expression profile which maximizes their effect on the pathogen, while minimizing negative effects on host fitness or behaviours.

Recently, the Gal4/UAS system has also been used to generate a driver line in *Drosophila* that allows for the tissue-specific expression of Cas9 endonuclease (Xue et al., 2014). Such a driver line removes the need to include Cas9 mRNA from any future gene editing experiments reducing the cost and labour needed for gene editing. The development of UAS-Cas9 mosquito strains that could be combined with existing or novel Gal4 drivers would give unprecedented control over tissue-specific gene disruption, and would be of great value to understanding the genetic basis of essential traits, particularly in regards to genes that are critical for both early developmental stages and adult stages, and thus are recalcitrant to ubiquitous disruption. Thus, characterization of novel promoters and generation of Gal4–UAS lines harnessing promoters already characterized would allow the mosquito genetics community to fully take advantage of this bipartite system (Fig. 5).

The Q system, identified in the fungus *Neurospora crassa*, utilizes genes associated with the *qa* gene cluster and is regulated by the concentration of glucose via the transcription factor (QF) and the transcription repressor (QS). While this system has not yet been used in mosquitoes, it is effective in *Drosophila* and in combination with the Gal4/UAS system, may lead to

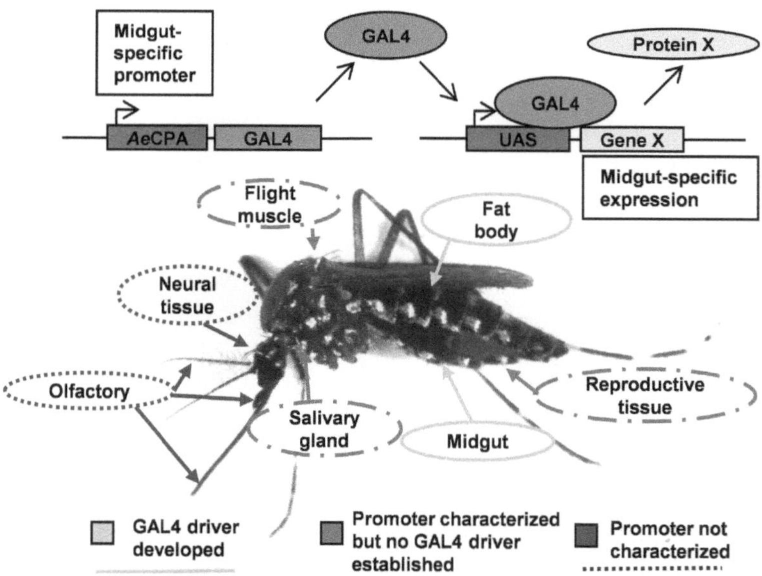

Fig. 5 The GAL4–UAS system allows for tissue-specific expression of effector genes in mosquitoes. The upper portion of the figure shows a typical driver-responder pair, in this case a midgut-specific promoter (*Ae*CPA) driving the expression of GAL4; GAL4 in turn interacts with the upstream-associated sequence or UAS to promote the production of a second gene X (gene of interest). The bipartite nature of the GAL4–UAS system allows for the experimenter to interchange either the promoter of GAL4 expression or the gene being promoted by the UAS promoter. This allows for quick and easy crosses to be made to test the impact of a specific effector gene within a targeted tissue. The bottom portion of the figure shows specific tissue types in which the GAL4–UAS system has been successfully utilized in mosquitoes (*yellow; solid line*), tissue types in which possible promoters have been identified but the GAL4–UAS system has not been utilized yet (*red; dash-dot line*), and tissue types in which no promoters have been identified and the GAL4–UAS system has not been utilized yet (*blue; dotted line*). (See the colour plate.)

even more precise control of transgenes (Potter and Luo, 2011). The ability to utilize both the GAL4–UAS system and Q system in conjunction could vastly expand the toolbox of bipartite systems in mosquitoes allowing for the testing of two independent genes of interest.

Lastly, tetracycline-controllable gene expression systems are very well established in mosquitoes in both Tet-Off and Tet-On forms (Lycett et al., 2004). Both of these systems rely on an external effector molecule (tetracycline or one of its analogues) which can be added or removed from the mosquito's diet. Lycett et al. (2004) was able to generate both Tet-On and Tet-Off transgenic lines in *An. stephensi* using the tet analogue doxycycline to regulate the expression of β-galactosidase with its expression being promoted in the transgenic Tet-On line and its expression being repressed in the corresponding Tet-Off lines. Following this initial proof of principle, the Tet-Off system was used in transgenic strains of *Ae. aegypti*, *Ae. albopictus* and *An. stephensi* (Fu et al., 2010; Labbé et al., 2012; Marinotti et al., 2013; Phuc et al., 2007; Wise de Valdez et al., 2011). Phuc et al. (2007) generated transgenic strains of *Ae. aegypti* carrying a gene which when expressed lead to lethality late in development (dominant lethality). In the presence of tetracycline the transgenic lines yielded survival rates comparable to wild-type controls (87–89%) but in its absence survival rates at adult emergence were severely reduced (2.6–4.2%). The late acting effect of the transgene allowed released transgenics to compete with the wild population, which provides an advantage when compared to traditional sterile insect techniques. The best of the strains developed by Phuc et al. (2007) were tested in field releases in the Cayman Islands (Harris et al., 2012), with eventual suppression rates calculated at around 80%. Currently, these techniques are being utilized by in Brazil to great effect, as Carvalho et al. (2015) found that release of transgenic male mosquitoes (OX513A) led to a 95% reduction of *Ae. aegypti* in the release area (Carvalho et al., 2015). Initially, 2800 transgenic males per hectare per week were released, but upon conducting a rangefinder study, production was optimized to release many more transgenic males than previously hypothesized (25,000 transgenic males per hectare per week). Given the recent outbreak of the Zika virus in South America (Imperato, 2016), the expedient use of these transgenic lines has become increasingly important. Fu et al. (2010) attempted a similar strategy using the Tet-Off system under the control of a female-specific indirect flight muscle promoter (Actin-4) to produce female *Ae. aegypti* which lacks the ability to fly. Follow-up cage trials indicated some success (Wise de Valdez et al., 2011). In a similar design, Labbé et al. (2012) and Marinotti et al. (2013) generated transgenic strains with female-specific flightless phenotypes in *Ae. albopictus* and *An. stephensi*, respectively. These experiments reinforce the utility of the Tet-Off system to control the expression of deleterious genes capable of conferring useful phenotypes in mosquitoes.

4. PROGRESS IN φC31 AND RECOMBINASE-MEDIATED CASSETTE EXCHANGE RECOMBINATION IN MOSQUITOES

The use of "docking" mechanisms allows the experimenter to reutilize a useful position within a genome to introduce additional genetic constructs. The insertion of a docking location can be accomplished via the random initial insertion of a TE, or a prespecified location utilizing HDR. The φC31 system (isolated from the φC31 bacteriophage; Thorpe and Smith, 1998) has been shown to be particularly useful in mosquitoes, and is based on the recombination reaction between attachment sites in the phage (*att*P) and target bacterium (*att*B). (For detailed background information on the φC31 system, see Fogg et al., 2014; Knapp et al., 2015; Smith et al., 2010).

First adapted for use in human cells and in *Drosophila* (Groth et al., 2000; Groth et al., 2004), the φC31system has been adapted for use in several mosquito species including *Ae. aegypti*, *Ae. albopictus*, *An. stephensi* and *An. gambiae* (Amenya et al., 2010; Labbe et al., 2010; Meredith et al., 2011; Nimmo et al., 2006). Nimmo et al. (2006) generated five *att*P φC31 docking strains in *Ae. aegypti* utilizing *piggyBac*, with between one and four docking sites per strain. Of the five strains generated, four were tested for site-specific integration by insertion of an *att*B construct containing a red fluorescent marker (DsRed); integration was observed at rates between 16.7% and 31.8% (average 23%). Labbe et al. (2010) utilized the φC31 docking system in *Ae. albopictus*, using a *piggyBac* element to first insert an *att*P docking site. After the recovery of five independent docking lines, three were chosen for integration of DsRed via an *att*B docking site, with reported integration rates between 2% and 6%. While significantly lower than that reported by Nimmo et al. (2006), these rates were similar to that observed in *Ae. aegypti* by Franz et al. (2011), who found integration rates of an EGFP marker gene of 1.7–4.9%. It is possible that the location of the docking site integration may play a large role in φC31 integration rates and that rates obtained by Labbe et al. (2010) and Franz et al. (2011) may be more typical of what can be expected.

Amenya et al. (2010) generated a cohort of integrated φC31 docking sites into *An. stephensi* to measure the impact of docking sites (in conjunction with a fluorescent marker gene) on mosquito fitness (transformation efficiency not reported). These researchers monitored fecundity, percentage of females that did not lay eggs, pupal development time, larval viability, male and female longevity and male and female wingspan. While there were individual differences between strains, Amenya et al. (2010) found no statistically significant difference, under the parameters measured, between mosquitoes containing the docking site and those without. This suggests that φC31 docking sites can be inserted into the genome without causing inherent fitness costs to the mosquito host. Meredith et al. (2011) utilized a φC31 docking site to integrate Vida3, an antimalarial peptide, in *An. gambiae*. Four independent docking strains were generated via *piggyBac*, a single strain was selected for integration of the corresponding *att*B sequence which was accompanied by the Vida3 antimalarial peptide and the DsRed marker gene, with reported efficiency of 10%. Windbichler et al. (2011) utilized a φC31 docking strain to provide the first proof of principle of a homing endonuclease (HEG) GD system (Windbichler et al., 2011). Using the docking site, they generated donor, reporter and target strains of *An. gambiae*, all at the same genetic locus. The donor strain contained EGFP interrupted with an HEG gene, the reporter contained an interrupted EGFP open-reading frame (which could be correctly expressed after incurring frameshift indels via the NHEJ pathway) and the target strain contained EGFP not interrupted via an HEG. Since the genetic constructs where driven by a male specific promoter, female crosses were used as a control. When males were crossed with WT females, a homing rate of 56% was observed with 97% of progeny containing the HEG cassette. The ability to utilize the φC31 docking system to generate a variety of transgenic lines within transgenic constructs in homologous portions of the chromosome, provide a unique, a useful asset for assessing possible future GD components as well as interrogating mechanisms of DSB repair.

Hammond et al. (2016) also utilized the φC31 docking system to improve their ability to generate complex multigene insertions for GD applications. Docking sites were initially integrated utilizing HDR in conjunction with an EGFP marker gene following DSB induction by a site-specific nuclease. This allowed for identification of lines with the docking site, and allowing for confirmation of the position of the docking site via PCR as well as for the characterization of the targeted genes with respect to their suspected role in female fertility. Once their role in female fertility was established, the GD construct was injected along with the recombinase construct allowing for recombination-mediated cassette exchange to occur. This relatively simple work flow allows for a smaller construct (the marker gene and φC31 docking site) to be inserted before introduction of a larger GD construct. An advantage of this technique could be the avoidance of the requirement to generate the large homologous flanks needed to insert a HDR construct which carries extensive of genetic cargo. It would also be advantageous for genetic targets which have only a small region which is targetable for HDR insertion (possibly due to repetitive elements).

Other recombinases such as the cre recombinase system and the FLP/FRT have been postulated for use in mosquitoes based on their successful use in *Drosophila* (Oberstein et al., 2005; Venken and Bellen, 2012). Jasinskiene et al. (2003) showed that the cre system catalyses transgene excision at extremely high rates (up to 99.4%) but was unable to exchange cassettes (Jasinskiene et al., 2003). Recently the FLP and cre systems have been used in conjunction with the φC31 system in *Ae. aegypti* successfully (Haghighat-Khah et al., 2015). Haghighat-Khah et al. (2015) generated a system which they termed integrase–recombinase-mediated cassette exchange (iRMCE) which utilizes the integration ability of the φC31 system and the excision abilities of the FLP and cre systems. This iRMCE technique allows for the removal of any unnecessary genetic components found in the plasmid backbone of the integrase. Utilizing this system, Haghighat-Khah et al. (2015) observed integration rates of 4.1–4.8% and minimum excision rates of 2.4% for cre-mediated excision and 1.3% for FLP-mediated excision in *Ae. aegypti*. These docking mechanisms provide for a unique tool which ultimately can be used in conjunction with all other previous tools and techniques mentioned in this chapter. The φC31 system may be used in association with TEs to insert transgenes throughout the mosquito genome to assess expression and stability and can also be utilized with HDR to insert docking sites in highly specific locations. The φC31, FLP and cre systems can also be utilized in conjunction with bipartite systems to achieve recombinase-mediated cassette exchange (RMCE) of new promoter sequences and effector genes into docking based lines. While the φC31 system is relatively well established in mosquitoes, more research and development is encouraged on the FLP and cre systems so that they can be utilized independently or in conjunction with φC31 to facilitate more highly efficient RMCE. For an excellent description of current best practices and protocols regarding the use of φC31 in mosquitoes, please see Pondeville et al. (2014).

5. CONSIDERATIONS AND OUTLOOK

The toolkit used for the manipulation and genetic modification of mosquitoes has expanded tremendously in the past two decades (Volohonsky et al., 2015). First TEs opened the door for the introduction of transgenic constructs. Next came in the introduction of docking systems such as φC31, which increased the specificity of genetic manipulation. Then came the first wave of specific nucleases including homing endonucleases, zinc finger nucleases and TALENS (these were not covered here due to recent advancements in CRISPR) in conjunction with utilization of HDR. Lastly, the most recent advancements have come with the development of CRISPR technology. CRISPR stands to continually revolutionize gene editing not only in mosquitoes but also in other model organisms of interest. In parallel with the advancements in the variety and efficiency of CRISPR a continued advancement in the understanding of how DNA double-stranded break repair promises to enhance the specificity and control associated with genetic modifications of mosquitoes. While older techniques such as the use of TEs and docking sites will continue to play a role in years to come, the CRISPR revolution and associated advancements stand to change gene editing and transgenesis in crucial disease vectors.

ACKNOWLEDGEMENTS

This project was supported by Grants (AI085091 and AI113643) from NIAID and its contents are solely the responsibility of the authors and do not necessarily represent the official views of the NIAID.

REFERENCES

Adelman, Z.N., Tu, Z., 2016. Control of mosquito-borne infectious diseases: sex and gene drive. Trends Parasitol. 32 (3), 219–229.

Adelman, Z.N., Basu, S., Myles, K.M., 2015. Gene insertion and deletion in mosquitoes. In: Adelman, Z.N. (Ed.), Genetic Control of Malaria and Dengue. Elsevier, London, UK/San Diego, CA/Waltham, MA/Oxford, UK, pp. 139–168.

Amenya, D.A., Bonizzoni, M., Isaacs, A.T., Jasinskiene, N., Chen, H., Marinotti, O., Yan, G., James, A.A., 2010. Comparative fitness assessment of *Anopheles stephensi* transgenic lines receptive to site-specific integration. Insect Mol. Biol. 19 (2), 263–269.

Awata, H., Watanabe, T., Hamanaka, Y., Mito, T., Noji, S., Mizunami, M., 2015. Knockout crickets for the study of learning and memory: dopamine receptor Dop1 mediates aversive but not appetitive reinforcement in crickets. Sci. Rep. 5, 15885.

Bassett, A.R., Tibbit, C., Ponting, C.P., Liu, J.L., 2013. Highly efficient targeted mutagenesis of Drosophila with the CRISPR/Cas9 system. Cell Rep. 4 (1), 220–228.

Basu, S., Aryan, A., Overcash, J.M., Samuel, G.H., Anderson, M.A., Dahlem, T.J., Myles, K.M., Adelman, Z.N., 2015. Silencing of end-joining repair for efficient site-specific gene insertion after TALEN/CRISPR mutagenesis in *Aedes aegypti*. Proc. Natl. Acad. Sci. U.S.A. 112 (13), 4038–4043.

Bernardini, F., Galizi, R., Menichelli, M., Papathanos, P.A., Dritsou, V., Marois, E., Crisanti, A., Windbichler, N., 2014. Site-specific genetic engineering of the *Anopheles gambiae* Y chromosome. Proc. Natl. Acad. Sci. U.S.A. 111 (21), 7600–7605.

Beumer, K.J., Trautman, J.K., Bozas, A., Liu, J.L., Rutter, J., Gall, J.G., Carroll, D., 2008. Efficient gene targeting in Drosophila by direct embryo injection with zinc-finger nucleases. Proc. Natl. Acad. Sci. U.S.A. 105 (50), 19821–19826.

Brazelton, V.A., Zarecor, S., Wright, D.A., Wang, Y., Liu, J., Chen, K., Yang, B., Lawrence-Dill, C.J., 2015. A quick guide to CRISPR sgRNA design tools. GM Crops Food 6 (4), 266–276.

Brinkman, E.K., Chen, T., Amendola, M., van Steensel, B., 2014. Easy quantitative assessment of genome editing by sequence trace decomposition. Nucleic Acids Res. 42 (22), e168.

Carvalho, D.O., McKemey, A.R., Garziera, L., Lacroix, R., Donnelly, C.A., Alphey, L., Malavasi, A., Capurro, M.L., 2015. Suppression of a field population of *Aedes aegypti* in Brazil by sustained release of transgenic male mosquitoes. PLoS Negl. Trop. Dis. 9 (7), e0003864.

Catteruccia, F., Nolan, T., Loukeris, T.G., Blass, C., Savakis, C., Kafatos, F.C., Crisanti, A., 2000. Stable germline transformation of the malaria mosquito *Anopheles stephensi*. Nature 405 (6789), 959–962.

Champer, J., Buchman, A., Akbari, O.S., 2016. Cheating evolution: engineering gene drives to manipulate the fate of wild populations. Nat. Rev. Genet. 17 (3), 146–159.

Chu, V.T., Weber, T., Wefers, B., Wurst, W., Sander, S., Rajewsky, K., Kuhn, R., 2015. Increasing the efficiency of homology-directed repair for CRISPR-Cas9-induced precise gene editing in mammalian cells. Nat. Biotechnol. 33 (5), 543–548.

Coates, C.J., Jasinskiene, N., Miyashiro, L., James, A.A., 1998. Mariner transposition and transformation of the yellow fever mosquito, *Aedes aegypti*. Proc. Natl. Acad. Sci. U.S.A. 95 (7), 3748–3751.

Cong, L., Ran, F.A., Cox, D., Lin, S., Barretto, R., Habib, N., Hsu, P.D., Wu, X., Jiang, W., Marraffini, L.A., Zhang, F., 2013. Multiplex genome engineering using CRISPR/Cas systems. Science 339 (6121), 819–823.

Corby-Harris, V., Drexler, A., Watkins de Jong, L., Antonova, Y., Pakpour, N., Ziegler, R., Ramberg, F., Lewis, E.E., Brown, J.M., Luckhart, S., Riehle, M.A., 2010. Activation of Akt signaling reduces the prevalence and intensity of malaria parasite infection and lifespan in *Anopheles stephensi* mosquitoes. PLoS Pathog. 6 (7), e1001003.

Dahlem, T.J., Hoshijima, K., Jurynec, M.J., Gunther, D., Starker, C.G., Locke, A.S., Weis, A.M., Voytas, D.F., Grunwald, D.J., 2012. Simple methods for generating and detecting locus-specific mutations induced with TALENs in the zebrafish genome. PLoS Genet. 8 (8), e1002861.

Dong, S., Lin, J., Held, N.L., Clem, R.J., Passarelli, A.L., Franz, A.W., 2015. Heritable CRISPR/Cas9-mediated genome editing in the yellow fever mosquito, *Aedes aegypti*. PLoS One 10 (3), e0122353.

Doudna, J.A., Charpentier, E., 2014. Genome editing. The new frontier of genome engineering with CRISPR-Cas9. Science 346 (6213), 1258096.

Esvelt, K.M., Smidler, A.L., Catteruccia, F., Church, G.M., 2014. Concerning RNA-guided gene drives for the alteration of wild populations. Elife. e03401.

Fogg, P.C.M., Colloms, S., Rosser, S., Stark, M., Smith, M.C.M., 2014. New applications for phage integrases. J. Mol. Biol. 426 (15), 2703–2716.

Franz, A.W., Jasinskiene, N., Sanchez-Vargas, I., Isaacs, A.T., Smith, M.R., Khoo, C.C., Heersink, M.S., James, A.A., Olson, K.E., 2011. Comparison of transgene expression in Aedes aegypti generated by mariner Mos1 transposition and PhiC31 site-directed recombination. Insect Mol. Biol. 20 (5), 587–598.

Fu, G., Lees, R.S., Nimmo, D., Aw, D., Jin, L., Gray, P., Berendonk, T.U., White-Cooper, H., Scaife, S., Kim Phuc, H., Marinotti, O., Jasinskiene, N., James, A.A., Alphey, L., 2010. Female-specific flightless phenotype for mosquito control. Proc. Natl. Acad. Sci. U.S.A. 107 (10), 4550–4554.

Gagnon, J.A., Valen, E., Thyme, S.B., Huang, P., Akhmetova, L., Pauli, A., Montague, T.G., Zimmerman, S., Richter, C., Schier, A.F., 2014. Efficient mutagenesis by Cas9 protein-mediated oligonucleotide insertion and large-scale assessment of single-guide RNAs. PLoS One 9 (5), e98186.

Galizi, R., Doyle, L.A., Menichelli, M., Bernardini, F., Deredec, A., Burt, A., Stoddard, B.L., Windbichler, N., Crisanti, A., 2014. A synthetic sex ratio distortion system for the control of the human malaria mosquito. Nat. Commun. 5, 3977.

Gantz, V.M., Bier, E., 2015. Genome editing. The mutagenic chain reaction: a method for converting heterozygous to homozygous mutations. Science 348 (6233), 442–444.

Gantz, V.M., Bier, E., 2016. The dawn of active genetics. Bioessays 38 (1), 50–63.

Gantz, V.M., Jasinskiene, N., Tatarenkova, O., Fazekas, A., Macias, V.M., Bier, E., James, A.A., 2015. Highly efficient Cas9-mediated gene drive for population modification of the malaria vector mosquito *Anopheles stephensi*. Proc. Natl. Acad. Sci. U.S.A. 112 (49), E6736–E6743.

Garneau, J.E., Dupuis, M.E., Villion, M., Romero, D.A., Barrangou, R., Boyaval, P., Fremaux, C., Horvath, P., Magadan, A.H., Moineau, S., 2010. The CRISPR/Cas bacterial immune system cleaves bacteriophage and plasmid DNA. Nature 468 (7320), 67–71.

Gilles, A.F., Schinko, J.B., Averof, M., 2015. Efficient CRISPR-mediated gene targeting and transgene replacement in the beetle *Tribolium castaneum*. Development 142 (16), 2832–2839.

Gratz, S.J., Ukken, F.P., Rubinstein, C.D., Thiede, G., Donohue, L.K., Cummings, A.M., O'Connor-Giles, K.M., 2014. Highly specific and efficient CRISPR/Cas9-catalyzed homology-directed repair in Drosophila. Genetics 196 (4), 961–971.

Gregory, M., Alphey, L., Morrison, N.I., Shimeld, S.M., 2016. Insect transformation with *piggyBac*: getting the number of injections just right. Insect Mol. Biol. 25 (3), 259–271. http://dx.doi.org/10.1111/imb.12220.

Griffin, R., Binari, R., Perrimon, N., 2014. Genetic odyssey to generate marked clones in Drosophila mosaics. Proc. Natl. Acad. Sci. U.S.A. 111 (13), 4756–4763.

Grossman, G.L., Rafferty, C.S., Clayton, J.R., Stevens, T.K., Mukabayire, O., Benedict, M.Q., 2001. Germline transformation of the malaria vector, *Anopheles gambiae*, with the *piggyBac* transposable element. Insect Mol. Biol. 10 (6), 597–604.

Groth, A.C., Olivares, E.C., Thyagarajan, B., Calos, M.P., 2000. A phage integrase directs efficient site-specific integration in human cells. Proc. Natl. Acad. Sci. U.S.A. 97 (11), 5995–6000.

Groth, A.C., Fish, M., Nusse, R., Calos, M.P., 2004. Construction of transgenic Drosophila by using the site-specific integrase from phage phiC31. Genetics 166 (4), 1775–1782.

Haghighat-Khah, R.E., Scaife, S., Martins, S., John, O. St, Matzen, K.J., Morrison, N., Alphey, L., 2015. Site-specific cassette exchange systems in the *Aedes aegypti* mosquito and the *Plutella xylostella* moth. PLoS One 10 (4), e0121097.

Hall, A.B., Basu, S., Jiang, X., Qi, Y., Timoshevskiy, V.A., Biedler, J.K., Sharakhova, M.V., Elahi, R., Anderson, M.A., Chen, X.G., Sharakhov, I.V., Adelman, Z.N., Tu, Z., 2015. A male-determining factor in the mosquito *Aedes aegypti*. Science 348 (6240), 1268–1270.

Hammond, A., Galizi, R., Kyrou, K., Simoni, A., Siniscalchi, C., Katsanos, D., Gribble, M., Baker, D., Marois, E., Russell, S., Burt, A., Windbichler, N., Crisanti, A., Nolan, T., 2016. A CRISPR-Cas9 gene drive system targeting female reproduction in the malaria mosquito vector *Anopheles gambiae*. Nat. Biotechnol. 34 (1), 78–83.

Harris, A.F., McKemey, A.R., Nimmo, D., Curtis, Z., Black, I., Morgan, S.A., Oviedo, M.N., Lacroix, R., Naish, N., Morrison, N.I., Collado, A., Stevenson, J., Scaife, S., Dafa'alla, T., Fu, G., Phillips, C., Miles, A., Raduan, N., Kelly, N., Beech, C., Donnelly, C.A., Petrie, W.D., Alphey, L., 2012. Successful suppression of a field mosquito population by sustained release of engineered male mosquitoes. Nat. Biotechnol. 30 (9), 828–830.

Imperato, P.J., 2016. The convergence of a virus, mosquitoes, and human travel in globalizing the Zika epidemic. J. Community Health 41 (3), 674–679.

Jasinskiene, N., Coates, C.J., Benedict, M.Q., Cornel, A.J., Rafferty, C.S., James, A.A., Collins, F.H., 1998. Stable transformation of the yellow fever mosquito, *Aedes aegypti*, with the *Hermes* element from the housefly. Proc. Natl. Acad. Sci. U.S.A. 95 (7), 3743–3747.

Jasinskiene, N., Coates, C.J., Ashikyan, A., James, A.A., 2003. High efficiency, site-specific excision of a marker gene by the phage P1 cre-loxP system in the yellow fever mosquito, *Aedes aegypti*. Nucleic Acids Res. 31 (22), e147.

Jinek, M., Chylinski, K., Fonfara, I., Hauer, M., Doudna, J.A., Charpentier, E., 2012. A programmable dual-RNA-guided DNA endonuclease in adaptive bacterial immunity. Science 337 (6096), 816–821.

Kim, D., Bae, S., Park, J., Kim, E., Kim, S., Yu, H.R., Hwang, J., Kim, J.-I., Kim, J.-S., 2015. Digenome-seq: genome-wide profiling of CRISPR-Cas9 off-target effects in human cells. Nat. Methods 12 (3), 237–243.

Kistler, K.E., Vosshall, L.B., Matthews, B.J., 2015. Genome engineering with CRISPR-Cas9 in the mosquito *Aedes aegypti*. Cell Rep. 11 (1), 51–60.

Kleinstiver, B.P., Prew, M.S., Tsai, S.Q., Topkar, V.V., Nguyen, N.T., Zheng, Z., Gonzales, A.P., Li, Z., Peterson, R.T., Yeh, J.R., Aryee, M.J., Joung, J.K., 2015. Engineered CRISPR-Cas9 nucleases with altered PAM specificities. Nature 523 (7561), 481–485.

Kleinstiver, B.P., Pattanayak, V., Prew, M.S., Tsai, S.Q., Nguyen, N.T., Zheng, Z., Joung, J.K., 2016. High-fidelity CRISPR-Cas9 nucleases with no detectable genome-wide off-target effects. Nature 529 (7587), 490–495.

Knapp, J.-M., Chung, P., Simpson, J.H., 2015. Generating customized transgene landing sites and multi-transgene arrays in drosophila using phiC31 integrase. Genetics 199 (4), 919–934.

Kokoza, V.A., Raikhel, A.S., 2011. Targeted gene expression in the transgenic *Aedes aegypti* using the binary Gal4-UAS system. Insect Biochem. Mol. Biol. 41 (8), 637–644.

Kokoza, V., Ahmed, A., Wimmer, E.A., Raikhel, A.S., 2001. Efficient transformation of the yellow fever mosquito *Aedes aegypti* using the piggyBac transposable element vector pBac[3xP3-EGFP afm]. Insect Biochem. Mol. Biol. 31 (12), 1137–1143.

Kokoza, V., Ahmed, A., Woon Shin, S., Okafor, N., Zou, Z., Raikhel, A.S., 2010. Blocking of plasmodium transmission by cooperative action of cecropin a and defensin a in transgenic *Aedes aegypti* mosquitoes. Proc. Natl. Acad. Sci. U.S.A. 107 (18), 8111–8116.

Kung, S.H., Retchless, A.C., Kwan, J.Y., Almeida, R.P., 2013. Effects of DNA size on transformation and recombination efficiencies in *Xylella fastidiosa*. Appl. Environ. Microbiol. 79 (5), 1712–1717.

Labbe, G.M., Nimmo, D.D., Alphey, L., 2010. *Piggybac-* and PhiC31-mediated genetic transformation of the Asian tiger mosquito, *Aedes albopictus* (Skuse). PLoS Negl. Trop. Dis. 4 (8), e788.

Labbé, G.M.C., Scaife, S., Morgan, S.A., Curtis, Z.H., Alphey, L., 2012. Female-specific flightless (fsRIDL) phenotype for control of *Aedes albopictus*. PLoS Negl. Trop. Dis. 6 (7), e1724.

Lamarche, B.J., Orazio, N.I., Weitzman, M.D., 2010. The MRN complex in double-strand break repair and telomere maintenance. FEBS Lett. 584 (17), 3682–3695.

Leenay, R.T., Maksimchuk, K.R., Slotkowski, R.A., Agrawal, R.N., Gomaa, A.A., Briner, A.E., Barrangou, R., Beisel, C.L., 2016. Identifying and visualizing functional PAM diversity across CRISPR-Cas systems. Mol. Cell 62 (1), 137–147.

Li, F., Scott, M.J., 2016. CRISPR/Cas9-mediated mutagenesis of the white and sex lethal loci in the invasive pest, *Drosophila suzukii*. Biochem. Biophys. Res. Commun. 469 (4), 911–916.

Liesch, J., Bellani, L.L., Vosshall, L.B., 2013. Functional and genetic characterization of Neuropeptide Y-like receptors in *Aedes aegypti*. PLoS Negl. Trop. Dis. 7 (10), e2486.

Lucas, K.J., Roy, S., Ha, J., Gervaise, A.L., Kokoza, V.A., et al., 2015. MicroRNA-8 targets the wingless signaling pathway in the female mosquito fat body to regulate reproductive processes. Proc. Natl. Acad. Sci. U.S.A. 112 (5), 1440–1445.

Lycett, G.J., Kafatos, F.C., Loukeris, T.G., 2004. Conditional expression in the malaria mosquito anopheles stephensi with Tet-On and Tet-Off systems. Genetics 167 (4), 1781–1790.

Lynd, A., Lycett, G.J., 2012. Development of the bi-partite Gal4-UAS system in the African malaria mosquito, *Anopheles gambiae*. PLoS One 7 (2), e31552.

Ma, S., Chang, J., Wang, X., Liu, Y., Zhang, J., Lu, W., Gao, J., Shi, R., Zhao, P., Xia, Q., 2014. CRISPR/Cas9 mediated multiplex genome editing and heritable mutagenesis of BmKu70 in *Bombyx mori*. Sci. Rep. 4, 4489.

Maeder, M.L., Linder, S.J., Cascio, V.M., Fu, Y., Ho, Q.H., Joung, J.K., 2013. CRISPR RNA-guided activation of endogenous human genes. Nat. Methods 10 (10), 977–979.

Mali, P., Aach, J., Stranges, P.B., Esvelt, K.M., Moosburner, M., Kosuri, S., Yang, L., Church, G.M., 2013a. CAS9 transcriptional activators for target specificity screening and paired nickases for cooperative genome engineering. Nat. Biotechnol. 31 (9), 833–838.

Mali, P., Esvelt, K.M., Church, G.M., 2013b. Cas9 as a versatile tool for engineering biology. Nat. Methods 10 (10), 957–963.

Mansour, W.Y., Schumacher, S., Rosskopf, R., Rhein, T., Schmidt-Petersen, F., Gatzemeier, F., Haag, F., Borgmann, K., Willers, H., Dahm-Daphi, J., 2008. Hierarchy of nonhomologous end-joining, single-strand annealing and gene conversion at site-directed DNA double-strand breaks. Nucleic Acids Res. 36 (12), 4088–4098.

Marinotti, O., Jasinskiene, N., Fazekas, A., Scaife, S., Fu, G., Mattingly, S.T., Chow, K., Brown, D.M., Alphey, L., James, A.A., 2013. Development of a population suppression strain of the human malaria vector mosquito, *Anopheles stephensi*. Malar. J. 12, 142.

Markert, M.J., Zhang, Y., Enuameh, M.S., Reppert, S.M., Wolfe, S.A., Merlin, C., 2016. Genomic access to monarch migration using TALEN and CRISPR/Cas9-mediated targeted mutagenesis. G3 (Bethesda) 6 (4), 905–915.

Marraffini, L.A., Sontheimer, E.J., 2010. CRISPR interference: RNA-directed adaptive immunity in bacteria and archaea. Nat. Rev. Genet. 11 (3), 181–190.

Maruyama, T., Dougan, S.K., Truttmann, M.C., Bilate, A.M., Ingram, J.R., Ploegh, H.L., 2015. Increasing the efficiency of precise genome editing with CRISPR-Cas9 by inhibition of nonhomologous end joining. Nat. Biotechnol. 33 (5), 538–542.

McMeniman, C.J., Corfas, R.A., Matthews, B.J., Ritchie, S.A., Vosshall, L.B., 2014. Multimodal integration of carbon dioxide and other sensory cues drives mosquito attraction to humans. Cell 156 (5), 1060–1071.

Meredith, J.M., Basu, S., Nimmo, D.D., Larget-Thiery, I., Warr, E.L., Underhill, A., McArthur, C.C., Carter, V., Hurd, H., Bourgouin, C., Eggleston, P., 2011. Site-specific integration and expression of an anti-malarial gene in transgenic *Anopheles gambiae* significantly reduces Plasmodium infections. PLoS One 6 (1), e14587.

Murr, R., Loizou, J.I., Yang, Y.G., Cuenin, C., Li, H., Wang, Z.Q., Herceg, Z., 2006. Histone acetylation by Trrap-Tip60 modulates loading of repair proteins and repair of DNA double-strand breaks. Nat. Cell Biol. 8 (1), 91–99.

Nimmo, D.D., Alphey, L., Meredith, J.M., Eggleston, P., 2006. High efficiency site-specific genetic engineering of the mosquito genome. Insect Mol. Biol. 15 (2), 129–136.

Oberstein, A., Pare, A., Kaplan, L., Small, S., 2005. Site-specific transgenesis by Cre-mediated recombination in Drosophila. Nat. Methods 2 (8), 583–585.

O'Brochta, D.A., Alford, R.T., Pilitt, K.L., Aluvihare, C.U., Harrell, R.A., 2011. *piggyBac* transposon remobilization and enhancer detection in Anopheles mosquitoes. Proc. Natl. Acad. Sci. U.S.A. 108 (39), 16339–16344.

O'Brochta, D.A., Pilitt, K.L., Harrell 2nd, R.A., Aluvihare, C., Alford, R.T., 2012. Gal4-based enhancer-trapping in the malaria mosquito *Anopheles stephensi*. G3 (Bethesda) 2 (11), 1305–1315.

Overcash, J.M., Aryan, A., Myles, K.M., Adelman, Z.N., 2015. Understanding the DNA damage response in order to achieve desired gene editing outcomes in mosquitoes. Chromosome Res. 23 (1), 31–42.

Palavesam, A., Esnault, C., O'Brochta, D.A., 2013. Post-integration silencing of *piggyBac* transposable elements in *Aedes aegypti*. PLoS One 8 (7), e68454.

Pasay, C., Arlian, L., Morgan, M., Vyszenski-Moher, D., Rose, A., Holt, D., Walton, S., McCarthy, J., 2008. High-resolution melt analysis for the detection of a mutation associated with permethrin resistance in a population of scabies mites. Med. Vet. Entomol. 22 (1), 82–88.

Peng, R., Lin, G., Li, J., 2016. Potential pitfalls of CRISPR/Cas9-mediated genome editing. FEBS J. 283 (7), 1218–1231.

Phuc, H.K., Andreasen, M.H., Burton, R.S., Vass, C., Epton, M.J., Pape, G., Fu, G., Condon, K.C., Scaife, S., Donnelly, C.A., Coleman, P.G., White-Cooper, H., Alphey, L., 2007. Late-acting dominant lethal genetic systems and mosquito control. BMC Biol. 5, 11.

Pinder, J., Salsman, J., Dellaire, G., 2015. Nuclear domain 'knock-in' screen for the evaluation and identification of small molecule enhancers of CRISPR-based genome editing. Nucleic Acids Res. 43 (19), 9379–9392.

Pondeville, E., Puchot, N., Meredith, J.M., Lynd, A., Vernick, K.D., Lycett, G.J., Eggleston, P., Bourgouin, C., 2014. Efficient ΦC31 integrase–mediated site-specific germline transformation of *Anopheles gambiae*. Nat. Protoc. 9 (7), 1698–1712.

Port, F., Bullock, S.L., 2016. Expansion of the CRISPR toolbox in an animal with tRNA-flanked Cas9 and Cpf1 gRNAs. bioRxiv doi: http://dxdoi.org/10.1101/046417.

Potter, C.J., Luo, L., 2011. Using the Q system in Drosophila melanogaster. Nat. Protoc. 6 (8), 1105–1120.

Qi, L.S., Larson, M.H., Gilbert, L.A., Doudna, J.A., Weissman, J.S., Arkin, A.P., Lim, W.A., 2013. Repurposing CRISPR as an RNA-guided platform for sequence-specific control of gene expression. Cell 152 (5), 1173–1183.

Ren, X., Sun, J., Housden, B.E., Hu, Y., Roesel, C., Lin, S., Liu, L.P., Yang, Z., Mao, D., Sun, L., Wu, Q., Ji, J.Y., Xi, J., Mohr, S.E., Xu, J., Perrimon, N., Ni, J.Q., 2013. Optimized gene editing technology for *Drosophila melanogaster* using germ line-specific Cas9. Proc. Natl. Acad. Sci. U.S.A. 110 (47), 19012–19017.

Ren, X., Yang, Z., Xu, J., Sun, J., Mao, D., Hu, Y., Yang, S.J., Qiao, H.H., Wang, X., Hu, Q., Deng, P., Liu, L.P., Ji, J.Y., Li, J.B., Ni, J.Q., 2014. Enhanced specificity and efficiency of the CRISPR/Cas9 system with optimized sgRNA parameters in Drosophila. Cell Rep. 9 (3), 1151–1162.

Reyon, D., Tsai, S.Q., Khayter, C., Foden, J.A., Sander, J.D., Joung, J.K., 2012. FLASH assembly of TALENs for high-throughput genome editing. Nat. Biotechnol. 30 (5), 460–465.

Richardson, C.D., Ray, G.J., DeWitt, M.A., Curie, G.L., Corn, J.E., 2016. Enhancing homology-directed genome editing by catalytically active and inactive CRISPR-Cas9 using asymmetric donor DNA. Nat. Biotechnol. 34 (3), 339–344.

Sethuraman, N., Fraser Jr., M.J., Eggleston, P., O'Brochta, D.A., 2007. Post-integration stability of piggyBac in Aedes aegypti. Insect Biochem. Mol. Biol. 37 (9), 941–951.

Shmakov, S., Abudayyeh, O.O., Makarova, K.S., Wolf, Y.I., Gootenberg, J.S., Semenova, E., Minakhin, L., Joung, J., Konermann, S., Severinov, K., Zhang, F., Koonin, E.V., 2015. Discovery and functional characterization of diverse class 2 CRISPR-Cas systems. Mol. Cell 60 (3), 385–397.

Smith, M.C., Brown, W.R., Brown, W.R., McEwan, A.R., Rowley, P.A., 2010. Site-specific recombination by φC31 integrase and other large serine recombinases. Biochem. Soc. Trans. 38 (2), 388–394.

The Centers for Disease Control and Prevention, 2014. CDC and Malaria.

Thorpe, H.M., Smith, M.C., 1998. In vitro site-specific integration of bacteriophage DNA catalyzed by a recombinase of the resolvase/invertase family. Proc. Natl. Acad. Sci. U.S.A. 95 (10), 5505–5510.

Tsai, S.Q., Zheng, Z., Nguyen, N.T., Liebers, M., Topkar, V.V., Thapar, V., Wyvekens, N., Khayter, C., Iafrate, A.J., Le, L.P., Aryee, M.J., Joung, J.K., 2015. GUIDE-seq enables genome-wide profiling of off-target cleavage by CRISPR-Cas nucleases. Nat. Biotechnol. 33 (2), 187–197.

Vaucheret, H., Fagard, M., 2001. Transcriptional gene silencing in plants: targets, inducers and regulators. Trends Genet. 17 (1), 29–35.

Venken, K.J., Bellen, H.J., 2012. Genome-wide manipulations of *Drosophila melanogaster* with transposons, Flp recombinase, and PhiC31 integrase. Methods Mol. Biol. 859, 203–228.

Volohonsky, G., Terenzi, O., Soichot, J., Naujoks, D.A., Nolan, T., Windbichler, N., Kapps, D., Smidler, A.L., Vittu, A., Costa, G., Steinert, S., Levashina, E.A., Blandin, S.A., Marois, E., 2015. Tools for *Anopheles gambiae* transgenesis. G3 (Bethesda) 5 (6), 1151–1163.

Wang, T., Wei, J.J., Sabatini, D.M., Lander, E.S., 2014. Genetic screens in human cells using the CRISPR-Cas9 system. Science 343 (6166), 80–84.

Wei, W., Xin, H., Roy, B., Dai, J., Miao, Y., Gao, G., 2014. Heritable genome editing with CRISPR/Cas9 in the silkworm, *Bombyx mori*. PLoS One 9 (7), e101210.

World Health Organization, 2014a. Dengue and severe dengue.

World Health Organization, 2014b. Malaria.

Williams, G.J., Hammel, M., Radhakrishnan, S.K., Ramsden, D., Lees-Miller, S.P., Tainer, J.A., 2014. Structural insights into NHEJ: building up an integrated picture of the dynamic DSB repair super complex, one component and interaction at a time. DNA Repair (Amst.) 17, 110–120.

Wilson, R., Orsetti, J., Klocko, A.D., Aluvihare, C., Peckham, E., Atkinson, P.W., Lehane, M.J., O'Brochta, D.A., 2003. Post-integration behavior of a *Mos*1 mariner gene vector in *Aedes aegypti*. Insect Biochem. Mol. Biol. 33 (9), 853–863.

Windbichler, N., Menichelli, M., Papathanos, P.A., Thyme, S.B., Li, H., Ulge, U.Y., Hovde, B.T., Baker, D., Monnat, R.J., Burt, A., Crisanti, A., 2011. A synthetic homing endonuclease-based gene drive system in the human malaria mosquito. Nature 473 (7346), 212–215.

Wise de Valdez, M.R., Nimmo, D., Betz, J., Gong, H.F., James, A.A., Alphey, L., Black, W.C.t., 2011. Genetic elimination of dengue vector mosquitoes. Proc. Natl. Acad. Sci. U.S.A. 108 (12), 4772–4775.

Wu, X., Kriz, A.J., Sharp, P.A., 2014a. Target specificity of the CRISPR-Cas9 system. Quant. Biol. 2 (2), 59–70.

Wu, X., Scott, D.A., Kriz, A.J., Chiu, A.C., Hsu, P.D., Dadon, D.B., Cheng, A.W., Trevino, A.E., Konermann, S., Chen, S., Jaenisch, R., Zhang, F., Sharp, P.A., 2014b. Genome-wide binding of the CRISPR endonuclease Cas9 in mammalian cells. Nat. Biotechnol. 32 (7), 670–676.

Wu, B., Luo, L., Gao, X.J., 2016. Cas9-triggered chain ablation of cas9 as a gene drive brake. Nat. Biotechnol. 34 (2), 137–138.

Xue, Z., Wu, M., Wen, K., Ren, M., Long, L., Zhang, X., Gao, G., 2014. CRISPR/Cas9 mediates efficient conditional mutagenesis in Drosophila. G3 (Bethesda) 4 (11), 2167–2173.

Ye, L., You, Z., Qian, Q., Zhang, Y., Che, J., Song, J., Zhong, B., 2015. TAL effectors mediate high-efficiency transposition of the piggyBac transposon in silkworm *Bombyx mori* L. Sci. Rep. 5, 17172.

Yu, C., Liu, Y., Ma, T., Liu, K., Xu, S., Zhang, Y., Liu, H., La Russa, M., Xie, M., Ding, S., Qi, L.S., 2015. Small molecules enhance CRISPR genome editing in pluripotent stem cells. Cell Stem Cell 16 (2), 142–147.

Zeng, B., Zhan, S., Wang, Y., Huang, Y., Xu, J., Liu, Q., Li, Z., Huang, Y., Tan, A., 2016. Expansion of CRISPR targeting sites in *Bombyx mori*. Insect Biochem. Mol. Biol. 72, 31–40.

Zetsche, B., Gootenberg, J.S., Abudayyeh, O.O., Slaymaker, I.M., Makarova, K.S., Essletzbichler, P., Volz, S.E., Joung, J., van der Oost, J., Regev, A., Koonin, E.V., Zhang, F., 2015. Cpf1 is a single RNA-guided endonuclease of a class 2 CRISPR-Cas system. Cell 163 (3), 759–771.

Zhang, Y., Heidrich, N., Ampattu, B.J., Gunderson, C.W., Seifert, H.S., Schoen, C., Vogel, J., Sontheimer, E.J., 2013. Processing-independent CRISPR RNAs limit natural transformation in *Neisseria meningitidis*. Mol. Cell 50 (4), 488–503.

Zhao, B., Kokoza, V.A., Saha, T.T., Wang, S., Roy, S., Raikhel, A.S., 2014. Regulation of the gut-specific carboxypeptidase: a study using the binary Gal4/UAS system in the mosquito *Aedes aegypti*. Insect Biochem. Mol. Biol. 54, 1–10.

Zhu, L., Mon, H., Xu, J., Lee, J.M., Kusakabe, T., 2015. CRISPR/Cas9-mediated knock-out of factors in non-homologous end joining pathway enhances gene targeting in silkworm cells. Sci. Rep. 5, 18103.

CHAPTER TWO

Sex Determination in Mosquitoes

J.K. Biedler, Z. Tu
Department of Biochemistry, The Fralin Life Science Institute, Virginia Tech, Blacksburg, VA, United States

Contents

Abstract

Sexual reproduction is critical to the survival and evolution of many organisms including mosquitoes. Sex determination in insects is often initiated by a fast-evolving primary signal(s) that switches on or off a cascade of molecular events that lead to sex-specific splicing of the transcripts of *doublesex* (*dsx*) and *fruitless* (*fru*), two genes that encode highly conserved transcription factors. The sex-specific isoforms of the DSX and FRU proteins programme sexual dimorphism in *Drosophila melanogaster* and other insects, and sex-specific splicing of *dsx* and *fru* is the molecular output of the process of sex determination. Sex-biassed splice isoforms of *dsx* and *fru* have been identified in mosquitoes although what regulates their splicing is unknown. Previous genetic evidence

Advances in Insect Physiology, Volume 51
ISSN 0065-2806
http://dx.doi.org/10.1016/bs.aiip.2016.05.005

suggests, depending on the mosquito species, that a dominant male-determining factor (M-factor) from either the Y chromosome or a male-determining locus (M-locus) is the primary signal that initiates male development and controls sex determination. In this chapter, we focus on recent progress in deciphering the molecular mechanisms of sex determination in mosquitoes, including the discovery of complex and distinct *dsx* splicing patterns in three mosquito genera, and the discovery and characterization of the first M-factor in *Aedes aegypti* and additional M-factor candidates in divergent mosquito species. We discuss recent discoveries of the genetic content and rapid remodelling of the Y chromosomes in the *Anopheles gambiae* species complex and the molecular consequence of Y chromosomal decay. We end by discussing the remaining challenges in illustrating the entire sex determination pathway in mosquitoes, and in developing novel mosquito control strategies based on reducing the number of blood-sucking females or converting them into harmless males.

1. INTRODUCTION

Sex determination refers to the developmental process by which the sex of an individual is determined. It is one of the most fundamental processes in biology as sex is critical to the survival and evolution of sexually reproducing organisms. Sexual reproduction generally results in offspring being composed of 50% of each sex, male and female. What is hidden behind this apparently simplistic phenotypic binary output is a myriad variety of mechanisms. Sex is determined genetically in most insects and sex-specific splicing of the primary transcripts of *doublesex* (*dsx*) and *fruitless* (*fru*), two genes that encode highly conserved transcription factors, programmes sexual dimorphism in *Drosophila melanogaster* and other insects (Herpin and Schartl, 2015; Salz and Erickson, 2010). Thus, the *dsx* and *fru* genes are at the bottom of the sex determination pathway, and their sex-specific splicing is the output of the process of sex determination (Fig. 1). The primary signals that determine sex are the protein or RNA molecules that serve as the master switches that control the sex-specific splicing of *dsx/fru*, either directly or through intermediates. Interestingly, these primary signals and the sex chromosome systems appear to evolve rapidly (Table 1, Fig. 2). Thus, information from model species such as *D. melanogaster* is of limited help when investigating the sex determination pathways in mosquitoes and other divergent insects.

Sex-biassed splice isoforms of *dsx* and *fru* have been identified in mosquitoes, although what regulates their splicing is unknown. In mosquitoes, previous genetic evidence suggests, depending on the species, that a

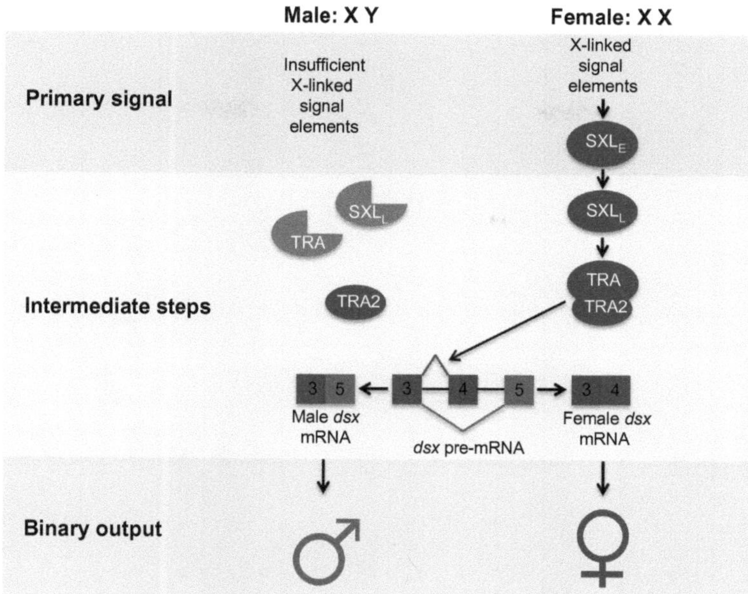

Fig. 1 The sex determination pathway in *Drosophila melanogaster*. *Ovals* indicate proteins in the pathway where colours *red* and *green* represent female-specific and nonsex-specific proteins, respectively. SXL_E and SXL_L refer to the early and late sex lethal proteins, respectively; TRA and TRA2 refer to the transformer and transformer-2 proteins, respectively. In males, insufficient X-linked signal elements result in the absence of SXL_E expression, which subsequently results in a truncated nonfunctional SXL_L protein (shown in *grey*). The lack of functional SXL_L results in a truncated and nonfunctional TRA (shown in *grey*). The lack of a functional TRA results in the male-specific splicing of *dsx* mRNA, excluding the female-specific exon 4. In females, a functional TRA and hence TRA/TRA2 complex directs female-specific splicing of *dsx* by activating a weak splice acceptor upstream of exon 4. Only select exons are shown that are alternatively spliced by TRA/TRA2 regulation. Exons are colour-coded *red* (female-specific), *blue* (male-specific), and *green* (nonsex-specific). TRA2 is expressed in both males and females. *fru* is also subjected to sex-specific alternative splicing by TRA but is not shown. (See the colour plate.)

dominant male-determining factor (M-factor) from either the Y chromosome or a male-determining locus (M-locus) initiates male development and is the primary signal for sex determination. Aside from fundamental biology, research in mosquito sex determination is also motivated by the potential applications for controlling mosquito-borne infectious diseases, because it is only the female sex that feeds on vertebrate blood and in doing so transmits disease-causing pathogens. Readers are encouraged to consult

Table 1 Diversity of Sex Chromosome Systems in Endopterygota (Holometabola)

Order	Genus	Species	Sex Chromosomes (Female/Male)	Male (M)/ Female (F)- Factor	M/F- Factor Location	Common Name	References
Coleoptera	*Tribolium*	*castaneum*	XX/XY	M	Y	Flour beetle	Shukla and Palli (2013, 2014)
Hymenoptera	*Apis*	*mellifera*	2n/n[a]	F (*csd*)	Chr 8	Honey bee	Beye et al. (1996, 2003)
Trichoptera	*Polycentropus*	*flavomaculatus*	ZO/ZZ	?	?	Caddis fly	Marec and Novak (1998)
Lepidoptera	*Micropterix*	*calthella*	ZO/ZZ	?	?	Marsh marigold moth	Traut and Marec (1997)
	Bombyx	*mori*	ZW/ZZ	F (*fem*)	W	Silkworm	Kiuchi et al. (2014), Abe et al. (2008), Tanaka (1916) and Hashimoto (1933)
Diptera	*Drosophila*	*melanogaster*	XX/XY	F (*Sxl*)	X	Common fruit fly	Bridges (1916) and Cline (1993)
	Ceratitis	*capitata*	XX/XY	M	Y	Medfly	Willhoeft and Franz (1996)
	Tephritis	*arnicae*	XX/XO	?	?	Fruit fly/gall fly	Bush (1966)
	Chrysotrypanea	*trifasciata*	ZW/ZZ	?	?	Fruit fly/gall fly	Bush (1966)

	Musca	*domestica*	XX/XY	M	Y	House fly	Dubendorfer et al. (2002)
	Anopheles	*stephensi*	XX/XY	M	Y		Rishikesh (1959)
	Anopheles	*gambiae*	XX/XY	M	Y		Clements (1992) and Marin and Baker (1998)
	Culex	*quinquefasciatus*	mm/Mm[b]	M	Chr 1		Jost and Laven (1971)
	Culex	*tarsalis*	mm/Mm[b]	M	Chr 3		Venkatesan et al. (2009) and McDonald et al. (1978)
	Culex	*tritaeniorhynchus*	mm/Mm[b]	M	Chr 3		Baker and Sakai (1976), Baker et al. (1977) and Mori et al. (2001)
	Culex	*tritaeniorhynchus*	mm/Mm[b]	M	Chr 1		Baker et al. (1971) and Selinger (1972)
	Armigeres	*subalbatus*	mm/Mm[b]	M	Chr 3		Ferdig et al. (1998)
	Aedes	*aegypti*	mm/Mm[b]	M (*Nix*)	Chr 1		McDonald and Rai (1970)

[a] *Apis mellifera* are haplodiploids where heterozygosity of the *csd* gene results in female development and hemizygosity of the *csd* gene results in male development.
[b] Homomorphic chromosomes (M/m) where M contains a Y-like region harbouring the M-factor.

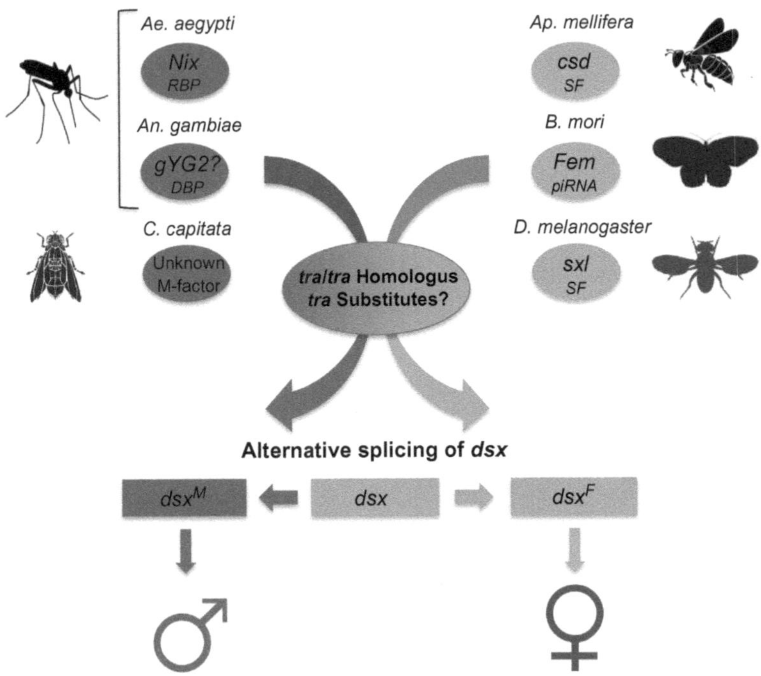

Fig. 2 Comparison of sex determination pathways of select holometabolous insects. *Blue-* and *pink-coloured circles* represent masculinizing (M-factors) and feminizing primary sex-determining signals, respectively. These primary signals determine the sex-specific spliceform of *transformer* or *transformer* substitutes produced. When these signals are not present (in the other sex), the 'default' splicing of *transformer* (or *transformer* substitutes, if present), and hence 'default' splicing of *dsx* take place, instructing development of the other sex. The type of molecule is indicated below the primary sex-determining signal gene name. In the case of *C. capitata* an M-factor is present but unidentified. In some species with an M-factor such as *Nix* in *Ae. aegypti*, the M-factor may directly modulate *dsx/fru* splicing. RNA-binding protein, *RBP*; DNA-binding protein, *DBP*; splicing factor, *SF*; Piwi-interacting RNA, *piRNA*. (See the colour plate.)

recent reviews on this topic (Adelman and Tu, 2016; Alphey, 2014; Biedler et al., 2016). However, the focus of this chapter is on the fundamental biology of sex determination in mosquitoes. We begin by reviewing the molecular mechanisms of sex determination in *D. melanogaster* and the diversity of the chromosome systems and primary signals that determine sex in divergent insect species. We then discuss the homomorphic and heteromorphic sex chromosomes in mosquitoes and the characteristics and rapid remodelling of the Y chromosomes in the *Anopheles gambiae* species complex. We describe the molecular mechanisms of sex determination in mosquitoes, including the discovery and characterization of the first M-factor found in the yellow fever mosquito, *Aedes aegypti* and progress towards the discovery of the M-factor in *Anopheles* mosquitoes. We end by discussing the remaining challenges in deciphering the entire sex determination pathway in mosquitoes, and in developing novel sex separation and mosquito control strategies based on reducing the number of blood-sucking females or even converting them into harmless males.

2. SEX DETERMINATION IN *D. melanogaster*

Sex determination has been extensively studied in the fruit fly *D. melanogaster* (reviewed in Salz and Erickson, 2010; Verhulst and Van De Zande, 2015). *D. melanogaster* contains heteromorphic or well-differentiated XY chromosomes; XX individuals are females while XY individuals are males. Although the *D. melanogaster* Y chromosome does not contain a dominant M-factor that determines sex, the dozen or so single-copy genes that remain on the Y (Carvalho et al., 2015a) are important for spermatogenesis, mating behaviour and modulation of chromatin and gene expression (Carvalho et al., 2000; Lemos et al., 2008; Zhou et al., 2012). Sex is determined in *D. melanogaster* by the balance between the X chromosome(s) and the autosomes (Bridges, 1916), through sensing the X:Autosome (X:A) ratio, where X:A=1 results in female development and X:A=0.5 results in male development (Bridges, 1916, 1921; Cline, 1993; Salz and Erickson, 2010). Thus, *D. melanogaster* mutants with a triploid sex chromosome XXY and diploid autosomes develop as infertile females. A slightly different view is that *D. melanogaster* determines sex by X dosage or by counting the number of X chromosomes (Erickson and Quintero, 2007). In any case, having two X chromosomes provides two sets of X-linked signal elements, which encode X-linked transcription factors (Fig. 1) that turn on the transcription of the *sex lethal* (*sxl*) gene from its early zygotic promoter, leading to production of the SXL_E protein, which serves as the master regulator or primary signal of sex determination. SXL_E is a splicing factor and triggers a cascade of molecular events that lead to the female-specific splicing of *dsx/fru*, and thus female development. First, the presence of SXL_E in the females ensures female-specific splicing of the *sxl* transcripts produced from the late promoter of the *sxl* gene, leading to the translation of a functional late SXL protein (SXL_L). SXL_L is also a splicing factor that ensures the production of the functional isoform (and protein) of *transformer* (TRA) in females. Subsequently, TRA complexes with the nonsex-specific TRA2 protein and ensures female-specific splicing of pre-mRNAs for the transcription factors *dsx* and *fru*. DSX and FRU proteins are at the bottom of the sex determination hierarchy and are responsible for instructing a sex-specific programme and sexual dimorphism in each cell where required. Indeed, the two protein isoforms DSX^F and DSX^M recognize the same response elements in genes since they share the same DNA-binding domain. However, the sex-specific C-terminal domains confer sex-specific gene regulation (Clough et al., 2014). SXL_E is only produced in the early embryonic stage yet this transient primary signal is sufficient to trigger the female-specific cascade of events that are then maintained through development.

3. SEX DETERMINATION IN INSECTS: CHROMOSOMAL SYSTEMS AND MASTER SWITCHES CHANGE, WHILE ENDPOINT EFFECTORS *dsx/fru* REMAIN

Sex determination genes are thought to have unusual evolutionary dynamics, where the most downstream genes of the pathway, *dsx* and *fru*, are highly conserved while genes that directly or indirectly regulate *dsx* and *fru* are fast evolving (Graham et al., 2003). Indeed, DSX is part of a highly conserved DM-domain protein family that regulates sexual differentiation in most if not all animals (Bachtrog et al., 2014; Bopp et al., 2014; Gailey et al., 2006; Geuverink and Beukeboom, 2014; Picard et al., 2015; Verhulst and van de Zande, 2015). Sex-specific splicing of *dsx* and *fru* has been demonstrated in diverse insect species.

What regulates this apparently conserved binary output of *dsx/fru* splicing is a variety of molecular mechanisms and chromosomal systems that could change dramatically among divergent, and sometimes closely related taxa (Bachtrog et al., 2014; Gempe and Beye, 2011; Herpin and Schartl, 2015). This is apparent when looking at sample insect species from the superorder Endopterygota (Holometabola) and within the order Diptera (Table 1, Fig. 2). In the order Diptera, for example, sex-determining chromosome systems encompass a wide karyotypic spectrum including XX/XY in *Anopheles* mosquitoes and *Drosophila* fruit flies, ZW/ZZ and XX/XO in some Tephritidae flies, and homomorphic sex-determining chromosomes in Culicinae mosquitoes (Davies and Roderick, 2005; Marin and Baker, 1998). Likewise, the primary sex-determining signals in insects are highly divergent and rapidly evolving. In honeybees, heterozygosity of the *complementary sex determiner* (*csd*) gene initiates female development (Hasselmann et al., 2008). In silkworms a W chromosome-linked piRNA gene has been shown to be the female-determining factor (Kiuchi et al., 2014). In *Drosophila* as described earlier, an X:Autosome ratio sensing mechanism determines sex (Bridges, 1916, 1921; Cline, 1993; Salz and Erickson, 2010).

Several non-*Drosophila* flies contain a dominant M-factor on their Y chromosomes. In the medfly, *Ceratitis capitata*, the functional TRA protein is produced as a result of splicing by a maternally deposited TRA/TRA2 complex, leading to the female sex; unless a yet-to-be-discovered M-factor interrupts the self-sustaining loop of *tra* splicing and function, as in males (Pane et al., 2002) (Table 1, Fig. 2). RNAi knockdown of *tra* in *C. capitata* produced fully fertile XX males, suggesting that the Y chromosome is not needed for sexual development other than providing the M-factor (Pane et al., 2002). A similar type of sex determination is also shown in the house fly, *Musca domestica* (Hediger et al., 2010). Intriguingly, the M-factor appears to be mobile among different strains or populations of the house fly, sometimes found on one of the autosomes instead of the Y chromosome (Feldmeyer et al., 2008).

In the silkworm *Bombyx mori*, sex determination is controlled by ZW sex chromosomes, where females are the heterogametic sex and the W chromosome has a female-determining factor, a piRNA-producing gene named *Feminizer* or *Fem* (Kiuchi et al., 2014). This was the first example of a molecule other than a protein being the initial sex-determining signal in genetic sex determination systems. A target for *Fem*-produced piRNAs was found on the Z chromosome in the ninth exon of a gene called *Masculinizer* (*Masc*) that potentially encodes a zinc finger protein. *Masc* was found to be necessary for male development and inhibition of *Masc* by *Fem* is needed for female development. When piRNAs produced from the precursor *Fem* RNA were inhibited in female embryos the male spliceform of *dsx* ($Bmdsx^{M}$) was detected. However, knockdown of *Masc* mRNA resulted in resulted in male-specific embryonic lethality instead of sex conversion, possibly due to dosage compensation as discussed in Section 4.3. As in mosquitoes, no *transformer* homologue has been discovered (Geuverink and Beukeboom, 2014).

In the honey bee *Apis mellifera*, which are haplodiploids, sex determination is governed by the *csd* gene (Beye et al., 2003). Unfertilized haploid eggs are hemizygous for *csd*, and develop as males, but fertilized eggs that are heterozygous for *csd* develop as females. *Csd* encodes an arginine–serine-rich protein, like many splicing factors, and has similarities to TRA. A second gene found to be necessary in sex determination is *feminizer* (*fem*), which acts downstream of *csd* and is also an arginine–serine-rich protein similar to TRA (Hasselmann et al., 2008).

4. SEX CHROMOSOMES IN MOSQUITOES

4.1 Homomorphic and Heteromorphic Sex Chromosomes

In mosquitoes, sex determination is carried out via at least two types of sex chromosome karyotypes (Fig. 3). One is heteromorphic as *Anopheles* mosquitoes, where the sex chromosomes (X and Y) are morphologically or karyotypically distinct. The other is homomorphic as in Culicinae mosquitoes, where the sex-determining chromosomes appear like autosomes, but contain a Y-like sex-determining region called the M-locus (Newton et al., 1974). More than a century ago it was discovered that *Anopheles punctipennis* contained heteromorphic chromosomes (Stevens, 1911). In contrast, in this same study it was determined that *Culex quinquefasciatus*, *Culex tarsalis* and *Theobaldia incidens* had homomorphic sex chromosomes. The current picture that has emerged is that of the two subfamilies of the mosquito family Culicidae, Anophelinae mosquitoes contain heteromorphic sex

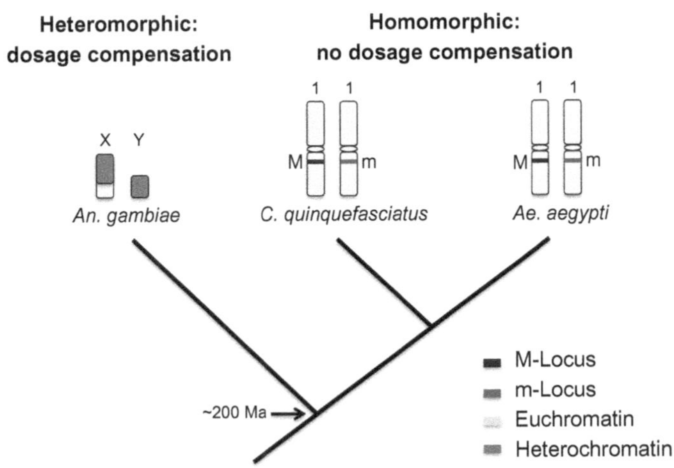

Fig. 3 Sex chromosomes and dosage compensation in species from three genera of mosquitoes. Cladogram shows relationship between the three species. Heteromorphic systems require dosage compensation while homomorphic systems do not. Chromosome figures show male sex chromosome set. *Blue* (*dark grey* in the print version) *bar* indicates M-locus and presence of a dominant male-determining factor (M-factor). *Pink* (*grey* in the print version) bar indicates the M-locus, the counterpart to the M-locus. Chromosomes are not drawn to scale and are labelled at top. Divergence time for Anophelinae/Culicinae is indicated in millions of years (Ma) (Chen et al., 2015; Krzywinski et al., 2006b; Reidenbach et al., 2009). Note that the euchromatic region of the *An. gambiae* X chromosome harbours hundreds of functional genes.

chromosomes, while Culicinae mosquitoes, which include both the *Aedes* and *Culex* genera, contain homomorphic sex chromosomes. These two subfamilies diverged approximately 200 Ma (Chen et al., 2015; Krzywinski et al., 2006b; Reidenbach et al., 2009), which makes an interesting scenario for the study of the evolution of sex determination mechanisms.

Regardless of the sex chromosome karyotype, what appears to be in common in all mosquitoes is the presence of an M-factor at the top of the sex determination hierarchy that initiates male development. In *Anopheles culifacies*, it was found that individuals with the XXY genotype developed as males (Baker and Sakai, 1979), suggesting the presence of a dominant M-factor on the Y chromosome. This is in contrast to *D. melanogaster* where there is no M-factor and the XXY genotype develops as females due to having two X chromosomes and therefore maintaining the X:A ratio (Bridges, 1921; Cline, 1993; Marin and Baker, 1998). The *Anopheles* Y chromosome has been shown to also control stenogamy–eurygamy mating behaviour (Fraccaro et al., 1977).

Studies in species of the *Aedes* and *Culex* genera also indicated the presence of an M-factor (Gilchrist and Haldane, 1947; McClelland, 1962), where *Mm* and *mm* are males and females, respectively (Motara and Rai, 1978). There are three pairs of chromosomes in *Ae. aegypti* (Rai, 1963). The M-locus was mapped to chromosome 1, band 1q21 (McClelland, 1962), and evidence suggests that the M-locus is nonrecombining (Severson et al., 2002; Toups and Hahn, 2010). There are clearly distinguishable cytological differences between the M- and m-loci, although the rest of the two chromosome 1 homologs are nondistinguishable (Motara and Rai, 1977, 1978). It is very interesting that the M-locus has been observed in noncanonical locations in species of both the *Aedes* and *Culex* genera (Table 1), which could indicate either translocation or turnover of the sex-determining locus. However, the M-locus is linked to the same markers in *Ae. aegypti* and *Culex pipiens*, suggesting a common origin dating back to at least their common ancestor (Malcolm et al., 1998; Mori et al., 1999). Therefore, the sex-determining chromosomes may have remained homomorphic since that time. Other studies suggest that the homomorphic sex chromosome of *Ae. aegypti* is ancestral within Culicidae and homomorphic sex chromosomes may have existed since the common ancestor of *Culicine and Anopheline* mosquitoes, approximately 150 My (Toups and Hahn, 2010). Evolutionary theories suggest that homomorphic sex chromosomes may eventually progress into heteromorphic sex chromosomes (Charlesworth et al., 2005). After the acquisition of a sex-determining locus, linkage between sexually antagonistic genes and the sex-determining locus is favoured and may lead to suppression of recombination. Subsequently progressive expansion of the nonrecombining region leads to the formation of heteromorphic sex chromosomes (Bachtrog, 2013; Bachtrog et al., 2011; Bull, 1983; Charlesworth and Mank, 2010; Charlesworth et al., 2005). The nascent sex-determining homomorphic chromosomes are often referred to as the proto-X and proto-Y (Wang et al., 2012). A recent study of the emu genome suggests that sex-biassed gene expression could resolve sexual antagonism and alleviate the need for a large nonrecombining region, thus maintaining the homomorphic sex-determining chromosomes in evolution (Vicoso et al., 2013). The homomorphic sex-determining chromosomes in *Aedes* and *Culex* mosquitoes provide an opportunity to glean new insights into sex chromosome evolution (Hall et al., 2014).

4.2 Cracking the Y Chromosome and a Case for Its Rapid Remodelling in the *An. gambiae* Species Complex

With the exception of evolutionarily nascent or newly formed Y chromosomes, repeat-rich heterochromatin dominates the Y chromosome and creates a bottleneck for Y chromosome analysis (Bachtrog, 2013). Not surprisingly, it has been challenging to uncover the molecular nature of the *Anopheles* Y chromosome (Krzywinski et al., 2004, 2006a), likely due to the difficulty in working with repeat-rich Y sequences. Integrating next-generation sequencing and a bioinformatics method named Chromosome Quotient, several Y chromosome genes have recently been discovered in *Anopheles stephensi* and *An. gambiae* (Criscione et al., 2013; Hall et al., 2013). The Chromosomal Quotient method finds male-specific DNA or RNA sequences by comparing them to separately acquired male and female genomic sequence databases, consisting of Illumina sequences of 25 × or higher genome coverage (Hall et al., 2013). Using single-molecule genome sequencing and the Chromosome Quotient method, the sequence and structure of the male-specific Y (MSY) of *An. gambiae* was determined, which is entirely heterochromatic (Hall et al., 2016). This was the first systematic characterization of a mosquito Y chromosome. Approximately 91% of the MSY are highly amplified structures consisting of several transposable elements and satellite sequences arranged in a particular order. The region of the tandemly repeated *zanzibar* retrotransposons, which accounts for 43% of the *An. gambiae* MSY, is nearly undetectable in the sibling species *Anopheles merus*. The highly Y-enriched satellites AgY477 and AgY373, which occupy another 43% of the *An. gambiae* MSY, are either present on both X and Y or absent altogether in the other sibling species of the *An. gambiae* species complex, namely *An. arabiensis*, *Anopheles quadriannulatus* and *An. merus*. Thus up to 86% of the MSY turns over in less than 2 million years, through changes of a few repetitive sequences. Such a structural dynamism is also observed between *An. gambiae* populations and among individuals within populations. These results provide new insights into the sequence, structure and evolution of heterochromatic Y chromosomes.

4.3 Dosage Compensation as a Consequence of Y Chromosome Decay

In sex chromosome systems with a heterogametic sex (eg, XY) there is a biological challenge in the heterogametic sex to equalize the gene expression of a monosomic X with that of autosomes (Straub and Becker, 2007). This is because the X chromosomes retain hundreds of functional genes while the Y chromosomes are mostly heterochromatic and only harbour a very small number of genes (Fig. 3). Complete dosage compensation, a mechanism that serves to hyperexpress the entire X (or Z chromosome in ZW chromosome systems) in the heterogametic sex (Mank, 2013; Ohno, 1967), results in hyperexpression of the entire X chromosome in *D. melanogaster* (Gelbart and Kuroda, 2009; Straub and Becker, 2007). Complete dosage compensation was determined to be in effect by RNA-sequencing analysis of male and female samples of *An. stephensi* (Jiang et al., 2015) and *An. gambiae* (Rose et al., 2016). In contrast, in the Culicinae subfamily where the M-factor resides at the M-locus on homomorphic sex chromosomes, dosage compensation is not expected as the remainder of the sex chromosome is much like other autosomes with many genes (Timoshevskiy et al., 2014).

Dosage compensation and sex determination are not regulated independently. *Sxl*, *Fem/Masc* and *xo-lethal 1*, the primary signals of sex determination in *D. melanogaster*, *B. mori* and *Caenorhabditis elegans*, respectively, also regulate dosage compensation (Kiuchi et al., 2014; Schutt and Nothiger, 2000; Thomas et al., 2012). Loss-of-function *sxl* mutations and knockdown of *Masc* cause female embryonic lethality in *D. melanogaster* and *B. mori*, respectively, likely due to misregulation of dosage compensation (Biedler et al., 2016; Cline, 1978; Kiuchi et al., 2014). It is not known if the M-factor(s) in *Anopheles* mosquitoes also regulates dosage compensation but if so this may make it difficult to disentangle the pathways of sex determination and dosage compensation in this group of mosquitoes. However, the presence of downstream intermediate players dedicated to one pathway may exist, which may be explored for mechanistic studies and for achieving sex conversion instead of sex-specific lethality (see Section 6.2).

5. MOLECULAR MECHANISMS OF SEX DETERMINATION IN MOSQUITOES

5.1 *Dsx* and *fru* in Mosquitoes: Complex and Distinct Splicing Patterns in Divergent Mosquito Genera

Dsx acts as the final binary switch at the bottom of the sex determination pathway (Verhulst and van de Zande, 2015). It is the most conserved factor in the pathway and is part of the large Doublesex/Male-abnormal-3-Related Transcription factor (*Dmrt*) gene superfamily (Picard et al., 2015). The *dsx* gene has been characterized in species of three mosquito genera: *Ae. aegypti*, *C. quinquefasciatus* and *An. gambiae* (Price et al., 2015; Salvemini et al., 2011; Scali et al., 2005). Interestingly, even though *dsx* is a highly conserved endpoint effector of the sex determination pathway, comparative analysis reveals conserved but also complex and varied splicing regulatory mechanisms between species of the three different mosquito genera (Fig. 4), and between mosquitoes and *Drosophila* (Price et al., 2015). In *Drosophila*, a weak 3′ splice acceptor of intron 3 is activated in order to retain the female-specific exon 4 (Lynch and Maniatis, 1995). In contrast, in *An. gambiae*, it is the promotion of a 5′ splice donor of intron 5 that is required to retain the female-specific exon 5 (homologous to *Drosophila* exon 4) (Scali et al., 2005), and this mechanism is similar to *fru* splicing in *Drosophila* (Lam et al., 2003). In *Ae. aegypti*, exon 5 has gained an intron resulting in what have been named exons 5a and 5b. The activation of a weak 3′ splice acceptor upstream of exon 5b produces the two female-specific isoforms, one that includes both exons 5a and 5b, and the other that has only exon 5b (Salvemini et al., 2011). Furthermore, the splice acceptor of exon 5a is ‘strong’ and may require a repressor for male-specific splicing. More recently, *dsx* has been characterized in *C. quinquefasciatus* (Price et al., 2015) where exon 5 is similar to that in *An. gambiae*, without the intron gain identified in *Ae. aegypti*. However, a weak 3′ splice acceptor identified in exon 6 is a unique characteristic in mosquito *dsx* gene structure. It is unknown if this is utilized for splicing the female, male or both isoforms. What all mosquito *dsx* genes have in common are the predicted TRA/TRA2 binding sites in the 3′ end of exon 5 (5b in *Ae. aegypti*). The complex and divergent sex-specific *dsx* splicing utilized by different mosquitoes may suggest the action of different regulatory mechanisms. However, biochemical studies are needed to test the earlier hypotheses.

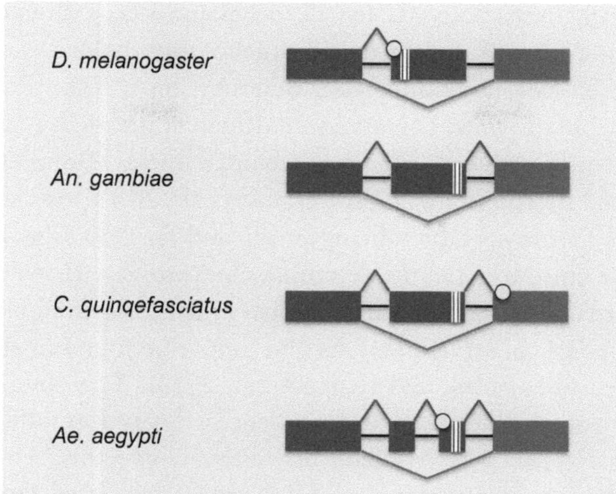

Fig. 4 Alternative splicing of *doublesex* pre-mRNA in *D. melanogaster* and mosquitoes. Simplified model of *doublesex* exon splicing is shown with *coloured rectangles* indicating homologous exons. *Red* exon in middle is the female-specific exon. In *Ae. aegypti*, the female-specific exon has gained an intron resulting in two exons. *Red* lines on top and *blue* lines on bottom connecting exons represent female-specific and male-specific splicing, respectively. *White* triplicate *vertical lines* indicate TRA/TRA2 binding sites and *yellow circles* indicate predicted weak splice sites (Price et al., 2015). Only select exons are shown that are spliced in a sex-specific manner. (See the colour plate.)

Like *dsx*, *fru* is an endpoint effector in the sex determination pathway and a transcription factor that programmes sexual dimorphism. In *Drosophila*, *fru* is subjected to sex-specific splicing and the male spliceform of *fru* directs neural development responsible for male courtship behaviour (Demir and Dickson, 2005). Analysis of *Ae. aegypti* and *An. gambiae fru* genes indicates similar sex-specific alternative splicing and conserved function (Gailey et al., 2006; Salvemini et al., 2013). *Ae. aegypti* and *An. gambiae fru* generate multiple nonsex-specific transcripts, with at least one transcript undergoing sex-specific splicing. The female-specific exon has predicted TRA/TRA-2-binding sites in the 3′ end that presumably mediate switching from default male-specific splicing to the female-specific isoform, including the female-specific exon. This exon has a stop codon early in the transcript, likely rendering the protein nonfunctional in females.

5.2 Is There a *Transformer* in Mosquitoes?

A generalized model of sex determination in holometabolous insects (except for *B. mori* which is in the order Lepidoptera) is one that puts TRA at the centre (Fig. 2), instructing the female developmental pathway (Verhulst et al., 2010). Although found outside of holometabolous insects, TRA apparently does not function in sex determination (Kato et al., 2010). TRA is arguably the most conserved fundamental instructive signal in the sex determination pathway of holometabolous insects (Bopp et al., 2014; Geuverink and Beukeboom, 2014; Verhulst et al., 2010). As a splicing factor, it directs female-specific splicing of *dsx* and *fru*, downstream effectors that are at the bottom of the sex determination pathway. However, despite having such an important role and being ancestral to holometabolous insects, TRA is much less conserved than *dsx* and cannot be found in all species of insects studied (Geuverink and Beukeboom, 2014). This absence may be due to divergent evolution beyond sequence recognition while retaining the functional capacity to regulate downstream effectors. For example, because of the low conservation of TRA, researchers used synteny of *tra* and another conserved gene in *Drosophila* to identify a candidate in *C. capitata* (Pane et al., 2002). This *C. capitata* homolog encodes an RS-rich protein that was indeed TRA but only had 32–40% identity over 120 amino acids with the *D. melanogaster* TRA. The *csd* and *fem* genes of the honeybee *Ap. mellifera* are RS-rich proteins with similarities to TRA (Beye et al., 2003), and comparative analysis of these genes with *tra* orthologs suggests a common evolutionary origin (Hasselmann et al., 2008). Interestingly, *csd* is believed to have arisen from a gene duplication of *fem* between 10 and 70 Ma.

Alternatively, in some species the ancestral TRA was lost and other proteins have been recruited to replace the necessary sex-specific splicing function of TRA, as dsx^M and dsx^F splice forms exist in the species where TRA has not been found. In *B. mori* of the order Lepidoptera where no *tra*-containing species have been identified, sex-specific splicing of *dsx* is carried out by BmIMP and BmPSI (Suzuki et al., 2010). BmIMP is a homolog of IGF-II mRNA-binding protein, and BmPSI is a homolog of P-element somatic inhibitor.

Tra is absent or partly absent from four insect orders (Geuverink and Beukeboom, 2014). It is absent from Lepidoptera (Mita et al., 2004) and Strepsiptera (but with only one genome considered), and partly absent from Coleoptera and basal Diptera (Salvemini et al., 2013). The loss of *tra* from basal Diptera is interesting as this includes the mosquito family and *tra* has not been found in *Ae. aegypti* or *An. gambiae* (Gailey et al., 2006; Nene et al., 2007). It is currently unknown if *tra* was replaced, or simply evolved beyond recognition. However, potential TRA/TRA-2 binding sites similar to those in *Drosophila* have been identified in *Ae. aegypti*, *C. quinquefasciatus* and *An. gambiae dsx* and *fru* genes, indicating the involvement of TRA-2 and possibly TRA-like factors (Gailey et al., 2006; Price et al., 2015; Salvemini et al., 2013; Scali et al., 2005).

5.3 The Discovery of an M-Factor and M-Factor Candidates in Mosquitoes

Recently an M-locus gene, called *Nix*, was discovered in *Ae. aegypti* (Hall et al., 2015) and was shown to be the first M-factor to be characterized in insects. NIX protein is predicted to have RNA-binding properties and is expressed at the onset of the maternal-to-zygotic transition, before sex is determined. When *Nix* was knocked out using clustered regularly interspersed palindromic repeats (CRISPR)/Cas9, genotypic males were feminized and had female antennae and genitals. When *Nix* was ectopically expressed, genotypic females developed male morphological features including male testis and external genitalia. Therefore, *Nix* was determined to be both necessary and sufficient to initiate male development. It was also determined that NIX functions by regulating the sex-specific splicing of *dsx/fru*. It is important to note that in this study CRISPR/Cas9 was not expected to knockout 100% of the target *Nix* gene in all cells due to limitations inherent in administering CRISPR/Cas9 by embryonic injection. This was supported by sequencing affected individuals and explains the resulting mosaic and variable nature of the observed phenotypes. *Nix* is a distant homologue of TRA-2 (Hall et al., 2015). One could speculate that female *dsx* splicing is the default in *Ae. aegypti* and NIX binds the TRA2-ISS sequences upstream of the 5a exon (Price et al., 2015; Salvemini et al., 2011) precluding inclusion of exon 5 and promoting generation of the male spliceform (Adelman and Tu, 2016; Salvemini et al., 2011, 2013). However, this hypothesis alone does not explain the splicing of exon 5b in males. It remains to be determined how NIX functions to either directly or indirectly control *dsx/fru* splicing. Although *Nix* homologue was found in the Asian tiger mosquito *Ae. albopictus*, it has not been identified in more divergent Culicinae species such as *C. quinquefasciatus* (Hall et al., 2015).

In *An. culifacies*, it was found that individuals with the XXY genotype developed as males (Baker and Sakai, 1979), suggesting the presence of a dominant M-factor on the Y chromosome. In *An. stephensi*, four Y genes were discovered including a candidate M-factor named *Guy1* (Criscione et al., 2013; Hall et al., 2013). *Guy1* (Gene unique to the Y) is expressed at the onset of the maternal-to-zygotic transition, the earliest of all known Y chromosome genes in *An. stephensi*. Furthermore, expression of *Guy1* does not depend on any other factors from the Y chromosome, making it a strong candidate for the primary signal of sex determination (Criscione et al., 2013). A few Y chromosome genes were also recently found in *An. gambiae*, including *gYG2* (Hall et al., 2013), which is expressed at the onset of the maternal-to-zygotic transition. Interestingly, this *gYG2* (*g* refers to *An. gambiae* and *YG* refers to Y gene) gene was the only Y chromosome gene that is shared among all species within the *An. gambiae* species complex (Hall et al., 2016), making it a strong candidate for the M-factor. Although *Guy1* and *gYG2* do not have easily detectable sequence homology, they both encode a 56 aa small protein rich in basic amino acid residues. Without demonstration of function, *Guy1* and *gYG2* remain just candidates for the M-factor. As mentioned earlier, if the M-factor in *Anopheles* mosquitoes also regulates dosage compensation, it may complicate the efforts to demonstrate the effect of the candidate M-factor on sex determination, as manipulations of such a factor may result in sex-specific lethality (Biedler et al., 2016).

5.4 Assumptions and Additional Considerations

Although recent years witnessed rapid progress in our understanding of the molecular basis of sex determination in mosquitoes, much remains to be investigated and it is important to consider some unstated assumptions. First of all, although sex-biassed splicing of *dsx/fru* has been clearly demonstrated as described earlier, and RNAi knockdown of *dsx*^{F} resulted in female lethality in *Ae. aegypti* (Whyard et al., 2015), there has been no direct demonstration that *dsx*^{M} and *dsx*^{F} leads to male and female differentiation, respectively. Although it is reasonable to assume that the sex-specific splicing of *dsx* and *fru* programmes sexual development in mosquitoes, based on conservation of the functions of the two transcription factors, this assumption remains to be tested. In addition, recent data from *D. melanogaster* suggest the existence of a *dsx/fru*-independent pathway downstream of *tra* in the adult intestinal stem cells (Hudry et al., 2016). Although *tra*/*tra*-2-independent branches of sex determination have been documented both in species that use *tra*/*tra*-2, and in species that appear to have lost the *tra* gene (Mita et al., 2004; Sun et al., 2015), the demonstration of a *dsx/fru*-independent pathway challenges the dogma that *dsx/fru* are the ultimate or sole output for sex determination.

6. CONTROL OF MOSQUITO-BORNE INFECTIOUS DISEASES THROUGH MANIPULATION OF GENES IN THE SEX DETERMINATION PATHWAY

6.1 Nature's Example: Meiotic Drive in Mosquitoes

Meiotic drive is a process that results in a distortion of segregation of genetic material during meiosis that can result in under- or overrepresentation in the gametes, and therefore deviates from Mendelian inheritance. It was discovered that some strains of *Ae. aegypti* had a paternal meiotic drive resulting in sex-ratio distortion, producing a higher percentage of males than females in offspring (Craig et al., 1960, 1961; Wood, 1961). The effect persisted with the male offspring for subsequent generations when tested. Some strains produced as few as ~20–30% females and it was shown that the male bias was not a result of postgametic mortality of females. Later studies produced lines having ~0–15% female offspring (Hickey and Craig, 1966). The locus responsible for the sex-ratio distortion was termed Distorter (*D*) and the distorter effect was found to be always associated with males (Hickey and Craig, 1966). The meiotic drive was determined to be associated with 'X' (non *M*-bearing) chromosome breakage during meiosis in males and mainly involved four sites (Newton et al., 1976). Cytological mapping of *D*, along with the *M*-locus showed that they were closely linked and had 1.2% recombination between them (Newton et al., 1978). Later studies showed 5.2% recombination (Shin et al., 2012). A sex-ratio distorter was also discovered in *C. quinquefasciatus* that produced an excessive number of male progeny (Sweeny and Barr, 1978). As in *Ae. aegypti* it was determined to only be active in the male during meiosis, but in contrast to *Ae. aegypti*, the distorter effect is explained by a recessive allele termed *d*. This discovery of the Distorter spawned ideas of harnessing the sex-ratio distorter trait for mosquito control (Craig, 1963; Craig et al., 1960; Huang et al., 2007; Mori et al., 2004). Although there were some initial promising results with *Ae. aegypti* (Cha et al., 2006; Curtis et al., 1976), development of this strategy has been hampered partly due to challenges including concerns about natural resistance and varying sensitivities to the distorter found in wild populations (Suguna et al., 1977; Wood, 1976; Wood and Ouda, 1987).

6.2 Applications for Disease Control

For a long time it was envisioned that vector control could be implemented using genetic strategies that either reduces mosquito populations or replaces competent vector populations with disease-resistant populations. As it is the female mosquitoes that bite and transmit the disease-causing agents, releasing only males is an important consideration for any genetic approach designed to control mosquito-borne diseases (Alphey, 2014; Benedict and Robinson, 2003; Black et al., 2011; Carvalho et al., 2015b; Collins, 1994; Fu et al., 2010; Galizi et al., 2014; Windbichler et al., 2008). Sex separation is still a major bottleneck, especially for *Anopheles* (Gilles et al., 2014; Papathanos et al., 2009). Genetic manipulations that result in all male progeny will be very helpful in this regard. Such transgenic lines may be maintained as homozygous if the sex-biassing transgene is under the control of a conditional promoter that is suppressed in the laboratory. Due to the introduction of male bias in subsequent generations, release of such transgenic males can achieve more effective population suppression than the classic SIT (Schliekelman et al., 2005; Thomas et al., 2000).

Targeting various components in the sex determination pathway can offer powerful strategies for vector control by altering sex ratios through female to male conversion or by causing female lethality. In *Ae. aegypti*, the feeding of larvae with dsRNA targeting the female isoform of *dsx* resulted in female lethality (Whyard et al., 2015). Targeting *dsx* in *Ae. aegypti* females by injection of siRNAs into pupae resulted in alteration of female morphological characteristics that resulted in lower fecundity (Mysore et al., 2015). Knockdown of *Ae. aegypti tra-2* using RNAi results in a dramatic increase in male/female offspring ratio by a segregation distortion (meiotic drive) mechanism (Hoang et al., 2016). Recent identification of the M-factor *Nix* in *Ae. aegypti* provides an exciting opportunity to improve understanding of the molecular pathways of sex determination in mosquitoes and will therefore facilitate the development of multiple genetic strategies to control mosquito-borne diseases.

Population reduction may be inflicted by causing female lethality, female sterility or converting females to males (Fig. 5). At the same time, the reduced number of females means less individuals that bite and transmit disease-causing pathogens. It is possible to significantly increase the efficiency of this approach by coupling a transgene that introduces male bias with a gene drive, a mechanism that increases the probability of inheritance of the transgene. Recently, a gene drive has been engineered using the CRISPR/Cas9 gene-editing system (Gantz et al., 2015) that utilized the mutagenic chain reaction strategy (Gantz and Bier, 2015). Combining a germline-specific CRISPR/Cas9 gene drive with an anti-*Plasmodium falciparum* dual effector gene cassette, it was demonstrated that these

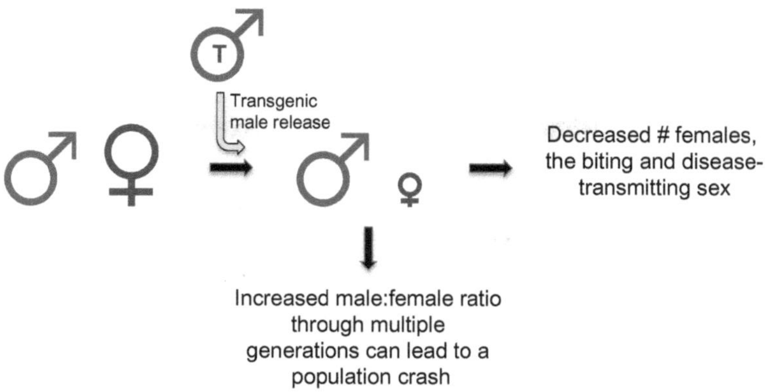

Fig. 5 A schematic of the strategy to control mosquito-borne infectious diseases by an engineered sex-ratio distorter. Coupling the transgene with a gene drive (Adelman and Tu, 2016; Biedler et al., 2016) will likely greatly improve the efficiency of the approach. Even with a drive mechanism multiple releases may be necessary to achieve desired effect. Effective vector control may be achieved without a drive mechanism by multiple and successive releases.

transgenes could be spread to offspring with greater than 99.5% efficiency. Similarly, using a CRISPR/Cas9-based gene drive, it was demonstrated that targeting three genes that confer a recessive sterility phenotype in the African malaria mosquito *An. gambiae*, gene disruption was transmitted to 91.4–99.6% of progeny (Hammond et al., 2016). Cage experiments in this study yielded promising results showing a rapid increase in allele frequency of the transgene in just a few generations. Therefore, driving 'maleness' into the mosquito population may represent an effective and efficient strategy to control mosquito-infectious diseases (Adelman and Tu, 2016; Biedler et al., 2016).

7. CONCLUDING REMARKS

Sexual reproduction is critical to the survival and evolution of many organisms including mosquitoes. Sex determination in mosquitoes captures significant attention beyond just basic science because sex separates the harmless nonbiting males from the deadly females that transmit pathogens that cause malaria, dengue and Zika fever. Despite strong interest, information on the molecular signals and mechanisms of sex determination has long been scarce in mosquitoes. Thus, it is rewarding that recent years have witnessed several breakthroughs that either brought novel insights, or raised important and exciting questions, or established enabling methodologies.

First, coupling next-generation sequencing with the Chromosome Quotient method, genes that were previously hidden in the male-specific but repeat-rich Y chromosome or M-locus have been discovered in *An. stephensi*, *An. gambiae* and *Ae. aegypti* (Criscione et al., 2013; Hall et al., 2013, 2014, 2015, 2016). This made it possible to study the function of these important genes in sex determination and male development, and will enable evolutionary analysis in a diverse array of mosquito species. Second, the discovery of the M-factor *Nix* in *Ae. aegypti* (Hall et al., 2015) has revealed for the first time the molecular identity of a primary signal that controls male development in this important vector of arboviruses. Although substantial challenges remain to deciphering the sex determination pathway in *Ae. aegypti* and other mosquitoes, the characterization of the primary signal fills a major gap and will facilitate the discovery of downstream players in sex determination. Third, the first systematic characterization of an *Anopheles* Y chromosome revealed the scarcity of genes and the predominance of a few repeat sequences ordered in tandem arrays (Hall et al., 2016). The Y chromosome content changed rapidly within the *An. gambiae* populations and within sister species of the *An. gambiae* complex. Although significant challenges remain, including the possible complication due to joint regulation between sex determination and dosage compensation, the discovery of candidates for the M-factors (Criscione et al., 2013; Hall et al., 2013, 2016) in *An. stephensi* and *An. gambiae* provides hope for improving understanding of both sex determination and dosage compensation in *Anopheles* mosquitoes. Fourth, complex and divergent patterns of sex-specific alternative splicing of *dsx* have been demonstrated in species of three mosquito genera (Price et al., 2015; Salvemini et al., 2011; Scali et al., 2005), suggesting possibly different regulatory mechanisms and calling for biochemical studies of the mechanisms of *dsx* splicing. Finally, significant progress has been made in demonstrating that sex ratio may be altered by targeting various components in the sex determination pathway (Hoang et al., 2016; Mysore et al., 2015; Whyard et al., 2015). The discovery of *Nix* in *Ae. aegypti* provides an opportunity to manipulate the primary sex determination signal and may lead to additional targets for manipulation. Coupling 'maleness' genes with recently developed CRISPR/cas9-based gene drives (Gantz et al., 2015; Hammond et al., 2016) may represent an effective and efficient strategy to control mosquito-borne infectious diseases (Adelman and Tu, 2016; Biedler et al., 2016). It is clear that major challenges are ahead. However, armed with newly acquired knowledge and emerging technologies, we are much better equipped than only a few years ago to make progress in understanding the molecular mechanisms and evolution of sex determination in mosquitoes and realizing the potential for novel genetic control of mosquito-borne infectious diseases.

ACKNOWLEDGEMENTS

We thank Brantley Hall and Janet Webster for critical review of the manuscript. This work is supported by NIH Grants AI121284, AI123338, AI113643, AI077680 and AI105575 to Z.T., by the Virginia Agricultural Experimental Station and the Virginia Tech Fralin Life Science Institute.

REFERENCES

Abe, H., Fujii, T., Tanaka, N., Yokoyama, T., Kakehashi, H., Ajimura, M., Mita, K., Banno, Y., Yasukochi, Y., Oshiki, T., Nenoi, M., Ishikawa, T., Shimada, T., 2008. Identification of the female-determining region of the W chromosome in Bombyx mori. Genetica 133, 269–282.

Adelman, Z.N., Tu, Z., 2016. Control of mosquito-borne infectious diseases: sex and gene drive. Trends Parasitol. 32, 219–229.

Alphey, L., 2014. Genetic control of mosquitoes. Annu. Rev. Entomol. 59, 205–224.

Bachtrog, D., 2013. Y-chromosome evolution: emerging insights into processes of Y-chromosome degeneration. Nat. Rev. Genet. 14, 113–124.

Bachtrog, D., Kirkpatrick, M., Mank, J.E., McDaniel, S.F., Pires, J.C., Rice, W., Valenzuela, N., 2011. Are all sex chromosomes created equal? Trends Genet. 27, 350–357.

Bachtrog, D., Mank, J.E., Peichel, C.L., Kirkpatrick, M., Otto, S.P., Ashman, T.L., Hahn, M.W., Kitano, J., Mayrose, I., Ming, R., Perrin, N., Ross, L., Valenzuela, N., Vamosi, J.C., Consortium, T.O.S., 2014. Sex determination: why so many ways of doing it? PLoS Biol. 12, e1001899.

Baker, R.H., Sakai, R.K., 1976. Male determining factor on chromosome 3 in the mosquito, Culex tritaeniorhynchus. J. Hered. 67, 289–294.

Baker, R.H., Sakai, R.K., 1979. Triploids and male determination in the mosquito, Anopheles culicifacies. J. Hered. 70, 345–346.

Baker, R.H., Sakai, R.K., Mian, A., 1971. Linkage group-chromosome correlation in Culex tritaeniorhynchus. Science 171, 585–587.

Baker, R.H., Umaima, T., Uddin, S., Sakai, R.K., 1977. Variations in the linkage of the sex allele in laboratory colonies of the mosquito Culex tritaeniorhynchus. Jpn. J. Genet. 42, 425–430.

Benedict, M.Q., Robinson, A.S., 2003. The first releases of transgenic mosquitoes: an argument for the sterile insect technique. Trends Parasitol. 19, 349–355.

Beye, M., Moritz, R.F.A., Crozier, R.H., Crozier, Y.C., 1996. Mapping the sex locus of the honeybee (Apis mellifera). Naturwissenschaften 83, 424–426.

Beye, M., Hasselmann, M., Fondrk, M.K., Page, R.E., Omholt, S.W., 2003. The gene csd is the primary signal for sexual development in the honeybee and encodes an SR-type protein. Cell 114, 419–429.

Biedler, J.K., Hall, A.B., Jiang, X., Tu, Z., 2016. Exploring the Sex-Determination Pathway for Control of Mosquito-Borne Infectious Diseases. In: Adelman, Z.N. (Ed.), Elsevier, San Diego, CA.

Black, W.C.T., Alphey, L., James, A.A., 2011. Why RIDL is not SIT. Trends Parasitol. 27, 362–370.

Bopp, D., Saccone, G., Beye, M., 2014. Sex determination in insects: variations on a common theme. Sex Dev. 8, 20–28.
Bridges, C.B., 1916. Non-disjunction as proof of the chromosome theory of heredity. Genetics 1, 1–52.
Bridges, C.B., 1921. Triploid intersexes in Drosophila melanogaster. Science 54, 252–254.
Bull, J.J., 1983. Evolution of Sex Determining Mechanisms. The Benjamin/Cummings Publishing Company, Inc., San Francisco, CA.
Bush, G.L., 1966. Female heterogamety in the family tephritidae (acalyptratae, diptera). Am. Nat. 100, 119–126.
Carvalho, A.B., Lazzaro, B.P., Clark, A.G., 2000. Y chromosomal fertility factors kl-2 and kl-3 of Drosophila melanogaster encode dynein heavy chain polypeptides. Proc. Natl. Acad. Sci. U.S.A. 97, 13239–13244.
Carvalho, A.B., Vicoso, B., Russo, C.A., Swenor, B., Clark, A.G., 2015a. Birth of a new gene on the Y chromosome of Drosophila melanogaster. Proc. Natl. Acad. Sci. U.S.A. 112, 12450–12455.
Carvalho, D.O., Mckemey, A.R., Garziera, L., Lacroix, R., Donnelly, C.A., Alphey, L., Malavasi, A., Capurro, M.L., 2015b. Suppression of a field population of Aedes aegypti in Brazil by sustained release of transgenic male mosquitoes. PLoS Negl. Trop. Dis. 9, e0003864.
Cha, S.J., Mori, A., Chadee, D.D., Severson, D.W., 2006. Cage trials using an endogenous meiotic drive gene in the mosquito Aedes aegypti to promote population replacement. Am. J. Trop. Med. Hyg. 74, 62–68.
Charlesworth, D., Mank, J.E., 2010. The birds and the bees and the flowers and the trees: lessons from genetic mapping of sex determination in plants and animals. Genetics 186, 9–31.
Charlesworth, D., Charlesworth, B., Marais, G., 2005. Steps in the evolution of heteromorphic sex chromosomes. Heredity (Edinb) 95, 118–128.
Chen, X.G., Jiang, X., Gu, J., Xu, M., Wu, Y., Deng, Y., Zhang, C., Bonizzoni, M., Dermauw, W., Vontas, J., Armbruster, P., Huang, X., Yang, Y., Zhang, H., He, W., Peng, H., Liu, Y., Wu, K., Chen, J., Lirakis, M., Topalis, P., Van Leeuwen, T., Hall, A.B., Jiang, X., Thorpe, C., Mueller, R.L., Sun, C., Waterhouse, R.M., Yan, G., Tu, Z.J., Fang, X., James, A.A., 2015. Genome sequence of the Asian tiger mosquito, Aedes albopictus, reveals insights into its biology, genetics, and evolution. Proc. Natl. Acad. Sci. U.S.A. 112, E5907–E5915.
Clements, A.N., 1992. The biology of mosquitoes. Development, Nutrition and Reproduction, vol. 1. Chapman & Hall, London.
Cline, T.W., 1978. Two closely linked mutations in Drosophila melanogaster that are lethal to opposite sexes and interact with daughterless. Genetics 90, 683–698.
Cline, T.W., 1993. The Drosophila sex determination signal: how do flies count to two? Trends Genet. 9, 385–390.
Clough, E., Jimenez, E., Kim, Y.A., Whitworth, C., Neville, M.C., Hempel, L.U., Pavlou, H.J., Chen, Z.X., Sturgill, D., Dale, R.K., Smith, H.E., Przytycka, T.M., Goodwin, S.F., Van Doren, M., Oliver, B., 2014. Sex- and tissue-specific functions of Drosophila doublesex transcription factor target genes. Dev. Cell 31, 761–773.
Collins, F.H., 1994. Prospects for malaria control through the genetic manipulation of its vectors. Parasitol. Today 10, 370–371.
Craig, G.B., 1963. Prospects for vector control through genetic manipulation of populations. Bull. World Health Organ. 29 (Suppl.), 89–97.
Craig Jr., G.B., Hickey, W.A., Vandehey, R.C., 1960. An inherited male-producing factor in Aedes aegypti. Science 132, 1887–1889.
Craig Jr., G.B., Vandehey, R.C., Hickey, W.A., 1961. Genetic variability in populations of Aedes aegypti. Bull. World Health Organ. 24, 527–539.

Criscione, F., Qi, Y., Saunders, R., Hall, B., Tu, Z., Criscione, F., Qi, Y., Saunders, R., Hall, B., Tu, Z., 2013. A unique Y gene in the Asian malaria mosquito Anopheles stephensi encodes a small lysine-rich protein and is transcribed at the onset of embryonic development. Insect Mol. Biol. 22, 433–441.

Curtis, C.F., Grover, K.K., Suguna, S.G., Uppal, D.K., Dietz, K., Agarwal, H.V., Kazmi, S.J., 1976. Comparative field cage tests of the population suppressing efficiency of three genetic control systems for Aedes Aegypti. Heredity (Edinb) 36, 11–29.

Davies, N., Roderick, G.K., 2005. Dipteran sex chromosomes in evolutionary developmental biology. In: Yeates, D.K., Wegmann, B.M. (Eds.), The Evolutionary Biology of Flies. Columbia University Press, New York.

Demir, E., Dickson, B.J., 2005. Fruitless splicing specifies male courtship behavior in Drosophila. Cell 121, 785–794.

Dubendorfer, A., Hediger, M., Burghardt, G., Bopp, D., 2002. Musca domestica, a window on the evolution of sex-determining mechanisms in insects. Int. J. Dev. Biol. 46, 75–79.

Erickson, J.W., Quintero, J.J., 2007. Indirect effects of ploidy suggest X chromosome dose, not the X:a ratio, signals sex in *Drosophila*. PLoS Biol. 5, e332.

Feldmeyer, B., Kozielska, M., Kuijper, B., Weissing, F.J., Beukeboom, L.W., Pen, I., 2008. Climatic variation and the geographical distribution of sex-determining mechanisms in the housefly. Evol. Ecol. Res. 10, 797–809.

Ferdig, M.T., Taft, A.S., Severson, D.W., Christensen, B.M., 1998. Development of a comparative genetic linkage map for Armigeres subalbatus using Aedes aegypti RFLP markers. Genome Res. 8, 41–47.

Fraccaro, M., Tiepolo, L., Laudani, U., Marchi, A., Jayakar, S.D., 1977. Y chromosome controls mating behaviour on Anopheles mosquitoes. Nature 265, 326–328.

Fu, G., Lees, R.S., Nimmo, D., Aw, D., Jin, L., Gray, P., Berendonk, T.U., White-Cooper, H., Scaife, S., Kim Phuc, H., Marinotti, O., Jasinskiene, N., James, A.A., Alphey, L., 2010. Female-specific flightless phenotype for mosquito control. Proc. Natl. Acad. Sci. U.S.A. 107, 4550–4554.

Gailey, D.A., Billeter, J.C., Liu, J.H., Bauzon, F., Allendorfer, J.B., Goodwin, S.F., 2006. Functional conservation of the fruitless male sex-determination gene across 250 Myr of insect evolution. Mol. Biol. Evol. 23, 633–643.

Galizi, R., Doyle, L.A., Menichelli, M., Bernardini, F., Deredec, A., Burt, A., Stoddard, B.L., Windbichler, N., Crisanti, A., 2014. A synthetic sex ratio distortion system for the control of the human malaria mosquito. Nat. Commun. 5, 3977.

Gantz, V.M., Bier, E., 2015. Genome editing. The mutagenic chain reaction: a method for converting heterozygous to homozygous mutations. Science 348, 442–444.

Gantz, V.M., Jasinskiene, N., Tatarenkova, O., Fazekas, A., Macias, V.M., Bier, E., James, A.A., 2015. Highly efficient Cas9-mediated gene drive for population modification of the malaria vector mosquito Anopheles stephensi. Proc. Natl. Acad. Sci. U.S.A. 112, E6736–E6743.

Gelbart, M.E., Kuroda, M.I., 2009. Drosophila dosage compensation: a complex voyage to the X chromosome. Development 136, 1399–1410.

Gempe, T., Beye, M., 2011. Function and evolution of sex determination mechanisms, genes and pathways in insects. Bioessays 33, 52–60.

Geuverink, E., Beukeboom, L.W., 2014. Phylogenetic distribution and evolutionary dynamics of the sex determination genes doublesex and transformer in insects. Sex Dev. 8, 38–49.

Gilchrist, B.M., Haldane, J.B.S., 1947. Sex linkage and sex determination in a mosquito, Culex molestus. Hereditas 33, 175–190.

Gilles, J.R.L., Schetelig, M.F., Scolari, F., Marec, F., Capurro, M.L., Franz, G., Bourtzis, K., 2014. Towards mosquito sterile insect technique programmes: exploring genetic,

molecular, mechanical and behavioural methods of sex separation in mosquitoes. Acta Trop. 132 (Suppl), S178–S187.
Graham, P., Penn, J.K., Schedl, P., 2003. Masters change, slaves remain. Bioessays 25, 1–4.
Hashimoto, H., 1933. The role of the W-chromosome in the sex determination of Bombyx mori. Jpn. J. Genet. 8, 245–247.
Hall, A.B., Qi, Y., Timoshevskiy, V., Sharakhova, M.V., Sharakhov, I.V., Tu, Z., 2013. Six novel Y chromosome genes in Anopheles mosquitoes discovered by independently sequencing males and females. BMC Genomics 14, 273.
Hall, A.B., Timoshevskiy, V.A., Sharakhova, M.V., Jiang, X., Basu, S., Anderson, M.A., Hu, W., Sharakhov, I.V., Adelman, Z.N., Tu, Z., 2014. Insights into the preservation of the homomorphic sex-determining chromosome of Aedes aegypti from the discovery of a male-biased gene tightly linked to the M-locus. Genome Biol. Evol. 6, 179–191.
Hall, A.B., Basu, S., Jiang, X., Qi, Y., Timoshevskiy, V.A., Biedler, J.K., Sharakhova, M.V., Elahi, R., Anderson, M.A., Chen, X.G., Sharakhov, I.V., Adelman, Z.N., Tu, Z., 2015. A male-determining factor in the mosquito Aedes aegypti. Science 348, 1268–1270.
Hall, A.B., Papathanos, P.-A., Sharma, A., Cheng, C., Akbari, O.S., Assour, L., Bergman, N.H., Cagnetti, A., Crisanti, A., Dottorini, T., Fiorentini, E., Galizi, R., Hnath, J., Jiang, X., Koren, S., Nolan, T., Radune, D., Sharakhova, M.V., Steele, A., Timoshevskiy, V.A., Windbichler, N., Zhang, S., Hahn, M.W., Phillippy, A.M., Emrich, S.J., Sharakhov, I.V., Tu, Z.J., Besansky, N.J., 2016. Radical remodeling of the Y chromosome in a recent radiation of malaria mosquitoes. Proc. Natl. Acad. Sci. U.S.A. 113, E2114–E2123.
Hammond, A., Galizi, R., Kyrou, K., Simoni, A., Siniscalchi, C., Katsanos, D., Gribble, M., Baker, D., Marois, E., Russell, S., Burt, A., Windbichler, N., Crisanti, A., Nolan, T., 2016. A CRISPR-Cas9 gene drive system targeting female reproduction in the malaria mosquito vector Anopheles gambiae. Nat. Biotechnol. 34, 78–83.
Hasselmann, M., Gempe, T., Schiøtt, M., Nunes-Silva, C.G., Otte, M., Beye, M., 2008. Evidence for the evolutionary nascence of a novel sex determination pathway in honeybees. Nature 454, 519–522.
Hediger, M., Henggeler, C., Meier, N., Perez, R., Saccone, G., Bopp, D., 2010. Molecular characterization of the key switch F provides a basis for understanding the rapid divergence of the sex-determining pathway in the housefly. Genetics 184, 155–170.
Herpin, A., Schartl, M., 2015. Plasticity of gene-regulatory networks controlling sex determination: of masters, slaves, usual suspects, newcomers, and usurpators. EMBO Rep. 16, 1260–1274.
Hickey, W.A., Craig JR., G.B., 1966. Genetic distortion of sex ratio in a mosquito, Aedes aegypti. Genetics 53, 1177–1196.
Hoang, K.P., Teo, T.M., Ho, T., Le, V.S., 2016. Mechanisms of sex determination and transmission ratio distortion in Aedes aegypti. Parasit. Vectors 9, 49.
Huang, Y.X., Magori, K., Lloyd, A.L., Gould, F., 2007. Introducing desirable transgenes into insect populations using Y-linked meiotic drive—a theoretical assessment. Evolution 61, 717–726.
Hudry, B., Khadayate, S., Miguel-Aliaga, I., 2016. The sexual identity of adult intestinal stem cells controls organ size and plasticity. Nature 530, 344–348.
Jiang, X., Biedler, J.K., Qi, Y., Hall, A.B., Tu, Z., 2015. Complete dosage compensation in anopheles stephensi and the evolution of Sex-biased genes in mosquitoes. Genome Biol. Evol. 7, 1914–1924.
Jost, E., Laven, H., 1971. Meiosis in translocation heterozygotes in the mosquito Culex pipiens. Chromosoma 35, 184–205.
Kato, Y., Kobayashi, K., Oda, S., Tatarazako, N., Watanabe, H., Iguchi, T., 2010. Sequence divergence and expression of a transformer gene in the branchiopod crustacean, Daphnia magna. Genomics 95, 160–165.

Kiuchi, T., Koga, H., Kawamoto, M., Shoji, K., Sakai, H., Arai, Y., Ishihara, G., Kawaoka, S., Sugano, S., Shimada, T., Suzuki, Y., Suzuki, M.G., Katsuma, S., 2014. A single female-specific piRNA is the primary determiner of sex in the silkworm. Nature 509, 633–636.

Krzywinski, J., Nusskern, D.R., Kern, M.K., Besansky, N.J., 2004. Isolation and characterization of Y chromosome sequences from the African malaria mosquito Anopheles gambiae. Genetics 166, 1291–1302.

Krzywinski, J., Chrystal, M.A., Besansky, N.J., 2006a. Gene finding on the Y: fruitful strategy in Drosophila does not deliver in Anopheles. Genetica 126, 369–375.

Krzywinski, J., Grushko, O.G., Besansky, N.J., 2006b. Analysis of the complete mitochondrial DNA from Anopheles funestus: an improved dipteran mitochondrial genome annotation and a temporal dimension of mosquito evolution. Mol. Phylogenet. Evol. 39, 417–423.

Lam, B.J., Bakshi, A., Ekinci, F.Y., Webb, J., Graveley, B.R., Hertel, K.J., 2003. Enhancer-dependent 5′-splice site control of fruitless pre-mRNA splicing. J. Biol. Chem. 278, 22740–22747.

Lemos, B., Araripe, L.O., Hartl, D.L., 2008. Polymorphic Y chromosomes harbor cryptic variation with manifold functional consequences. Science 319, 91–93.

Lynch, K.W., Maniatis, T., 1995. Synergistic interactions between two distinct elements of a regulated splicing enhancer. Genes Dev. 9, 284–293.

Malcolm, C.A., Bourguet, D., Ascolillo, A., Rooker, S.J., Garvey, C.F., Hall, L.M.C., Pasteur, N., Raymond, M., 1998. A sex-linked Ace gene, not linked to insensitive acetylcholinesterase-mediated insecticide resistance in Culex pipiens. Insect Mol. Biol. 7, 107–120.

Mank, J.E., 2013. Sex chromosome dosage compensation: definitely not for everyone. Trends Genet. 29, 677–683.

Marec, F., Novak, K., 1998. Absence of sex chromatin corresponds with a sex-chromosome univalent in females of Trichoptera. Eur. J. Entomol. 95, 197–209.

Marin, I., Baker, B.S., 1998. The evolutionary dynamics of sex determination. Science 281, 1990–1994.

McClelland, G.A.H., 1962. Sex-linkage in Aedes aegypti. Trans. R. Soc. Trop. Med. Hyg. 56, 4.

McDonald, P.T., Rai, K.S., 1970. Correlation of linkage groups with chromosomes in the mosquito, Aedes aegypti. Genetics 66, 475–485.

McDonald, P.T., Asman, S.M., Terwedow Jr., H.A., 1978. Sex-linked translocations in Culex tarsalis:chromosome—linkage group correlation and segregation patterns. J. Hered. 69, 282–360.

Mita, K., Kasahara, M., Sasaki, S., Nagayasu, Y., Yamada, T., Kanamori, H., Namiki, N., Kitagawa, M., Yamashita, H., Yasukochi, Y., Kadono-Okuda, K., Yamamoto, K., Ajimura, M., Ravikumar, G., Shimomura, M., Nagamura, Y., Shin, I.T., Abe, H., Shimada, T., Morishita, S., Sasaki, T., 2004. The genome sequence of silkworm, Bombyx mori. DNA Res. 11, 27–35.

Mori, A., Severson, D.W., Christensen, B.M., 1999. Comparative linkage maps for the mosquitoes (Culex pipiens and Aedes aegypti) based on common RFLP loci. J. Hered. 90, 160–164.

Mori, A., Tomita, T., Hidoh, O., Kono, Y., Severson, D.W., 2001. Comparative linkage map development and identification of an autosomal locus for insensitive acetylcholinesterase-mediated insecticide resistance in Culex tritaeniorhynchus. Insect Mol. Biol. 10, 197–203.

Mori, A., Chadee, D.D., Graham, D.H., Severson, D.W., 2004. Reinvestigation of an endogenous meiotic drive system in the mosquito, Aedes aegypti (Diptera: Culicidae). J. Med. Entomol. 41, 1027–1033.

Motara, M.A., Rai, K.S., 1977. Chromosomal differentiation in two species of Aedes and their hybrids revealed by Giemsa C-banding. Chromosoma 64, 125–132.

Motara, M.A., Rai, K.S., 1978. Giemsa C-banding patterns in Aedes (Stegomyia) mosquitoes. Chromosoma 70, 51–58.

Mysore, K., Sun, L., Tomchaney, M., Sullivan, G., Adams, H., Piscoya, A.S., Severson, D.W., Syed, Z., Duman-Scheel, M., 2015. siRNA-mediated silencing of doublesex during female development of the dengue vector mosquito Aedes aegypti. PLoS Negl. Trop. Dis. 9, e0004213.

Nene, V., Wortman, J.R., Lawson, D., Haas, B., Kodira, C., Tu, Z.J., Loftus, B., Xi, Z., Megy, K., Grabherr, M., Ren, Q., Zdobnov, E.M., Lobo, N.F., Campbell, K.S., Brown, S.E., Bonaldo, M.F., Zhu, J., Sinkins, S.P., Hogenkamp, D.G., Amedeo, P., Arensburger, P., Atkinson, P.W., Bidwell, S., Biedler, J., Birney, E., Bruggner, R.V., Costas, J., Coy, M.R., Crabtree, J., Crawford, M., Debruyn, B., Decaprio, D., Eiglmeier, K., Eisenstadt, E., El-Dorry, H., Gelbart, W.M., Gomes, S.L., Hammond, M., Hannick, L.I., Hogan, J.R., Holmes, M.H., Jaffe, D., Johnston, J.S., Kennedy, R.C., Koo, H., Kravitz, S., Kriventseva, E.V., Kulp, D., Labutti, K., Lee, E., Li, S., Lovin, D.D., Mao, C., Mauceli, E., Menck, C.F., Miller, J.R., Montgomery, P., Mori, A., Nascimento, A.L., Naveira, H.F., Nusbaum, C., O'Leary, S., Orvis, J., Pertea, M., Quesneville, H., Reidenbach, K.R., Rogers, Y.H., Roth, C.W., Schneider, J.R., Schatz, M., Shumway, M., Stanke, M., Stinson, E.O., Tubio, J.M., Vanzee, J.P., Verjovski-Almeida, S., Werner, D., White, O., Wyder, S., Zeng, Q., Zhao, Y., Zhao, Q., Hill, C.A., Raikhel, A.S., Soares, M.B., Knudson, D.L., Lee, N.H., Galagan, J., Salzberg, S.L., Paulsen, I.T., Dimopoulos, G., Collins, F.H., Birren, B., Fraser-Liggett, C.M., Severson, D.W., 2007. Genome sequence of Aedes aegypti, a major arbovirus vector. Science 316, 1718–1723.

Newton, M.E., Southern, D.I., Wood, R.J., 1974. X and Y chromosomes of Aedes aegypti (L.) distinguished by Giemsa C-banding. Chromosoma 49, 41–49.

Newton, M.E., Wood, R.J., Southern, D.I., 1976. Cytogenetic analysis of meiotic drive in mosquito, Aedes-aegypti (L). Genetica 46, 297–318.

Newton, M.E., Wood, R.J., Southern, D.I., 1978. Cytological mapping of the M and D loci in the mosquito, Aedes aegypti (L.). Genetica 48, 137–143.

Ohno, S., 1967. Sex Chromosomes and Sex-Linked Genes. Springer-Verlag, Berlin, New York, etc.

Pane, A., Salvemini, M., Delli Bovi, P., Polito, C., Saccone, G., 2002. The transformer gene in Ceratitis capitata provides a genetic basis for selecting and remembering the sexual fate. Development 129, 3715–3725.

Papathanos, P.A., Bossin, H.C., Benedict, M.Q., Catteruccia, F., Malcolm, C.A., Alphey, L., Crisanti, A., 2009. Sex separation strategies: past experience and new approaches. Malar. J. 8 (Suppl. 2), S5.

Picard, M.A., Cosseau, C., Mouahid, G., Duval, D., Grunau, C., Toulza, E., Allienne, J.F., Boissier, J., 2015. The roles of Dmrt (double sex/male-abnormal-3 related transcription factor) genes in sex determination and differentiation mechanisms: ubiquity and diversity across the animal kingdom. C. R. Biol. 338, 451–462.

Price, D.C., Egizi, A., Fonseca, D.M., 2015. Characterization of the doublesex gene within the Culex pipiens complex suggests regulatory plasticity at the base of the mosquito sex determination cascade. BMC Evol. Biol. 15, 108.

Rai, K.S., 1963. A comparative study of mosquito karyotypes. Ann. Entomol. Soc. Am. 56, 160–170.

Reidenbach, K.R., Cook, S., Bertone, M.A., Harbach, R.E., Wiegmann, B.M., Besansky, N.J., 2009. Phylogenetic analysis and temporal diversification of mosquitoes (Diptera: Culicidae) based on nuclear genes and morphology. BMC Evol. Biol. 9, 298.

Rishikesh, N., 1959. Chromosome behaviour during spermatogenesis of Anopheles stephensi sensu stricto. Cytologia 24, 447–458.

Rose, G., Krzywinska, E., Kim, J., Revuelta, L., Ferretti, L., Krzywinski, J., 2016. Dosage compensation in the African malaria mosquito Anopheles gambiae. Genome Biol. Evol. 8, 411–425.

Salvemini, M., Mauro, U., Lombardo, F., Milano, A., Zazzaro, V., Arcà, B., Polito, L.C., Saccone, G., 2011. Genomic organization and splicing evolution of the doublesex gene, a Drosophila regulator of sexual differentiation, in the dengue and yellow fever mosquito Aedes aegypti. BMC Evol. Biol. 11, 41.

Salvemini, M., D'Amato, R., Petrella, V., Aceto, S., Nimmo, D., Neira, M., Alphey, L., Polito, L.C., Saccone, G., 2013. The orthologue of the fruitfly sex behaviour gene fruitless in the mosquito Aedes aegypti: evolution of genomic organisation and alternative splicing. PLoS One 8, e48554.

Salz, H.K., Erickson, J.W., 2010. Sex determination in Drosophila: the view from the top. Fly (Austin) 4, 60–70.

Scali, C., Catteruccia, F., Li, Q., Crisanti, A., 2005. Identification of sex-specific transcripts of the Anopheles gambiae doublesex gene. J. Exp. Biol. 208, 3701–3709.

Schliekelman, P., Ellner, S., Gould, F., 2005. Pest control by genetic manipulation of sex ratio. J. Econ. Entomol. 98, 18–34.

Schutt, C., Nothiger, R., 2000. Structure, function and evolution of sex-determining systems in Dipteran insects. Development 127, 667–677.

Selinger, R., 1972. Inherited semisterility for control of harmful insects. V. Translocations in Culex tritaenio-rhynchus. Experientia 28, 481–482.

Severson, D.W., Meece, J.K., Lovin, D.D., Saha, G., Morlais, I., 2002. Linkage map organization of expressed sequence tags and sequence tagged sites in the mosquito, Aedes aegypti. Insect Mol. Biol. 11, 371–378.

Shin, D., Mori, A., Severson, D.W., 2012. Genetic mapping a meiotic driver that causes sex ratio distortion in the mosquito Aedes aegypti. J. Hered. 103, 303–307.

Shukla, J.N., Palli, S.R., 2013. Tribolium castaneum Transformer-2 regulates sex determination and development in both males and females. Insect Biochem. Mol. Biol. 43, 1125–1132.

Shukla, J.N., Palli, S.R., 2014. Production of all female progeny: an evidence for the presence of male sex determination factor on Y chromosome. J. Exp. Biol. 217, 1653–1655.

Stevens, N.M., 1911. Further studies on heterochromosomes in mosquitoes. Biol. Bull. 20, 109–120.

Straub, T., Becker, P.B., 2007. Dosage compensation: the beginning and end of generalization. Nat. Rev. Genet. 8, 47–57.

Suguna, S.G., Wood, R.J., Curtis, C.F., Whitelaw, A., Kazmi, S.J., 1977. Resistance to meiotic drive at the MD locus in an Indian wild population of Aedes aegypti. Genet. Res. 29, 123–132.

Sun, X., Yang, H., Sturgill, D., Oliver, B., Rabinow, L., Samson, M.L., 2015. Sxl-dependent, tra/tra2-independent alternative splicing of the Drosophila melanogaster X-linked gene found in neurons. G3 (Bethesda) 5, 2865–2874.

Suzuki, M.G., Imanishi, S., Dohmae, N., Asanuma, M., Matsumoto, S., 2010. Identification of a male-specific RNA binding protein that regulates sex-specific splicing of Bmdsx by increasing RNA binding activity of BmPSI. Mol. Cell. Biol. 30, 5776–5786.

Sweeny, T.L., Barr, A.R., 1978. Sex ratio distortion caused by meiotic drive in a mosquito, Culex pipiens L. Genetics 88, 427–446.

Tanaka, Y., 1916. Genetic studies in the silkworm. J. Coll. Agric. Sapporo. 6, 1–33.

Thomas, D.D., Donnelly, C.A., Wood, R.J., Alphey, L.S., 2000. Insect population control using a dominant, repressible, lethal genetic system. Science (New York, NY) 287, 2474–2476.

Thomas, C.G., Woodruff, G.C., Haag, E.S., 2012. Causes and consequences of the evolution of reproductive mode in Caenorhabditis nematodes. Trends Genet. 28, 213–220.

Timoshevskiy, V.A., Kinney, N.A., Debruyn, B.S., Mao, C., Tu, Z., Severson, D.W., Sharakhov, I.V., Sharakhova, M.V., 2014. Genomic composition and evolution of Aedes aegypti chromosomes revealed by the analysis of physically mapped supercontigs. BMC Biol. 12, 27.

Toups, M.A., Hahn, M.W., 2010. Retrogenes reveal the direction of sex-chromosome evolution in mosquitoes. Genetics 186, 763–766.

Traut, W., Marec, F., 1997. Sex chromosome differentiation in some species of Lepidoptera (Insecta). Chromosome Res. 5, 283–291.

Venkatesan, M., Broman, K.W., Sellers, M., Rasgon, J.L., 2009. An initial linkage map of the West Nile Virus vector Culex tarsalis. Insect Mol. Biol. 18, 453–463.

Verhulst, E.C., Van De Zande, L., 2015. Double nexus-Doublesex is the connecting element in sex determination. Brief. Funct. Genomics 14, 396–406.

Verhulst, E.C., Van De Zande, L., Beukeboom, L.W., 2010. Insect sex determination: it all evolves around transformer. Curr. Opin. Genet. Dev. 20, 376–383.

Vicoso, B., Kaiser, V.B., Bachtrog, D., 2013. Sex-biased gene expression at homomorphic sex chromosomes in emus and its implication for sex chromosome evolution. Proc. Natl. Acad. Sci. U.S.A. 110, 6453–6458.

Wang, J., Na, J.K., Yu, Q., Gschwend, A.R., Han, J., Zeng, F., Aryal, R., Vanburen, R., Murray, J.E., Zhang, W., Navajas-Perez, R., Feltus, F.A., Lemke, C., Tong, E.J., Chen, C., Wai, C.M., Singh, R., Wang, M.L., Min, X.J., Alam, M., Charlesworth, D., Moore, P.H., Jiang, J., Paterson, A.H., Ming, R., 2012. Sequencing papaya X and Yh chromosomes reveals molecular basis of incipient sex chromosome evolution. Proc. Natl. Acad. Sci. U.S.A. 109, 13710–13715.

Whyard, S., Erdelyan, C.N.G., Partridge, A.L., Singh, A.D., Beebe, N.W., Capina, R., 2015. Silencing the buzz: a new approach to population suppression of mosquitoes by feeding larvae double-stranded RNAs. Parasit. Vectors 8, 96.

Willhoeft, U., Franz, G., 1996. Identification of the sex-determining region of the Ceratitis capitata Y chromosome by deletion mapping. Genetics 144, 737–745.

Windbichler, N., Papathanos, P.A., Crisanti, A., 2008. Targeting the X chromosome during spermatogenesis induces Y chromosome transmission ratio distortion and early dominant embryo lethality in *Anopheles gambiae*. PLoS Genet. 4, e1000291.

Wood, R.J., 1961. Biological and genetic studies on sex ratio in DDT resistant and susceptible strains of Aedes aegypti Linn. Genet. Agrar. 8–4, 287–307.

Wood, R.J., 1976. Between family variation in sex-ratio in trinidad (T-30) strain of Aedes-aegypti (L) indicating differences in sensitivity to meiotic drive gene Md. Genetica 46, 345–361.

Wood, R.J., Ouda, N.A., 1987. The genetic basis of resistance and sensitivity to the meiotic drive gene D in the mosquito Aedes aegypti L. Genetica 72, 69–79.

Zhou, J., Sackton, T.B., Martinsen, L., Lemos, B., Eickbush, T.H., Hartl, D.L., 2012. Y chromosome mediates ribosomal DNA silencing and modulates the chromatin state in Drosophila. Proc. Natl. Acad. Sci. U.S.A. 109, 9941–9946.

CHAPTER THREE

Sexual Selection and the Evolution of Mating Systems in Mosquitoes

A. South, F. Catteruccia
Harvard T.H. Chan School of Public Health, Boston, MA, United States

Contents

Abstract

While Darwin first conceptualized the connection between morphological trait variation across related taxa and mating system evolution nearly 150 years ago, a mechanistic understanding of the relationship between genes, physiology and behaviour across species has only recently become possible through remarkable progress in genetics and molecular biology. With their extensive diversification, ecological variation and burgeoning genomic resources, mosquitoes are an ideal group to identify patterns of trait variation and suggest causal evolutionary mechanisms. Here we discuss the mating systems of *Anopheles gambiae* and *Aedes aegypti* mosquitoes, vectors of major infectious diseases that afflict humanity. In this review, our emphasis is on interspecific comparisons of traits relevant to mechanisms of pre- and postcopulatory selection, including mating behaviour, molecular pathways of reproduction, postmating physiology and morphological traits. Based on a wealth of recent studies in mosquito genomics and phylogenetics, we provide an evolutionary context to examine adaptive changes in mosquito reproduction, their possible impact on pathogen transmission and their potential consequences for disease control.

Advances in Insect Physiology, Volume 51
ISSN 0065-2806
http://dx.doi.org/10.1016/bs.aiip.2016.04.004

1. INTRODUCTION

One-sixth of the world's total infection-associated morbidity and mortality is attributable to vector borne diseases, with more than 90% due to infectious agents transmitted by mosquitoes (McGraw and O'Neill, 2013). *Aedes* and *Anopheles* mosquitoes have the dubious honour of being the most efficient and prolific vectors of the two most virulent mosquito borne diseases, dengue and malaria, respectively. An estimated 100 million new cases of dengue and over 200 million new cases of malaria occur every year, with over the half the world's population at risk of infection (Bhatt et al., 2013; McGraw and O'Neill, 2013; WHO, 2014). Without the development of viable vaccines, strategies that directly target the mosquito vector will likely remain the most effective method to combat these diseases. However, current mosquito control methods based on the use of long-lasting insecticide-treated nets (LLINs) and indoor residual sprays (IRS) for *Anopheles*, and space spraying and larvicide treatment for *Aedes*, are severely threatened by the rapid spread of insecticide resistance in mosquito populations around the globe (Deming et al., 2016; Hemingway, 2014; Marcombe et al., 2012; Toe et al., 2014). Drug treatments are also losing their efficacy, with elevated levels of artemisinin resistance appearing in the malaria-causing *Plasmodium* parasites (Ariey et al., 2014; Miotto et al., 2013). The potential impending failure of these control strategies would put millions of people in disease endemic areas at risk, making the development of new approaches for vector control a crucial priority (malERA Group, 2011). Studying mosquito biology offers multiple opportunities to develop more effective approaches, either by leading directly to the production of new technology or by informing existing methods.

In this review, we focus on the reproductive ecology of two mosquito species, *Anopheles gambiae* and *Aedes aegypti*, both major threats to human health. *An. gambiae* and their sister species *An. coluzzii* are the principal vectors of human malaria in Africa, while *Ae. aegypti* transmits a number of widespread arboviruses including dengue, yellow fever and Zika, in tropical and subtropical countries throughout the globe. Despite their medical and economic importance, basic elements of aedine and anopheline reproductive ecology remain unknown, hampering progress in the development of new vector control tools. Others have noted that these gaps in our knowledge primarily stem from a lack of attention paid to male mating biology (Diabate and Tripet, 2015; Howell and Knols, 2009). However, we would argue that there is limited basic biological knowledge regarding both sexes, paralleled by a lack of perspective on the operational mechanisms of sexual selection. Specifically, little consideration or acknowledgment is given to realities such as phenotypic plasticity and reaction norms (the range of phenotypic expressions that can arise from a single genotype in response to different environmental conditions) in these species.

Understanding mosquito reproductive ecology is a critical prerequisite for developing new and effective vector control strategies. In particular, this knowledge is key to the successful deployment of genetic control strategies that rely on mating between genetically modified males and field females (Catteruccia et al., 2009). At a time when large-scale trial releases of genetically modified *Ae. aegypti* mosquitoes are currently underway in different countries, a deeper grasp of relevant aspect of mosquito ecology could improve the chances of this strategy (Alphey et al., 2011; Bargielowski et al., 2011; Carvalho et al., 2015). Modelling the release of genetically modified mosquitoes has revealed that the successful implementation of various strategies requires understanding elements of mosquito ecology. For instance, introducing modified mosquitoes refractory to infection and determining the most appropriate genetic drive mechanism to propagate genetic modification requires a deep understanding of evolutionary trade-offs that can only be accurately understood in the context of reproductive ecology and associated life-history traits (Alphey et al., 2011; Boette and Koella, 2002; Harshman and Zera, 2007; Koella and Boete, 2003). Relatively simple mechanisms of sexual selection, such as identifying male traits that females use in mate choice decisions, can be the difference between the success or failure of such vector control strategies. Indeed, numerous attempts to control vector populations through the use of sterile insect technique failed largely due to a lack of competiveness of released males vs their wild-type rivals, most notably in multiple species of *Anopheles* (Benedict and Robinson, 2003). As genetic approaches become more and more sophisticated and powerful in their ability to introduce desired modification into the mosquito genome, it is imperative that vector biologists glean what conclusions they can from past failures in order to optimize the development and deployment of new vector control strategies.

Secondarily, we wish to highlight the relevance of *An. gambiae* and *Ae. aegypti* as model systems for studying the evolution of mating systems. Determining the secondary sexual characters under selection provides an opportunity to understand the selective pressures that shape those characters. *An. gambiae* and *Ae. aegypti* females are predominantly monandrous, a mating system that is certainly the exception rather than the rule across arthropod taxa (Arnqvist and Nilsson, 2000), while males at least have the opportunity to be polygynous. The highly male-biased operational sex ratio (in this instance, the ratio of sexually active males to females in a population, hereafter referred to as OSR) provides an excellent opportunity to validate theoretical predictions underlying mating system evolution, such as the conditional nature of alternative mating strategies. Thus, it is our hope that this review will inform research in *Anopheles* and *Aedes* mosquitoes from both a basic and applied science perspective (see Table 1 for a comparison of the reproductive systems of these mosquito species).

Table 1 Comparison of the Known Mating System Components of *Ae. aegypti* and *An. gambiae* Mosquitoes

Reproductive Character	*Aedes aegypti*	*Anopheles gambiae*
Precopulatory		
Operational sex ratio	Male biased	Male biased
Mode of mating	Female: Monandrous; context-dependent polyandry	Female: Monandrous
	Male: Polygynous	Male: Polygynous
Mating strategy	Microswarms; single pair mating	Swarming
Swarm location	Near host	Swarm marker
Mode of attraction to swarm	Aggregation pheromone; wingbeat flight tones	Unknown
Method of mate choice	Harmonic convergence	Harmonic convergence
Postcopulatory		
Coagulated ejaculate	No	Yes
Hormone in ejaculate	Juvenile hormone	20-Hydroxyecdysone
Female postmating response trigger	Unknown	20-Hydroxyecdysone
Female refractory	Yes	Yes
Fecundity increased	Yes	Yes
Ovulation and oviposition induced	Yes	Yes

2. THE RELEVANCE OF SEXUAL SELECTION RESEARCH FOR MOSQUITO CONTROL STRATEGIES

Sexual selection is among the most powerful drivers of evolutionary change, shaping a diverse array of physiological, morphological and behavioural traits that impact the reproductive success of both sexes. This evolutionary process is predicated upon differential reproductive success stemming from intrasexual competition or intersexual choice. It is now widely accepted that these mechanisms continue to be active following the initiation of mating (Simmons, 2001; Sirot and Wolfner, 2015), meaning that reproductive fitness cannot be understood without considering how sexual selection operates before (precopulatory) and after (postcopulatory) mating. Here we are highlighting what is known about the critical pre- and postcopulatory elements of the *Anopheles* and *Aedes* mating system, while also emphasizing the components that require additional research efforts (Table 1).

Mating system classifications define how an individual obtains a mate, describing mating behaviours in addition to the number of mates that each sex has over a defined period of time. The pioneering work of Bateman (1948), Trivers (1972) as well as Emlen and Oring (1977) demonstrated that the presence of a limiting sex (in mosquitoes, the male-biased OSR means females are the limiting sex) creates a very strong gradient of selection, one that shapes secondary sexual characters that are the underlying determinants of reproductive output. Considering such fitness-related traits is a central component of vectorial capacity, here defined as the interaction of entomological components shaping the ability of insect vectors to transmit disease (Macdonald, 1957). As such, these traits represent potential targets for control strategies. Extrinsic factors can also influence the OSR and spatiotemporal distribution of the limiting sex, as predation, temperature and character displacement (trait evolution that stems from selection imposed by a related species with similar ecological requirements) can alter sex ratios and/or reduce access to mates.

The nature and strength of sexual selection is highly dynamic, with natural feedback occurring between episodes of selection as a function of alterations in the OSR, with implications for a species' mating system. As an example, if males of a particular species have invested heavily into adaptations that controls female mating rate (for instance through the transfer of specific ejaculate components (Perry et al., 2013) or mate guarding (Alcock,

1994)), thereby ensuring paternity, then it becomes increasingly important for males to find and mate with virgin females before their rivals. This increases the selective gradient of precopulatory selection (Bateman, 1948), driving male investment into traits that maximize mating success. Given that sexually selected traits across episodes of selection could be subject to resource trade-offs (Harshman and Zera, 2007), males will have an inevitable decision to make: how will their finite resource pool be divided between traits active in both pre- and postcopulatory selection? This could result in a decrease in male investment into adaptations that ensure monandry, leading to an increase in the number of females that remate, thereby increasing the strength of postcopulatory selection.

This dynamic nature of selection is important to consider in the context of vector control. By their very nature, some vector control interventions have the potential to dramatically alter the strength of selection at the population and species level. Both *Ae. aegypti* and *An. gambiae* are known to have an OSR that is highly biased towards males. This biased OSR is likely maintained by a combination of males emerging from larval breeding sites first and the relative rapidity with which adult females drop out of breeding population (because of mating-induced refractoriness to additional mating attempts—see later for more details). Given this OSR inequality, selection is expected to favour the evolution of physiological or behavioural adaptations that increase the likelihood of males encountering and/or successfully copulating with virgin females. Consistent with this theory, both species utilize a swarm-based mating strategy, aggregating the sexes for the purposes of mating (see later for more detail).

As mating system theory predicts, it is the interaction between the OSR and the strength of selection that determines the behavioural, morphological and physiological traits that compose a species' mating system. But how could this central feature of mating system evolution affect vector control? As a salient example, we consider currently deployed methods of IRS and LLINs that are almost exclusively focused towards females (as females preferentially enter human homes to find a host for blood feeding), removing them from the mating pool and sharpening the OSR and gradient of sexual selection. Such alterations in OSR have important implications for mating system evolution, as they are associated with profound changes in male behavioural strategies in other species (Emlen and Oring, 1977; Enders, 1993; Weir et al., 2010). Increases in OSR can lead to concomitant increases in male aggression, with males engaged in intrasexual competition at the expense of courting females (Enders, 1993; Weir et al., 2010), or to the emergence of alternative mating strategies such as nonaggressive 'sneaker' males (Dubois et al., 2004; Sinervo and Lively, 1996; Utne-Palm et al., 2015; Zatz et al., 2011). Such alternative strategies can be condition dependent, implemented only under particular circumstances (Gross, 1996; Tomkins and Hazel, 2007). In mosquitoes, the presence of these alternative mating strategies could complicate the implementation of vector control methods, especially those that are critically dependent upon targeting or exploiting aspects of mating behaviour. Evidence exists to suggest that within the span of only a few years, various *Anopheles* species have responded to the selective pressure of IRS and LLINs with a shift in biting and host-seeking behaviours (Gatton et al., 2013; Russell et al., 2011). Therefore, it is possible that similar selective pressures could be operating on components of sexual selection. In fact, Dao et al. (2008) report evidence of *An. coluzzii* males and females mating, most likely during the day, inside of human homes rather than in crepuscular swarms as normally observed. It is conceivable that indoor mating may be emerging as an alternative mating strategy driven by the strong selection gradient imposed by a male-biased OSR. The authors speculate that this form of mating could be utilizing a different series of reproductive characters during the process of courtship and pair formation. It is conceivable that the presence of such an alternative mating strategy could reduce the impact of control strategies that utilize males who have been modified to maximize their competitiveness in a swarm-based mating strategy, potentially leaving an indoor mode of mating untargeted and leading to its increase in frequency. Thus, it becomes even more imperative to not only understand current mechanisms of selection but also monitor and understand how they could be potentially changing in the face of various interventions.

3. PRECOPULATORY ELEMENTS OF SEXUAL SELECTION

3.1 Swarm Formation and Mate Choice in *Anopheles* Mosquitoes

Many species of *Anopheles* mosquitoes are typified by their use of swarm aggregations as a mating strategy (Diabate and Tripet, 2015; Howell and Knols, 2009; Yuval, 2006). In general, swarming behaviour is seemingly far more of an obligatory part of the anopheline mating system relative to *Aedes* mosquitoes, but exceptions do exist (Howell and Knols, 2009; Yuval, 2006). *An. gambiae* and *An. coluzzii* males form crepuscular swarms at specific sites, with *An. coluzzii* forming swarms generally over some sort of landmark known as a swarm marker, while *An. gambiae* form swarms mostly over bare ground (Diabate et al., 2009, 2011). Swarms are formed repeatedly over the same site within a single season and even across seasons (Diabate et al., 2009, 2011). Swarms can vary greatly in size, ranging from a few dozen males to thousands (Howell and Knols, 2009). While the mechanisms of swarm site detection remain unknown, visual cues seem to play an important role. Swarm markers are generally an object that serves as a dark–light contrast point on the ground or the horizon and can be used for orientation (Howell and Knols, 2009), while height and shape are fairly plastic, possibly a reflection of environmental constraints at the swarm site (Charlwood et al., 2002). Females enter the swarms after initiation has already occurred, and generally leave the swarm in copula. Ample evidence exists to classify *An. gambiae* and *An. coluzzii* swarms as leks, as they fit most, if not all, of the criteria used to define leks (Alcock, 1987): (1) swarming occurs over sites that are devoid of any resources usable by females and are used solely as a location for mating; (2) females have the opportunity to choose among the males present in the swarm; (3) males do not contribute to parental care; and (4) extensive male–male interactions are occurring within the swarm, likely as a mechanism allowing them to compete for females. Indeed, recent evidence suggests that females have the opportunity to evaluate multiple males before forming a mating pair (Butail et al., 2012; Shishika et al., 2014).

Despite the role that visual cues play in swarm site selection and formation, relatively little is known about this critical behaviour. No mechanism of site recognition has yet been identified, and the use of additional sensory cues remains an open possibility. Given the variety of ecological parameters in the life-history of *Anopheles* mosquitoes that are reliant upon the perception of chemical signals (Pitts et al., 2014; Takken and Knols, 1999), some have postulated aggregation or recruitment pheromones could play a role in both swarm formation and localization by females (Diabate and Tripet, 2015). Although to date attempts to isolate such a chemical have not yet been successful (Pitts et al., 2014), the identification of long-range aggregation pheromones could provide an additional tool for vector control strategies as it would facilitate the development of traps to attract and kill male and female mosquitoes looking for a mate. What is clear about swarm formation is that mechanisms of swarm segregation can be an effective reproductive isolating mechanism. For example, partial swarm segregation between *An. coluzzii* and *An. gambiae* as well as between *An. coluzzii* and *An. melas* has been reported in various parts of West Africa (Assogba et al., 2014; Diabate et al., 2006, 2009, 2011). Additionally, different subtaxa and populations of *Anopheles* stratify their swarms by height over the ground (Assogba et al., 2014). Although mixed swarms do occur at a certain low-level frequency, swarm segregation still appears to serve as an initial barrier to hybridization between cryptic species or populations.

Through intensive field observations of swarms and the application of 3D video tracking, the basis of sexual selection within the context of a mating swarm is being slowly unravelled. Diabate et al. (2011) concluded that the mating dynamics within swarms in Burkina Faso were consistent with the 'hotshot' model of lek formation, with males not only initiating swarms but also adopting particular positions within the swarm to maximize their chances of interacting with arriving females. Shishika et al. (2014) reports the presence of extensive male–male interactions within a swarm, with males engaging in parallel flights, behaviours that may function to improve their mating success. Once females enter the swarm, the process of mating pair formation occurs relatively rapidly, with copulation lasting only 15–20 s (Charlwood and Jones, 1979). Female *An. gambiae* as well as all *Anopheles* species tested so far mate only a single time (Tripet et al., 2003; Yuval, 2006), but it remains unknown if females in the field ever return to the mating swarm or encounter males on subsequent nights. The rapid nature of mating pair formation may suggest that neither the male nor the female engages in the process of courtship or choice. However, considering that females mate only a single time and the entirety of their reproductive fitness is dependent upon their single mate, a complete lack of female choice seems extremely unlikely. Laboratory-based assays using tethered females have documented instances of females kicking males in order to resist copulation attempts (Charlwood and Jones, 1979), a possible indication and mechanism of choice. More in depth behavioural studies are needed to better understand and characterize choice in free flying females.

Multiple mechanisms have been proposed to explain the process of sex recognition and choice within the context of swarming. The use of close contact chemical cues, such as cuticular hydrocarbons (CHCs), has been suggested as a likely mechanism of choice (Diabate and Tripet, 2015; Pitts et al., 2014; Tripet et al., 2004). Polerstock et al. (2002) reported a change in CHC profiles following mating, indicating a possible role of CHCs in mediating aspects of mating, but other investigations into the CHC profiles of *Anopheles* mosquitoes have not found similar patterns (Caputo et al., 2007; Pitts et al., 2014). Ample evidence does exist for the usage of flight tones as a mechanism for conspecific recognition and mate choice. Mosquito wingbeats produce a flight tone composed of a fundamental frequency and multiple harmonics perceived by the Johnston's organ in the pedicel of the antenna (Clements, 1999). Experiments with *An. gambiae* have indicated that female and male mosquitoes can adjust their wingbeat frequencies in a mutually matching fashion so that they converge on a common frequency, a process known as harmonic convergence (Gibson et al., 2010). Additional experiments utilizing *An. gambiae* and *An. coluzzii* have determined that harmonic convergence is more likely to occur between members of the same species (Pennetier et al., 2010). Moreover, playback experiments utilizing mosquitoes exposed to recorded flight tones typical of larger or smaller members of the opposite sex demonstrated differences in time to harmonic convergence as a function of the body size of the potential mate (Cator et al., 2010). These studies represent the best evidence for a mechanism underlying conspecific recognition and mate choice in *Anopheles* mosquitoes, yet it remains unclear if harmonic convergence occurs as a post hoc consequence of a decision between a male and a female to mate, or if this is a critical mechanism underlying the process of precopulatory mate choice. Moreover, it is important to note that these experiments have been performed utilizing tethered individuals, which may lead to unnatural flight patterns and lower wingbeat frequencies in a variety of insects (Bartussek et al., 2013; Dudley and Ellington, 1990; Kutsch et al., 1999). Thus, methods that allow for free flight in the context of a swarm while still allowing for the recording of wingbeat frequencies will be critical to more fully explore the role of flight tones in the process of sexual selection.

Additional factors could also influence mating success. Size is often an important indicator of fitness in mosquitoes, especially in females where a larger body size is generally associated with higher fecundity (Briegel, 1990; Lyimo and Takken, 1993). The importance of body size in mediating the mating success of male *Anopheles* has produced contrasting results, but the majority of data from both field and laboratory studies suggest that males of intermediate size are often the most successful (Diabate and Tripet, 2015), potentially as a result of inherent trade-offs between agility and ability to maintain sufficient nutrient stores that would manifest as variation in body size. Swarming is a calorically costly behaviour (Yuval et al., 1994), and Gary et al. (2009) demonstrated that limited access to sugar feeding negatively correlates with male *An. gambiae* mating success. Interestingly enough, there is some laboratory-based evidence of mate choice by male *An. gambiae*, as they have been observed to preferentially mate with larger females (Okanda et al., 2002), although based on the highly skewed OSR the extent that males would have an opportunity to exert choice in the field is unknown.

3.2 Aggregation and Mate Choice in *Aedes* Mosquitoes

Mechanisms of precopulatory sexual selection in *Ae. aegypti* mosquitoes exhibit distinct differences and similarities when compared to *An. gambiae*. *Ae. aegypti* also has a male-biased OSR, with females for the most part being considered monandrous (but see later) and males at least having the potential to be polygynous. The process of courtship and mating in *Ae. aegypti* is relatively far more plastic than what has been reported for *Anopheles*, with mating occurring via an array of possible strategies (Oliva et al., 2014). Mating can occur in the context of an isolated single pair, as individual males have been observed to intercept a female in flight to mate (Downes, 1969), but mating swarms are common, with formation of swarms of both *Ae. aegypti* and *Ae. albopictus* frequently observed in close proximity to a vertebrate host (Cator et al., 2011; Gubler and Bhattacharya, 1972; Hartberg, 1971; Yuval, 2006). While the formation of swarms and mating pairs can occur throughout the day, peaks of activity at dusk and immediately following dawn have been reported (Oliva et al., 2014). Given the plasticity of how mating occurs in *Ae. aegypti*, their mating system is relatively difficult to categorize, although some have referred to the system as lek-like (Cabrera and Jaffe, 2007). However, swarming is facultative, occurs in microswarms (Cabrera and Jaffe, 2007; Oliva et al., 2014) and is principally associated with the presence of a vertebrate host and thus occurs at a spot that could be considered as resource-associated.

Both chemical signals and acoustic flight tones have been associated with attraction, conspecific recognition and mate choice in *Aedes* mosquitoes. Behavioural studies concluded that males likely produced a volatile aggregation pheromone capable of stimulating flight activity and attracting both sexes (Cabrera and Jaffe, 2007), but others have reported that these volatile aggregation cues only attract females, while males aggregate to swarms in response to same sex flight tones (Fawaz et al., 2014). Beyond a role for these tones in mediating aggregation to a swarm, ample evidence demonstrates that the process of courtship and mate choice in both sexes primarily relies on the usage of flight tones (Oliva et al., 2014), although Nijhout and Craig (1971) suggested that *Ae. albopictus* males could be using a chemical contact cue to identify conspecific females through receptors on their tarsi. Similar to the mechanism of harmonic convergence observed in *An. gambiae*, male and female *Ae. aegypti* actively modulate their flight tones in order to achieve harmonic convergence (Cator et al., 2009). Not only has this been demonstrated to function in the process of mate choice, but there is evidence that females could be gaining an indirect benefit by discriminating between males based on their flight tones, as the ability of males to achieve harmonic convergence (and subsequent relatively higher mating success) is heritable (Cator and Harrington, 2011). Thus, it is possible that the dynamics of flight tone production and perception could be an honest signal of male quality.

As discussed earlier for *An. gambiae*, the size in *Ae, aegypti* is also correlated with reproductive success. Large females are more fecund (Briegel, 1990), and male size is a predictor of sperm production capacity, with a positive correlation between male size, sperm number and the number of sperm transferred to the female (Ponlawat and Harrington, 2007, 2009). Smaller males also become sperm depleted faster than larger males when subject to repeated rematings (Helinski and Harrington, 2011).

4. POSTCOPULATORY ELEMENTS OF SEXUAL SELECTION

4.1 Degree of Multiple Mating in Female *Anopheles* and *Aedes* Mosquitoes

The formation of a mating pair does not signal the end of sexual selection. Inside of female reproductive tracts, diverse molecules manufactured by each sex play key roles not only in prolonged sexual interactions but in altering resource budgets in a manner that can impact fecundity, lifespan and immune response (Schwenke et al., 2016; Simmons, 2001; Sirot and Wolfner, 2015). In general, female *Anopheles* and *Aedes* mosquitoes are largely considered to be monandrous (Yuval, 2006), with females becoming refractory to mating as a consequence of substances transferred in the male ejaculate. In *An. gambiae*, a microsatellite parentage analysis of field caught females found a rate of polyandry <3% (Tripet et al., 2003), while in aedines the extent to which female mate multiply is far more variable. Parentage analysis of field caught *Ae. albopictus* females found evidence of multiple matings in 26% of the families analysed. In *Ae. aegypti*, 14% of females were found to be polyandrous in an experiment that utilized semifield enclosures (Helinski et al., 2012), while Richardson et al. (2015) concluded that the degree of polyandry present in a natural population was at least 6.5%, and possibly as high as 14.6%. Degner and Harrington (2016) demonstrated that the degree of polyandry varies as a function of the postmating time interval, with the majority of females (76%) becoming fully refractory to additional mating attempts 2 h following mating, and rates of remating dropping to 3% 20 h following the first mating. Remating in *Aedes* mosquitoes may therefore be context dependent and occur at a higher rate than what is seen in *An. gambiae*. As speculated by in Degner and Harrington (2016), *Ae. aegypti* females could use a trading-up strategy, with the probability of remating tied to female assessment of male cues that could be linked to overall male quality. While no increase in the fecundity of singly vs multiply mated females was found (Helinski and Harrington, 2011), females could also be accruing indirect benefits in a context- or condition-dependent fashion, a benefit that, to our knowledge, has not been tested. Indeed, a monandrous mating system is considered the exception rather than the rule, as a single mating generally does not maximize female fitness (Arnqvist and Nilsson, 2000; South and Lewis, 2011).

4.2 Postmating Response in *Anopheles* Females

A growing body of evidence across multiple taxa points to the receipt of sexually transferred seminal secretions (proteins, hormones, transcripts, lipids, free amino acids and carbohydrates) as the mediators of a range of behavioural and physiological changes in the female, including the induction of monandry (Avila et al., 2011; Perry et al., 2013). *Anopheles* and *Aedes* are no exception, with a variety of changes in the females being linked to ejaculate transfer. Both *Anopheles* and *Aedes* males manufacture in their single pair of male accessory glands (MAGs) a varied and potent cocktail of seminal substances that is transferred to the female during mating. The male reproductive tracts of these species are markedly similar, yet the content of their ejaculate is not (Table 1).

Males of multiple *Anopheles* species transfer their seminal secretions as a coagulated structure known as the mating plug (Giglioli and Mason, 1966; Mitchell et al., 2015). In *An. gambiae* and other anophelines, this mating plug is a rod-shaped gelatinous structure composed of a mixture of proteins, lipids and steroid hormones that is deposited into the female atrium and digested over the next 24–48 h (Giglioli and Mason, 1966; Mitchell et al., 2015; Pondeville et al., 2008; Rogers et al., 2009). Dottorini's et al. (2007) bioinformatic comparison of *An. gambiae* and *Drosophila melanogaster* MAGs was the first of several recent efforts to determine the composition of *An. gambiae* MAG secretions, which collectively have identified ~120 MAG-enriched proteins (Baldini et al., 2012; Rogers et al., 2009). These investigations have been critical in identifying the basis for mating plug coagulation (Le et al., 2013; Rogers et al., 2009), highlighting a key role for a MAG-specific transglutaminase 3 (TG3) enzyme in cross-linking other MAG proteins and forming this gelatinous rod. Correct mating plug coagulation and transfer are linked to important reproductive functions, including the proper storage of sperm in the female (Rogers et al., 2009). Different degrees of coagulation are observed across anophelines (Mitchell et al., 2015). Males from African anophelines such as *An. gambiae*, *An. arabiensis* and *An. funestus* transfer a highly coagulated plug, South East Asian species like *An. dirus* and *An. farauti* have an intermediate level of plug coagulation, and species such as the Central American *An. albimanus* show a complete lack of coagulation. This variation in mating plug phenotypes may be due to the function of the MAG-specific enzyme TG3 (Le et al., 2013; Rogers et al., 2009). When we compared the evolutionary rates of TG3 to that of the two other TG paralogs present in the genomes of 17 *Anopheles* species as well as in the fruit fly *D. melanogaster* and *Aedes* and *Culex* mosquitoes, we found that TG3 is present specifically in anophelines and is highly divergent within this genus, evolving far more rapidly than its two paralogs (Neafsey et al., 2015). This observation suggests that this enzyme has been subject to strong and taxon-specific selection, potentially leading to the observed variation in mating plug phenotypes described earlier.

Most studies on the female postmating response have been carried out in *An. gambiae*. In these mosquitoes, the receipt of the mating plug initiates a strong transcriptional response in the female reproductive tract, paralleled by a series of physiological and behavioural changes with profound implications for the reproductive fitness of both sexes (Gabrieli et al., 2014; Rogers et al., 2008). After mating, females become refractory to additional mating attempts, egg development is bolstered and egg laying is induced. A number of recent studies from our laboratory have demonstrated that the primary trigger of these physiological, behavioural and transcriptional changes is a MAG produced steroid hormone, 20-hydroxyecdysone (20E) (Baldini et al., 2013; Gabrieli et al., 2014; Mitchell et al., 2015; Shaw et al., 2014), which is packaged into the mating plug and transferred to females during mating (Baldini et al., 2013; Pondeville et al., 2008). In many arthropod species, 20E mediates moulting between juvenile stages during larval development (Yamanaka et al., 2013). In adult mosquitoes, a prohormone ecdysone is produced by the ovaries following a blood meal (Brown et al., 1998; Hagedorn et al., 1975). It is converted into an active hormone, 20E, in peripheral tissues such as fat body, where 20E initiates a cascade of transcription factors that leads to activation of yolk protein precursors genes and initiates vitellogenesis (Attardo et al., 2005; Raikhel et al., 2005; discussed in detail in chapter "Regulation of Reproductive Processes in Female Mosquitoes" by Raikhel). The relative contribution of MAG- and ovary-derived ecdysteroids to the female reproductive physiology remains to be determined.

Using a series of microarray experiments, our laboratory demonstrated an extremely high degree of overlap in the transcriptional responses of females that were either mated or injected with 20E, suggesting that it is this male-derived steroid hormone that is responsible for many of the postmating transcriptional changes (Gabrieli et al., 2014). Additionally, through mating females to males with experimentally depleted levels of 20E and 20E injections into virgin females, we identified 20E as the trigger for mating refractoriness and egg laying, demonstrating that this steroid hormone is a major modulator of female reproductive physiology (Gabrieli et al., 2014). Moreover, sexually transferred 20E regulates the mating-induced increase in egg development experienced by many anopheline species (Baldini et al., 2013; Mitchell et al., 2015) via its interaction with a 20E-induced female protein called MISO (mating-induced stimulator of oogenesis) specifically expressed in the atrium after mating (Baldini et al., 2013). Sexual transfer of 20E also positively affects fertility upon the upregulation of a spermathecal-specific heme peroxidase (HPX15) (Shaw et al., 2014).

Importantly, our laboratory is also moving beyond biochemical and physiological characterization of anopheline reproductive characters by exploring their evolution within a phylogenetic context. In Mitchell et al. (2015), we detected high variability in the ability of males in nine *Anopheles* species to synthesize 20E in their MAGs, similar to the divergence observed in mating plug phenotypes. Males from four species (three African malaria vectors, *An. gambiae*, *An. arabiensis* and *An. funestus*, and the Indian vector *An. stephensi*) have high levels of MAG 20E, while additional species from Europe, China and Southeast Asia have lower titres of MAG 20E, with 20E absent in the central American species *An. albimanus*. Based on this divergence in both mating plug phenotypes and 20E synthesis, ancestral state reconstruction across these species and related dipteran outgroups showed that MAG 20E synthesis and mating plug coagulation are both derived characters that have undergone a pattern of correlated evolution as the anopheline lineage has evolved (Mitchell et al., 2015). Moreover, possible coevolution between the sexes is suggested by the observation that evolution of a mating system based on 20E transfer has induced parallel coadaptation in the female reproductive protein MISO (Mitchell et al., 2015), as mentioned earlier a female protein responsible for translating male-transferred 20E into a fecundity-boosting signal in *An. gambiae* (Baldini et al., 2013). This series of results suggests that different *Anopheles* lineages may have been subject to different selective regimes, possibly resulting in divergent mating strategies. Consistent with this hypothesis, the plugless and 20E-less species *An. albimanus* is not known to utilize a swarm-based mating system as seen in other *Anopheles* (Yuval, 2006). The anopheline clade therefore is an excellent model system for future investigations into the interaction of mating systems and reproductive character evolution.

4.3 Postmating Response in *Aedes* Females

While similar female postmating responses occur in *Aedes* females, relatively little is known about the components of male ejaculate initiating this response. Seminal fluid is not coagulated into a structured plug, but still represents a complex mixture of peptides, hormones and transcripts (Alfonso-Parra et al., 2016; Clifton et al., 2014; Sirot et al., 2011). Recent investigations have focused on identifying and characterizing the seminal fluid proteome of both *Ae. albopictus* and *Ae. aegypti* in order to identify the possible factor(s) responsible for triggering the female postmating response (Boes et al., 2014; Sirot et al., 2011). Sirot et al. (2011) and Boes et al. (2014) identified ~100 and ~200 seminal fluid proteins (SFPs) in *Ae. aegypti* and *Ae. albopictus*, respectively. Interestingly, while only 36.4% of *Ae. albopictus* SFPs had orthologues in *Ae. aegypti*, comparisons of the evolutionary rate of these shared proteins showed no evidence of positive selection, indicating a high degree of conservation (Boes et al., 2014). In parallel to the ejaculate of *An. gambiae* males, hormones have also been found in ejaculate of both *Ae. aegypti* and *Ae. albopictus*. The presence of a putative male adipokinetic hormone, a hormone with a hypothesized role in sperm protection and/or egg production, was detected in *Ae. albopictus* (Boes et al., 2014). Moreover, *Ae. aegypti* males transfer juvenile hormone to females, triggering an increase in female fecundity (Clifton et al., 2014). As these two hormones play a role in regulating a large variety of biochemical processes critical to reproductive success (Clifton and Noriega, 2011, 2012; Kodrik, 2008), it is tempting to speculate on selective pressures that could have led to the evolution of a mating strategy whereby male mosquitoes supply females with fitness boosting hormones, potentially in return for a reduction in female mating rate. Indeed, theoretical models of mating system evolution predict that alterations in how a male's ejaculate impact female fitness and female mating rate generate selection that impacts the evolutionary trajectory of these traits (Alonzo and Pizzari, 2010, 2013). If the male-transferred hormones in *Aedes* mosquitoes have equivalent impacts on female fitness as 20E in *An. gambiae*, then these hormone pathways could represent an important target for vector control strategies. Given the demonstrated role of *Ae. aegypti* MAG extracts in inducing female refractoriness to mating (Fuchs and Hiss, 1970; Fuchs et al., 1968, 1969) and for fertility (Adlakha and Pillai, 1975), it is only a question of time before the male-derived factors critical for inducing postmating responses in *Aedes* females are identified.

Mating in *Ae. aegypti* is also known to influence oocyte development, oviposition behaviour, blood feeding and even the digestion of a bloodmeal (Downe, 1975; Edman, 1970; Klowden and Chambers, 1991; Klowden and Lea, 1979; Lavoipierre, 1958; Leahy and Craig, 1965). MAG contents have even been implicated in affecting female longevity, as females mated to males with depleted levels of ejaculate suffered a reduction in lifespan (Helinski and Harrington, 2011). Alfonso-Parra et al. (2016) adopted a transcriptomic approach to investigate the changes in the *Ae. aegypti* female reproductive tract after mating. They found a total of 364 differentially expressed transcripts across their treatments, with surprising evidence for the transfer of male transcripts in the ejaculate. Based on the observed transcriptional response, they suggest that the male ejaculate serves a general function of 'priming' the female for the subsequent processes of blood feeding and the initiation of the first gonotrophic cycle, possibly connected to the structure of the *Ae. aegypti* mating system in which males swarm around vertebrate hosts. Thus, the components of the male ejaculate and female postmating response could have been shaped by mating, bloodfeeding and egg development occurring sequentially within a short interval of time. A comparison of transcriptional postmating responses with *An. gambiae* females shows a conserved upregulation of the immune system 24 h after mating, while both species also experience downregulation of genes related to general metabolism, possibly indicating a conserved process of shunting resources towards reproduction as a consequence of mating (Alfonso-Parra et al., 2016; Gabrieli et al., 2014). Further studies comparing the transcriptomic responses of both *Aedes* and *Anopheles* mosquitoes to mating and with additional elements of their ecology, such as bloodfeeding or infection status, should allow for the identification of both conserved and divergent pathways shaping the postmating response of these mosquitoes.

The reproductive tracts of female *Ae. aegypti* and *An. gambiae* demonstrate some profound morphological differences (shown in Fig. 1). Most notable are the multiple sites of sperm storage in *Aedes* females compared to the single spermatheca in *An. gambiae*. In *Ae. aegypti*, a structure known

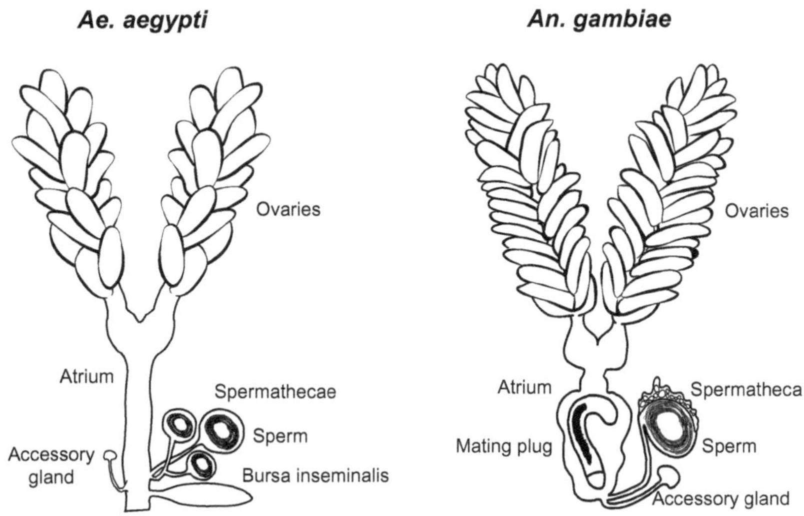

Fig. 1 The reproductive tracts of female *Ae. aegypti* and *An. gambiae* mosquitoes. *Ae. aegypti* are characterized by the presence of multiple spermathecae, a bursa inseminalis and a more tubular atrium. By comparison, *An. gambiae* females have only a single spermatheca with a rounder atrium, shown here with a male transferred mating plug. (See the colour plate.)

as the bursa inseminalis (BI) is the first, but only temporary, site of sperm storage. Following the termination of copulation, sperm begin to migrate from the BI to the three spermathecae, one larger central structure and two lateral ones. The presence of multiple sites of sperm storage is a feature more commonly associated with insect species with polyandrous females, a mating system with higher degrees of postcopulatory selection than monandrous species (Simmons, 2001). The lack of a coagulated mating plug, the relatively higher degree of polyandry, the multiple sites of sperm storage and the plastic nature of mating strategies employed by *Ae. aegypti* mosquitoes are all notable differences from *An. gambiae* and could represent a suite of reproductive characters that have evolved together. The most parsimonious explanation for these differences is that *Aedes* has retained reproductive characters from a common dipteran ancestor, while the anopheline lineage has been subject to a different selective regime responsible for producing its extant mating system characters. Of particular interest is *An. albimanus*, a species in the Nyssorhynchus lineage that seems to share more features of its mating system with *Aedes* mosquitoes than with other *Anopheles*. After the divergence of Anophelinae from Culicinae approximately 150 million years ago, Nyssorhynchus was the first lineage to split from the rest of the

Anophelinae (Neafsey et al., 2015). Molecular and behavioural comparisons of *An. albimanus* to *An. gambiae* and *Ae. aegypti* should give some fascinating insights into the interplay between mating system structure and reproductive characters while also providing insights into the anopheline ancestral mating system.

5. CONCLUSIONS

The past decade has seen a meteoric increase in the application of genomic and molecular biology to understanding the biology of mosquito vectors. In the case of *Ae. aegypti*, this has facilitated the development and deployment of genetically modified mosquitoes in order to combat dengue fever (Alphey, 2014; Carvalho et al., 2015), while the hope of similar progress in the development of modified *Anopheles* mosquitoes to control malaria is ever growing. The importance of ecological studies to the successful development of such novel means of vector control has been increasingly recognized, but progress in conducting that research requires additional effort. In part, relevant ecological studies lag behind due to the absence of a framework to focus and unite these efforts into translational, integrative studies to bridge the gap between genomics and ecology. Knowledge of multiple aspects of reproductive biology and ecology will be crucial to the generation of novel vector control strategies, as it will contribute to an effective implementation of genetic modification and mass release strategies and may lead to the production of new methods to attract, trap or kill male and female mosquitoes.

ACKNOWLEDGEMENTS

We are grateful to Douglas Paton, Ethan Degner, Laura Sirot and Manuela Bernardi for information, comments and assistance with the figure and table. A.S. and F.C. were supported by a grant from the National Institutes of Health, R01 AI104956-04.

REFERENCES

Adlakha, V., Pillai, M.K., 1975. Involvement of male accessory gland substance in the fertility of mosquitoes. J. Insect Physiol. 21, 1453–1455.
Alcock, J., 1987. Leks and hilltopping in insects. J. Nat. Hist. 21, 319–328.
Alcock, J., 1994. Postinsemination associations between males and females in insects: the mate-guarding hypothesis. Annu. Rev. Entomol. 39, 1–21.
Alfonso-Parra, C., Ahmed-Braimah, Y.H., Degner, E.C., Avila, F.W., Villarreal, S.M., Pleiss, J.A., Wolfner, M.F., Harrington, L.C., 2016. Mating-induced transcriptome changes in the reproductive tract of female *Aedes aegypti*. PLoS Negl. Trop. Dis. 10, e0004451.

Alonzo, S.H., Pizzari, T., 2010. Male fecundity stimulation: conflict and cooperation within and between the sexes: model analyses and coevolutionary dynamics. Am. Nat. 175, 174–185.
Alonzo, S.H., Pizzari, T., 2013. Selection on female remating interval is influenced by male sperm competition strategies and ejaculate characteristics. Philos. Trans. R. Soc. Lond. B Biol. Sci. 368, 20120044.
Alphey, L., 2014. Genetic control of mosquitoes. Annu. Rev. Entomol. 59, 205–224.
Alphey, N., Bonsall, M., Alphey, L., 2011. Modeling resistance to genetic control of insects. J. Theor. Biol. 270, 42–55.
Ariey, F., Witkowski, B., Amaratunga, C., Beghain, J., Langlois, A., Khim, N., Kim, S., Duru, V., Bouchier, C., Ma, L., Lim, P., Leang, R., Duong, S., Sreng, S., Suon, S., Chuor, C., Bout, D.M., Menard, S., Rogers, W.O., Genton, B., Fandeur, T., Miotto, O., Ringwald, P., Le Bras, J., Berry, A., Barale, J., Fairhurst, R.M., Benoit-Vical, F., Mercereau-Puijalon, O., Menard, D., 2014. A molecular marker of artemisinin-resistant *Plasmodium falciparum* malaria. Nature 505, 50–55.
Arnqvist, G., Nilsson, T., 2000. The evolution of polyandry: multiple mating and female fitness in insects. Anim. Behav. 60, 145–164.
Assogba, B.S., Djogbenou, L., Saizonou, J., Diabate, A., Dabire, K.R., Moiroux, N., Gilles, J.R.L., Makautode, M., Baldet, T., 2014. Characterization of swarming and mating behaviour between *Anopheles coluzzii* and *Anopheles melas* in a sympatry area of Benin. Acta Trop. 132, S53–S63.
Attardo, G.M., Hansen, I.A., Raikhel, A.S., 2005. Nutritional regulation of vitellogenesis in mosquitoes: implications for anautogeny. Insect Biochem. Mol. Biol. 35, 661–675.
Avila, F.W., Sirot, L.K., Laflamme, B.A., Rubinstein, C.D., Wolfner, M.F., 2011. Insect seminal fluid proteins: identification and function. Annu. Rev. Entomol. 56, 21–40.
Baldini, F., Gabrieli, P., Rogers, D.W., Catteruccia, F., 2012. Function and composition of male accessory gland secretions in *Anopheles gambiae*: a comparison with other insect vectors of infectious diseases. Pathog. Glob. Health 106, 82–93.
Baldini, F., Gabrieli, P., South, A., Valim, C., Mancini, F., Catteruccia, F., 2013. The interaction between a sexually transferred steroid hormone and a female protein regulates oogenesis in the malaria mosquito *Anopheles gambiae*. PLoS Biol. 11, e1001695.
Bargielowski, I., Nimmo, D., Alphey, L., Koella, J.C., 2011. Comparison of life history characteristics of the genetically modified OX513A line and a wild type strain of *Aedes aegypti*. PLoS One 6, e20699.
Bartussek, J., Mutlu, A.K., Zaptocky, M., Fry, S.N., 2013. Limit-cycle-based control of the myogenic wingbeat rhythm in the fruit fly Drosophila. J. R. Soc. Interface 10, 20121013.
Bateman, A.J., 1948. Intra-sexual selection in *Drosophila*. Heredity 2, 349–368.
Benedict, M.Q., Robinson, A.S., 2003. The first releases of transgenic mosquitoes: an argument for the sterile insect technique. Trends Parasitol. 19, 349–355.
Bhatt, S., Gething, P.W., Brady, O.J., Messina, J.P., Farlow, A.W., Moyes, C.L., Drake, J.M., Brownstein, J.S., Hoen, A.G., Sankoh, O., Myers, M.F., George, D.B., Jaenisch, T., Wint, G.R.W., Simmons, C.P., Scott, T.W., Farrar, J.J., Hay, S.I., 2013. The global distribution and burden of dengue. Nature 496, 504–507.
Boes, K.E., Ribeiro, J.M., Wong, A., Harrington, L.C., Wolfner, M.F., Sirot, L.K., 2014. Identification and characterization of seminal fluid proteins in the Asian tiger mosqutio, Aedes, albopictus. PLoS Negl. Trop. Dis. 8, e2946.
Boette, C., Koella, J., 2002. A theoretical approach to predicting the success of genetic manipulation of malaria mosquitoes in malaria control. Malar. J. 1, 3.
Briegel, H., 1990. Metabolic relationship between female body size, reserves, and fecundity of *Aedes aegypti*. J. Insect Physiol. 36, 165–172.
Brown, M.R., Graf, R., Swiderek, K.M., Fendley, D., Stracker, T.H., Champagne, D.E., Lea, A.O., 1998. Identification of a steroidogenic neurohormone in female mosquitoes. J. Biol. Chem. 273, 3967–3971.

Butail, S., Manoukis, N., Diallo, M., Ribeiro, J.M., Lehmann, T., Paley, D.A., 2012. Reconstructing the flight kinematics of swarming and mating in wild mosquitoes. J. R. Soc. Interface 9, 2624–2638.

Cabrera, M., Jaffe, K., 2007. An aggregation pheromone modulates Lekking behavior in the vector mosquito *Aedes Aegypti* (Diptera: Culicidae). J. Am. Mosq. Control Assoc. 23, 1–10.

Caputo, B., Dani, F.R., Horne, G.L., N'fale, S., Diabate, A., Turillazzi, S., Coluzzi, M., Constantini, C., Priestman, A.A., Petrarca, V., Della Torre, A., 2007. Comparative analysis of epicuticular lipid profiles of sympatric and allopatric field populations of *Anopheles gambiae* s.s. molecular forms and *An. arabiensis* from Burkina Faso (West Africa). Insect Biochem. Mol. Biol. 37, 389–398.

Carvalho, D.O., Mckemey, A.R., Garziera, L., Lacroix, R., Donnelly, C.A., Alphey, L., Malavasi, A., Capurro, M.L., 2015. Suppression of a field population of *Aedes aegypti* in Brazil by sustained release of transgenic male mosquitoes. PLoS Negl. Trop. Dis. 9, e0003864.

Cator, L.J., Harrington, L.C., 2011. The harmonic convergence of fathers predicts the mating success of sons in the yellow fever mosquito. Anim. Behav. 82, 627–633.

Cator, L.J., Arthur, B.J., Harrington, L.C., Hoy, R.R., 2009. Harmonic convergence in the love songs of the dengue vector mosquito. Science 323, 1077–1079.

Cator, L.J., Ng'habi, K.R., Hoy, R.R., Harrington, L.C., 2010. Sizing up a mate: variation in production and response to acoustic signals in *Anopheles gambiae*. Behav. Ecol. 21, 1033–1039.

Cator, L.J., Arthur, B.J., Ponlawat, A., Harrington, L.C., 2011. Behavioral observations and sound recordings of free-flight mating swarms of *Ae. aegypti* (Diptera: Culicidae) in Thailand. J. Med. Entomol. 48, 941–946.

Catteruccia, F., Crisanti, A., Wimmer, E.A., 2009. Transgenic technologies to induce sterility. Malar. J. 8, S7.

Charlwood, J.D., Jones, M.D.R., 1979. Mating behaviour in the mosquito, *Anopheles gambiae* s.l. I. Close range and contact behaviour. Physiol. Entomol. 4, 111–120.

Charlwood, J.D., Pinto, J., Sousa, C.A., Ferreira, C., Do Rosario, V.E., 2002. Male size does not affect mating success (of *Anopheles gambiae* in Sao Tome). Med. Vet. Entomol. 16, 109–111.

Clements, A.N., 1999. The Biology of Mosquitoes. CABI Publishing, Wallingford, United Kingdom.

Clifton, M.E., Noriega, F.G., 2011. Nutrient limitation results in juvenile hormone-mediated resorption of previtellogic ovarian follicles in mosquitoes. J. Insect Physiol. 57, 1274–1281.

Clifton, M.E., Noriega, F.G., 2012. The fate of follicles after a blood meal is dependent on previtellogenic nutrition and juvenile hormone in *Aedes aegypti*. J. Insect Physiol. 58, 1007–1019.

Clifton, M.E., Correa, S., Rivera-Perez, C., Nouzova, M., Noriega, F.G., 2014. Male *Aedes aegypti* mosquitoes use JH III transferred during copulation to influence previtellogenic ovary physiology and affect the reproductive output of female mosquitoes. J. Insect Physiol. 64, 40–47.

Dao, A., Adamou, A., Yaro, A.S., Maiga, H., Kassogue, Y., Traore, S., Lehmann, T., 2008. Assessment of alternative mating strategies in *Anopheles gambiae*: does mating occur indoors? J. Med. Entomol. 45, 643–652.

Degner, E.C., Harrington, L.C., 2016. Polyandry depends on postmating time interval in the dengue vector Aedes aegypti. Am. J. Trop. Med. Hyg. 94 (4), 780–785.

Deming, R., Manrique-Saide, P., Barreiro, A.M., Cardena, E.U.K., Che-Mendoza, A., Jones, B., Liebman, K., Vizcaino, L., Vasquez-Prokopec, G., Lenhart, A., 2016. Spatial variation of insecticide resistance in the dengue vector *Aedes aegypti* presents unique vector control challenges. Parasit. Vectors 9, 67.

Diabate, A., Tripet, F., 2015. Targeting male mosquito mating behaviour for malaria control. Parasit. Vectors 8, 347.
Diabate, A., Dabire, R.K., Kengne, P., Brengues, C., Baldet, T., Ouari, A., Simard, F., Lehmann, T., 2006. Mixed swarms of the molecular M and S forms of *Anopheles gambiae* (Diptera: Culicidae) in sympatric area from Burkina Faso. J. Med. Entomol. 43, 480–483.
Diabate, A., Dao, A., Yaro, A.S., Adamou, A., Gonzalez, R., Manoukis, N.C., Traore, S.F., Gwadz, R.W., Lehmann, T., 2009. Spatial swarm segregation and reproductive isolation between the molecular forms of *Anopheles gambiae*. Proc. Biol. Sci. 276, 4215–4222.
Diabate, A., Yaro, A.S., Dao, A., Diallo, M., Huestis, D.L., Lehmann, T., 2011. Spatial distribution and male mating success of *Anopheles gambiae* swarms. BMC Evol. Biol. 11, 184.
Dottorini, T., Nicolaides, L., Ranson, H., Rogers, D.W., Crisanti, A., Catteruccia, F., 2007. A genome-wide analysis in *Anopheles gambiae* mosquitoes reveals 46 male accessory gland genes, possible modulators of female behavior. Proc. Natl. Acad. Sci. U.S.A. 104, 16215–16220.
Downe, A.E., 1975. Internal regulation of rate of digestion of blood meals in the mosquito *Aedes aegypti*. J. Insect Physiol. 21, 1835–1839.
Downes, J.A., 1969. The swarming and mating flight of Diptera. Annu. Rev. Entomol. 14, 271–298.
Dubois, F., Giraldeau, L.A., Hamilton, I.M., Grant, J.W., Lefebvre, L., 2004. Distraction sneakers decrease the expected level of aggression with groups: a game-theoretic model. Am. Nat. 164, E32–E45.
Dudley, R., Ellington, C.P., 1990. Mechanics of forward flight in bumblebees: I. Kinematics and morphology. J. Exp. Biol. 148, 19–52.
Edman, J.D., 1970. Rate of digestion of vertebrate blood in *Aedes aegypti* (L.). Effect of age, mating, and parity. Am. J. Trop. Med. Hyg. 19, 1031–1033.
Emlen, S.T., Oring, L.W., 1977. Ecology, sexual selection and the evolution of mating systems. Science 197, 215–233.
Enders, M.M., 1993. The effect of male size and operational sex ratio on male mating success in the common spider mite, *Tetranychus urticae* Koch (Acari: Tetranychidae). Anim. Behav. 46, 835–846.
Fawaz, E.Y., Allan, S.A., Bernier, U.R., Obernauer, P.J., Diclaro II, J.W., 2014. Swarming mechanisms in the yellow fever mosquito: aggregation pheromones are involved in the mating behavior of *Aedes aegypti*. J. Vector Ecol. 39, 347–354.
Fuchs, M.S., Hiss, E.A., 1970. The partial purification and separation of the protein components of matrone from *Aedes aegypti*. J. Insect Physiol. 16, 931–939.
Fuchs, M.S., Craig, G.B., Hiss, E.A., 1968. The biochemical basis of female monogamy in mosquitoes. I. Extraction of the active principle from *Aedes aegypti*. Life Sci. 7, 835–839.
Fuchs, M.S., Craig, G.B., Despommier, D.D., 1969. The protein nature of the substance inducing female monogamy in *Aedes aegypti*. J. Insect Physiol. 15, 701–709.
Gabrieli, P., Kakani, E., Mitchell, S.N., Mameli, E., Want, E.J., Serrao, A., Baldini, F., Catteruccia, F., 2014. Sexual transfer of the steroid hormone 20E induces the post-mating switch in *Anopheles gambiae* mosquitoes. Proc. Natl. Acad. Sci. U.S.A. 111 (46), 16353–16358.
Gary, R.E., Cannon, J.W., Foster, W.A., 2009. Effect of sugar on male *Anopheles gambiae* mating performance, as modified by temperature, space and body size. Parasit. Vectors 2, 19.
Gatton, M.L., Chitnis, N., Churcher, T., Donnelly, M.J., Ghani, A.C., Godfray, H.C.J., Gould, F., Hastings, I., Marshall, J., Ranson, H., Rowland, M., Shaman, J., Lindsay, S.W., 2013. The importance of mosquito behavioural adaptations to malaria control in Africa. Evolution 67, 1218–1230.

Gibson, G., Warren, B., Russell, I.J., 2010. Humming in tune: sex and species recognition by mosquitoes on the wing. J. Assoc. Res. Otolaryngol. 11, 527–540.
Giglioli, M., Mason, G., 1966. The mating plug of anopheline mosquitoes. Proc. R. Ent. Soc. Lond. 41, 123–129.
Gross, M.R., 1996. Alternative reproductive strategies and tactics: diversity within the sexes. Trends Ecol. Evol. 11, 92–98.
Gubler, D.J., Bhattacharya, N.C., 1972. Swarming and mating of *Aedes (S.) albopictus* in nature. Mosq. News 32, 219–223.
Hagedorn, H.H., O'connor, J.D., Fuchs, M.S., Sage, B., Schlaeger, D.A., Bohm, M.K., 1975. The ovary as a source of alpha-ecdysone in an adult mosquito. Proc. Natl. Acad. Sci. U.S.A. 72, 3255–3259.
Harshman, L.G., Zera, A.J., 2007. The cost of reproduction: the devil in the details. Trends Ecol. Evol. 22, 80–86.
Hartberg, W.K., 1971. Observations on the mating behaviour of *Aedes aegypti* in nature. Bull. World Health Organ. 45, 847–850.
Helinski, M.E.H., Harrington, L.C., 2011. Male mating history and body size influence female fecundity and longevity of the dengue vector *Aedes aegypti*. J. Med. Entomol. 48, 202–211.
Helinski, M.E.H., Valerio, L., Facchinelli, L., Scott, T.W., Ramsey, J., Harrington, L.C., 2012. Evidence of polyandry for *Aedes aegypti* in semifield enclosures. Am. J. Trop. Med. Hyg. 86, 635–641.
Hemingway, J., 2014. The role of vector control in stopping the transmission of malaria: threats and opportunities. Philos. Trans. R. Soc. Lond. B Biol. Sci. 369, 20130431.
Howell, P.I., Knols, B.G., 2009. Male mating biology. Malar. J. 8 (Suppl. 2), S8.
Klowden, M.J., Chambers, G.M., 1991. Male accessory gland substances activate egg development in nutritionally stressed *Aedes aegypti* mosquitoes. J. Insect Physiol. 37, 721–726.
Klowden, M.J., Lea, A.O., 1979. Humoral inhibition of host-seeking in *Aedes aegypti* during oocyte maturation. J. Insect Physiol. 25, 231–235.
Kodrik, D., 2008. Adipokinetic hormone functions that are not associated with insect flight. Physiol. Entomol. 33, 171–180.
Koella, J.C., Boete, C., 2003. A model for the coevolution of immunity and immune evasion in vector-borne diseases with implications for the epidemiology of malaria. Am. Nat. 161, 698–707.
Kutsch, W., Van Der Wall, M., Fischer, H., 1999. Analysis of free forward flight of *Schistocerca gregaria* employing telemetric transmission of muscle potentials. J. Exp. Zool. 284, 119–129.
Lavoipierre, M.M., 1958. Biting behavior of mated and unmated females of an African strain of *Aedes aegypti*. Nature 181, 1781–1782.
Le, B.V., Nguyen, J.B., Logarajah, S., Wang, B., Marcus, J., Williams, H.P., Catteruccia, F., Baxter, R.H.G., 2013. Characterization of *Anopheles gambiae* transglutaminase 3 (AgTG3) and its native substrate plugin. J. Biol. Chem. 288, 4844–4853.
Leahy, M., Craig, G., 1965. Accessory gland substance as a stimulant for oviposition in *Aedes aegypti* and *Ae. albopictus*. Mosq. News 25, 448–452.
Lyimo, E.O., Takken, W., 1993. Effects of body size on fecundity and pre-gravid rate of *Anopheles gambiae* females in Tanzania. Med. Vet. Entomol. 7, 328–332.
Macdonald, G., 1957. The Epidemiology and Control of Malaria. Oxford University Press, London.
malERA Consultative Group on Vector Control, 2011. A research agenda for malaria eradication: vector control. PLoS Med. 8, e1000401.
Marcombe, S., Mathieu, R.B., Pocquet, N., Riaz, M., Poupardin, R., Selior, S., Darriet, F., Reynaud, S., Yebakima, A., Corbel, V., David, J., Chandre, F., 2012. Insecticide resistance in the dengue vector *Aedes aegypti* from Martinique: distribution, mechanisms and relations with environmental factors. PLoS One 7, e30989.

Mcgraw, E.A., O'neill, S.L., 2013. Beyond insecticides: new thinking on an ancient problem. Nat. Rev. Microbiol. 11, 181–193.

Miotto, O., Almagro-Garcia, J., Manske, M., Macinnis, B., Campino, S., Rockett, K.A., Amaratunga, C., LIM, P., Suon, S., Sreng, S., Anderson, J.M., Duong, S., Nguon, C., Chuor, C.M., Saunders, D., Se, Y., Lon, C., Fukuda, M.M., Amenga-Etego, L., Hodgson, A.V.O., Asoala, V., Imwong, M., Takala-Harrison, S., Nosten, F., Su, X.-Z., Ringwald, P., Ariey, F., Dolecek, C., Hien, T.T., Boni, M.F., Thai, C.Q., Amambua-Ngwa, A., Conway, D.J., Djimdé, A.A., Doumbo, O.K., Zongo, I., Ouedraogo, J.-B., Alcock, D., Drury, E., Auburn, S., Koch, O., Sanders, M., Hubbart, C., Maslen, G., Ruano-Rubio, V., Jyothi, D., Miles, A., O'brien, J., Gamble, C., Oyola, S.O., Rayner, J.C., Newbold, C.I., Berriman, M., Spencer, C.C.A., Mcvean, G., Day, N.P., White, N.J., Bethell, D., Dondorp, A.M., Plowe, C.V., Fairhurst, R.M., Kwiatkowski, D.P., 2013. Multiple populations of artemisinin-resistant *Plasmodium falciparum* in Cambodia. Nat. Genet. 45, 648–655.

Mitchell, S.N., Kakani, E.G., South, A., Howell, P.I., Waterhouse, R.M., Catteruccia, F., 2015. Evolution of sexual traits influencing vectorial capacity in anopheline mosquitoes. Science 347, 985–988.

Neafsey, D.E., Waterhouse, R.M., Abai, M.R., Aganezov, S.S., Alekseyev, M.A., Allen, J.E., Amon, J., Arca, B., Arensburger, P., Artemov, G., Assour, L.A., Basseri, H., Berlin, A., Birren, B.W., Blandin, S.A., Brockman, A.I., Burkot, T.R., Burt, A., Chan, C.S., Chauve, C., Chiu, J.C., Christensen, M., Costantini, C., Davidson, V.L., Deligianni, E., Dottorini, T., Dritsou, V., Gabriel, S.B., Guelbeogo, W.M., Hall, A.B., Han, M.V., Hlaing, T., Hughes, D.S., Jenkins, A.M., Jiang, X., Jungreis, I., Kakani, E.G., Kamali, M., Kemppainen, P., Kennedy, R.C., Kirmitzoglou, I.K., Koekemoer, L.L., Laban, N., Langridge, N., Lawniczak, M.K., Lirakis, M., Lobo, N.F., Lowy, E., Maccallum, R.M., Mao, C., Maslen, G., Mbogo, C., Mccarthy, J., Michel, K., Mitchell, S.N., Moore, W., Murphy, K.A., Naumenko, A.N., Nolan, T., Novoa, E.M., O'loughlin, S., Oringanje, C., Oshaghi, M.A., Pakpour, N., Papathanos, P.A., Peery, A.N., Povelones, M., Prakash, A., Price, D.P., Rajaraman, A., Reimer, L.J., Rinker, D.C., Rokas, A., Russell, T.L., Sagnon, N., Sharakhova, M.V., Shea, T., Simao, F.A., Simard, F., Slotman, M.A., Somboon, P., Stegniy, V., Struchiner, C.J., Thomas, G.W., Tojo, M., Topalis, P., Tubio, J.M., Unger, M.F., Vontas, J., Walton, C., Wilding, C.S., Willis, J.H., Wu, Y.C., YAN, G., Zdobnov, E.M., Zhou, X., Catteruccia, F., Christophides, G.K., Collins, F.H., Cornman, R.S., et al., 2015. Mosquito genomics. Highly evolvable malaria vectors: the genomes of 16 Anopheles mosquitoes. Science 347, 1258522.

Nijhout, H.F., Craig, G.B., 1971. Reproductive isolation in *Stegomyia* mosquitoes. III Evidence for a sexual pheromone. Entomol. Exp. Appl. 14, 399–412.

Okanda, F., Dao, A., Njiru, B., Arija, J., Akelo, H., Toure, Y., Odulaja, A., Beier, J.C., Githure, J.I., Yan, G., Gouagna, L.C., Knols, B., Killeen, G.F., 2002. Behavioural determinants of gene flow in malaria vector populations: *Anopheles gambiae* males select large females as mates. Malar. J. 1, 10.

Oliva, C.F., Damiens, D., Benedict, M.Q., 2014. Male reproductive biology of *Aedes* mosquitoes. Acta Trop. 1325, S12–S19.

Pennetier, C., Warren, B., Dabire, K.R., Russell, I.J., Gibson, G., 2010. "Singing on the wing" as a mechanism for species recognition in the malarial mosquito *Anopheles gambiae*. Curr. Biol. 20, 131–136.

Perry, J.C., Sirot, L., Wigby, S., 2013. The seminal symphony: how to compose an ejaculate. Trends Ecol. Evol. 28, 414–422.

Pitts, R.J., Mozuraitis, R., Gauvin-Bialecki, A., Lemperiere, G., 2014. The roles of kairomones, synomones and pheromones in the chemically-mediated behaviour of male mosquitoes. Acta Trop. 132 (Suppl.), S26–S34.

Polerstock, A.R., Eigenbrode, S.D., Klowden, M.J., 2002. Mating alters the cuticular hydrocarbons of female Anopheles gambiae sensu stricto and Aedes aegypti (Diptera: Culicidae). J. Med. Entomol. 39 (3), 545–552.

Pondeville, E., Maria, A., Jacques, J.C., Bourgouin, C., Dauphin-Villemant, C., 2008. *Anopheles gambiae* males produce and transfer the vitellogenic steroid hormone 20-hydroxyecdysone to females during mating. Proc. Natl. Acad. Sci. U.S.A. 105, 19631–19636.

Ponlawat, A., Harrington, L.C., 2007. Age and body size influence male sperm capacity of the dengue vector *Aedes aegypti* (Diptera: Culicidae). J. Med. Entomol. 44, 422–426.

Ponlawat, A., Harrington, L.C., 2009. Factors associated with male mating success of the dengue vector mosquito, *Aedes aegypti*. Am. J. Trop. Med. Hyg. 80, 395–400.

Raikhel, A.S., Brown, M., Belles, X., 2005. Endocrine control of reproductive processes. In: Gilbert, L., Gill, S., Latrou, I. (Eds.), Comprehensive Molecular Insect Science. Elsevier Press, Oxford.

Richardson, J.B., Jameson, S.B., Gloria-Soria, A., Wesson, D.M., Powell, J., 2015. Evidence of limited polyandry in a natural population of *Aedes aegypti*. Am. J. Trop. Med. Hyg. 93, 189–193.

Rogers, D.W., Whitten, M.M., Thailayil, J., Soichot, J., Levashina, E.A., Catteruccia, F., 2008. Molecular and cellular components of the mating machinery in *Anopheles gambiae* females. Proc. Natl. Acad. Sci. U.S.A. 105, 19390–19395.

Rogers, D.W., Baldini, F., Battaglia, F., Panico, M., Dell, A., Morris, H.R., Catteruccia, F., 2009. Transglutaminase-mediated semen coagulation controls sperm storage in the malaria mosquito. PLoS Biol. 7, e1000272.

Russell, T.L., Govella, N.J., Azizi, S., Drakeley, C.J., Kachur, S.P., Killeen, G.F., 2011. Increased proportions of outdoor feeding among residual malaria vector populations following increased use of insecticide-treated nets in rural Tanzania. Malar. J. 10, 80.

Schwenke, R.A., Lazzaro, B.P., Wolfner, M.F., 2016. Reproduction-immunity trade-offs in insects. Annu. Rev. Entomol. 61, 239–256.

Shaw, W.R., Teodori, E., Mitchell, S.N., Baldini, F., Gabrieli, P., Rogers, D.W., Catteruccia, F., 2014. Mating activates the heme peroxidase HPX15 in the sperm storage organ to ensure fertility in *Anopheles gambiae*. Proc. Natl. Acad. Sci. U.S.A. 111, 5854–5859.

Shishika, D., Manoukis, N.C., Butail, S., Paley, D.A., 2014. Male motion coordination in anopheline mating swarms. Sci. Rep. 4, 6318.

Simmons, L.W., 2001. Sperm Competition and Its Evolutionary Consequences in the Insects. Princeton University Press, New Jersey.

Sinervo, B., Lively, C., 1996. The rock-paper-scissors game and the evolution of alternative mating strategies. Nature 380, 240–243.

Sirot, L.K., Wolfner, M.F., 2015. Who's zooming who? Seminal fluids and cryptic female choice in Diptera. In: Peretti, A.V., Aisenberg, A. (Eds.), Cryptic Female Choice in Arthropods. Springer International Publishing, Switzerland.

Sirot, L.K., Hardstone, M.C., Helinski, M.E., Ribeiro, J.M., Kimura, M., Deewatthanawong, P., Wolfner, M.F., Harrington, L.C., 2011. Towards a semen proteome of the dengue vector mosquito: protein identification and potential functions. PLoS Negl. Trop. Dis. 5, e989.

South, A., Lewis, S.M., 2011. The influence of male ejaculate quantity on female fitness: a meta-analysis. Biol. Rev. Camb. Philos. Soc. 86, 299–309.

Takken, W., Knols, B., 1999. Odor-mediated behavior of Afrotropical malaria mosquitoes. Annu. Rev. Entomol. 44, 131–157.

Toe, K.H., Jones, C.M., N'fale, S., Ismail, H.M., Dabire, K.R., Ranson, H., 2014. Increased pyrethroid resistance in malaria vectors and decreased bed net effectiveness, Burkina Faso. Emerg. Infect. Dis. 20, 1691–1696.
Tomkins, J.L., Hazel, W., 2007. The status of the conditional evolutionary stable strategy. Trends Ecol. Evol. 22, 522–528.
Tripet, F., Toure, Y.T., Dolo, G., Lanzaro, G.C., 2003. Frequency of multiple inseminations in field-collected *Anopheles gambiae* females revealed by DNA analysis of transferred sperm. Am. J. Trop. Med. Hyg. 68, 1–5.
Tripet, F., Dolo, G., Traore, S., Lanzaro, G.C., 2004. The wingbeat hypothesis of reproductive isolation between members of the *Anopheles gambiae* complex (Diptera: Culicidae) does not fly. J. Med. Entomol. 41, 375–384.
Trivers, R., 1972. Parental investment and sexual selection. In: Cambell, B. (Ed.), Sexual Selection and the Descent of Man. Aldine Press, Chicago, pp. 1871–1971.
Utne-Palm, A.C., Eduard, K., Jensen, K.H., Mayer, I., Jakobsen, P.J., 2015. Size dependent male reproductive tactic in the two-spotted goby (*Gobiusculus flavescens*). PLoS One 10, e0143487.
Weir, L.K., Grant, J.W.A., Hutchings, J., 2010. Patterns of aggression and operational sex ratio with alternative male phenotypes in Atlantic Salmon. Ethology 116, 166–175.
WHO, 2014. World Malaria Report. World Health Organization, Geneva, Switzerland.
Yamanaka, N., Rewitz, K.F., O'connor, M.B., 2013. Ecdysone control of developmental transitions: lessons from *Drosophila* research. Annu. Rev. Entomol. 58, 497–516.
Yuval, B., 2006. Mating systems of blood-feeding flies. Annu. Rev. Entomol. 51, 413–440.
Yuval, B., Holliday-Hanson, L., Washino, R.K., 1994. Energy budget of swarming male mosquitoes. Ecol. Entomol. 19, 74–78.
Zatz, C., Werneck, R.M., Macias-Ordonez, R., Machado, G., 2011. Alternative mating tactics in dimorphic males of the harvestman *Longiperna concolor* (Arachnida: Opiliones). Behav. Ecol. Sociobiol. 65, 995–1005.

CHAPTER FOUR

The Role of Juvenile Hormone in Mosquito Development and Reproduction

J. Zhu*, F.G. Noriega†
*Virginia Tech, Blacksburg, VA, United States
†Florida International University, Miami, FL, United States

Contents

Abstract

Juvenile hormone (JH) is a key regulator of development and reproduction in mosquitoes. JH delays metamorphosis until larvae have attained an appropriate stage and size. At that point, a drop in JH titre permits a metamorphic moult. As the antimetamorphic role of JH comes to an end, the late pupa becomes again "competent" to synthesize JH, which plays an essential role orchestrating reproductive maturation. JH synthesis is controlled by the rate of flux of isoprenoid precursors, with a complex interplay of changes in precursor pools, enzyme levels and nutritional and developmental modulators, such as 20-hydroxyecdysone, ecdysis-triggering hormone, insulin and allatostatin-C. JH uses multiple molecular mechanisms to exert its pleiotropic functions at different stages of the mosquito life cycle. JH acts via an unidentified membrane receptor and an

Advances in Insect Physiology, Volume 51
ISSN 0065-2806
http://dx.doi.org/10.1016/bs.aiip.2016.04.005

intracellular receptor, methoprene-tolerant protein. The latter functions as a transcription factor and directly regulates the expression of JH target genes.

1. INTRODUCTION

Juvenile hormone (JH) regulates development and reproductive maturation in insects (Goodman and Cusson, 2012); therefore, JH has long been considered as a target for the development of novel insecticides (Cusson et al., 2013). This chapter will mostly emphasize the work that has been done on biosynthesis, mechanism of action and roles of JH in controlling development and reproduction in the mosquito *Aedes aegypti*. The importance of *Ae. aegypti* as a vector of diseases has attracted the interest of scientists and funding agencies for many years. Consequently, there is ample information available on biological, ecological, anatomical and physiological aspects of this mosquito. Vector base is an excellent web-resource available for genomic analysis (Lawson et al., 2009). Molecular tools such as DNA microarrays (Xi et al., 2008), RNA interference (RNAi) (Perez-Hedo et al., 2013), germline transformation (Kokoza and Raikhel, 2011) and high-throughput transcriptome sequencing (Poelchau et al., 2013) are readily available. All these factors have contributed to make *Ae. aegypti* an excellent model for the study of JH biology.

2. REGULATION OF JH TITRES

2.1 JH Titres in Mosquitoes

JH titres are regulated by the balance between biosynthesis and release of the hormone from the *corpora allata* (CA), as well as its degradation and clearance from the haemolymph by tissue uptake and excretion (Feyereisen, 1985; Goodman and Cusson, 2012). JH titre in mosquitoes is fundamentally determined by the rate of biosynthesis in the CA (Hernandez-Martinez et al., 2015). JH is not stored in the CA, and therefore the amount of JH "released" to the haemolymph or the incubation medium represents the amount of JH synthesized (Li et al., 2003a).

JH-III is the primary form of JH that is found in *Ae. aegypti* (Baker et al., 1983). The biosynthetic pathway of JH-III (hereafter, JH) in the CA of mosquitoes involves 13 discrete enzymatic reactions and is conventionally divided into early (mevalonic acid pathway or MVAP) and late (JH-branch) steps (Nouzova et al., 2011; Rivera-Perez et al., 2014). The 13 enzymes of the JH biosynthesis pathway have been characterized in *Ae. aegypti* (Noriega, 2014). The main trunk of the MVAP consists of multiple enzymatic steps through which Acetyl-CoA is gradually transformed into the 5-carbon compound isopentenyl-pyrophosphate (IPP), and later on to the 15-carbon molecule farnesyl-pyrophosphate (FPP). In the CA of mosquitoes, FPP is sequentially transformed to farnesol (FOL), farnesal (FAL), farnesoic acid (FA), methyl farnesoate (MF) and JH (Nouzova et al., 2011).

The nutritionally dependent development of ovaries is an excellent physiological framework to understand the dynamic changes in JH synthesis during the gonotrophic cycle of female *Ae. aegypti* (Fig. 1). A decrease in JH synthesis in the last instar larvae allows pupation and metamorphosis to proceed. As the antimetamorphic role of JH comes to an end, the "*inactive*" CA

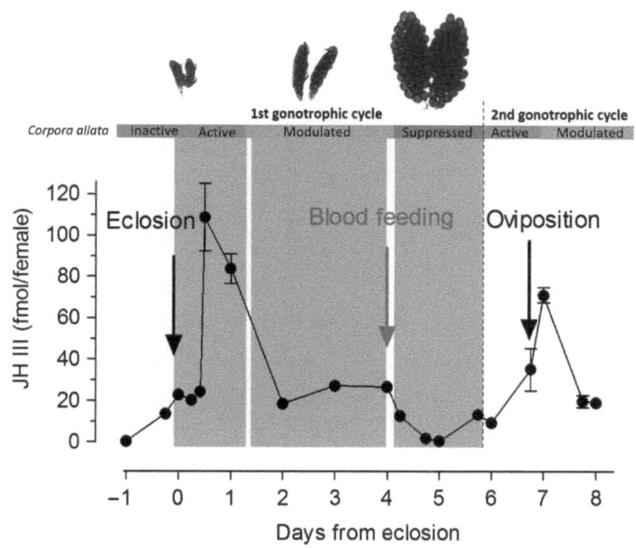

Fig. 1 Dynamic changes in JH levels in haemolymph are related to four distinct CA conditions and linked to the female reproductive physiology. Panels show (1) the development of the ovaries, (2) the four distinct CA conditions (inactive, active, modulated and suppressed) and (3) changes in JH titers in haemolymph in pupae and the first two gonotrophic cycles of *Ae. aegypti* females. The CA is *activated* at the end of the pupal stage and the rise in JH titres controls the previtellogenic development of ovarian follicles. A nutritionally dependent *modulation* of JH synthesis controls resorption of ovarian follicles during the state of arrest in mature sugar-fed females. After a blood meal there is a *suppression* and *reactivation* of JH synthesis, essential for the proper completion of vitellogenic follicle development and the initiation of the second gonotrophic cycle.

of the late pupa recovers its "competence" to synthesize JH. The "*activation*" of JH synthesis and the rise of haemolymph titre after adult eclosion occur in two independent steps; there is a developmentally regulated twofold increase in JH synthesis at eclosion, which is followed by a major nutritionally dependent increase in JH synthesis around 12 h postecdysis (Hernandez-Martinez et al., 2015; Rivera-Perez et al., 2014). Decapitation during these first 12 h of adult life precludes this major increase of JH synthesis and prevents the activation of previtellogenic reproductive maturation (Hernandez-Martinez et al., 2007). This suggests the brain plays a key role sensing nutritional status, and stimulates the CA to its maximum activity only if teneral nutrients are above a threshold (Caroci et al., 2004). After the previtellogenic maturation has been completed (48–72 h), mosquitoes enter the ovarian resting stage and are capable of "*modulating*" JH synthesis at rates proportional to the supply of Acetyl-CoA units derived from sugar-meals (Rivera-Perez et al., 2014). Blood feeding results in active "*suppression*" of JH synthesis. Reduction of JH titre during the vitellogenic stage is critical for the appropriate completion of oogenesis in mosquitoes (Patterson, 1974). As oogenesis proceeds, JH synthesis increases again to prepare females for the second gonotrophic cycle.

An important question that remains partially unanswered is if the changes in JH synthesis described in adult female *Ae. aegypti* are a reflection of the titre in other mosquitoes. Although less comprehensive, analysis of JH biosynthesis in *Culex pipiens* (Readio et al., 1988, 1999) and *Anopheles gambiae* (S. Hernandez-Martinez et al., unpublished) showed overall similar profiles: with low JH biosynthesis at adult eclosion, high in the sugar-fed adults, a dramatic decrease after blood ingestion and a new peak afterwards, as females get ready for a second gonotrophic cycle. Considering that JH titres in mosquitoes are fundamentally determined by the rate of biosynthesis in the CA (Hernandez-Martinez et al., 2015), it is reasonable to postulate that haemolymph titres of JH in *Culex* and *Anopheles* species should show similar trends to those in *Ae. aegypti*.

The potential role of MF as a "JH" in insect preimaginal stages is a controversial issue that is just starting to be addressed (Wen et al., 2015). MF is abundant in the haemolymph of immature stages of several insects (Teal et al., 2014), including *Ae. aegypti* 4th instar larvae mosquito (Hernandez-Martinez et al., 2015). MF is the immediate biosynthetic precursor of JH in mosquitoes, and therefore is very abundant in CA extracts (Rivera-Perez et al., 2014). On the other hand, MF is not released by the CA of adult *Ae. aegypti*, and it is undetectable in the haemolymph of adult female mosquito (Hernandez-Martinez et al., 2015). Detailed temporal profiles of JH titres have never been investigated in embryos, larvae and adult male mosquitoes.

2.2 Regulation of CA Activity

Periodic pulses of 20-hydroxyecdysone (20E) and JH orchestrate the timing of organism-wide developmental transitions in insects (Riddiford, 2012; Yamanaka et al., 2013). High levels of JH prevent metamorphosis in immature stages and stimulate oogenesis in adult mosquitoes. Growth and reproduction are nutrient-dependent processes; JH is therefore an important part of a transduction mechanism that connects developmental changes and nutritional status with the activation of specific physiological events during development and reproduction. Consequently, CA activity is tightly regulated by the concerted action of multiple factors linked to developmental and nutritional signals (Fig. 2).

The brain plays a critical role in the integration of information that governs CA activity; stimulatory and inhibitory effects of brain factors have been

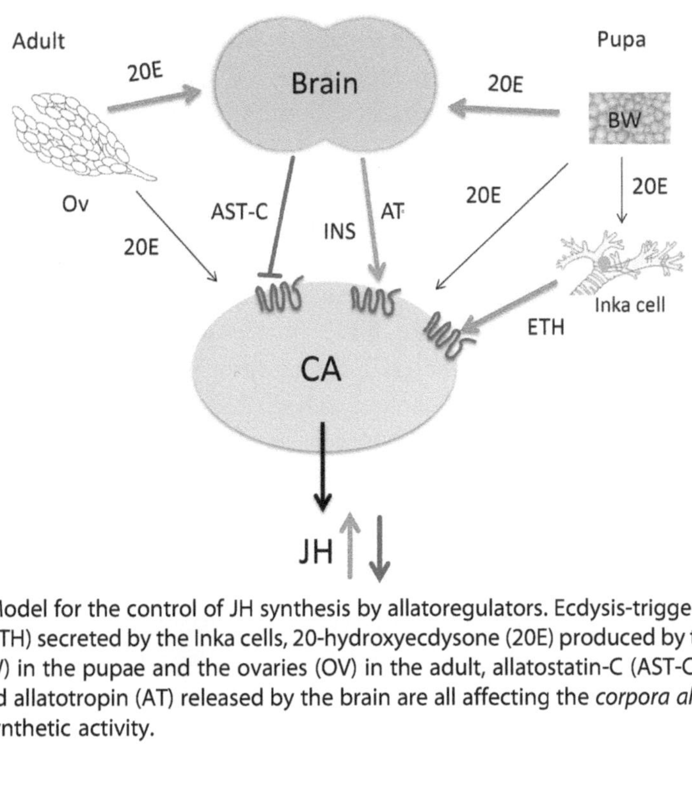

Fig. 2 Model for the control of JH synthesis by allatoregulators. Ecdysis-triggering hormone (ETH) secreted by the Inka cells, 20-hydroxyecdysone (20E) produced by the body wall (BW) in the pupae and the ovaries (OV) in the adult, allatostatin-C (AST-C), insulin (INS) and allatotropin (AT) released by the brain are all affecting the *corpora allata* (CA) JH biosynthetic activity.

described in mosquitoes (Hernandez-Martinez et al., 2007; Li et al., 2004). Separation of the CA from the brain (denervation) results in a remarkable activation of JH synthesis in early pupae (Areiza et al., 2015). In contrast, denervation prevents the 10-fold activation of JH synthesis that occurs 12 h after adult eclosion (Hernandez-Martinez et al., 2007). In sugar-fed and blood-fed females, denervation causes a significant increase in JH synthesis (Li et al., 2004). These effects are likely mediated by different modulators that are present in the brain of *Ae. aegypti*, such as allatostatin-C (AST-C) and allatotropin (AT) (Hernandez-Martinez et al., 2005). They both modify JH synthesis in vitro (Li et al., 2003b, 2006) and their receptors are expressed in the *corpora allata–corpora cardiaca* complex (CA–CC) (Mayoral et al., 2010; Nouzova et al., 2012).

For pharate adult mosquitoes the exact timing of CA activation is critical, and it has to be developmentally coordinated with eclosion. Neuroendocrine factors involved in the regulation of ecdysis, such as ecdysis-triggering hormone (ETH) and 20E, are perfect molecules to synchronize CA activation with moulting to adult. The increase of 20E titre provides temporal cues for the execution of a CA maturation programme; 20E acts as a developmental signal that ensures proper reactivation of JH synthesis in the mosquito pupae (Areiza et al., 2015). Before the rise of 20E titres in late pupa, the CA activation programme has been progressively assembled, but it is kept on hold by a putative brain repressor. 20E stimulates JH synthesis by increasing juvenile hormone acid methyl transferase (JHAMT) activity, which catalyses the conversion of FA into MF in the CA (Areiza et al., 2015).

ETH is a small C-terminally amidated peptide that is released into the haemolymph to activate preecdysis and ecdysis motor programmes in the central nervous system (Žitňan and Adams, 2012). Specialized endocrine cells called Inka cells synthesize ETH in mosquitoes (Dai and Adams, 2009). In *Ae. aegypti* pupae levels of ETH receptor transcripts in the CA rise in synchronism with 20E levels before ecdysis (Areiza et al., 2014). ETH acts as an allatotropic regulator of CA activity, ensuring the proper timing of JH synthesis in pharate adult mosquitoes (Areiza et al., 2014). CA from late pupae stimulated with ETH show increases in JHAMT activity and JH synthesis.

Before a blood meal is taken, nutritional regulators adjust JH synthesis at rates proportional to the reserves derived from sugar-meals. AST-C plays a key role in this nutritional modulation, exerting a strong, rapid and reversible inhibition of JH synthesis that can be overridden by addition of any of the 13 JH precursors, indicating that the AST-C target is located before entry of Acetyl-CoA into the JH biosynthetic pathway (Nouzova et al., 2015). Stimulation experiments using different sources of carbon (glucose, pyruvate, acetate and citrate) revealed that AST-C acts after pyruvate is converted to citrate in the mitochondria (Nouzova et al., 2015). AST-C inhibits JH synthesis by blocking the citrate carrier (CIC) that transports citrate from the mitochondria to the cytosol, obstructing the production of cytoplasmic Acetyl-CoA that sustains JH synthesis in the CA of mosquitoes. In vitro inhibition of the CIC transporter mimics the effect of AST-C, and can be overridden by addition of citrate or acetate (Nouzova et al., 2015).

The insulin/TOR pathways control growth, development, stress resistance, metabolic homeostasis, reproduction and lifespan in animals (Howell and Manning, 2011; Siddle, 2012). In *Ae. aegypti*, starvation decreases JH synthesis via a reduction in insulin signalling in the CA (Perez-Hedo et al., 2013). Starvation-induced upregulation of the insulin receptor; increases CA insulin sensitivity and "primes" the gland to respond rapidly to rises in insulin levels. During this response to starvation the synthetic potential of the CA remains unaffected, and the gland rapidly and efficiently responds to insulin stimulation by increasing JH synthesis to rates similar to those of CA from nonstarved females (Perez-Hedo et al., 2014). A balance between the actions of AST-C and insulin might modulate JH synthesis in sugar-fed adult mosquitoes.

2.3 Flux Control of JH Synthesis Rate

The JH biosynthetic rate is influenced by a complex interplay of changes in precursor pools, enzyme levels and external modulators such as nutrients and allatoregulatory factors (Noriega, 2014; Rivera-Perez et al., 2014). There is a coordinated expression of the 13 JH biosynthetic enzymes in pupae and adult female mosquitoes (Nouzova et al., 2011; Rivera-Perez et al., 2014). In early pupae (−24 h), transcript levels of the synthetic enzymes are low. Steady-state mRNA levels of all enzymes increase in sugar-fed females and decrease again after blood feeding. Genes operating in the early and late steps of the pathway (MVAP and JH-branch) are transcriptionally coregulated as a single unit, and catalytic activities for enzymes of the MVAP and JH-branch also change in a coordinated fashion in "active" and "inactive" CA (Rivera-Perez et al., 2014).

JH synthesis is controlled by the rate of flux of isoprenoids; therefore, JH precursor pool concentrations and fluxes (which are flows into and out of pools) are critical variables in JH regulation (Nouzova et al., 2011). Global fluctuations in the intermediate pool sizes in the MVAP and JH-branch are not functioning as a unit, but behave inversely (Rivera-Perez et al., 2014). Principal component analysis of the metabolic pools indicated that in reproductive female mosquitoes, at least four developmental switches alter JH synthesis by modulating the flux of isoprenoids at distinct points. Metabolic analysis delimited four distinct CA physiological conditions that were named: inactive, active, modulated and suppressed CA (Rivera-Perez et al., 2014) (Fig. 1). The molecular bases for regulation of JH synthesis, as well as the roles of brain factors or other endocrine regulators change during these four phases. Nutritional modulators like AST-C and insulin control the availability of precursors, such as cytoplasmic Acetyl-CoA, without affecting the synthetic potential of the CA (Nouzova et al., 2015) (Fig. 3). On the contrary, developmental regulators such as ETH and 20E, tend to modulate the activity of key enzymes like JHAMT (Areiza et al., 2014). Enzyme levels need to surpass a minimum threshold to achieve a net flux of precursors through the JH biosynthetic pathway; an increase in JHAMT enzymatic activity is critical for the increases in JH biosynthesis in pharate adult and newly eclosed females (Areiza et al., 2014; Rivera-Perez et al., 2014).

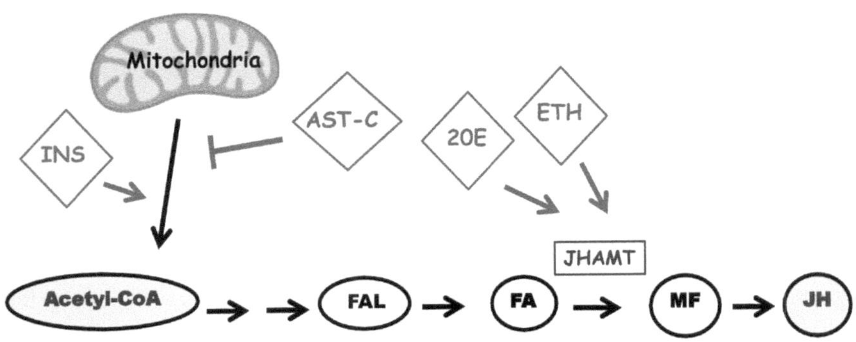

Fig. 3 Model for the control of the flux of precursors in the JH biosynthetic pathway. Precursor pools are represented by *ovals* and connected by *arrows*. Enzymes are in *blue* (*grey* in the print version) *boxes*. Modulators are in *diamonds*. Precursor pools (Acetyl-CoA, FAL, FA and MF) Enzyme: JHAMT (juvenile hormone acid methyl transferase). Regulatory factors affecting precursor pool sizes or enzymatic activities: *AS-C*, allatostatin-C; *INS*, insulin; *20E*, 20-OH-ecdysone and *ETH*, ecdysis-triggering hormone.

3. THE ROLE OF JH IN MOSQUITO DEVELOPMENT

JH plays an antimetamorphic role during development, maintaining the status quo in preimaginal stages and preventing immature insects from precociously turning into adults (Jindra et al., 2013). JH titres are high while the larva is feeding and growing, but drop to permit metamorphosis (Goodman and Cusson, 2012). A recent study on the silkworm *Bombyx mori* indicated that JH is not essential during embryonic and early larval stages (Daimon et al., 2015). *B. mori* mutants were generated with null mutations in the JHAMT gene or the two JH receptor genes (*Met1* and *Met2*). Deprivation of JH or disruption of JH signalling during the embryonic stages had no adverse effect on embryonic growth and morphogenesis, but significantly reduced hatchability of eggs (Daimon et al., 2015). In the JH-deficient *B. mori* mutants, precocious metamorphosis occurred only in the third and fourth larval instars, but did not affect first- and second-instar larvae, suggesting that JH is responsible for maintaining the status quo only during the late larval stages when larvae have already gained competence to enter metamorphosis (Daimon et al., 2015; Smykal et al., 2014). This hypothesis has not been tested in mosquitoes.

The functions of JH during preimaginal stages in mosquitoes have been mostly extrapolated from studies on other insects or deduced from the phenotypic changes induced by topical application of JH or its analogues (JHA). Methoprene and pyriproxyfen affect the development of mosquitoes at multiple stages, and therefore have been widely used as pesticides for mosquito control (Marcombe et al., 2014). Methoprene blocks the embryonic development of *Ae. aegypti* and inhibit egg hatching (Spielman and Williams, 1966). JHAs also interfere with metamorphosis and prevent the emergence of adults (Spielman and Williams, 1966). Methoprene is particularly effective against 4th instar mosquito larvae. A majority of the methoprene-treated larvae died during the pupal stage (Henrick, 2007). Larval exposure to JHA affects metamorphic midgut remodelling and adult cuticle formation in mosquitoes (Nishiura et al., 2003; Spielman and Skaff, 1967; Spielman and Williams, 1966; Wu et al., 2006). The metamorphic midgut remodelling includes two processes, formation of pupal/adult midguts through the division and differentiation of imaginal diploid cells, and programmed death of larval polytene midgut cells (Nishiura, 2002); treating 4th instar *Ae. aegypti* larvae with methoprene blocks degeneration of larval midgut epithelium (Nishiura et al., 2003). Methoprene-treated larvae pupate, but the pupae developed from treated larvae contain two midgut epithelial layers, larval midgut and the pupal/adult midgut (Wu et al., 2006). This methoprene effect is presumably achieved through modulating the action of 20E, as the methoprene treatment causes dysregulation of many genes involved in the stage-specific response to 20E, including *ecdysone receptor, ultraspiracle, broad, E93* and *ftz-f1* (Wu et al., 2006).

4. THE ROLE OF JH IN MOSQUITO REPRODUCTION

Organisms are subject to "trade-offs" between the energetic demands of reproduction and the energy required to survive (Williams, 1966). JH is the central hormonal regulator of life-history trade-offs in many insects, including mosquitoes (Clifton and Noriega, 2011, 2012; Noriega, 2004). Insects must not only allocate nutrients properly within each developmental stage but must also consider the effects of immediate resource allocations on future reproduction and overall fitness (Boggs, 1981, 1992).

Three major periods can be defined in the development of the ovaries during a gonotrophic cycle in *Ae. aegypti* mosquitoes: previtellogenesis (PVG), ovarian resting stage (ORS) and vitellogenesis (VG) (Klowden, 1997). Females emerge with 40 μm immature primary follicles that grow into 100 μm mature PVG follicles in the next 48–72 h (Feinsod and Spielman, 1980; Gwadz and Spielman, 1973). Oocytes remain in a dynamic "state of arrest" and will enter VG only after a blood meal (Hagedorn et al., 1977; Klowden, 1997). Three sources of nutrients provide energy and building blocks for the three distinct phases of ovarian development in *Ae. aegypti*. Preimaginal reserves are partially consumed during PVG, nectar-feeding adds critical reserves during the ORS, and later a blood meal triggers VG (Briegel, 1985, 1990; Klowden, 1997; Noriega, 2004; Zhou et al., 2004a,b).

Ae. aegypti females can laid over 120 eggs in a gonotrophic cycle, therefore a tightly regulated control of nutrient allocations to the ovaries is critical. JH directly controls nutrient allocation into the ovaries in the PVG and ORS phases, and indirectly influences the fate of VG follicles after a blood meal (Clifton and Noriega, 2011, 2012). The ovarian resting stage in *Ae. aegypti* is a period marked by constant adjustment of the reproductive output based on nutritional status; this adjustment occurs through follicular resorption by apoptosis (Clifton and Noriega, 2011). By resorbing excess reproductive tissues, mosquitoes can alter previous reproductive decisions by redirecting resources away from reproduction in favour of competing physiological activities. Resorption occurs in response to poor nutrition/starvation and can be prevented by application of JH (Clifton and Noriega, 2011). The allocation of incoming resources has often been described using a Y-model of resource allocation whereby resources entering at the stem of the "Y" are allocated towards either reproductive physiology on one branch and/or somatic physiology on the other (Boggs, 2009). Under this model, excessive allocations toward somatic physiology or reductions in incoming nutrition often result in reductions to reproductive output (Fig. 4).

The developmental fate of the remaining follicles (those not resorbing) is also sensitive to nutrition and JH. Nutrients and transcripts related to VG (lipophorin receptor, vitellogenin receptor, etc.) accumulated in mosquitoes treated with the JH analogue methoprene or fed 20% sucrose (Clifton and Noriega, 2011); these mosquitoes have increased fecundity and decreased follicle resorption after a blood meal. These results suggest that the molecular and energetic contents of a follicle during the PVG stage are related to that follicle's fate after a blood meal. In other words, nutrition and JH during the ORS alter a follicle's intrinsic ability to develop after a blood meal.

In addition to nutrition, mating triggers many changes in female mosquitoes that enhance reproductive success (Klowden and Chambers, 1991; Shaw et al., 2015). Many of the phenotypic, molecular and biochemical changes observed when manipulating nutrition and hormonal status (ie, increased follicle size, increased triglycerides, transcripts accumulation) also occurred in response to mating (Clifton et al., 2014). Male *Ae. aegypti* mosquitoes might alter the female's resource allocation priorities towards

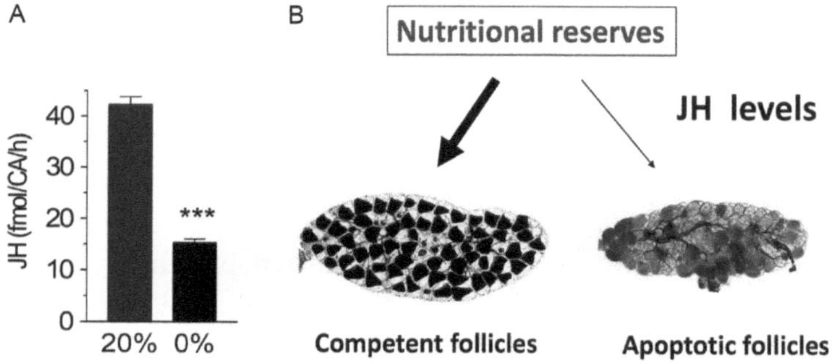

Fig. 4 Role of JH on controlling resource allocations in female mosquitoes. (A) JH biosynthesis levels depend on nutrient reserves (0% and 20% sugar meals). (B) Nutrients availability determine the levels of JH synthesis and reserve allocations into the ovaries. More nutrients result in higher JH (+) and decreased follicle resorption. Poor nutrition/starvation results in lower JH titres (−) and an increase of follicular resorption by apoptosis. (See the colour plate.)

reproduction by transferring JH-III during copulation; reducing the rate of previtellogenic resorption and increasing the amount of stored ovarian lipids (Clifton et al., 2014). In *An. gambiae*, similar results were obtained with the hormone 20E. In this species, high levels of 20E are synthesized and stored in the accessory gland before being transferred to the female during copulation (Pondeville et al., 2008). The interaction between sexually transferred 20E and female proteins results in an increase in the number of eggs produced (Baldini et al., 2013).

Topical application of JH analogues reduced fecundity in *Ae. aegypti* (Brabant and Dobson, 2013) and *An. gambiae* (Bai et al., 2010); and pyriproxyfen-impregnated bed nets have been tested for malaria control (Harris et al., 2013; Kawada et al., 2014). JHA-sterilizing effects have been described as mediated by suppressing the expression of ecdysone-regulated genes in female mosquitoes (Bai et al., 2010); stressing the importance of the endocrine balance between JH and 20E during the gonotrophic cycle.

5. MECHANISM OF ACTION OF JH

5.1 The JH Receptor Methoprene Tolerant

Many effects of JH in mosquitoes are mediated by the methoprene-tolerant (Met) protein, an intracellular JH receptor (Li et al., 2011; Zou et al., 2013). Met belongs to the basic helix–loop–helix (bHLH)-Per-Arnt-Sim (PAS) family of transcription factors (Ashok et al., 1998). The HLH domain promotes formation of homo- or heterodimers and generally bind to DNA with a hexanucleotide core called E-box (CANNTG) (Sailsbery et al., 2012). The PAS domain functions as a protein dimerization motif and consists of two similar hydrophobic repeats, termed PAS-A and PAS-B (Kewley et al., 2004). PAS domains in signal transduction proteins serve as a versatile sensor to detect a wide variety of chemical and physical stimuli, and to regulate the activity of effector domains covalently linked to the PAS domains (Moglich et al., 2009).

In vitro-synthesized *Ae. aegypti* Met (AaMet) protein binds to [^{3}H]-labelled JH-III with high affinity ($K_d = 4.4 \pm 1.9$ nM) (Li et al., 2014). The JH-binding activity resides primarily in the PAS-B domain (Charles et al., 2011). A three-dimensional model generated by homology modelling has revealed a potential JH-binding pocket in the AaMet PAS-B domain (Fig. 5). Site-directed mutagenesis of selected amino acid residues, which are predicted to be in close contact with the docked JH-III in the PAS-B domain, substantially reduces the binding of AaMet to JH-III and two JH

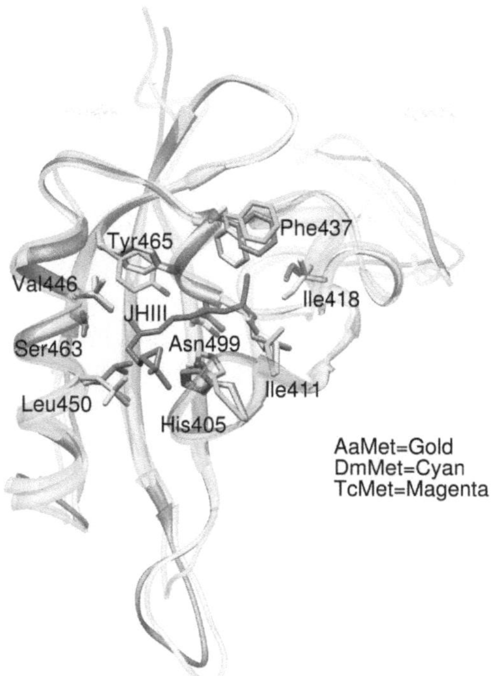

Fig. 5 Cartoon representation of the modelled AaMet PAS-B domain with the lowest-energy docked conformation of JH-III (shown as *red sticks*). *AaMet, Aedes aegypti* Met; *DmMet, Drosophila melanogaster* Met; *TcMet, Tribolium castaneum* Met. (See the colour plate.)

analogues (methoprene and pyriproxyfen), providing solid support to the current structural model (Li et al., 2014).

5.2 Transcriptional Regulation by the JH Receptor

In the previtellogenic adult *Ae. aegypti* mosquitoes, JH modulates the expression of a large number of genes (Zhu et al., 2010; Zou et al., 2013). This JH action relies on the Met protein. RNAi-mediated depletion of Met severely hinders the JH-induced growth of primary ovarian follicles in the newly emerged female adults (Zou et al., 2013). Compared with control mosquitoes, depletion of Met leads to upregulation of 1385 genes and downregulation of 1169 genes in the fat body alone at 72 h after eclosion (Zou et al., 2013). Chromatin immunoprecipitation experiments have demonstrated the binding of AaMet in vivo to the promoter of the early trypsin gene (*AaET*), a well-characterized JH target gene (Noriega et al., 1997), providing explicit evidence that AaMet directly participates in the transcriptional regulation to mediate the JH response (Li et al., 2011). JH response elements (JHREs) have been identified in the regulatory regions of *AaET* and other JH-responsive genes (Li et al., 2011; Shin et al., 2012; Zou et al., 2013). Each JHRE contains an E-box like sequence (5′-GCACGTG-3′) (Li et al., 2014).

Binding of AaMet to the identified JHREs involves another bHLH-PAS protein, Taiman (Tai), a p160 steroid receptor coactivator. RNAi-mediated knockdown of AaTai in newly emerged adult mosquitoes considerably undermines JH-induced gene expression, suggesting that AaTai is a component of the JH signalling pathway (Li et al., 2011). Formation of the AaMet–AaTai complex is induced by JH and the two proteins dimerize through their HLH and PAS domains (Li et al., 2011). While AaMet or AaTai alone cannot bind to the JHREs identified in the JH-regulated genes, the AaMet–AaTai complex is sufficient to bind to the JHREs. This binding requires participation of the DNA-binding domains of both AaMet and AaTai, indicating that AaTai acts as the DNA-binding partner of AaMet (Li et al., 2014). AaMet has also been reported to bind to cycle (CYC) in a JH-dependent manner (Shin et al., 2012). The AaMet–AaCYC dimer also recognizes the E-box-like JHREs in the JH target genes and mediates circadian regulation of the JH-induced gene expression (Shin et al., 2012). Multiple copies of JHREs have been identified in each of the JH-regulated promoters (Li et al., 2014; Shin et al., 2012). It remains to be tested whether the AaMet–AaTai and AaMet–AaCYC complexes act collaboratively on the same promoter to mediate the JH-regulated gene expression.

5.3 Other Transcription Factors in the JH Signalling Pathway

JH induces rapid expression of several transcription factors, including Hairy and Kruppel homolog 1 (Kr-h1) (Shin et al., 2012; Zhu et al., 2010). These proteins act downstream of AaMet and also contribute significantly to regulation of gene expression in response to JH. A recent study has shown that Hairy binds to the promoters of some *Ae. aegypti* genes the expression of which is repressed by JH and recruits a transcriptional corepressor (Groucho) to execute the repressive action of JH (Saha et al., 2016). RNAi-mediated knockdown of either AaMet or Hairy substantially alleviates the repression by JH (Saha et al., 2016). Likewise, Kr-h1 is found to be associated with the regulatory regions of a few Met-dependent genes in adult female mosquitoes (R. Ojani et al., unpublished). Depletion of Kr-h1 after eclosion considerably upregulates the expression of some of the putative

Kr-h1 target genes but downregulates the expression of other Kr-h1 target genes, suggesting that Kr-h1 can activate or repress gene transcription in a promoter-specific manner (R. Ojani et al., unpublished).

5.4 Plasma Membrane-Initiated JH Signalling

The function of Met as a transcription regulator is modulated not only by the intracellular binding of JH but also by a JH signalling pathway that is initiated by proteins associated with plasma membrane (Fig. 6). In *Ae. aegypti* mosquitoes, JH activates phospholipase C (PLC) and quickly increases the levels of inositol 1,4,5-trisphosphate, diacylglycerol and intracellular calcium, leading to activation of protein kinase C (PKC) and calcium/calmodulin-dependent protein kinase II (CaMKII) (Liu et al., 2015; Ojani et al., 2015). The events culminate in phosphorylation of both AaMet and AaTai (Liu et al., 2015). In newly emerged female mosquitoes, RNAi-mediated depletion of PLC, PKC or CaMKII significantly reduces the binding of AaMet and AaTai to the promoters of the JH-regulated genes and

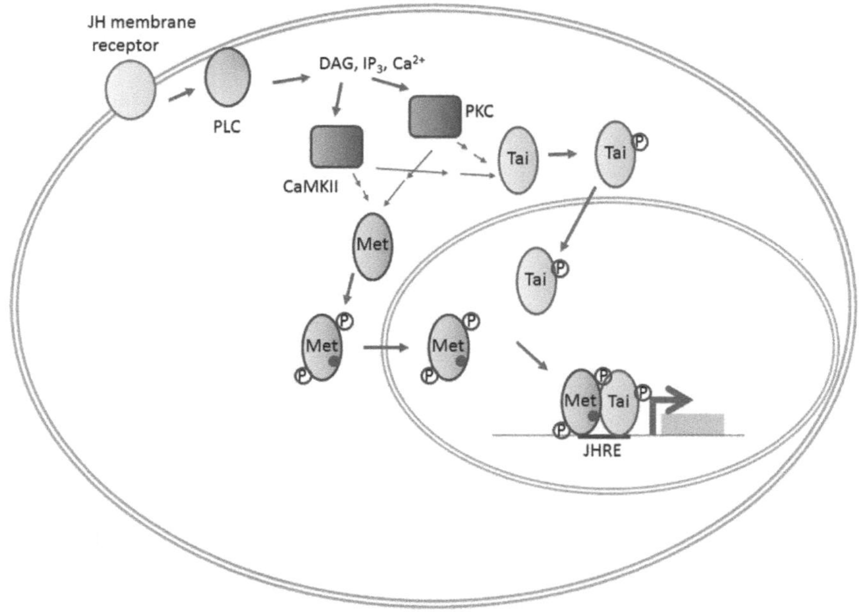

Fig. 6 Juvenile hormone signalling pathway. In the current model, JH binds to an unidentified JH receptor that is associated with the plasma membrane. As a result, PLC is activated, causing increases in DAG, IP_3 and Ca^{2+}. The second messengers activate CaMKII and PKC, resulting in phosphorylation of Met and Tai which in turn potentiates DNA binding of those two transcription factor to the JH-regulated genes. (See the colour plate.)

consequently dampens the JH-induced gene expression (Liu et al., 2015; Ojani et al., 2015). In vitro studies using nuclear proteins extracted from adult mosquitoes indicate that dephosphorylation of AaMet and AaTai eliminates the binding of AaMet and AaTai to JHRE (Liu et al., 2015). These results together suggest that the membrane-initiated signalling pathway modifies the DNA-binding activity of AaMet and AaTai via phosphorylation and thus modulates the genomic responses to JH. This additional layer of regulation may allow for variable and heterogeneous cell responses to JH.

6. CONCLUDING NOTES

Our understanding of the regulation of JH titres and mechanisms of action has significantly changed over the past few years. The potential for genetic manipulation made *Drosophila melanogaster*, *Tribolium castaneum* and *B. mori* suitable insect model organisms to investigate the molecular mechanisms of action of JH (Jindra et al., 2013, 2015a, 2015b). Although the general features of JH biosynthesis and signalling seem to be conserved in most insects, there is clearly diversity in aspects such as the presence of particular JH homologues, the order of the final biosynthetic enzymatic steps, the role of allatoregulators, the number of receptor paralogs, etc.; therefore, JH biosynthesis studies need to be extended beyond the classic model insects. As a matter of fact, recent studies on *Ae. aegypti* have significantly contributed to our overall understanding of the synthesis and action of JH in insects (Areiza et al., 2014, 2015; Li et al., 2011, 2014; Liu et al., 2015; Saha et al., 2016; Zou et al., 2013). Further studies will be necessary to understand the regulation of JH synthesis and mechanisms of action in preimaginal stages and adult male mosquitoes; as well as to expand the research in other mosquito species; in particular, those that are relevant for disease transmission.

ACKNOWLEDGEMENTS

We would like to express our thanks to Mark Clifton for comments on the manuscript. This work was supported by NIH Grant No. AI 45545 to F.G.N. and NIH Grant No. AI099250 to J.Z.

REFERENCES

Areiza, M., Nouzova, M., Rivera-Perez, C., Noriega, F.G., 2014. Ecdysis triggering hormone ensures proper timing of juvenile hormone biosynthesis in pharate adult mosquitoes. Insect Biochem. Mol. Biol. 54, 98–105.

Areiza, M., Nouzova, M., Rivera-Perez, C., Noriega, F.G., 2015. 20-hydroxyecdysone stimulation of juvenile hormone biosynthesis by the mosquito *corpora allata*. Insect Biochem. Mol. Biol. 64, 100–105.

Ashok, M., Turner, C., Wilson, T.G., 1998. Insect juvenile hormone resistance gene homology with the bHLH-PAS family of transcriptional regulators. Proc. Natl. Acad. Sci. U.S.A. 95, 2761–2766.

Bai, H., Gelman, D.B., Palli, S.R., 2010. Mode of action of methoprene in affecting female reproduction in the African malaria mosquito, *Anopheles gambiae*. Pest Manag. Sci. 66, 936–943.

Baker, F.C., Hagedorn, H.H., Schooley, D.A., Wheelock, G., 1983. Mosquito juvenile hormone: identification and bioassay activity. J. Insect Physiol. 29, 465–470.

Baldini, F., Gabrieli, P., South, A., Valim, C., Mancini, F., Catteruccia, F., 2013. The interaction between a sexually transferred steroid hormone and a female protein regulates oogenesis in the malaria mosquito *Anopheles gambiae*. PLoS Biol. 10, e1001695.

Boggs, C.L., 1981. Nutritional and life-history determinants of resource allocation in holometabolous insects. Am. Nat. 117, 692–709.

Boggs, C.L., 1992. Resource allocation: exploring connections between foraging and life history. Funct. Ecol. 6, 508–518.

Boggs, C.L., 2009. Understanding insect life histories and senescence through a resource allocation lens. Funct. Ecol. 23, 927–937.

Brabant, P.J., Dobson, S.D., 2013. Methoprene effects on survival and reproductive performance of adult female and male *Aedes aegypti*. J. Am. Mosq. Control Assoc. 29, 369–375.

Briegel, H., 1985. Mosquito reproduction: incomplete utilization of the blood meal protein for oogenesis. J. Insect Physiol. 31, 15–21.

Briegel, H., 1990. Metabolic relationship between female body size, reserves and fecundity of *Aedes aegypti*. J. Insect Physiol. 36, 165–172.

Caroci, A., Li, Y., Noriega, F.G., 2004. Reduced juvenile hormone synthesis in mosquitoes with low teneral reserves reduces ovarian previtellogenic development in *Aedes aegypti*. J. Exp. Biol. 207, 2685–2690.

Charles, J.P., Iwema, T., Epa, V.C., Takaki, K., Rynes, J., Jindra, M., 2011. Ligand-binding properties of a juvenile hormone receptor, Methoprene-tolerant. Proc. Natl. Acad. Sci. U.S.A. 108, 21128–21133.

Clifton, M.E., Noriega, F.G., 2011. Nutrient limitation results in juvenile hormone-mediated resorption of previtellogenic ovarian follicles in mosquitoes. J. Insect Physiol. 57, 1274–1281.

Clifton, M.E., Noriega, F.G., 2012. The fate of follicles after a blood meal is dependent on previtellogenic nutrition and juvenile hormone in *Aedes aegypti*. J. Insect Physiol. 58, 1007–1019.

Clifton, M.E., Correa, S., Rivera-Perez, C., Nouzova, M., Noriega, F.G., 2014. Male *Aedes aegypti* mosquitoes use JH III transferred during copulation to influence previtellogenic ovary physiology and affect the reproductive output of female mosquitoes. J. Insect Physiol. 64, 40–47.

Cusson, M., Sen, S.E., Shinoda, T., 2013. Juvenile hormone biosynthetic enzymes as targets for insecticide discovery. In: Ishayya, I., Palli, S.R., Horowitz, A.R. (Eds.), Advanced Technologies for Managing Insect Pests. Springer, Netherlands, pp. 31–55.

Dai, L., Adams, M.E., 2009. Ecdysis triggering hormone signaling in the yellow fever mosquito *Aedes aegypti*. Gen. Comp. Endocrinol. 162, 43–51.

Daimon, T., Uchibori, M., Nakao, H., Sezutsu, H., Shinoda, T., 2015. Knockout silkworms reveal a dispensable role for juvenile hormones in holometabolous life cycle. Proc. Natl. Acad. Sci. U.S.A. 112, 4226–4235.

Feinsod, F.M., Spielman, A., 1980. Nutritional-mediated juvenile hormone secretion in mosquitoes. J. Insect Physiol. 26, 113–117.
Feyereisen, R., 1985. Regulation of juvenile hormone titer: synthesis. In: Kerkut, G.A., Gilbert, L.I. (Eds.), Comprehensive Insect Physiology Biochemistry and Pharmacology, vol. 7. Pergamon Press, Oxford, pp. 391–430.
Goodman, W.G., Cusson, M., 2012. The juvenile hormones. In: Gilbert, L.I. (Ed.), Insect Endocrinology. Academic Press, San Diego, pp. 310–365.
Gwadz, R.W., Spielman, A., 1973. Corpus allatum control of ovarian development in *Aedes aegypti*. J. Insect Physiol. 19, 1441–1448.
Hagedorn, H.H., Turner, S., Hagedorn, E.A., Pontecorvo, D., Greenbaum, P., Pfeiffer, D., Wheelock, G., Flanagan, T.R., 1977. Post emergence growth of the ovarian follicles of *Aedes aegypti*. J. Insect Physiol. 23, 203–206.
Harris, C.L., Wetoijera, D.W., Dongus, S., Matowo, N.S., Lorenz, L.M., Devine, G.J., Majambere, S., 2013. Sterilizing effects of pyriproxyfen in *Anopheles arabiensis* and its potential use in malaria control. Parasit. Vectors 6, 144–151.
Henrick, C.A., 2007. Methoprene. J. Am. Mosq. Control. Assoc. 23, 225–239.
Hernandez-Martinez, S., Li, Y., Rodriguez, M.H., Lanz-Mendoza, H., Noriega, F.G., 2005. Allatotropin and PISCF- and YXFGL-amide-allatostatins distribution in *Aedes aegypti* and *Anopheles albimanus* mosquitoes. Cell Tissue Res. 321, 105–113.
Hernandez-Martinez, S., Mayoral, J.G., Li, Y., Noriega, F.G., 2007. Role of juvenile hormone and allatotropin on nutrient allocation, ovarian development and survivorship in mosquitoes. J. Insect Physiol. 53, 230–234.
Hernandez-Martinez, S., Rivera-Perez, C., Nouzova, M., Noriega, F.G., 2015. Coordinated changes in JH biosynthesis and JH hemolymph titers in *Aedes aegypti* mosquitoes. J. Insect Physiol. 72, 22–27.
Howell, J.J., Manning, B.D., 2011. mTOR couples cellular nutrient sensing to organismal metabolic homeostasis. Trends Endocrinol. Metab. 22, 94–102.
Jindra, M., Palli, S.R., Riddiford, L.M., 2013. The juvenile hormone signaling pathway in insect development. Annu. Rev. Entomol. 58, 181–204.
Jindra, M., Belles, X., Shinoda, T., 2015a. Molecular basis of juvenile hormone signaling. Curr. Opin. Insect Sci. 11, 39–46.
Jindra, M., Uhlirova, M., Charles, J.-P., Smykal, V., Hill, R.J., 2015b. Genetic evidence for function of the bHLH-PAS protein Gce/Met as a juvenile hormone receptor. PLoS Genet. 7, e1005394.
Kawada, H., Dida, G.O., Ohashi, K., Kawashima, E., Sonye, G., Njenga, S.M., Mwandawiro, C., Minakawa, N., 2014. A small-scale field trial of pyriproxyfen-impregnated bed nets against pyrethroid-resistant *Anopheles gambiae* s.s. in Western Kenya. PLoS One 9, e111195.
Kewley, R.J., Whitelaw, M.L., Chapman-Smith, A., 2004. The mammalian basic helix-loop-helix/PAS family of transcriptional regulators. Int. J. Biochem. Cell Biol. 36, 189–204.
Klowden, M.J., 1997. Endocrine aspects of mosquito reproduction. Arch. Insect Biochem. Physiol. 35, 491–512.
Klowden, M.J., Chambers, G.M., 1991. Male accessory gland substances activate egg development in nutritionally stressed *Aedes aegypti* mosquitoes. J. Insect Physiol. 37, 721–726.
Kokoza, V.A., Raikhel, A.S., 2011. Targeted gene expression in the transgenic *Aedes aegypti* using the binary Gal4-UAS system. Insect Biochem. Mol. Biol. 41, 637–644.
Lawson, D., Arensburger, P., Atkinson, P., et al., 2009. VectorBase: a data resource for invertebrate genomics. Nucleic Acids Res. 37, D583–D587.
Li, Y., Unnithan, C., Hernandez-Martinez, S., Feyereisen, R., Noriega, F.G., 2003a. Activity of the corpora allata of adult female *Aedes aegypti*: effects of mating and feeding. Insect Biochem. Mol. Biol. 33, 1307–1315.

Li, Y., Unnithan, C., Veenstra, J., Feyereisen, R., Noriega, F.G., 2003b. Stimulation of juvenile hormone biosynthesis by the *corpora allata* of adult *Aedes aegypti* in vitro: effect of farnesoic acid and Aedes allatotropin. J. Exp. Biol. 206, 1825–1832.

Li, Y., Hernandez-Martinez, S., Noriega, F.G., 2004. Inhibition of juvenile hormone biosynthesis in mosquitoes: effect of allatostatic head-factors, PISCF- and YXFGL-amide-allatostatins. Regul. Pept. 118, 175–182.

Li, Y., Martinez-Hernandez, S., Fernandez, F., Mayoral, J.G., Topalis, P., Priestap, H., Perez, M., Navarete, A., Noriega, F.G., 2006. Biochemical, molecular and functional characterization of PISCF-allatostatin, a regulator of juvenile hormone biosynthesis in the mosquito *Aedes aegypti*. J. Biol. Chem. 281, 34048–34055.

Li, M., Mead, E.A., Zhu, J., 2011. Heterodimer of two bHLH-PAS proteins mediates juvenile hormone-induced gene expression. Proc. Natl. Acad. Sci. U.S.A. 108, 638–643.

Li, M., Liu, P., Wiley, J.D., Ojani, R., Bevan, D.R., Li, J., Zhu, J., 2014. A steroid receptor coactivator acts as the DNA-binding partner of the methoprene-tolerant protein in regulating juvenile hormone response genes. Mol. Cell. Endocrinol. 394, 47–58.

Liu, P., Peng, H.J., Zhu, J., 2015. Juvenile hormone-activated phospholipase C pathway enhances transcriptional activation by the methoprene-tolerant protein. Proc. Natl. Acad. Sci. U.S.A. 112, 1871–1879.

Marcombe, S., Farajollahi, A., Healy, S.P., Clark, G.G., Fonseca, D.M., 2014. Insecticide resistance status of United States populations of *Aedes albopictus* and mechanisms involved. PLoS One 9, e101992.

Mayoral, J.G., Nouzova, M., Brockhoff, A., Goodwin, M., Hernandez-Martinez, S., Richter, D., Meyerhof, W., Noriega, F.G., 2010. Allatostatin-C receptors in mosquitoes. Peptides 31, 442–450.

Moglich, A., Ayers, R.A., Moffat, K., 2009. Structure and signaling mechanism of Per-ARNT-Sim domains. Structure 17, 1282–1294.

Nishiura, J.T., 2002. Coordinated morphological changes in midgut, imaginal discs, and respiratory trumpets during metamorphosis of Aedes aegypti (Diptera: Culicidae). Ann. Entomol. Soc. Am. 95, 498–504.

Nishiura, J.T., Ho, P., Ray, K., 2003. Methoprene interferes with mosquito midgut remodeling during metamorphosis. J. Med. Entomol. 40, 498–507.

Noriega, F.G., 2004. Nutritional regulation of JH synthesis: a mechanism to control reproductive maturation in mosquitoes? Insect Biochem. Mol. Biol. 34, 687–693.

Noriega, F.G., 2014. Juvenile hormone biosynthesis in insects: what is new, what do we know, what questions remain? ISRN 2014, 1–16. http://dx.doi.org/10.1155/2014/967361.

Noriega, F.G., Shaa, D., Wells, M.A., 1997. Juvenile Hormone controls early trypsin gene expression in the midgut of *Aedes aegypti*. Insect Mol. Biol. 6, 63–66.

Nouzova, M., Edwards, M., Mayoral, J.M., Noriega, F.G., 2011. A coordinated expression of biosynthetic enzymes controls the flux of juvenile hormone precursors in the *corpora allata* of mosquitoes. Insect Biochem. Mol. Biol. 41, 660–669.

Nouzova, M., Mayoral, J.M., Brockhoff, A., Goodwin, M., Meyerhof, W., Noriega, F.G., 2012. Functional characterization of an allatotropin receptor expressed in the *corpora allata* of mosquitoes. Peptides 34, 201–208.

Nouzova, M., Rivera-Perez, C., Noriega, F.G., 2015. Allatostatin-C reversibly blocks the transport of citrate out of the mitochondria and inhibits juvenile hormone synthesis in mosquitoes. Insect Biochem. Mol. Biol. 57, 20–26.

Ojani, R., Liu, P., Fu, X., Zhu, J., 2015. Protein kinase C modulates transcriptional activation by the juvenile hormone receptor methoprene-tolerant. Insect Biochem. Mol. Biol. 70, 44–52.

Patterson, J.W., 1974. A comparison of the morphogenetic and sterilizing activities of juvenile hormone mimics on *Aedes aegypti*. J. Insect Physiol. 20, 2095–2106.

Perez-Hedo, M., Rivera-Perez, C., Noriega, F.G., 2013. The insulin/TOR signal transduction pathway is involved in the nutritional regulation of juvenile hormone synthesis in *Aedes aegypti*. Insect Biochem. Mol. Biol. 43, 495–500.

Perez-Hedo, M., Rivera-Perez, C., Noriega, F.G., 2014. Starvation increases insulin sensitivity and reduces juvenile hormone synthesis in mosquitoes. PLoS One 9, e86183.

Poelchau, M.F., Reynolds, J.A., Elsik, C.G., Denlinger, D.L., Armbruster, P.A., 2013. Deep sequencing reveals complex mechanisms of diapause preparation in the invasive mosquito, *Aedes albopictus*. Proc. Biol. Sci. 280, 20130143.

Pondeville, E., Maria, A., Jacques, J.C., Bourgouin, C., Dauphin-Villemant, C., 2008. *Anopheles gambiae* males produce and transfer the vitellogenic steroid hormone 20-hydroxyecdysone to females during mating. Proc. Natl. Acad. Sci. U.S.A. 105, 19631–19636.

Readio, J., Peck, K., Meola, R., Dahm, K.H., 1988. Corpus allatum activity (in vitro) in female Culex pipiens during adult life cycle. J. Insect Physiol. 34, 131–135.

Readio, J., Chen, M.H., Meola, R., 1999. Juvenile hormone biosynthesis in diapausing and nondiapausing Culex pipiens (Diptera: Culicidae). J. Med. Entomol. 36, 355–360.

Riddiford, L.M., 2012. How does juvenile hormone control insect metamorphosis and reproduction? Gen. Comp. Endocrinol. 179, 477–484.

Rivera-Perez, C., Nouzova, M., Lamboglia, I., Noriega, F.G., 2014. Metabolomics reveals dynamic changes in the mevalonate and juvenile hormone synthesis pathways linked to the mosquito reproductive physiology. Insect Biochem. Mol. Biol. 51, 1–9.

Saha, T.T., Shin, S.W., Dou, W., Roy, S., Zhao, B., Hou, Y., Wang, X.L., Zou, Z., Girke, T., Raikhel, A.S., 2016. Hairy and Groucho mediate the action of juvenile hormone receptor Methoprene-tolerant in gene repression. Proc. Natl. Acad. Sci. U.S.A. 113, 735–743.

Sailsbery, J.K., Atchley, W.R., Dean, R.A., 2012. Phylogenetic analysis and classification of the fungal bHLH domain. Mol. Biol. Evol. 29, 1301–1318.

Shaw, W.R., Attardo, G.M., Aksoy, S., Catteruccia, F., 2015. A comparative analysis of reproductive biology of insect vectors of human disease. Curr. Opin. Insect Sci. 10, 142–148.

Shin, S.W., Zou, Z., Saha, T.T., Raikhel, A.S., 2012. bHLH-PAS heterodimer of methoprene-tolerant and Cycle mediates circadian expression of juvenile hormone-induced mosquito genes. Proc. Natl. Acad. Sci. U.S.A. 109, 16576–16581.

Siddle, K., 2012. Molecular basis of signaling specificity of insulin and IGF receptors: neglected corners and recent advances. Front. Endocrinol. 3, 1–24.

Smykal, V., Daimon, T., Kayukawa, T., Takaki, K., Shinoda, T., Jindra, M., 2014. Importance of juvenile hormone signaling arises with competence of insect larvae to metamorphose. Dev. Biol. 390, 221–230.

Spielman, A., Skaff, V., 1967. Inhibition of metamorphosis and of ecdysis in mosquitoes. J. Insect Physiol. 13, 1087–1095.

Spielman, A., Williams, C.M., 1966. Lethal effects of synthetic juvenile hormone on larvae of the yellow fever mosquito, *Aedes aegypti*. Science 154, 1043–1044.

Teal, P.E., Jones, D., Jones, G., Torto, B., Nyasembe, V., Borgemeister, C., Alborn, H.T., Kaplan, F., Boucias, D., Lietze, V.U., 2014. Identification of methyl farnesoate from the hemolymph of insects. J. Nat. Prod. 77, 402–405.

Wen, D., Rivera-Perez, C., Abdou, M., Jia, Q., He, Q., Zyaan, O., Bendena, W.B., Tobe, S.S., Noriega, F.G., Palli, S.R., Wang, J., Li, S., 2015. Methyl farnesoate plays a dual role in regulating drosophila metamorphosis. PLoS Genet 11 (3), e1005038.

Williams, G.C., 1966. Natural selection, the costs of reproduction, and a refinement of Lack's principle. Am. Nat. 100, 687–690.

Wu, Y., Parthasarathy, R., Bai, H., Palli, S.R., 2006. Mechanisms of midgut remodeling: juvenile hormone analog methoprene blocks midgut metamorphosis by modulating ecdysone action. Mech. Dev. 123, 530–547.

Xi, Z., Ramirez, J.L., Dimopoulos, G., 2008. The *Aedes aegypti* toll pathway controls dengue virus infection. PLoS Pathog. 7, e1000098.

Yamanaka, N., Rewitz, K.F., O'Connor, M.B., 2013. Ecdysone control of developmental transitions: lessons from Drosophila research. Annu. Rev. Entomol. 58, 497–516.

Zhou, G., Flowers, M., Friedrich, M.K., Horton, J., Pennington, J., Wells, M.A., 2004a. Metabolic fate of [14C]-labeled meal protein amino acids in *Aedes aegypti* mosquitoes. J. Insect Physiol. 50, 337–349.

Zhou, G., Pennington, J., Wells, M.A., 2004b. Utilization of pre-existing energy stores of female *Aedes aegypti* mosquitoes. Insect Biochem. Mol. Biol. 50, 337–349.

Zhu, J., Busche, J.M., Zhang, X., 2010. Identification of juvenile hormone target genes in the adult female mosquitoes. Insect Biochem. Mol. Biol. 40, 23–29.

Žitňan, D., Adams, M.E., 2012. Neuroendocrine regulation of ecdysis. In: Gilbert, L.I. (Ed.), Insect Endocrinology. Academic Press, San Diego, pp. 253–309.

Zou, Z., Saha, T.T., Roy, S., Shin, S.W., Backman, T.W., Girke, T., White, K.P., Raikhel, A.S., 2013. Juvenile hormone and its receptor, methoprene-tolerant, control the dynamics of mosquito gene expression. Proc. Natl. Acad. Sci. U.S.A. 110, 2173–2181.

CHAPTER FIVE

Regulation of Reproductive Processes in Female Mosquitoes

S. Roy*, V. Smykal*, L. Johnson†, T.T. Saha*, Z. Zou‡, A.S. Raikhel*

*Department of Entomology and Institute of Integrative Genome Biology, University of California Riverside, Riverside, CA, United States

†Graduate Program in Molecular, Cell and Developmental Biology, University of California Riverside, Riverside, CA, United States

‡State Key Laboratory of Integrated Management of Pest Insects and Rodents, Institute of Zoology, Chinese Academy of Sciences, Beijing, China

Contents

Abstract

Females of the vector mosquito species require vertebrate blood for egg development and consequently transmit numerous disease pathogens of humans and animals. Previous studies have determined that a steroid hormone, 20-hydroxyecdysone (20E), is the major hormone regulating egg maturation in female mosquitoes. Its action has been particularly well studied in the fat body, a metabolic tissue functionally

Advances in Insect Physiology, Volume 51
ISSN 0065-2806
http://dx.doi.org/10.1016/bs.aiip.2016.05.004

analogous to the vertebrate liver. This chapter discusses the regulation of ecdysteroid synthesis and the molecular basis of 20E action in reproducing female mosquitoes. Research has also revealed that in addition to 20E, a number of other regulators, such as insulin-like peptides and amino acids, are involved in directing sequential gene expression and the progression of a reproductive cycle. The role of these factors has been reviewed in detail. The concluding section describes recent studies related to differential expression of genes and their temporal regulation, which have been possible due to the availability of mosquito genomes and new and improved molecular techniques.

1. INTRODUCTION

Reproduction in haematophagous insects, like mosquitoes, is linked to blood intake. Mosquitoes such as *Aedes aegypti* and *Anopheles gambiae* are the most important disease vectors. They are anautogenous with an absolute requirement for blood feeding. The females develop and lay a batch of eggs only when they acquire a large enough blood meal. Therefore, reproduction in female mosquitoes is cyclic and serves as a foundation for pathogen transmission. The female mosquito reproductive cycle can be divided into two distinct phases, previtellogenic (posteclosion in the first cycle, PE) and vitellogenic (postblood meal, PBM), during which female mosquitoes undergo dramatic shifts in feeding, alternating between consumption of sugars (nectar) and proteins (blood), respectively. This biphasic reproductive cycle is regulated mainly by alternating peaks of juvenile hormone (JH) and 20-hydroxyecdysone (20E) (Fig. 1). Female mosquitoes undergo a JH-controlled PE phase during which their reproductive tissues mature for subsequent vitellogenesis. At about 72 h PE, females enter the state of reproductive arrest, while seeking a host. A blood meal triggers the onset of a series of swift and synchronized reproductive events, required for 20E-controlled vitellogenesis and the completion of egg development and oviposition within a 72-h period. Following the fall of ecdysteroid titre, vitellogenesis is terminated and the reproductive tissues go through a preparatory stage for the next gonadotrophic cycle. Understanding the regulatory mechanisms governing egg developmental cycles is of paramount importance for exploration of novel control approaches and has been a focus of researchers for decades.

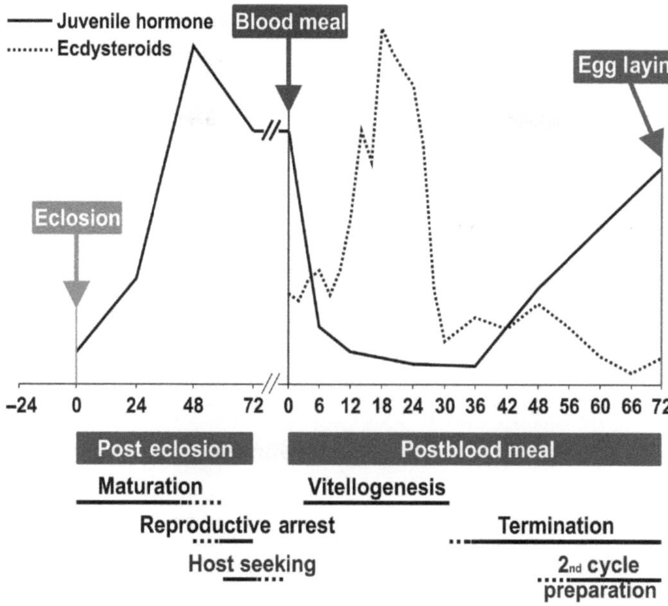

Fig. 1 A schematic diagram of events during the first reproductive cycle of the *Aedes aegypti* female. After eclosion, a rise in juvenile hormone titre promotes the growth of primary ovarian follicles and overall maturation of the female. Around 72 h posteclosion (PE), a female actively seeks a host for a blood meal. Ingestion of a blood meal shuts down JH synthesis in the corpora allata and concurrently triggers synthesis of ecdysteroids in the ovary. Ecdysteroids are the main regulator of vitellogenesis in the fat body. The titre rises at 12 h postblood meal (PBM), peaks at 18 h PBM and sharply declines at 30 h PBM. Low ecdysteroid level is a signal for the cessation of vitellogenesis and start of the terminal phase of the reproductive cycle when the eggs develop fully and are ready to be laid around 72 h PBM. JH titre increases again around 36 h PBM to prepare the female for the next reproductive cycle. JH and ecdysteroid titres are from Shapiro et al. (1986) and Hagedorn et al. (1975), respectively. See chapter "The Role of Juvenile Hormone in Mosquito Development and Reproduction" by Zhu and Noriega for details on JH synthesis.

2. EARLY STUDIES OF ENDOCRINE REGULATION OF MOSQUITO REPRODUCTION

Ever since the classic studies of the blood-sucking bug *Rhodnius prolixus* pioneered by the insect physiologist Sir Vincent Wigglesworth (Wigglesworth, 1947, 1951, 1957), researchers have been searching for factors regulating mosquito reproduction. The role of endocrine signalling in the link between ovarian development and blood feeding in the mosquito *Anopheles maculipennis* was first recognized by Detinova (1945). Studies of the yellow fever mosquito *Ae. aegypti* revealed that a factor from the mosquito head was required for normal egg development (Clements, 1956; Gillet, 1956, 1957; Larsen and Bodenstein, 1959). In a series of elegant microsurgical experiments, Lea (1967) showed that the source of this factor is brain medial neurosecretory cells (NSCs); removal of these cells led to arrested egg maturation, which was restored by reimplantation of the cells. Lea (1972) also demonstrated that an endocrine factor from the cells is stored in the corpus cardiacum and is released only after a mosquito takes a blood meal.

The source of JH is the corpora allata (CA), a paired gland that resides in the mosquito prothorax. Removal of this gland (allatectomy) from the *Ae. aegypti* female at eclosion arrested egg development (Lea, 1969). However, if allatectomy was performed at 72 h PE, it did not prevent egg maturation, indicating the importance of the CA for the PE phase. Gwadz and Spielman (1973) demonstrated that the CA is responsible for the PE growth of primary ovarian follicles; reimplantation of the CA restored follicular development previously arrested by allatectomy at eclosion. Hagedorn et al. (1977) further established that ovarian growth is controlled by JH. This hormone has also been implicated in the development of ovarian responsiveness to a brain factor (Shapiro and Hagedorn, 1982).

Early studies also established the role of ecdysteroids in mosquito reproduction. Injection of ecdysteroids into female *Ae. aegypti* stimulated egg development, suggesting involvement of this hormone in female reproduction (Spielman et al., 1971). In a breakthrough study, Hagedorn and coworkers found that the ovaries of female *Ae. aegypti* secrete ecdysone (E) after a blood meal (Hagedorn et al., 1975). E secretion reached its maximal level at 16 h PBM and rapidly declined thereafter. Hagedorn et al. (1979) further showed that a brain extract stimulates E production by follicle cells of the ovary. The significance of this discovery has been highlighted by numerous follow-up studies demonstrating that female reproduction in all Diptera and some Lepidoptera species is controlled by ovary-derived ecdysteroids (reviewed in Brown et al., 2009).

Vitellogenesis is the central event of female reproduction. It involves production and secretion of yolk protein precursors (YPPs) such as vitellogenin (Vg) by the fat body and internalization of YPPs by developing oocytes through receptor-mediated endocytosis (reviewed in Raikhel, 2005). Hagedorn and Fallon (1973) established that Vg synthesis occurs in the *Ae. aegypti* fat body and further observed that it is controlled by 20E at the transcriptional level (Fallon et al., 1974; Hagedorn et al., 1973, 1975). These early studies, collectively, laid a foundation for our understanding of the network of endocrine factors controlling sequential periods of the female mosquito reproductive cycle.

3. REGULATION OF ECDYSTEROIDOGENESIS IN THE OVARY

3.1 E Biosynthesis

The prothoracic glands (PGs) synthesize and release ecdysteroids during immature stages of insects; in female dipterans, however, the ovary is the major source of these hormones. Ecdysteroid biosynthesis is conserved in both of these tissues (reviewed in Brown et al., 2009; Rewitz et al., 2007). A unique feature of insect metabolism is the inability to synthesize cholesterol de novo; therefore, it must be acquired from food (reviewed in Brown et al., 2009). In adult mosquitoes, cholesterol is obtained directly from blood. Cholesterol delivery from the gut to the site of E biosynthesis by a lipid carrier protein, lipophorin, has been observed in lepidopteran insects (Gilbert et al., 2002; Jouni et al., 2002, 2003). Although the transport of cholesterol to the ovary has not been demonstrated in mosquitoes, the ovarian lipophorin receptor has already been characterized (Cheon et al., 2001). In the ecdysteroidogenic tissues, the first step is the conversion of cholesterol to 7-dehydrocholesterol, catalysed by 7,8-dehydrogenase (Fig. 2). This is followed by multiple biosynthetic steps mediated by P450 enzymes encoded by "Halloween genes" that are localized in the endoplasmic reticulum (ER) and mitochondria (reviewed in Gilbert, 2004; Niwa and Niwa, 2014). The product of ecdysteroid biosynthesis is a prohormone, E, which is secreted into the haemolymph and internalized by peripheral tissues. There, the final step in conversion of E to an active hormone, 20E, is mediated by the cytochrome P450 enzyme, CYP314A1 (Petryk et al., 2003). Several ecdysteroidogenic enzymes have been identified in ovaries of *Ae. aegypti* and *An. gambiae* females (Pondeville et al., 2013; Sieglaff et al., 2005). Their transcriptional upregulation positively correlates with the blood meal and the peak of ecdysteroids during the reproductive cycle, suggesting a link between the blood meal regulatory cascade and E biosynthesis in the ovaries (Sieglaff et al., 2005).

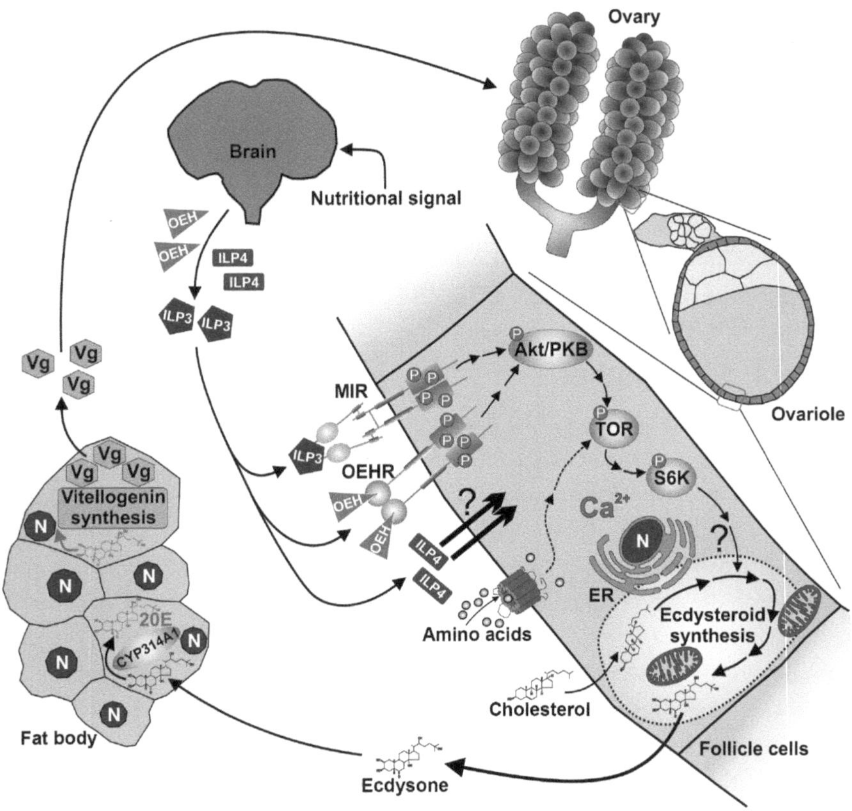

Fig. 2 Ecdysteroidogenesis in the mosquito ovary. Nutritional signals from the blood meal triggers release of neuropeptides, ovary ecdysteroidogenic hormone (OEH) and insulin-like peptides 3 and 4 (ILP3 and ILP4, respectively) from the brain of an *Aedes aegypti* female. These neuropeptides bind their corresponding receptors in the membrane of ovarian follicle cells: OEH binds ovary ecdysteroidogenic hormone receptor (OEHR), ILP3 binds mosquito insulin receptor (MIR) and ILP4 binds to an unknown receptor. After binding, OEHR and MIR are phosphorylated and along with amino acid signalling activates insulin and AA/TOR signalling pathways, manifested by phosphorylation of Akt/PKB, TOR and S6 kinases. Activation of these pathways drives the multistep synthesis of ecdysteroids from cholesterol in the endoplasmic reticulum (ER) and mitochondria of the follicle cells. Ca^{2+} flux seems to be an important factor for the ecdysteroidogenesis although its precise role remains unclear. Ecdysone secreted into the haemolymph is converted by the cytochrome P450 enzyme CYP314A1 to the active form of 20-hydroxyecdysone (20E) in the peripheral tissues. 20E triggers *vitellogenin* (Vg) expression in the fat body and, once released into the haemolymph, Vg is taken by the oocytes and packed in yolk granules. Abbreviations: *20E*, 20-hydroxyecdysone; *Akt/PKB*, Akt/protein kinase B; *ILP3*, insulin-like peptide 3; *ILP4*, insulin-like peptide 4; *MIR*, mosquito insulin receptor; *N*, Nucleus; *P*, phosphorylation; *OEH*, ovary ecdysteroidogenic hormone; *S6K*, p70 ribosomal S6 kinase; *TOR*, target of rapamycin; *Vg*, Vitellogenin. (See the colour plate.)

3.2 Endocrine Control of E Biosynthesis in the Ovary

The neuropeptide prothoracicotropic hormone (PTTH) is a major regulator of E biosynthesis in the PG (reviewed in Rewitz et al., 2013). PTTH is expressed in the lateral NSCs of the brain and stored and released from a pair of neurohaemal organs—corpora cardiaca in most insects or ring gland in higher Diptera—in response to developmental stimuli (Rybczynski, 2005). Once released, it binds Torso, a receptor tyrosine kinase (RTK) in the membrane of PG cells, and signals via the mitogen-activated protein kinase pathway to induce ecdysteroid biosynthesis (Rewitz et al., 2009). Neither recombinant *Ae. aegypti* nor *Bombyx mori* PTTH activated ecdysteroid production in *Ae. aegypti* ovaries in vitro, indicating that PTTH is not a regulator of ovarian ecdysteroidogenesis in mosquitoes (McKinney et al., 2016).

3.2.1 Ovary Ecdysteroidogenic Hormone

As discussed earlier, Hagedorn et al. (1979) described gonadotropic activity in brain extracts of blood-fed mosquitoes that induces ecdysteroid synthesis in *Ae. aegypti* ovaries. Brown et al. (1998) isolated a neuroparsin-like factor, named ovary ecdysteroidogenic hormone (OEH), from mosquito head extracts and identified it as an ecdysteroidogenic gonadotropin with highly specific activity in bioassays with *Ae. aegypti* ovaries (see chapter "Mosquito Peptide Hormones: Diversity, Production, and Function" by Strand et al. for additional details). Blood meal-originated nutrients, particularly amino acids (AAs), trigger the release of OEH from the medial NSCs in the brain of *Ae. aegypti* females (Brown et al., 2008; Wen et al., 2010). Recombinant *Ae. aegypti* OEH and 20E were shown to restore egg maturation in decapitated, facultatively autogenous *Georgecraigius atropalpus* female mosquitoes without a blood meal (Gulia-Nuss et al., 2012).

OEH signals via the recently characterized RTK AAEL001915 (Fig. 2) (Dhara et al., 2013; Gulia-Nuss et al., 2012; Vogel et al., 2015). This receptor is referred to as the OEH receptor (OEHR) and is conserved in mosquitoes. Its expression peaks at 12 h PBM, preceding the ecdysteroid peak in *Ae. aegypti*. Vogel et al. (2015) demonstrated that OEHR depletion via RNA interference (RNAi) attenuated both E production in the ovary and subsequent Vg synthesis in the fat body.

3.2.2 Insulin-Like Peptides

Riehle and Brown (1999) showed that inhibition of the insulin pathway with specific inhibitors sequesters ecdysteroidogenesis in the *Ae. aegypti* ovary. Later, it was demonstrated that two *Ae. aegypti* insulin-like peptides (ILPs), ILP3 and ILP4, exhibit gonadotrophic activity and stimulate ovaries to produce ecdysteroids (Fig. 2) (Brown et al., 2008; Wen et al., 2010). Like OEH, ILPs are produced by the medial NSCs in the brains of *Ae. aegypti* females (Brown et al., 2008; Wen et al., 2010). Gulia-Nuss et al. (2012) demonstrated that, in the same species, ILP3 stimulates serine protease activity in the midgut of blood-fed females, and thus contributes to the availability of AAs for Vg synthesis.

The mosquito insulin receptor (MIR) is present in both nurse and follicle cells in *Ae. aegypti* ovaries (Graf et al., 1997; Helbling and Graf, 1998; Riehle and Brown, 2002). MIR RNAi knockdown prevents ILP3, but not ILP4, from inducing ecdysteroid synthesis by the ovaries of *Ae. aegypti* (Brown et al., 2008; Wen et al., 2010). This suggests that, unlike ILP3, ILP4 signalling is independent of the MIR. Moreover, ILP3 binds specifically to MIR, while ILP4 binds to a 55-kDa membrane protein (Fig. 2) (Wen et al., 2010). However, functional insulin signalling is required for the gonadotropic activity of both ILP3 and ILP4.

It is also known that OEH and ILP3 signal through their respective receptors and, together with AAs, converge on activation of the insulin signalling and amino acid–Target of Rapamycin (AA/TOR) pathways to stimulate synthesis of ecdysteroids in *Ae. aegypti* ovaries (Dhara et al., 2013). This convergence is supported by independent activation of Akt/protein kinase B by ILP3 and OEH in the ovary. Furthermore, ILP3 and OEH with AAs costimulate phosphorylation of p70 S6 kinase (S6K) in the AA/TOR pathway (Fig. 2) (Dhara et al., 2013; Vogel et al., 2015). Recently, Yamanaka et al. (2015) characterized complex regulation of E release from the PG by calcium-stimulated, vesicle-mediated exocytosis in the ring gland of *Drosophila melanogaster.* Ca^{2+} signalling seems to play some role in the *Ae. aegypti* ovary as well, since it has been demonstrated that Ca^{2+} flux is likely contributing to, but not essential for, OEH and ILP3 activity (Fig. 2) (McKinney et al., 2016).

4. MOLECULAR BIOLOGY OF 20E ACTION IN FEMALE MOSQUITO REPRODUCTION

4.1 Studies of the 20E Regulatory Hierarchy

4.1.1 The Ecdysone Receptor

Since the early discoveries described earlier, significant progress has been achieved in understanding the action of 20E. We owe most of the present knowledge to genetic studies in *D. melanogaster* (reviewed in Yamanaka et al., 2013). However, despite the wealth of information provided by *Drosophila* studies, solving the 20E regulation of reproductive events activated by a blood meal in female mosquitoes represents a unique problem.

Ashburner et al. (1974), observing formation of 20E-responsive chromosomal puffing in salivary glands of *Drosophila* larvae, had predicted the existence of the 20E regulatory hierarchy, in which this hormone acts through its receptor causing transcription of so-called early responsive genes. The protein products of these genes in turn serve as transcription factors that activate a larger number of effector genes (Ashburner, 1990; Ashburner et al., 1974). The molecular events underlying the Ashburner model have been subsequently worked out using rapidly developing genetic and molecular methods.

As Ashburner had predicted, 20E was found to act through its cognate receptor (Koelle et al., 1991). However, the 20E functional receptor was identified as a heterodimer, composed of two nuclear receptors—ecdysone receptor (EcR) and Ultraspiracle (USP) (Thomas et al., 1993; Yao et al., 1992, 1993). EcR and USP are orthologs of the vertebrate farnesoid X receptor and the retinoid X receptor, respectively (reviewed in King-Jones and Thummel, 2005). Both EcR and USP possess a conserved structure characteristic of members of the nuclear receptor superfamily, consisting of the DNA-binding (DBD) and ligand-binding (LBD) domains, as well as amino-terminal and carboxyl-terminal transactivation domains (Fig. 3A). EcR binds 20E as its ligand with high affinity; in contrast, unliganded USP interacts with EcR (reviewed in Hill et al., 2013). Crystallographic studies have shown that EcR LBD consists of 12 alpha helices and a beta-sheet forming a ligand-binding pocket fitting 20E. In contrast, USP LBD has a closed configuration with a pocket filled with lipid (Billas et al., 2003; Hill et al., 2013). Upon binding of 20E, the EcR–USP heterodimer interacts with the ecdysone response element (EcRE) consisting of two hexamers (AGGTCA) forming an inverted or direct palindrome separated by a variable number of nucleotides (Fig. 3B) (reviewed in King-Jones and Thummel, 2005).

The two isoforms of EcR, EcR-A and EcR-B, identified in the *Ae. aegypti* mosquito, are differentially expressed during the PBM phase of the female reproductive cycle (Wang et al., 2002). In the fat body, the highest EcR-A transcript abundance corresponds to the 20E peak at 18 h PBM, while the EcR-B transcript is maximally present at the termination stage, 36–42 h PBM (Mane-Padros et al., 2012). Unlike *D. melanogaster*, in which

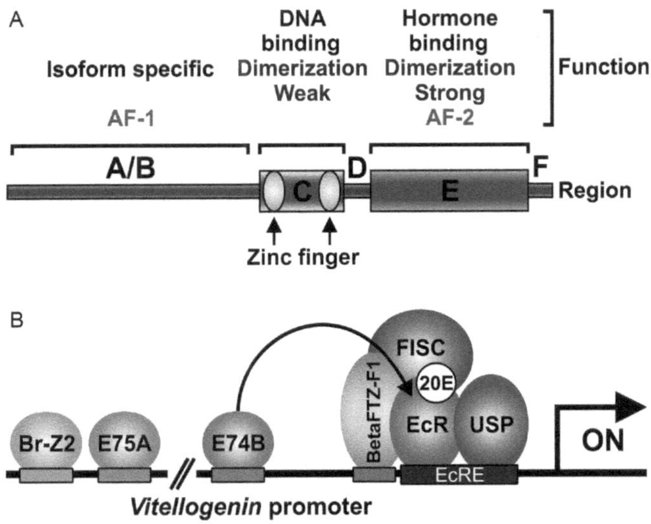

Fig. 3 Molecular mechanism of 20E action. (A) Schematic representation of nuclear receptor protein structure. Nuclear receptors share six conserved domains, A–F. The variable, N-terminal isoform-specific (A/B) domain containing the activation function 1 (AF-1) is responsible for relatively weak transactivation. The conserved DNA-binding domain (DBD) (C) with two zinc fingers specifically binds DNA sequences, called hormone response elements, in the gene promoter region and partially participates in receptor dimerization. The hinge-region (D) is variable in sequence and connects the DBD and the following ligand-binding domain (LBD). It also contributes to intracellular trafficking and subcellular distribution of the receptors. The moderately conserved LBD (E) covers a wide range of functions from ligand-binding and ligand-dependent transactivation through its activation function 2 (AF-2), receptor dimerization, to coactivator and corepressor binding. The last, C-terminal domain (F) is variable in sequence and flexible in structure. (B) Transactivation complex on the promoter of the *vitellogenin* (*Vg*) gene. The nuclear receptor, ecdysone receptor (EcR), binds 20E with high affinity, heterodimerizes with another nuclear receptor, ultraspiracle (USP) and the heterodimer binds ecdysone response elements (EcREs) in the *Vg* gene promoter in the fat body of the *Ae. aegypti* female. Nuclear receptor BetaFTZ-F1 binds the histone acetyltransferase p160/SRC (FISC) in a ligand-independent manner and the complex is recruited to the EcR/USP heterodimer in the presence of 20E. BetaFTZ-F1/FISC/EcR/USP complex upregulates *Vg* transcription which is further enhanced by the binding of Br-Z2, E75A and E74B transcription factors to the *Vg* promoter. (See the colour plate.)

there is only a single USP, in *Ae. aegypti*, two USP isoforms, USP-A and USP-B, were found (Kapitskaya et al., 1996). Multiple USP isoforms were later identified in other insects (Jindra et al., 1997; Lan et al., 1999). Similar to EcR isoforms, *Ae. aegypti* USPs are expressed differentially, with USP-B being constitutively present during the PBM phase and USP-A exhibiting a sharp expression peak at the termination stage (Mane-Padros et al., 2012).

20E has opposite effects on USP isoform transcripts; it is required for sustaining a high level of USP-B expression but downregulates expression of USP-A (Wang et al., 2000). USP-B, which has been implicated as the primary heterodimerization partner for EcR during vitellogenesis, binds the EcREs with a much higher affinity than USP-A (Wang et al., 2000). However, the precise functions of EcR and USP isoforms in *Ae. aegypti* gene regulation remain to be determined.

4.1.2 The Early Genes of the 20E Genetic Hierarchy

Molecular and genetic studies in *D. melanogaster* confirmed the Ashburner model by identifying genes corresponding to 20E-responsive early chromatin puffs (King-Jones and Thummel, 2005; Yamanaka et al., 2013). The *E74* gene is located within the E74EF puff and consists of two transcription units, E74A and E74B (Fletcher et al., 1995). As a result, there are two isoforms of the early factor E74 that belong to the ETS family of transcription factors. E74 is involved in control of stage- and tissue-specific expression of 20E-regulated genes in *Drosophila* (Fletcher and Thummel, 1995). In *Ae. aegypti*, two E74 isoforms play different roles in vitellogenesis. E74B is induced after a blood meal and expressed during the peak of vitellogenesis, while E74A is expressed during the termination stage (Sun et al., 2002). There are several E74 binding sites on the *Vg* gene promoter (Sun et al., 2005). E74B activates expression of the *Vg* gene, whereas E74A has no effect (Sun et al., 2004). E74B serves as an enhancer, synergistically acting with EcR/USP (Fig. 3B) (Sun et al., 2005). A direct protein–protein interaction between EcR and E74B on the *Vg* promoter was demonstrated by means of two-hybrid assays and coimmunoprecipitation analysis. A dominant negative E74 mutant abolished Vg activation in a cell transactivation assay (Sun et al., 2005).

The nuclear receptor E75, an insect ortholog of the vertebrate Rev-erb, is another classic component of the 20E genetic hierarchy (Segraves and Hogness, 1990). In *Drosophila*, E75 is represented by several isoforms that play different roles in coordinating the 20E response (reviewed in Thummel, 1997). In *Ae. aegypti*, there are three E75 isoforms—E75A, E75B and E75C—that are activated by a blood-meal-regulated hormonal cascade (Pierceall et al., 1999). RNAi-mediated silencing of E75 isoforms has revealed their distinct roles in regulating the level and timing of expression of key genes involved in vitellogenesis (Cruz et al., 2012). RNAi depletion of E75A results in repression of *Vg* expression, whereas depletion of E75C results in sustained elevation of Vg levels (Cruz et al., 2012). Haem serves as a ligand for E75, reversibly binding to its LBD (Reinking et al., 2005). During the process of blood digestion, mosquitoes accumulate haem. A combinatorial action of 20E and haem results in a very high level of *Vg* expression, providing an additional link between blood feeding and activation of vitellogenesis (Cruz et al., 2012). Thus, haem derived from a blood meal serves as a signalling molecule mediated by E75 and assures adequate nutrient availability.

Broad (Br) encodes a family of C2H2-type, zinc-finger DNA-binding transcription factors that are critical members of the 20E regulatory hierarchy and is required for expression of other early genes, E74 and E75 (Karim et al., 1993). Br plays a critical role in insect metamorphosis and serves as a pupal stage specifier (Bayer et al., 2003; Zhou and Riddiford, 2002). However, studies in adult mosquitoes have revealed that Br is essential for female reproduction (Chen et al., 2004; Zhu et al., 2007). Four Br isoforms have been identified in *Ae. aegypti*—Z1, Z2, Z3 and Z4—all of which play different roles in vitellogenesis. Expressions of Z1, Z2 and Z4 are stimulated after blood feeding and by 20E in vitro. The regulatory region of the *Vg* gene contains multiple binding sites for Br isoforms. Cell transfection and RNAi analyses showed that Z2 acts as an activator by directly binding to the *Vg* promoter, whereas Z1 and Z4 function as repressors (Chen et al., 2004; Zhu et al., 2007).

In *Ae. aegypti*, transcription of the *Vg* gene is controlled via blood-meal-triggered pathways, with 20E signalling being the central regulator. The *Vg* 5′-regulatory region contains several EcREs, providing evidence of direct control of this gene by EcR/USP (Kokoza et al., 2001; Martin et al., 2001b; Zhu et al., 2006). The authenticity of *Ae. aegypti* Vg EcREs has been confirmed using electrophoretic mobility shift and cell transfection assays, as well as by transgenesis (Kokoza et al., 2001; Martin et al., 2001a; Zhu et al., 2006). The presence of additional binding sites for early gene products of the 20E hierarchy in the *Vg* gene promoter indicates combinatorial control of its expression (Chen et al., 2004; Cruz et al., 2012; Sun et al., 2005).

4.2 Regulation of Developmental Transitions

Timing of 20E signalling during metamorphosis in *Drosophila* is controlled by a set of sequentially expressed nuclear receptors, eg, HR3 and betaFTZ-F1 (Broadus et al., 1999; Lam et al., 1999; White et al., 1997). HR3 is an orphan receptor that binds target genes as a monomer and serves as a developmental switch required for the prepupal–pupal transition (Lam et al.,

1999). HR3 activates the betaFTZ-F1 nuclear receptor that determines competence for stage-specific responses to a subsequent 20E pulse (Broadus et al., 1999). Surprisingly, this genetic signalling is reiteratively utilized during the reproductive cycle in female *Ae. aegypti*, demonstrating conservation of regulatory pathways (Kapitskaya et al., 2000; Li et al., 2000). HR3 is expressed in the late pupa and activates betaFTZ-F1 after the decrease of 20E titre, whereas betaFTZ-F1 continues to be expressed during the JH-controlled PE phase (Li et al., 2000).

Studies of *Ae. aegypti* have established the molecular basis of competency for 20E responsiveness (Fig. 3B). RNAi silencing of betaFTZ-F1 attenuated expression of early genes and the target *Vg* gene after either a blood meal or 20E treatment, providing proof that this nuclear receptor is the factor determining competence for the 20E response in females (Zhu et al., 2003). JH is critical for acquisition of responsiveness by reproductive organs to 20E during the PE phase because it has been implicated in the translational control of betaFTZ-F1 (Zhu et al., 2003). Translated betaFTZ-F1 protein binds the histone acetyltransferase p160/SRC (FISC) in a ligand-independent manner. In the presence of 20E, the betaFTZ-F1/FISC complex is recruited to the EcR/USP heterodimer, initiating loading of ECR/USP/FISC onto target gene promoters and leading to local histone acetylation and robust activation of 20E-responsive genes (Zhu et al., 2006).

A timely cessation of *YPP* gene expression is essential for transition to another gonadotrophic cycle. HR3 plays a central role in orchestrating developmental transitions during the termination stage (Mane-Padros et al., 2012) and is expressed again during the declining 20E titre at 24–36 h PBM, downregulating *Vg* expression. This negative effect of HR3 is a direct result of its binding to its cognate recognition site in the promoter of the *Vg* gene. HR3 is also essential for turning off TOR signalling and initiating programmed autophagy in the fat body (Bryant and Raikhel, 2011; Mane-Padros et al., 2012; Smykal and Raikhel, 2015). Similar to the pupal–adult transition, HR3 activates betaFTZ-F1, which in turn is critical for progression to the second gonadotrophic cycle (Mane-Padros et al., 2012).

5. AA/TOR AND INSULIN SIGNALLING IN MOSQUITO VITELLOGENESIS

Although ecdysteroids are major regulators of mosquito vitellogenesis (Hagedorn et al., 1975), other pathways contribute significantly to the precise regulation of female mosquito reproduction. The AA/TOR and insulin pathways control many aspects of insect life, including nutrient-dependent growth during development, vitellogenesis and oocyte growth (reviewed in Badisco et al., 2013; Hansen et al., 2014; Okamoto and Yamanaka, 2015; Smykal and Raikhel, 2015). They serve as check points, ensuring that sufficient nutrients are available for the vitellogenic phase. The relative contribution of AA/TOR and insulin pathways in the nutritional regulation of vitellogenesis is not completely clear, although simultaneous inhibition of both pathways has been shown to block egg production, suggesting their synergistic action (Gulia-Nuss et al., 2011).

5.1 The AA/TOR Signalling

A blood meal obtained by a female mosquito serves as a major source of AAs and other nutrients for the upcoming reproductive cycle. Hansen et al. (2004) identified a critical role of the AA/TOR pathway in PBM regulation of vitellogenesis in the female *Ae. aegypti* fat body (Fig. 4). AAs are transported by specific transporters that belong to a group of transceptors, which are involved in both AA binding and transport (Boudko et al., 2015; Kriel et al., 2011). Multiple studies have shown that the RNAi silencing of some of the AA transporters/transceptors lowers TOR signalling and mosquito fertility in *Ae. aegypti* females (Attardo et al., 2006; Boudko et al., 2015; Carpenter et al., 2012; Hansen et al., 2011).

AAs signal by means of a nutrient-sensitive TORC1 (hereafter TOR) branch of the TOR signalling pathway (Kim and Guan, 2011; Loewith and Hall, 2011; Wullschleger et al., 2006). Most signals upstream of TOR converge on tuberous sclerosis complex 1 and 2 (TSC1/2), a heterodimeric negative regulator of Rheb (Ras homolog enriched in brain) GTPase (Inoki et al., 2003; Tee et al., 2003). Hansen et al. (2004) showed that, in contrast to TOR knockdown, RNAi of TSC2 upregulated Vg expression in the female *Ae. aegypti* fat body. Roy and Raikhel (2011) reported that Rheb, a cell membrane protein localized at the lysosome, activates TOR protein kinase activity in response to AAs in *Ae. aegypti* and controls egg development. Rag A/B and Rag C/D, Ras-related small GTP-binding protein GTPases, also participate in transmitting the AA signal to TOR by localizing TOR protein to the lysosome and enabling its activation by Rheb (Groenewoud and Zwartkruis, 2013; Yang et al., 2013). Other critical components of the TOR signalling pathway have been connected to egg development in mosquito females. Once activated by a nutritional signal, TOR kinase phosphorylates p70 S6 kinase (S6K) and

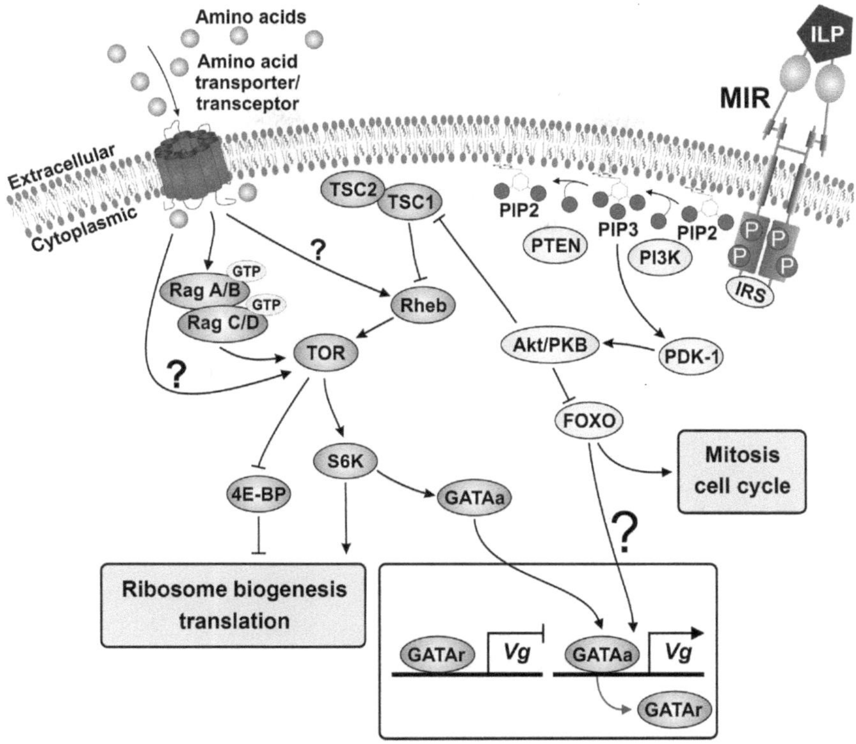

Fig. 4 Insulin and AA/TOR signalling pathways. Amino acids released during digestion of a blood meal are sensed and transported into the fat body cells of *Ae. aegypti* where they participate in activation of the target of rapamycin (TOR) kinase via Ras homolog enriched in brain (Rheb). Activated TOR kinase phosphorylates downstream proteins and thereby derepresses ribosome biogenesis and protein translation by inactivation of the translational inhibitor eukaryotic translation initiation factor 4E-binding protein (4E-BP) and by activation of p70 S6 kinase (S6K). Activated S6K elicits massive translation of GATAa transcription factor that replaces repressive GATAr in the *Vg* gene promoter and participates in *Vg* transcription. Insulin-like peptides (ILPs) bound to the mosquito insulin receptor (MIR) localized in the membrane of follicle cells results in its phosphorylation and activation. Activated MIR transduces its signalling through a series of phosphorylation events to activate phosphatidylinositol 3-kinase (PI3K). PI3K phosphorylates phosphatidylinositol-4,5-bisphosphate (PIP2) to generate phosphatidylinositol-3,4,5-trisphosphate (PIP3), which activates Akt/protein kinase B (Akt/PKB). Active Akt/PKB can stimulate TOR signalling pathway through repression of the TOR-negative regulators, tuberous sclerosis complex 1 and 2 (TSC1/2). Forkhead box O (FOXO) transcription factor regulates cell cycle progression and mitosis, and Akt/PKB phosphorylation keeps FOXO from its translocation to the nucleus and likely repression of *Vg* transcription. However, the role of FOXO in Vg expression needs verification. Phosphatase and tensin homolog (PTEN) negatively regulates insulin signalling by dephosphorylation of the PIP3 to PIP2. Other abbreviations: *GTP*, guanosine-5′-triphosphate; *IRS*, insulin receptor substrate; *P*, phosphate/phosphorylation; *PDK*-1, 3-phosphoinositide-dependent protein kinase-1; *Rag A/B* and *Rag C/D*, Ras-related small GTP-binding protein GTPases; *Vg*, Vitellogenin. (See the colour plate.)

eukaryotic translation initiation factor 4E binding protein (4E-BP), a repressor of cap-dependent mRNA translation (Hansen et al., 2005; Roy et al., 2007). Weng and Shiao (2015) suggested possible cross-talk between AA/TOR and Wnt signalling pathways, when silencing of Wnt membrane receptor Frizzled 2 decreased S6K phosphorylation and Vg expression.

Following TOR activation, phosphorylation of 4E-PB derepresses systemic translation inhibition and, together with translational activator S6K, permits both translation of GATA activator (GATAa) transcription factor in the *Ae. aegypti* fat body and *Vg* expression (Attardo et al., 2003; Park et al., 2006). GATAa replaces the GATA repressor in the *Vg* promoter, thus permitting expression (Attardo et al., 2003; Martin et al., 2001a; Park et al., 2006).

5.2 Insulin Signalling

In addition to the role in E biosynthesis described earlier, ILPs play a significant role in the nutritional regulation of vitellogenesis. ILPs are peptide hormones with three interchain disulfide bonds (Sajid et al., 2011) that are structurally and functionally analogous to vertebrate insulin (Luckhart and Riehle, 2007; Wu and Brown, 2006). In mosquitoes and other insects, the ILP and insulin-like growth factor signalling pathways play a major role in regulation of growth, development, lifespan and reproduction (Broughton et al., 2005; Giannakou and Partridge, 2007; Hyun, 2013; Okamoto and Yamanaka, 2015; Smykal and Raikhel, 2015; Taguchi and White, 2008; Wu and Brown, 2006).

Various numbers of ILPs have been identified in mosquito species: eight in *Ae. aegypti* (Riehle et al., 2006), seven in *An. gambiae* (Krieger et al., 2004), five in *Anopheles stephensi* (Marquez et al., 2011) and at least three in *Culex pipiens* (Sim and Denlinger, 2009). Brown et al. (2008) showed that injection of decapitated, blood-fed *Ae.* aegypti with ILP3-stimulated ovarian ecdysteroidogenesis and egg maturation. RNAi of ILP1 or insulin receptor expression in *C. pipiens* arrested ovarian development in the previtellogenic stage and mimicked a diapause-like state in nondiapausing females (Sim and Denlinger, 2008, 2011). Wen et al. (2010) demonstrated that although ILP3 and ILP4 likely signal via different receptors, they both stimulate ecdysteroidogenesis in the ovaries (see earlier). Interestingly, Kang et al. (2008) described activation of insulin signalling in the midgut by insulin and insulin-like growth factor ingested in a blood meal by female *An. stephensi.* ILPs activate phosphatidylinositol

3-kinase (PI3K), which phosphorylates phosphatidylinositol-4,5-bisphosphate (PIP2) to phosphatidylinositol-3,4,5-trisphosphate (PIP3). *Ae. aegypti* PI3K participates in dose-dependent ecdysteroid production by the ovaries and *Vg* expression in the fat body in vitro (Pri-Tal et al., 2008; Riehle and Brown, 1999; Roy et al., 2007). PIP3 activates the main effector Akt/protein kinase B and 3-phosphoinositide-dependent protein kinase-1 (PDK-1). Akt represses the negative regulators of TOR, TSC1/2 and permits TOR activation by phosphorylation (Loewith and Hall, 2011; Wullschleger et al., 2006). Akt expression has been detected in embryos, in fat bodies and in *Ae. aegypti* ovaries (Dhara et al., 2013; Riehle and Brown, 2003; Roy et al., 2007; Vogel et al., 2015). Bovine insulin can stimulate insulin signalling in the fat body of *Ae. aegypti* females, and insulin together with 20E triggers S6K phosphorylation and *Vg* expression in vitro, in the absence of AAs. RNAi of Akt or TOR expression blocks insulin- and 20E-mediated *Vg* expression (Hansen et al., 2004; Riehle and Brown, 1999; Roy et al., 2007).

In transgenic *Ae. aegypti* and *An. stephensi* females, fat body-specific over-expression of an active Akt increased *Vg* expression but not the number of eggs (Arik et al., 2015). Furthermore, fat body-targeted expression of active Akt in both mosquito species also activated the downstream signalling molecules Forkhead box O and S6 kinase, and increased mosquito survivorship PBM (Arik et al., 2015). Hansen et al. (2007) demonstrated that other members of the Forkhead box gene family seem to negatively affect *Vg* expression and egg production in *Ae. aegypti*, but the full understanding of their roles in mosquito reproduction needs further study. Phosphatase and tensin homolog (PTEN, a direct antagonist of PI3K) dephosphorylate PIP3 to PIP2 leading to inactivation of PI3K and shutdown of the insulin pathway. Arik et al. (2009) increased egg production in *Ae. aegypti* during the first reproductive cycle by knocking down PTEN splice variant 6 (PTEN6). Moretti et al. (2014) showed that knockdown or inhibition of another phosphatase, AAEL001919, an ortholog of protein-tyrosine phosphatase 1B (PTP1B), a negative regulator of the insulin pathway, decreased *Vg* expression and egg production in *Ae. aegypti*.

6. TEMPORAL COORDINATION OF GENE EXPRESSION IN MOSQUITO REPRODUCTION

Coordinated expression of regulatory and functional genes is essential for proper development and reproduction. Advances in genome sequencing, microarray and high-throughput RNA-seq techniques have profoundly influenced our understanding of differential gene expression in various organisms. These approaches have been frequently used in recent years to study genome-wide changes in gene expression related to haematophagy and mosquito reproductive cycles, mostly in *An. gambiae* and *Ae. aegypti* (Akbari et al., 2013; Bonizzoni et al., 2011; Dana et al., 2005; Holt et al., 2002; Hou et al., 2015; Marinotti et al., 2006; Price et al., 2011; Ribeiro, 2003; Roy et al., 2015; Zhu et al., 2010; Zou et al., 2013).

The first genomic-scale study of haematophagy was performed by Holt et al. (2002). They identified differential abundance of 168 ESTs between sugar- and blood-fed (24 h PBM) *An. gambiae* females. An additional 267 genes, which were differentially expressed PBM, were found by Ribeiro (2003). Dana et al. (2005) identified 359 more gene products that were differentially expressed, by examining seven different time points within the first 48 h PBM. They also determined the temporal patterns of gene expression and suggested that gene transcription may be influenced by changes in JH and 20E titres. Marinotti et al. (2006) conducted a detailed microarray-based study of differential gene expression in *An. gambiae*. These authors demonstrated global changes in transcript accumulation during five different time points PBM, during a gonadotrophic cycle (3–96 h PBM). They reported that almost 7500 genes (50% of those probed) showed significant variation in transcript accumulation during the first 72 h after a blood meal. Marinotti et al. (2006) also reported differential expression of genes in ovaries and fat bodies by comparing gene expression at 24 h PBM to that in tissue from nonblood-fed females.

All of the earlier mentioned studies were done in *An. gambiae* and were related to differential expression of genes PBM. In another microarray-based study, Zhu et al. (2010) analysed the transcriptomic changes in newly emerged female *Ae. aegypti*, after topical application of exogenous JH, which is essential for reproductive maturation. They identified a group of JH target genes and paved the way for elucidating genetic interactions in the JH-controlled maturation phase.

Zou et al. (2013) later demonstrated that over 6000 genes were differentially regulated during 72 h of the JH-controlled PE phase in the fat body of *Ae. aegypti* females. These authors identified temporal patterns of gene expression in the PE fat body. They also confirmed that gene transcription in the PE phase is influenced by changes in JH titres, as suggested by Dana et al. (2005). The role of JH and its receptor Methoprene-tolerant in the up- and downregulation of gene expression at different time points within the fat body during this period was experimentally validated. The authors further demonstrated that the functional identity of the gene clusters differs dramatically: genes belonging to metabolic processes, including carbohydrate and lipid metabolism, were upregulated during the first half of the PE phase, when the titre of JH is low; genes related to DNA replication, transcription and translation were highly expressed at a high titre of JH, late in the PE phase (Zou et al., 2013). They also reported that transcripts encoding ribosomal proteins represent a large portion of the fat body transcriptome in PE females, and these results concurred with those of Price et al. (2011).

Two studies seeking haematophagy-related differential expression of genes in *Ae. aegypti* examined changes in gene expression at 5 and 24 h PBM compared with that of nonblood-fed females. Bonizzoni et al. (2011) reported that ~6000 transcripts were differentially expressed between sugar- and blood-fed (5 h PBM) females on the basis of RNA-seq analysis. Price et al. (2011) summed up the results of 454-pyrosequencing by listing ~5000 genes that were differentially expressed in the fat body/abdomen tissue of female *Ae. aegypti*.

Organ- or tissue-specific transcriptome analyses have become more common in the last few years as they are particularly valuable for understanding differential gene expression. Akbari et al. (2013) provided a detailed investigation of ovary-specific genes transcribed over the gonadotrophic cycle in *Ae. aegypti*. While Hou et al. (2015) unravelled the expression dynamics of carbohydrate metabolism genes in the fat body using two separate time course microarray transcriptomes spanning the entire first gonadotrophic cycle, 0–6 h PE to 72 h PE, and 3–72 h PBM, respectively.

Roy et al. (2015) determined that about 7500 transcripts exhibited differential expression over nine time points during the 72-h PBM phase in the fat body of *Ae. aegypti* females. These numbers are similar to those from *An. gambiae*, reported by Marinotti et al. (2006). Roy et al. (2015) identified four major and sequential waves of gene expression over the 72 h PBM. Moreover, these results presented experimental evidence that AAs, 20E, HR3 and JH are the main regulators of differential regulation of the gene expression program during a blood-meal-activated gonadotrophic cycle. Each of these key regulators provides activating and repressive effects on one or more gene sets, thereby maintaining a strict temporal coordination of gene expression essential for sequential reproductive events (Roy et al., 2015). Taken together, these studies have contributed significantly to our understanding of temporal coordination of differential gene expression during mosquito gonadotrophic cycles at the genomic level.

7. CONCLUDING REMARKS

A clear understanding of the regulation of reproductive processes in female mosquitoes is of great importance as blood feeding and disease transmission are intrinsically linked to these processes. We have come a long way since the early studies of egg development, and now much more is known about the endocrine regulation of tissue-specific reproductive processes and their timing. To establish new and advanced control methods, unravelling of the molecular mechanisms behind these processes is of utmost importance. As a result, mosquito biologists have shifted their focus from studying 'what is happening' to 'how things are happening'. During the last decades, the perspective of studying mosquito reproduction from physiological and biochemical standpoints has moved towards molecular biology and genomics, yielding to new paradigms in the regulatory network controlling female mosquito reproduction. For example, studies employing molecular biology techniques coupled with bioinformatics tools have recently identified the roles of microRNAs involved in regulation of events related to blood feeding and reproduction (Bryant et al., 2010; Puthiyakunnon et al., 2013; Liu et al., 2014; Lucas et al., 2015a,b), thereby opening up an entirely new domain of research (see chapter "Functions of small RNAs in mosquitoes" by Hussain et al.).

The molecular biology of 20E action has been worked out in considerable detail. However, our current knowledge is based on studies of a limited number of genes. As a result, there are still important gaps in our understanding of the regulatory mechanisms of the mosquito reproductive cycle. Investigating the temporal regulation of gene expression in the vitellogenic fat body, Roy et al. (2015) postulated that in addition to activating some gene clusters, 20E differentially represses others. The intriguing question that remains to be investigated is the molecular mechanism of 20E repressive action. In this respect, it is important to elucidate which coregulators are involved in this pathway. Although previous studies have shown that AA/TOR and insulin pathways play important roles in controlling vitellogenesis, the relative contribution of each of these pathways remains unclear. Little is known about the complex networks of transcriptional factors and microRNAs responsible for differential regulation of gene expression during female reproduction. The role of posttranslational modifications and epigenetic events remains largely unexplored. Rapid development of novel molecular techniques and approaches and the exponential increase in genomics, metabolomics and bioinformatics studies should achieve a new level of understanding of the complex regulatory networks controlling female mosquito reproduction.

ACKNOWLEDGEMENTS

We are grateful to Mark Brown and Naoki Yamanaka for critical reading of the manuscript and helpful suggestions. This work was supported by a Grant RO1 AI036959 from the National Institutes of Health to A.S.R. and Strategic Priority Research Program of the Chinese Academy of Sciences (No. XDB11030600) to Z.Z.

REFERENCES

Akbari, O.S., Antoshechkin, I., Amrhein, H., Williams, B., Diloreto, R., Sandler, J., Hay, B.A., 2013. The developmental transcriptome of the mosquito Aedes aegypti, an invasive species and major arbovirus vector. G3 (Bethesda) 3, 1493–1509. http://dx.doi.org/10.1534/g3.113.006742.

Arik, A.J., Rasgon, J.L., Quicke, K.M., Riehle, M.A., 2009. Manipulating insulin signaling to enhance mosquito reproduction. BMC Physiol. 9, 15. http://dx.doi.org/10.1186/1472-6793-9-15.

Arik, A.J., Hun, L.V., Quicke, K., Piatt, M., Ziegler, R., Scaraffia, P.Y., Badgandi, H., Riehle, M.A., 2015. Increased Akt signaling in the mosquito fat body increases adult survivorship. FASEB J. 29, 1404–1413. http://dx.doi.org/10.1096/fj.14-261479.

Ashburner, M., 1990. Puffs, genes, and hormones revisited. Cell 61, 1–3. http://dx.doi.org/10.1016/0092-8674(90)90205-s.

Ashburner, M., Chihara, C., Meltzer, P., Richards, G., 1974. Temporal control of puffing activity in polytene chromosomes. Cold Spring Harb. Symp. Quant. Biol. 38, 655–662. http://dx.doi.org/10.1101/sqb.1974.038.01.070.

Attardo, G.M., Higgs, S., Klingler, K.A., Vanlandingham, D.L., Raikhel, A.S., 2003. RNA interference-mediated knockdown of a GATA factor reveals a link to an autogeny in the mosquito Aedes aegypti. Proc. Natl. Acad. Sci. U.S.A. 100, 13374–13379. http://dx.doi.org/10.1073/pnas.2235649100.

Attardo, G.M., Hansen, I.A., Shiao, S.-H., Raikhel, A.S., 2006. Identification of two cationic amino acid transporters required for nutritional signaling during mosquito reproduction. J. Exp. Biol. 209, 3071–3078. http://dx.doi.org/10.1242/jeb.02349.

Badisco, L., Van Wielendaele, P., Broeck, J.V., 2013. Eat to reproduce: a key role for the insulin signaling pathway in adult insects. Front. Physiol. 4, 202. http://dx.doi.org/10.3389/fphys.2013.00202.

Bayer, C., Zhou, X., Zhou, B., Riddiford, L.M., von Kalm, L., 2003. Evolution of the Drosophila broad locus: the Manduca sexta broad Z4 isoform has biological activity in Drosophila. Dev. Genes Evol. 213, 471–476. http://dx.doi.org/10.1007/s00427-003-0354-4.

Billas, I.M.L., Iwema, T., Garnier, J.-M., Mitschler, A., Rochel, N., Moras, D., 2003. Structural adaptability in the ligand-binding pocket of the ecdysone hormone receptor. Nature 426, 91–96. http://dx.doi.org/10.1038/nature02112.

Bonizzoni, M., Dunn, W., Campbell, C.L., Olson, K.E., Dimon, M.T., Marinotti, O., James, A.A., 2011. RNA-seq analyses of blood-induced changes in gene expression in the mosquito vector species, Aedes aegypti. BMC Genomics 12, 82. http://dx.doi.org/10.1186/1471-2164-12-82.

Boudko, D.Y., Tsujimoto, H., Rodriguez, S.D., Meleshkevitch, E.A., Price, D.P., Drake, L.L., Hansen, I.A., 2015. Substrate specificity and transport mechanism of amino-acid transceptor Slimfast from Aedes aegypti. Nat. Commun. 6, 8546. http://dx.doi.org/10.1038/ncomms9546.

Broadus, J., McCabe, J.R., Endrizzi, B., Thummel, C.S., Woodard, C.T., 1999. The Drosophila βFTZ-F1 orphan nuclear receptor provides competence for stage-specific responses to the steroid hormone ecdysone. Mol. Cell 3, 143–149. http://dx.doi.org/10.1016/S1097-2765(00)80305-6.

Broughton, S.J., Piper, M.D.W., Ikeya, T., Bass, T.M., Jacobson, J., Driege, Y., Martinez, P., Hafen, E., Withers, D.J., Leevers, S.J., Partridge, L., 2005. Longer lifespan, altered metabolism, and stress resistance in Drosophila from ablation of cells making insulin-like ligands. Proc. Natl. Acad. Sci. U.S.A. 102, 3105–3110. http://dx.doi.org/10.1073/pnas.0405775102.

Brown, M.R., Graf, R., Swiderek, K.M., Fendley, D., Stracker, T.H., Champagne, D.E., Lea, A.O., 1998. Identification of a steroidogenic neurohormone in female mosquitoes. J. Biol. Chem. 273, 3967–3971. http://dx.doi.org/10.1074/jbc.273.7.3967.

Brown, M.R., Clark, K.D., Gulia, M., Zhao, Z., Garczynski, S.F., Crim, J.W., Suderman, R.J., Strand, M.R., 2008. An insulin-like peptide regulates egg maturation and metabolism in the mosquito Aedes aegypti. Proc. Natl. Acad. Sci. U.S.A. 105, 5716–5721. http://dx.doi.org/10.1073/pnas.0800478105.

Brown, M.R., Sieglaff, D.H., Rees, H.H., 2009. Gonadal ecdysteroidogenesis in Arthropoda: occurrence and regulation. Annu. Rev. Entomol. 54, 105–125. http://dx.doi.org/10.1146/annurev.ento.53.103106.093334.

Bryant, B., Macdonald, W., Raikhel, A.S., 2010. microRNA miR-275 is indispensable for blood digestion and egg development in the mosquito *Aedes aegypti*. Proc. Natl. Acad. Sci. U.S.A. 107, 22391–22398. http://dx.doi.org/10.1073/pnas.1016230107.

Bryant, B., Raikhel, A.S., 2011. Programmed autophagy in the fat body of Aedes aegypti is required to maintain egg maturation cycles. PLoS One 6, e25502. http://dx.doi.org/10.1371/journal.pone.0025502.

Carpenter, V.K., Drake, L.L., Aguirre, S.E., Price, D.P., Rodriguez, S.D., Hansen, I.A., 2012. SLC7 amino acid transporters of the yellow fever mosquito Aedes aegypti and their role in fat body TOR signaling and reproduction. J. Insect Physiol. 58, 513–522. http://dx.doi.org/10.1016/j.jinsphys.2012.01.005.

Chen, L., Zhu, J., Sun, G., Raikhel, A.S., 2004. The early gene Broad is involved in the ecdysteroid hierarchy governing vitellogenesis of the mosquito Aedes aegypti. J. Mol. Endocrinol. 33, 743–761. http://dx.doi.org/10.1677/jme.1.01531.

Cheon, H.-M., Seo, S.-J., Sun, J., Sappington, T.W., Raikhel, A.S., 2001. Molecular characterization of the VLDL receptor homolog mediating binding of lipophorin in oocyte of the mosquito Aedes aegypti. Insect Biochem. Mol. Biol. 31, 753–760. http://dx.doi.org/10.1016/S0965-1748(01)00068-6.

Clements, A., 1956. Hormonal control of ovary development in mosquitoes. J. Exp. Biol. 33, 211–223.

Cruz, J., Mane-Padros, D., Zou, Z., Raikhel, A.S., 2012. Distinct roles of isoforms of the heme-liganded nuclear receptor E75, an insect ortholog of the vertebrate Rev-erb, in mosquito reproduction. Mol. Cell. Endocrinol. 349, 262–271. http://dx.doi.org/10.1016/j.mce.2011.11.006.

Dana, A.N., Hong, Y.S., Kern, M.K., Hillenmeyer, M.E., Harker, B.W., Lobo, N.F., Hogan, J.R., Romans, P., Collins, F.H., 2005. Gene expression patterns associated with blood-feeding in the malaria mosquito Anopheles gambiae. BMC Genomics 6, 5. http://dx.doi.org/10.1186/1471-2164-6-5.

Detinova, T.S., 1945. On the influence of the glands of internal secretion upon the ripening of the gonads and the imaginal diapause in Anopheles maculipennis (in Russian, with Eng. Summ.). Zool. Zhurnal. 24, 291–298.

Dhara, A., Eum, J.-H., Robertson, A., Gulia-Nuss, M., Vogel, K.J., Clark, K.D., Graf, R., Brown, M.R., Strand, M.R., 2013. Ovary ecdysteroidogenic hormone functions independently of the insulin receptor in the yellow fever mosquito, Aedes aegypti. Insect Biochem. Mol. Biol. 43, 1100–1108. http://dx.doi.org/10.1016/j.ibmb.2013.09.004.

Fallon, A.M., Hagedorn, H.H., Wyatt, G.R., Laufer, H., 1974. Activation of vitellogenin synthesis in the mosquito Aedes aegypti by ecdysone. J. Insect Physiol. 20, 1815–1823. http://dx.doi.org/10.1016/0022-1910(74)90211-x.

Fletcher, J., Thummel, C., 1995. The Drosophila E74 gene is required for the proper stage-specific and tissue-specific transcription of ecdysone-regulated genes at the onset of metamorphosis. Development 121, 1411–1421.

Fletcher, J., Burtis, K., Hogness, D., Thummel, C., 1995. The Drosophila E74 gene is required for metamorphosis and plays a role in the polytene chromosome puffing response to ecdysone. Development 121, 1455–1465.

Giannakou, M., Partridge, L., 2007. Role of insulin-like signalling in Drosophila lifespan. Trends Biochem. Sci. 32, 180–188. http://dx.doi.org/10.1016/j.tibs.2007.02.007.

Gilbert, L.I., 2004. Halloween genes encode P450 enzymes that mediate steroid hormone biosynthesis in Drosophila melanogaster. Mol. Cell. Endocrinol. 215, 1–10. http://dx.doi.org/10.1016/j.mce.2003.11.003.

Gilbert, L.I., Rybczynski, R., Warren, J.T., 2002. Control and biochemical nature of the ecdysteroidogenic pathway. Annu. Rev. Entomol. 47, 883–916. http://dx.doi.org/10.1146/annurev.ento.47.091201.145302.

Gillett, J.D., 1956. Initiation and promotion of ovarian development in the mosquito Aëdes (Stegomyia) Aegypti (Linnaeus). Ann. Trop. Med. Parasitol. 50, 375–380. http://dx.doi.org/10.1080/00034983.1956.11685779.

Gillett, J.D., 1957. Variation in the time of release of the ovarian development hormone in Aedes aegypti. Nature 180, 656–657. http://dx.doi.org/10.1038/180656a0.

Graf, R., Neuenschwander, S., Brown, M.R., Ackermann, U., 1997. Insulin-mediated secretion of ecdysteroids from mosquito ovaries and molecular cloning of the insulin receptor homologue from ovaries of bloodfed Aedes aegypti. Insect Mol. Biol. 6, 151–163. http://dx.doi.org/10.1111/j.1365-2583.1997.tb00083.x.

Groenewoud, M.J., Zwartkruis, F.J.T., 2013. Rheb and Rags come together at the lysosome to activate mTORC1. Biochem. Soc. Trans. 41, 951–955. http://dx.doi.org/10.1042/bst20130037.

Gulia-Nuss, M., Robertson, A.E., Brown, M.R., Strand, M.R., 2011. Insulin-like peptides and the target of rapamycin pathway coordinately regulate blood digestion and egg maturation in the mosquito Aedes aegypti. PLoS One 6, e20401. http://dx.doi.org/10.1371/journal.pone.0020401.

Gulia-Nuss, M., Eum, J.-H., Strand, M.R., Brown, M.R., 2012. Ovary ecdysteroidogenic hormone activates egg maturation in the mosquito Georgecraigius atropalpus after adult eclosion or a blood meal. J. Exp. Biol. 215, 3758–3767. http://dx.doi.org/10.1242/jeb.074617.

Gwadz, R.W., Spielman, A., 1973. Corpus allatum control of ovarian development in Aedes aegypti. J. Insect Physiol. 19, 1441–1448. http://dx.doi.org/10.1016/0022-1910(73)90174-1.

Hagedorn, H.H., Fallon, A.M., 1973. Ovarian control of vitellogenin synthesis by the fat body in Aedes aegypti. Nature 244, 103–105. http://dx.doi.org/10.1038/244103a0.

Hagedorn, H.H., Fallon, A.M., Laufer, H., 1973. Vitellogenin synthesis by the fat body of the mosquito Aedes aegypti. Dev. Biol. 31, 285–294. http://dx.doi.org/10.1016/0012-1606(73)90265-0.

Hagedorn, H.H., O'Connor, J.D., Fuchs, M.S., Sage, B., Schlaeger, D.A., Bohm, M.K., 1975. The ovary as a source of alpha-ecdysone in an adult mosquito. Proc. Natl. Acad. Sci. U.S.A. 72, 3255–3259. http://dx.doi.org/10.1073/pnas.72.8.3255.

Hagedorn, H.H., Turner, S., Hagedorn, E.A., Pontecorvo, D., Greenbaum, P., Pfeiffer, D., Wheelock, G., Flanagan, T.R., 1977. Postemergence growth of the ovarian follicles of Aedes aegypti. J. Insect Physiol. 23, 203–206. http://dx.doi.org/10.1016/0022-1910(77)90030-0.

Hagedorn, H.H., Shapiro, J.P., Hanaoka, K., 1979. Ovarian ecdysone secretion is controlled by a brain hormone in an adult mosquito. Nature 282, 92–94. http://dx.doi.org/10.1038/282092a0.

Hansen, I.A., Attardo, G.M., Park, J.-H., Peng, Q., Raikhel, A.S., 2004. Target of rapamycin-mediated amino acid signaling in mosquito anautogeny. Proc. Natl. Acad. Sci. U.S.A. 101, 10626–10631. http://dx.doi.org/10.1073/pnas.0403460101.

Hansen, I.A., Attardo, G.M., Roy, S.G., Raikhel, A.S., 2005. Target of rapamycin-dependent activation of S6 kinase is a central step in the transduction of nutritional signals during egg development in a mosquito. J. Biol. Chem. 280, 20565–20572. http://dx.doi.org/10.1074/jbc.m500712200.

Hansen, I.A., Sieglaff, D.H., Munro, J.B., Shiao, S.-H., Cruz, J., Lee, I.W., Heraty, J.M., Raikhel, A.S., 2007. Forkhead transcription factors regulate mosquito reproduction. Insect Biochem. Mol. Biol. 37, 985–997. http://dx.doi.org/10.1016/j.ibmb.2007.05.008.

Hansen, I.A., Boudko, D.Y., Shiao, S.-H., Voronov, D.A., Meleshkevitch, E.A., Drake, L.L., Aguirre, S.E., Fox, J.M., Attardo, G.M., Raikhel, A.S., 2011. AaCAT1 of the yellow fever mosquito, Aedes aegypti: a novel histidine-specific amino acid transporter from the SLC7 family. J. Biol. Chem. 286, 10803–10813. http://dx.doi.org/10.1074/jbc.m110.179739.

Hansen, I.A., Attardo, G.M., Rodriguez, S.D., Drake, L.L., 2014. Four-way regulation of mosquito yolk protein precursor genes by juvenile hormone-, ecdysone-, nutrient-, and insulin-like peptide signaling pathways. Front. Physiol. 5, 103. http://dx.doi.org/10.3389/fphys.2014.00103.

Helbling, P., Graf, R., 1998. Localization of the mosquito insulin receptor homolog (MIR) in reproducing yellow fever mosquitoes (Aedes aegypti). J. Insect Physiol. 44, 1127–1135. http://dx.doi.org/10.1016/s0022-1910(97)00039-5.

Hill, R.J., Billas, I.M.L., Bonneton, F., Graham, L.D., Lawrence, M.C., 2013. Ecdysone receptors: from the Ashburner model to structural biology. Annu. Rev. Entomol. 58, 251–271. http://dx.doi.org/10.1146/annurev-ento-120811-153610.

Holt, R.A., Subramanian, G.M., Halpern, A., Sutton, G.G., Charlab, R., Nusskern, D.R., Wincker, P., Clark, A.G., Ribeiro, J.M.C., Wides, R., Salzberg, S.L., Loftus, B., Yandell, M., Majoros, W.H., Rusch, D.B., Lai, Z., Kraft, C.L., Abril, J.F., Anthouard, V., Arensburger, P., Atkinson, P.W., Baden, H., de Berardinis, V., Baldwin, D., Benes, V., Biedler, J., Blass, C., Bolanos, R., Boscus, D., Barnstead, M., Cai, S., Center, A., Chaturverdi, K., Christophides, G.K., Chrystal, M.A., Clamp, M., Cravchik, A., Curwen, V., Dana, A., Delcher, A., Dew, I., Evans, C.A., Flanigan, M., Grundschober-Freimoser, A., Friedli, L., Gu, Z., Guan, P., Guigo, R., Hillenmeyer, M.E., Hladun, S.L., Hogan, J.R., Hong, Y.S., Hoover, J., Jaillon, O., Ke, Z., Kodira, C., Kokoza, E., Koutsos, A., Letunic, I., Levitsky, A., Liang, Y., Lin, J.-J., Lobo, N.F., Lopez, J.R., Malek, J.A., McIntosh, T.C., Meister, S., Miller, J., Mobarry, C., Mongin, E., Murphy, S.D., O'Brochta, D.A., Pfannkoch, C., Qi, R., Regier, M.A., Remington, K., Shao, H., Sharakhova, M.V., Sitter, C.D., Shetty, J., Smith, T.J., Strong, R., Sun, J., Thomasova, D., Ton, L.Q., Topalis, P., Tu, Z., Unger, M.F., Walenz, B., Wang, A., Wang, J., Wang, M., Wang, X., Woodford, K.J., Wortman, J.R., Wu, M., Yao, A., Zdobnov, E.M., Zhang, H., Zhao, Q., Zhao, S., Zhu, S.C., Zhimulev, I., Coluzzi, M., della Torre, A., Roth, C.W., Louis, C., Kalush, F., Mural, R.J., Myers, E.W., Adams, M.D., Smith, H.O., Broder, S., Gardner, M.J., Fraser, C.M., Birney, E., Bork, P., Brey, P.T., Venter, J.C., Weissenbach, J., Kafatos, F.C., Collins, F.H., Hoffman, S.L., 2002. The genome sequence of the malaria mosquito Anopheles gambiae. Science 298, 129–149. http://dx.doi.org/10.1126/science.1076181.

Hou, Y., Wang, X.-L., Saha, T.T., Roy, S., Zhao, B., Raikhel, A.S., Zou, Z., 2015. Temporal coordination of carbohydrate metabolism during mosquito reproduction. PLoS Genet. 11, e1005309. http://dx.doi.org/10.1371/journal.pgen.1005309.

Hyun, S., 2013. Body size regulation and insulin-like growth factor signaling. Cell. Mol. Life Sci. 70, 2351–2365. http://dx.doi.org/10.1007/s00018-013-1313-5.

Inoki, K., Li, Y., Xu, T., Guan, K.L., 2003. Rheb GTPase is a direct target of TSC2 GAP activity and regulates mTOR signaling. Genes Dev. 17, 1829–1834. http://dx.doi.org/10.1101/gad.1110003.

Jindra, M., Huang, J.-Y., Malone, F., Asahina, M., Riddiford, L.M., 1997. Identification and mRNA developmental profiles of two ultraspiracle isoforms in the epidermis and wings of Manduca sexta. Insect Mol. Biol. 6, 41–53. http://dx.doi.org/10.1046/j.1365-2583.1997.00153.x.

Jouni, Z.E., Yun, H.K., Wells, M.A., 2002. Cholesterol efflux from larval Manduca sexta fat body in vitro: high-density lipophorin as the acceptor. J. Insect Physiol. 48, 609–618. http://dx.doi.org/10.1016/S0022-1910(02)00081-1.

Jouni, Z.E., Takada, N., Gazard, J., Maekawa, H., Wells, M.A., Tsuchida, K., 2003. Transfer of cholesterol and diacylglycerol from lipophorin to Bombyx mori ovarioles in vitro: role of the lipid transfer particle. Insect Biochem. Mol. Biol. 33, 145–153. http://dx.doi.org/10.1016/S0965-1748(02)00102-9.

Kang, M.-A., Mott, T.M., Tapley, E.C., Lewis, E.E., Luckhart, S., 2008. Insulin regulates aging and oxidative stress in Anopheles stephensi. J. Exp. Biol. 211, 741–748. http://dx.doi.org/10.1242/jeb.012955.

Kapitskaya, M., Wang, S., Cress, D.E., Dhadialla, T.S., Raikhel, A.S., 1996. The mosquito ultraspiracle homologue, a partner of ecdysteroid receptor heterodimer: cloning and characterization of isoforms expressed during vitellogenesis. Mol. Cell. Endocrinol. 121, 119–132. http://dx.doi.org/10.1016/0303-7207(96)03847-6.

Kapitskaya, M.Z., Li, C., Miura, K., Segraves, W., Raikhel, A.S., 2000. Expression of the early-late gene encoding the nuclear receptor HR3 suggests its involvement in regulating the vitellogenic response to ecdysone in the adult mosquito. Mol. Cell. Endocrinol. 160, 25–37. http://dx.doi.org/10.1016/S0303-7207(99)00253-1.

Karim, F., Guild, G., Thummel, C., 1993. The drosophila broad-complex plays a key role in controlling ecdysone-regulated gene-expression at the onset of metamorphosis. Development 118, 977–988.

Kim, J., Guan, K.-L., 2011. Amino acid signaling in TOR activation. Annu. Rev. Biochem. 80, 1001–1032. http://dx.doi.org/10.1146/annurev-biochem-062209-094414.

King-Jones, K., Thummel, C.S., 2005. Nuclear receptors—a perspective from Drosophila. Nat. Rev. Genet. 6, 311–323. http://dx.doi.org/10.1038/nrg1581.

Koelle, M.R., Talbot, W.S., Segraves, W.A., Bender, M.T., Cherbas, P., Hogness, D.S., 1991. The drosophila EcR gene encodes an ecdysone receptor, a new member of the steroid receptor superfamily. Cell 67, 59–77. http://dx.doi.org/10.1016/0092-8674(91)90572-G.

Kokoza, V.A., Martin, D., Mienaltowski, M.J., Ahmed, A., Morton, C.M., Raikhel, A.S., 2001. Transcriptional regulation of the mosquito vitellogenin gene via a blood meal-triggered cascade. Gene 274, 47–65. http://dx.doi.org/10.1016/S0378-1119(01)00602-3.

Krieger, M.J.B., Jahan, N., Riehle, M.A., Cao, C., Brown, M.R., 2004. Molecular characterization of insulin-like peptide genes and their expression in the African malaria mosquito, Anopheles gambiae. Insect Mol. Biol. 13, 305–315. http://dx.doi.org/10.1111/j.0962-1075.2004.00489.x.

Kriel, J., Haesendonckx, S., Rubio-Texeira, M., Van Zeebroeck, G., Thevelein, J.M., 2011. From transporter to transceptor: signaling from transporters provokes re-evaluation of complex trafficking and regulatory controls. Bioessays 33, 870–879. http://dx.doi.org/10.1002/bies.201100100.

Lam, G., Hall, B.L., Bender, M., Thummel, C.S., 1999. DHR3 is required for the prepupal–pupal transition and differentiation of adult structures during Drosophila metamorphosis. Dev. Biol. 212, 204–216. http://dx.doi.org/10.1006/dbio.1999.9343.

Lan, Q., Hiruma, K., Hu, X., Jindra, M., Riddiford, L.M., 1999. Activation of a delayed-early gene encoding MHR3 by the ecdysone receptor heterodimer EcR-B1–USP-1 but not by EcR-B1–USP-2. Mol. Cell. Biol. 19, 4897–4906. http://dx.doi.org/10.1128/MCB.19.7.4897.

Larsen, J.R., Bodenstein, D., 1959. The humoral control of egg maturation in the mosquito. J. Exp. Zool. 140, 343–381. http://dx.doi.org/10.1002/jez.1401400208.

Lea, A.O., 1967. The medial neurosecretory cells and egg maturation in mosquitoes. J. Insect Physiol. 13, 419–429. http://dx.doi.org/10.1016/0022-1910(67)90082-0.

Lea, A.O., 1969. Egg maturation in mosquitoes not regulated by the corpora allata. J. Insect Physiol. 15, 537–541. http://dx.doi.org/10.1016/0022-1910(69)90251-0.

Lea, A.O., 1972. Regulation of egg maturation in the mosquito by the neurosecretory system: the role of the corpus cardiacum. Gen. Comp. Endocrinol. 3 (Suppl.), 602–608. http://dx.doi.org/10.1016/0016-6480(72)90191-8.

Li, C., Kapitskaya, M.Z., Zhu, J., Miura, K., Segraves, W., Raikhel, A.S., 2000. Conserved molecular mechanism for the stage specificity of the mosquito vitellogenic response to ecdysone. Dev. Biol. 224, 96–110. http://dx.doi.org/10.1006/dbio.2000.9792.

Liu, S., Lucas, K.J., Roy, S., Ha, J., Raikhel, A.S., 2014. Mosquito-specific microRNA-1174 targets serine hydroxymethyltransferase to control key functions in the gut. Proc. Natl. Acad. Sci. U.S.A. 111, 14460–14465. http://dx.doi.org/10.1073/pnas.1416278111.

Loewith, R., Hall, M.N., 2011. Target of rapamycin (TOR) in nutrient signaling and growth control. Genetics 189, 1177–1201. http://dx.doi.org/10.1534/genetics.111.133363.

Lucas, K.J., Roy, S., Ha, J., Gervaise, A.L., Kokoza, V.A., Raikhel, A.S., 2015a. MicroRNA-8 targets the wingless signaling pathway in the female mosquito fat body to regulate reproductive processes. Proc. Natl. Acad. Sci. U.S.A. 112, 1440–1445. http://dx.doi.org/10.1073/pnas.1424408112.

Lucas, K.J., Zhao, B., Roy, S., Gervaise, A.L., Raikhel, A.S., 2015b. Mosquito-specific microRNA-1890 targets the juvenile hormone-regulated serine protease JHA15 in the female mosquito gut. RNA Biol. 12, 1383–1390. http://dx.doi.org/10.1080/15476286.2015.1101525.

Luckhart, S., Riehle, M.A., 2007. The insulin signaling cascade from nematodes to mammals: insights into innate immunity of Anopheles mosquitoes to malaria parasite infection. Dev. Comp. Immunol. 31, 647–656. http://dx.doi.org/10.1016/j.dci. 2006.10.005.

Mane-Padros, D., Cruz, J., Cheng, A., Raikhel, A.S., 2012. A critical role of the nuclear receptor HR3 in regulation of gonadotrophic cycles of the mosquito Aedes aegypti. PLoS One 7, e45019. http://dx.doi.org/10.1371/journal.pone.0045019.

Marinotti, O., Calvo, E., Nguyen, Q.K., Dissanayake, S., Ribeiro, J.M.C., James, A.A., 2006. Genome-wide analysis of gene expression in adult Anopheles gambiae. Insect Mol. Biol. 15, 1–12. http://dx.doi.org/10.1111/j.1365-2583.2006.00610.x.

Marquez, A.G., Pietri, J.E., Smithers, H.M., Nuss, A., Antonova, Y., Drexler, A.L., Riehle, M.A., Brown, M.R., Luckhart, S., 2011. Insulin-like peptides in the mosquito Anopheles stephensi: identification and expression in response to diet and infection with Plasmodium falciparum. Gen. Comp. Endocrinol. 173, 303–312. http://dx.doi.org/10.1016/j.ygcen.2011.06.005.

Martin, D., Piulachs, M.-D., Raikhel, A.S., 2001a. A novel GATA factor transcriptionally represses yolk protein precursor genes in the mosquito Aedes aegypti via interaction with the CtBP corepressor. Mol. Cell. Biol. 21, 164–174. http://dx.doi.org/10.1128/mcb.21.1.164-174.2001.

Martin, D., Wang, S.-F., Raikhel, A.S., 2001b. The vitellogenin gene of the mosquito Aedes aegypti is a direct target of ecdysteroid receptor. Mol. Cell. Endocrinol. 173, 75–86. http://dx.doi.org/10.1016/s0303-7207(00)00413-5.

McKinney, D.A., Eum, J.-H., Dhara, A., Strand, M.R., Brown, M.R., 2016. Calcium influx enhances neuropeptide activation of ecdysteroid hormone production by mosquito ovaries. Insect Biochem. Mol. Biol. 70, 160–169. http://dx.doi.org/10.1016/j.ibmb.2016.01.001.

Moretti, D.M., Ahuja, L.G., Nunes, R.D., Cudischevitch, C.O., Daumas-Filho, C.R.O., Medeiros-Castro, P., Ventura-Martins, G., Jablonka, W., Gazos-Lopes, F., Senna, R., Sorgine, M.H.F., Hartfelder, K., Capurro, M., Atella, G.C., Mesquita, R.D., Silva-Neto, M.A.C., 2014. Molecular analysis of Aedes aegypti classical protein tyrosine phosphatases uncovers an ortholog of mammalian PTP-1B implicated in the control of egg production in mosquitoes. PLoS One 9, e104878. http://dx.doi.org/10.1371/journal.pone.0104878.

Niwa, R., Niwa, Y.S., 2014. Enzymes for ecdysteroid biosynthesis: their biological functions in insects and beyond. Biosci. Biotechnol. Biochem. 78, 1283–1292. http://dx.doi.org/10.1080/09168451.2014.942250.

Okamoto, N., Yamanaka, N., 2015. Nutrition-dependent control of insect development by insulin-like peptides. Curr. Opin. Insect Sci. 11, 21–30. http://dx.doi.org/10.1016/j.cois.2015.08.001.

Park, J.-H., Attardo, G.M., Hansen, I.A., Raikhel, A.S., 2006. GATA factor translation is the final downstream step in the amino acid/target-of-rapamycin-mediated vitellogenin gene expression in the anautogenous mosquito Aedes aegypti. J. Biol. Chem. 281, 11167–11176. http://dx.doi.org/10.1074/jbc.m601517200.

Petryk, A., Warren, J.T., Marques, G., Jarcho, M.P., Gilbert, L.I., Kahler, J., Parvy, J.-P., Li, Y., Dauphin-Villemant, C., O'Connor, M.B., 2003. Shade is the Drosophila P450 enzyme that mediates the hydroxylation of ecdysone to the steroid insect molting hormone 20-hydroxyecdysone. Proc. Natl. Acad. Sci. U.S.A. 100, 13773–13778. http://dx.doi.org/10.1073/pnas.2336088100.

Pierceall, W.E., Li, C., Biran, A., Miura, K., Raikhel, A.S., Segraves, W.A., 1999. E75 expression in mosquito ovary and fat body suggests reiterative use of ecdysone-regulated hierarchies in development and reproduction. Mol. Cell. Endocrinol. 150, 73–89. http://dx.doi.org/10.1016/s0303-7207(99)00022-2.

Pondeville, E., David, J.-P., Guittard, E., Maria, A., Jacques, J.-C., Ranson, H., Bourgouin, C., Dauphin-Villemant, C., 2013. Microarray and RNAi analysis of P450s in Anopheles gambiae male and female steroidogenic tissues: CYP307A1 is required for ecdysteroid synthesis. PLoS One 8, e79861. http://dx.doi.org/10.1371/journal.pone.0079861.

Price, D.P., Nagarajan, V., Churbanov, A., Houde, P., Milligan, B., Drake, L.L., Gustafson, J.E., Hansen, I.A., 2011. The fat body transcriptomes of the yellow fever mosquito Aedes aegypti, pre- and post- blood meal. PLoS One 6, e22573. http://dx.doi.org/10.1371/journal.pone.0022573.

Pri-Tal, B.M., Brown, J.M., Riehle, M.A., 2008. Identification and characterization of the catalytic subunit of phosphatidylinositol 3-kinase in the yellow fever mosquito Aedes aegypti. Insect Biochem. Mol. Biol. 38, 932–939. http://dx.doi.org/10.1016/j.ibmb.2008.07.004.

Puthiyakunnon, S., Yao, Y., Li, Y., Gu, J., Peng, H., Chen, X., 2013. Functional characterization of three microRNAs of the Asian tiger mosquito, *Aedes albopictus*. Parasit. Vectors 6, 230. http://dx.doi.org/10.1186/1756-3305-6-230.

Raikhel, A.S., 2005. Vitellogenesis of disease vectors, from physiology to genes. In: Marquardt, W.C. et al., (Ed.), Biology of Disease Vectors. Elsevier Academic Press, Burlington, San Diego, London, pp. 329–346.

Reinking, J., Lam, M.M.S., Pardee, K., Sampson, H.M., Liu, S., Yang, P., Williams, S., White, W., Lajoie, G., Edwards, A., Krause, H.M., 2005. The Drosophila nuclear receptor E75 contains heme and is gas responsive. Cell 122, 195–207. http://dx.doi.org/10.1016/j.cell.2005.07.005.

Rewitz, K.F., O'Connor, M.B., Gilbert, L.I., 2007. Molecular evolution of the insect Halloween family of cytochrome P450s: phylogeny, gene organization and functional conservation. Insect Biochem. Mol. Biol. 37, 741–753. http://dx.doi.org/10.1016/j.ibmb.2007.02.012.

Rewitz, K.F., Yamanaka, N., Gilbert, L.I., O'Connor, M.B., 2009. The insect neuropeptide PTTH activates receptor tyrosine kinase torso to initiate metamorphosis. Science 326, 1403–1405. http://dx.doi.org/10.1126/science.1176450.

Rewitz, K.F., Yamanaka, N., O'Connor, M.B., 2013. Developmental checkpoints and feedback circuits time insect maturation. In: Shi, Y.B. (Ed.), Animal Metamorphosis. Elsevier Academic Press Inc., San Diego, pp. 1–33

Ribeiro, J.M.C., 2003. A catalogue of Anopheles gambiae transcripts significantly more or less expressed following a blood meal. Insect Biochem. Mol. Biol. 33, 865–882. http://dx.doi.org/10.1016/s0965-1748(03)00080-8.

Riehle, M.A., Brown, M.R., 1999. Insulin stimulates ecdysteroid production through a conserved signaling cascade in the mosquito Aedes aegypti. Insect Biochem. Mol. Biol. 29, 855–860. http://dx.doi.org/10.1016/S0965-1748(99)00084-3.

Riehle, M., Brown, M., 2002. Insulin receptor expression during development and a reproductive cycle in the ovary of the mosquito Aedes aegypti. Cell Tissue Res. 308, 409–420. http://dx.doi.org/10.1007/s00441-002-0561-8.

Riehle, M.A., Brown, M.R., 2003. Molecular analysis of the serine/threonine kinase Akt and its expression in the mosquito Aedes aegypti. Insect Mol. Biol. 12, 225–232. http://dx.doi.org/10.1046/j.1365-2583.2003.00405.x.

Riehle, M.A., Fan, Y., Cao, C., Brown, M.R., 2006. Molecular characterization of insulin-like peptides in the yellow fever mosquito, Aedes aegypti: expression, cellular localization, and phylogeny. Peptides 27, 2547–2560. http://dx.doi.org/10.1016/j.peptides.2006.07.016.

Roy, S.G., Raikhel, A.S., 2011. The small GTPase Rheb is a key component linking amino acid signaling and TOR in the nutritional pathway that controls mosquito egg development. Insect Biochem. Mol. Biol. 41, 62–69. http://dx.doi.org/10.1016/j.ibmb.2010.10.001.

Roy, S.G., Hansen, I.A., Raikhel, A.S., 2007. Effect of insulin and 20-hydroxyecdysone in the fat body of the yellow fever mosquito, Aedes aegypti. Insect Biochem. Mol. Biol. 37, 1317–1326. http://dx.doi.org/10.1016/j.ibmb.2007.08.004.

Roy, S., Saha, T.T., Johnson, L., Zhao, B., Ha, J., White, K.P., Girke, T., Zou, Z., Raikhel, A.S., 2015. Regulation of gene expression patterns in mosquito reproduction. PLoS Genet. 11, e1005450. http://dx.doi.org/10.1371/journal.pgen.1005450.

Rybczynski, R., 2005. Prothoracicotropic hormone. In: Gilbert, L.I., Kostas, I., Gill, S.S. (Eds.), Comprehensive Molecular Insect Science. Elsevier, Oxford, pp. 61–123.

Sajid, W., Kulahin, N., Schluckebier, G., Ribel, U., Henderson, H.R., Tatar, M., Hansen, B.F., Svendsen, A.M., Kiselyov, V.V., Norgaard, P., Wahlund, P.-O., Brandt, J., Kohanski, R.A., Andersen, A.S., De Meyts, P., 2011. Structural and biological properties of the Drosophila insulin-like peptide 5 show evolutionary conservation. J. Biol. Chem. 286, 661–673. http://dx.doi.org/10.1074/jbc.m110.156018.

Segraves, W.A., Hogness, D.S., 1990. The E75 ecdysone-inducible gene responsible for the 75B early puff in Drosophila encodes two new members of the steroid receptor superfamily. Genes Dev. 4, 204–219. http://dx.doi.org/10.1101/gad.4.2.204.

Shapiro, J.P., Hagedorn, H.H., 1982. Juvenile hormone and the development of ovarian responsiveness to a brain hormone in the mosquito, Aedes aegypti. Gen. Comp. Endocrinol. 46, 176–183. http://dx.doi.org/10.1016/0016-6480(82)90199-x.

Shapiro, A., Wheelock, G., Hagedorn, H., Baker, F., Tsai, L., Schooley, D., 1986. Juvenile-hormone and juvenile-hormone esterase in adult females of the mosquito aedes-aegypti. J. Insect Physiol. 32, 867–877. http://dx.doi.org/10.1016/0022-1910(86)90102-2.

Sieglaff, D.H., Duncan, K.A., Brown, M.R., 2005. Expression of genes encoding proteins involved in ecdysteroidogenesis in the female mosquito, Aedes aegypti. Insect Biochem. Mol. Biol. 35, 471–490. http://dx.doi.org/10.1016/j.ibmb.2005.01.011.

Sim, C., Denlinger, D.L., 2008. Insulin signaling and FOXO regulate the overwintering diapause of the mosquito Culex pipiens. Proc. Natl. Acad. Sci. U.S.A. 105, 6777–6781. http://dx.doi.org/10.1073/pnas.0802067105.

Sim, C., Denlinger, D.L., 2009. A shut-down in expression of an insulin-like peptide, ILP-1, halts ovarian maturation during the overwintering diapause of the mosquito Culex pipiens. Insect Mol. Biol. 18, 325–332. http://dx.doi.org/10.1111/j.1365-2583.2009.00872.x.

Sim, C., Denlinger, D.L., 2011. Catalase and superoxide dismutase-2 enhance survival and protect ovaries during overwintering diapause in the mosquito Culex pipiens. J. Insect Physiol. 57, 628–634. http://dx.doi.org/10.1016/j.jinsphys.2011.01.012.

Smykal, V., Raikhel, A.S., 2015. Nutritional control of insect reproduction. Curr. Opin. Insect Sci. 11, 31–38. http://dx.doi.org/10.1016/j.cois.2015.08.003.

Spielman, A., Gwadz, R.W., Anderson, W.A., 1971. Ecdysone-initiated ovarian development in mosquitoes. J. Insect Physiol. 17, 1807–1814. http://dx.doi.org/10.1016/0022-1910(71)90125-9.

Sun, G., Zhu, J., Li, C., Tu, Z., Raikhel, A.S., 2002. Two isoforms of the early E74 gene, an Ets transcription factor homologue, are implicated in the ecdysteroid hierarchy governing vitellogenesis of the mosquito, Aedes aegypti. Mol. Cell. Endocrinol. 190, 147–157. http://dx.doi.org/10.1016/s0303-7207(01)00726-2.

Sun, G., Zhu, J., Raikhel, A.S., 2004. The early gene E74B isoform is a transcriptional activator of the ecdysteroid regulatory hierarchy in mosquito vitellogenesis. Mol. Cell. Endocrinol. 218, 95–105. http://dx.doi.org/10.1016/j.mce.2003.12.014.

Sun, G., Zhu, J., Chen, L., Raikhel, A.S., 2005. Synergistic action of E74B and ecdysteroid receptor in activating a 20-hydroxyecdysone effector gene. Proc. Natl. Acad. Sci. U.S.A. 102, 15506–15511. http://dx.doi.org/10.1073/pnas.0503501102.

Taguchi, A., White, M.F., 2008. Insulin-like signaling, nutrient homeostasis, and life span. Annu. Rev. Physiol. 70, 191–212. http://dx.doi.org/10.1146/annurev.physiol.70.113006.100533.

Tee, A.R., Manning, B.D., Roux, P.P., Cantley, L.C., Blenis, J., 2003. Tuberous sclerosis complex gene products, Tuberin and Hamartin, control mTOR signaling by acting as a GTPase-activating protein complex toward Rheb. Curr. Biol. 13, 1259–1268. http://dx.doi.org/10.1016/s0960-9822(03)00506-2.

Thomas, H.E., Stunnenberg, H.G., Stewart, A.F., 1993. Heterodimerization of the Drosophila ecdysone receptor with retinoid X receptor and ultraspiracle. Nature 362, 471–475. http://dx.doi.org/10.1038/362471a0.

Thummel, C.S., 1997. Dueling orphans-interacting nuclear receptors coordinate Drosophila metamorphosis. Bioessays 19, 669–672. http://dx.doi.org/10.1002/bies.950190806.

Vogel, K.J., Brown, M.R., Strand, M.R., 2015. Ovary ecdysteroidogenic hormone requires a receptor tyrosine kinase to activate egg formation in the mosquito Aedes aegypti. Proc. Natl. Acad. Sci. U.S.A. 112, 5057–5062. http://dx.doi.org/10.1073/pnas.1501814112.

Wang, S.-F., Li, C., Zhu, J., Miura, K., Miksicek, R.J., Raikhel, A.S., 2000. Differential expression and regulation by 20-hydroxyecdysone of mosquito ultraspiracle isoforms. Dev. Biol. 218, 99–113. http://dx.doi.org/10.1006/dbio.1999.9575.

Wang, S.-F., Li, C., Sun, G., Zhu, J., Raikhel, A.S., 2002. Differential expression and regulation by 20-hydroxyecdysone of mosquito ecdysteroid receptor isoforms A and B. Mol. Cell. Endocrinol. 196, 29–42. http://dx.doi.org/10.1016/s0303-7207(02)00225-3.

Wen, Z., Gulia, M., Clark, K.D., Dhara, A., Crim, J.W., Strand, M.R., Brown, M.R., 2010. Two insulin-like peptide family members from the mosquito Aedes aegypti exhibit differential biological and receptor binding activities. Mol. Cell. Endocrinol. 328, 47–55. http://dx.doi.org/10.1016/j.mce.2010.07.003.

Weng, S.-C., Shiao, S.-H., 2015. Frizzled 2 is a key component in the regulation of TOR signaling-mediated egg production in the mosquito Aedes aegypti. Insect Biochem. Mol. Biol. 61, 17–24. http://dx.doi.org/10.1016/j.ibmb.2015.03.010.
White, K.P., Hurban, P., Watanabe, T., Hogness, D.S., 1997. Coordination of Drosophila metamorphosis by two ecdysone-induced nuclear receptors. Science 276, 114–117. http://dx.doi.org/10.1126/science.276.5309.114.
Wigglesworth, V.B., 1947. The corpus allatum and the control of metamorphosis in insects. Nature 159, 872. http://dx.doi.org/10.1038/159872a0.
Wigglesworth, V.B., 1951. Source of moulting hormone in Rhodnius. Nature 168, 558. http://dx.doi.org/10.1038/168558b0.
Wigglesworth, V.B., 1957. The action of growth hormones in insects. Symp. Soc. Exp. Biol. 11, 204–227.
Wu, Q., Brown, M.R., 2006. Signaling and function of insulin-like peptides in insects. Annu. Rev. Entomol. 51, 1–24. http://dx.doi.org/10.1146/annurev.ento.51.110104.151011.
Wullschleger, S., Loewith, R., Hall, M.N., 2006. TOR signaling in growth and metabolism. Cell 124, 471–484. http://dx.doi.org/10.1016/j.cell.2006.01.016.
Yamanaka, N., Rewitz, K.F., O'Connor, M.B., 2013. Ecdysone control of developmental transitions: lessons from Drosophila research. Annu. Rev. Entomol. 58, 497–516.
Yamanaka, N., Marqués, G., O'Connor, M.B., 2015. Vesicle-mediated steroid hormone secretion in Drosophila melanogaster. Cell 163, 907–919. http://dx.doi.org/10.1016/j.cell.2015.10.022.
Yang, H., Gong, R., Xu, Y., 2013. Control of cell growth: Rag GTPases in activation of TORC1. Cell. Mol. Life Sci. 70, 2873–2885. http://dx.doi.org/10.1007/s00018-012-1195-y.
Yao, T.-P., Segraves, W.A., Oro, A.E., McKeown, M., Evans, R.M., 1992. Drosophila ultraspiracle modulates ecdysone receptor function via heterodimer formation. Cell 71, 63–72. http://dx.doi.org/10.1016/0092-8674(92)90266-f.
Yao, T.-P., Forman, B.M., Jiang, Z., Cherbas, L., Chen, J.-D., McKeown, M., Cherbas, P., Evans, R.M., 1993. Functional ecdysone receptor is the product of EcR and Ultraspiracle genes. Nature 366, 476–479. http://dx.doi.org/10.1038/366476a0.
Zhou, X.F., Riddiford, L.M., 2002. Broad specifies pupal development and mediates the "status quo" action of juvenile hormone on the pupal-adult transformation in Drosophila and Manduca. Development 129, 2259–2269.
Zhu, J., Chen, L., Raikhel, A.S., 2003. Posttranscriptional control of the competence factor FTZ-F1 by juvenile hormone in the mosquito Aedes aegypti. Proc. Natl. Acad. Sci. U.S.A. 100, 13338–13343. http://dx.doi.org/10.1073/pnas.2234416100.
Zhu, J., Chen, L., Sun, G., Raikhel, A.S., 2006. The competence factor Ftz-F1 potentiates ecdysone receptor activity via recruiting a p160/SRC coactivator. Mol. Cell. Biol. 26, 9402–9412. http://dx.doi.org/10.1128/MCB.01318-06.
Zhu, J., Chen, L., Raikhel, A.S., 2007. Distinct roles of Broad isoforms in regulation of the 20-hydroxyecdysone effector gene, Vitellogenin, in the mosquito Aedes aegypti. Mol. Cell. Endocrinol. 267, 97–105. http://dx.doi.org/10.1016/j.mce.2007.01.006.
Zhu, J., Busche, J.M., Zhang, X., 2010. Identification of juvenile hormone target genes in the adult female mosquitoes. Insect Biochem. Mol. Biol. 40, 23–29. http://dx.doi.org/10.1016/j.ibmb.2009.12.004.
Zou, Z., Saha, T.T., Roy, S., Shin, S.W., Backman, T.W.H., Girke, T., White, K.P., Raikhel, A.S., 2013. Juvenile hormone and its receptor, methoprene-tolerant, control the dynamics of mosquito gene expression. Proc. Natl. Acad. Sci. U.S.A. 110, E2173–E2181. http://dx.doi.org/10.1073/pnas.1305293110.

CHAPTER SIX

Mosquito Peptide Hormones: Diversity, Production, and Function

M.R. Strand, M.R. Brown, K.J. Vogel
University of Georgia, Athens, GA, United States

Contents

Abstract

Mosquitoes, like other insects, produce a diversity of peptide hormones that are processed from different precursor proteins and have a range of activities. Early studies relied on purification of bioactive peptides for hormone identification, but more recently genomic data have provided the information needed to more comprehensively identify peptide hormone genes and associated receptors. The first part of this chapter summarizes the known or predicted peptide hormones that are produced by mosquitoes. The second part of this chapter discusses the sources of these molecules and their functions.

1. INTRODUCTION

All multicellular organisms produce peptide hormones that are classified on the basis of shared structural features and/or function (Fricker, 2012). Most derive from precursors called prepropeptides that are produced and

Advances in Insect Physiology, Volume 51
ISSN 0065-2806
http://dx.doi.org/10.1016/bs.aiip.2016.05.003

enzymatically processed in different types of endocrine or neurosecretory cells. Endocrine cells usually store and directly release peptide hormones into circulation, whereas neurosecretory cells can either release hormones directly or transport them through axons to neurohemal organs or other locations where they are stored and released (Nässel and Winther, 2010; Szapiro and Barbour, 2009). Peptide hormones mediate activities by binding to membrane receptors that activate intracellular signalling and effector functions in different types of cells. A majority of peptide hormones bind receptors in the large G protein-coupled receptor (GPCR) family, but some bind protein kinase receptors (PKRs) or receptor guanylyl cyclases (RGCs) (Biarc et al., 2011; Lemmon and Schlessinger, 2010; Potter, 2011). The focus of this chapter is the peptide hormones of mosquitoes. We first discuss the known or predicted peptide hormones mosquitoes produce and the receptors they bind. We then examine the sources of these hormones and their functions. Insects, including mosquitoes, use some peptide hormones as neurotransmitters, while also producing other peptides that have important roles in development, locomotion, circadian rhythms, olfaction, and mating. We largely do not discuss these topics because of space constraints and/or the limited mosquito literature that is available.

2. PEPTIDE HORMONES AND RECEPTOR DISCOVERY IN MOSQUITOES

The first endocrine studies in mosquitoes began in the mid-20th century with experiments showing that a humoral factor released from the head of female mosquitoes stimulated development of eggs after blood feeding (Clements, 1956; Detinova, 1945; Gillett, 1956). This factor was initially named egg development neurosecretory hormone (EDNH) because experiments showed it was produced by brain medial secretory cells (Lea, 1967, 1972). Studies later established that EDNH was a protein or small peptide that stimulated ovaries of *Aedes aegypti* to produce the steroid hormone ecdysone, which was shown to be essential for yolk biosynthesis by the fat body, yolk uptake by primary oocytes and egg maturation (Hagedorn et al., 1975, 1979; Lea, 1972) (see chapter "Regulation of Reproductive Processes in Female Mosquitoes" by Roy et al.). This finding resulted in the renaming of EDNH as ovary ecdysteroidogenic hormone (OEH) (Matsumoto et al., 1989a). However, identification of OEH as a 140 amino acid peptide required many years of effort because large quantities of starting material from mosquito heads were needed to ultimately purify the bioactive peptide and determine its sequence (Brown et al., 1998).

Other studies during this period provided evidence that mosquitoes produce peptide hormones with functions in nutrient storage (Lea and Van Handel, 1970), water balance and diuresis (Maddrell and Phillips, 1978; Nijhout and Carrow, 1978), host seeking (Klowden and Lea, 1979), and blood meal digestion (Graf et al., 1998). Yet the small size of mosquitoes presented similar difficulties to those confronted with OEH. As a result, few of these factors were identified. Instead, most peptide hormones that were successfully purified and identified came from larger insects like certain moths, locusts, and cockroaches (Gäde et al., 1997; Hauser et al., 1997, 2006a,b; Raabe, 1989). In some cases, these sequence data were then used to identify homologous peptide sequences from mosquitoes (Veenstra, 1994, 1999; Veenstra et al., 1997a,b, 1999).

Approaches to studying peptide hormones and their receptors dramatically changed with the sequencing and annotation of the first insect genome from *Drosophila melanogaster* (Adams et al., 2000). This was soon followed by the sequencing of the mosquito, *Anopheles gambiae* (Holt et al., 2002), and several other mosquito species including *Ae. aegypti* and *Culex quinquefasciatus* (Arensburger et al., 2010; Jiang et al., 2014; Neafsey et al., 2015; Nene et al., 2007; Waterhouse, 2015). These data together with the genomes for several other insects, invertebrates, and vertebrates provided the foundation needed to identify candidate peptide hormone and receptor genes through homology-based searches (Hauser et al., 2010; Hewes and Taghert, 2001; Hummon et al., 2006; Li et al., 2008; Predel et al., 2010; Riehle et al., 2002; Roller et al., 2008; Southey et al., 2008; Veenstra et al., 2012).

2.1 Peptide Hormone Genes

Annotation of the *Ae. aegypti* and *An. gambiae* genomes originally identified 43 and 35 peptide hormone genes, respectively (Predel et al., 2010; Riehle et al., 2002). Additions to the literature since these studies were published together with annotation improvements increase this number to 50 for *Ae. aegypti*, 47 for *An. gambiae,* and 45 *C. quinquefasciatus*. We list these in alphabetical order in Table 1 along with their identifiers, which will be useful to readers because some of these genes are still difficult to find or correctly recognize in databases. We also list the 45 genes that have been identified from *D. melanogaster*, which is of value to this discussion from two perspectives. First, more peptide hormones have been functionally studied in *D. melanogaster* than in any mosquito species. Second, mosquitoes and *D. melanogaster* reside in families (Culicidae and Drosophilidae) that arose at different times in the evolutionary history of the order Diptera. The Culicidae is an early lineage that arose 225 million years ago (mya) while the Drosophilidae evolved much more recently (40–65 mya) in concert with other cyclorrhaphan flies (Wiegmann et al., 2003). Thus, the peptide hormone genes from mosquitoes and drosophilids together provide information across the breadth of the order.

2.1.1 Comparative Genomics

Most peptide hormone genes are single copy and produce mature hormones that are monomeric. However, differential processing can yield multiple bioactive forms for some peptide hormones with the most extreme example in mosquitoes being prepro-FMRFamide, which is processed into nine functional FMRFamide isoforms (Predel et al., 2010). Some peptide hormone genes along with their receptors are duplicated in mosquitoes and select other insects. These include adipokinetic hormone (AKH), corazonin, and AKH/corazonin peptide (ACP) (Table 1). Some of these duplications, however, have been lost in drosophilids including *D. melanogaster* (Hauser and Grimmelikhuijzen, 2014; Vogel et al., 2013). The only multimember peptide hormone genes in mosquitoes, drosophilids, and other insects are the insulin-like peptides (ILPs), but it is unclear all family members derive from a common ancestral gene (Grönke et al., 2010). Lastly, bursicon and glycoprotein A2/B5 are dimeric peptide hormones that are produced from cleavage products of different prepropeptides (bursicon/partner of bursicon and glyprotein hormone A2/A5), whose corresponding genes are also expressed in different neurosecretory cells (Table 1). Trypsin modulating oostatic factor (TMOF) is a proline rich, 10 amino acid peptide identified in *Ae. aegypti* that is often discussed as a peptidyl regulator of blood meal digestion and egg formation in mosquitoes (Borovsky et al., 1990; Verlinden et al., 2014). However, TMOF is not included in Table 1 or further discussed here because it matches a 10 amino acid sequence of the vitelline envelope protein (Lin et al., 1993), which clearly is not a peptide hormone precursor protein. In addition, no data compellingly supports that TMOF binds a specific receptor including any GPCR, PRK, or RGC, or activates signalling in any target cell that is consistent with it functioning as a peptide hormone.

Overall, the data in Table 1 show *Ae. aegypti*, *An. gambiae*, and *C. quinquefasciatus* encode the same peptide hormone genes with two exceptions. First the short neuropeptide (sNPF) gene in *Ae. aegypti* has duplicated, which we refer to here as sNPF1 and sNPF2 (see Section 4.1.16),

Table 1 Peptide Hormone Genes for Three Mosquito Species and *D. melanogaster*[a]

Hormone	Ae. aegypti	An. gambiae	C. quinquefasciatus	D. melanogaster	Mosquito Species with Reported Localization Data[b]
Adipokinetic hormone (AKH)	AAEL011996	AGAP008834	CPIJ000869	CG1171	Aa, Ag
Adipokinetic/corazonin peptide (ACP)	AAEL010950	AGAP002430	CPIJ001379	ND	Aa, Ag
Agatoxin-like neuropeptide	PA	AGAP007821	CPIJ000956	ND	
Allatostatin A (ASTA)	AAEL015251	AGAP003712	CPIJ008017	CG13633	Aa
Allatostatin C (ASTC)	AAEL005747	AGAP010157	partial unan.	CG14919	Aa
Allatotropin	AAEL009541	AGAP012130	CPIJ007896	ND	Aa
Apis-ITG-like	AAEL006369	AGAP008993	CPIJ005482	CG8216	
Bursicon	PA	AGAP002537	CPIJ009600	CG13419	Ag
Partner of bursicon	AAEL013722	AGAP004506	CPIJ012985	CG15284	Ag
CCHamide 1	AAGE02019353	BM585352	AAWU01008744	CG14358	
CCHamide 2	AAEL004890	AGAP004553	AAWU01038417	CG14375	
Corazonin	AAEL005252	AGAP012665	ND	CG3302	Aa
Crustacean cardioactive peptide (CCAP)	AAEL000630	AGAP009729	CPIJ005842	CG4910	Ag

Continued

Table 1 Peptide Hormone Genes for Three Mosquito Species and *D. melanogaster*—cont'd

Hormone	Ae. aegypti	An. gambiae	C. quinquefasciatus	D. melanogaster	Mosquito Species with Reported Localization Data
Diuretic hormone 31 (DH31)	AAEL008070	AGAP001382	ND	CG13094	Aa
Diuretic hormone 44 (DH44)	AAEL008292	AGAP003269	CPIJ008822	CG8348	Aa
Ecdysis triggering hormone (ETH)	AAEL001762	AGAP007062	CPIJ004945	CG18105	Aa
Eclosion hormone	AAEL011229	AGAP010437	CPIJ011911	CG5400	
FMRFamide	AAEL013645	AGAP005518	CPIJ000101	CG2346	Aa
Glycoprotein A2	BN001241	AGAP008301	CPIJ013236	CG17878	
Glycoprotein B5	AAEL001474	AGAP008302	CPIJ013235	CG40041	
Insulin-like peptide (ILP) 1	AAEL000937	AGAP010605	CPIJ018051	CG14173	
ILP2	AAEL000960	AY324308	CPIJ018049	CG8167	
ILP3	DQ845751	AY324309	CPIJ018050	CG14167	Aa, Ag, As
ILP4	AAEL000932	AY324310	ND	CG6736	
ILP5	AAEL003000	AGAP003927	CPIJ001698	CG33273	
ILP6	DQ845755	AY324313	CPIJ003329	CG14049	
ILP7	DQ845757	AY324314	ND	CG13317	
ILP8	DQ845754	ND	ND	CG14059	

Ion transport peptide	AAEL015332	AGAP005055	CPIJ003972	CG13586	
Limostatin 1	AAEL008355	AGAP013197	ND	CG8317	
Limostatin 2	AAEL008359	ND	ND	ND	
Kinin (leucokinin)	AAEL010172	AGAP013518	CPIJ010343	CG13480	Aa
Myoinhibitory peptide	AAEL012139	AGAP000833	CPIJ802231	CG6456	Aa
Myosuppressin	AAEL007294	AGAP001474	CPIJ012769	CG6440	
Natalisin	AAEL003260	AGAP005277	CPIJ001072	CG34388	
Neuropeptide F	AAEL002733	AGAP004642	PA	CG10342	Aa
Neuropeptide-like peptides (NPLPs)	AAEL014708	AGAP010366	CPIJ014175	CG3441	Aa
Orcokinin	AAEL010172	AGAP012220	CPIJ010343	CG13565	
Ovary ecdysteroidogenic hormone (OEH)	AAEL004155	AGAP000108	CPIJ010626	ND	Aa, Ag
Pigment dispersing hormone	AAEL001754	AGAP005776	CPIJ004895	CG6496	
Proctolin	ND	ND	ND	CG7105	
Prothoracicotropic hormone (PTTH)	PA	AGAP000859	CPIJ003196	CG13687	
Pyrokinin 1 (PK1)	AAEL012060	AGAP002292	CPIJ005970	CG6371	Aa, As, Cp

Continued

Table 1 Peptide Hormone Genes for Three Mosquito Species and *D. melanogaster*—cont'd

Hormone	**Ae. aegypti**	**An. gambiae**	**C. quinquefasciatus**	**D. melanogaster**	**Mosquito Species with Reported Localization Data**
Pyrokinin 2 (PK2)	AAEL005444	AGAP000347	partial unan.	CG15520	Aa, As, Cp
RYamide	AAEL011702	AGAP006765	CPIJ008988	CG40733	
Short neuropeptide F 1 (sNPF1)	AAEL012542	DQ437578	CPIJ009049	CG13968	Aa
Short neuropeptide F 2 (sNPF2)	AF155738.1	ND	ND	ND	Aa
SIFamide	AAEL009858	AGAP007056	CPIJ004953	CG33527	Aa
Sulfakinin	PA	AGAP009275	CPIJ004208	CG18090	Aa
Tachykinin	AAEL006644	AGAP010014	Partial unan.	CG14734	Aa, Cs
Trissin	AAEL008756	AGAP012496	CPIJ016124	CG14871	

[a]Gene entries for each species are by their VectorBase, GenBank, or FlyBase identifier. *PA*: identification of a partial exon in the genome that is unannotated. *ND*: gene not detected in databases.
[b]Aa, *Aedes aegypti*; Ag, *Anopheles gambiae*; As, *An stephensi*; Cp, *Culex pipiens*; and Cs, *C. salinarius*.

whereas *An. gambiae* and *C. quinquefasciatus* have only one sNPF gene. Second, *Ae. aegypti* has eight ILP genes while *An. gambiae* has seven and *C. quinquefasciatus* has five. Table 1 notes a few other instances of genes not being detected in *C. quinquefasciatus* but this is likely due to annotation issues. *D. melanogaster* lacks genes for ACP, allatotropin, and OEH but orthologs of all other peptide hormone genes in mosquitoes are present. Orthologs of most peptide hormone genes in mosquitoes and drosophilids are also present in insects in other orders. The proctolin gene is present in *D. melanogaster* and other insects but is absent in mosquitoes. A few other peptide hormone genes present in one or more insects from other orders are absent from both mosquitoes and drosophilids (Table 1). The latter could indicate these peptide hormones are absent from all of the Diptera.

2.1.2 Transcription and Peptidomics

Transcriptional profiling of individual genes has been reported for several peptide hormones in mosquitoes, but only a few studies report expression data in specific tissues or life stages. In *An. gambiae*, tissue and/or life stage expression data are available for AKH and ACP (Kaufmann and Brown, 2006, 2008), bursicon (Robertson et al., 2007), ILPs, (Arsic and Guerin, 2008; Krieger et al., 2004), neuropeptide F (NPF) (Garczynski et al., 2005), and sNPF (Garczynski et al., 2007). In *Ae. aegypti*, such data are available for AKH and ACP (Kaufmann et al., 2009), allatostatin A and C (ASTA and C) (Hernández-Martínez et al., 2005; Li et al., 2006; Veenstra et al., 1997a), allatotropin (Hernández-Martínez et al., 2005; Veenstra and Costes, 1999), ILPs (Brown et al., 2008; Riehle et al., 2006), kinin (Veenstra, 1994), NPF (Stanek et al., 2002), OEH (Brown et al., 1998), and sNPF2 (Stracker et al., 2002). Data are also available for what we name in this paper pyrokinin 2 (Choi et al., 2013) in *Ae. aegypti* (see Section 4.1.15) and prothoracicotropic hormone (PTTH) in *Culex pipiens* (Zhang and Denlinger, 2011). Profiling of multiple *An. gambiae* or *Ae. aegypti* genes through array based or RNAseq studies provide additional data on peptide hormone gene expression (Akbari et al., 2013; Baker et al., 2011; Matthews et al., 2016). The only expression data available for other mosquito species are for ILP genes in *C. pipiens* (Sim and Denlinger, 2009) and *Anopheles stephensi* (Marquez et al., 2011; Pietri et al., 2015), and tachykinins in *Culex salinaris* (Meola et al., 1998). Insights about prepropeptide processing are currently only available for *Ae. aegypti* through peptidomic analysis of the central nervous system and select other tissues (Predel et al., 2010; Siju et al., 2013). It is also important to note that the accuracy of the above data depends on the quality of the reference genomes that were available at the time the above studies were conducted.

2.2 Peptide Hormone Receptors

Insights into the known or predicted receptors for mosquito peptide hormones derive from two lines of investigation. The first is experimental data in one or more mosquito species showing that a particular receptor is activated or required for the biological activity of a given peptide hormone. The second is homology-based studies of mosquito receptor genes relative to *D. melanogaster*, other insects, or other animals where experimental data on orthologs are available. While several studies discuss GPCRs that bind peptide hormones (see later), Vogel et al. (2013) provides the only comprehensive summary of the receptors in mosquitoes and drosophilids that are known or predicted to bind peptide hormones. Key features of these receptors are summarized below.

2.2.1 GPCRs

GPCRs form a very large protein family that is present in all multicellular animals. However, the first studies showing that peptide hormones are often ligands for GPCRs derive from experiments first conducted in mammals. GPCRs range from 40 to 60 kDa in size and reside as monomers in the plasma membrane of cells. All exhibit a standard topology which consists of three regions: a variable extracellular N-terminal domain, a conserved transmembrane domain with seven hydrophobic α-helices that functions as a ligand binding pocket, and an intracellular C-terminal domain that mediates signalling through interactions with G proteins (Bockhaert and Pin, 1999). GPCRs are further subdivided into six classes (A–F) of which only Class A (rhodopsin-like) and Class B (secretin-like) contain members that bind peptide hormones (Attwood and Findlay, 1994; Hauser et al., 2006a,b).

The conserved transmembrane domain shared by GPCRs was used during annotation to interrogate the *D. melanogaster* genome. This led to identification of more than 200 predicted GPCR genes of which 44 were Class A and B members (Brody and Cravchik, 2000; Hewes and Taghert, 2001). A few functional studies of GPCRs preceded sequencing of the *D. melanogaster* genome but most of these predicted GPCRs were 'orphans' that were subsequently targeted for study by many groups using different experimental approaches. This resulted in 'deorphanization' and identification of GPCRs that bind several of the peptide hormones listed in Table 1 (see Johnson et al., 2003; Nässel and Winther, 2010). The same approach was used to annotate the GPCR genes of *An. gambiae*, *Ae. aegypti*, and *C. quinquefasciatus* (Arensburger et al., 2010; Hill et al., 2002; Nene et al., 2007). However, experimental studies identifying mosquito GPCRs that bind peptide hormones are currently restricted in *Ae. aegypti* to ASTC (Mayoral et al., 2010), allatotropin (Hayes et al., 1997; Nouzova et al., 2012), kinins (Kersch and Pietrantonio, 2011; Kwon et al., 2012; Pietrantonio et al., 2005; Taneja-Bageshwar et al., 2009), NPF (Liesch et al., 2013), pyrokinin 2 (Choi et al., 2013), and sNPFs (Liesch et al., 2013). In anopheline mosquitoes, GPCRs that bind AKH, crustacean cardioactive peptide (CCAP), and corazonin (Belmont et al., 2006), ACP (Hansen et al., 2010), myosuppressin (Scholler et al., 2005), NPF (Garczynski et al., 2005), and sNPFs (Garczynski et al., 2007) have been identified in *An. gambiae* while the GPCR receptor for a kinin has been identified from *A. stephensi* (Radford et al., 2004).

The likely GPCRs for other peptide hormones in mosquitoes were comparatively assessed by Vogel et al. (2013) using all of the Class A (245) and B (82) genes from *Ae. aegypti*, *An. gambiae, C. quinquefasciatus, D. melanogaster,* and *D. mojavensis*. This produced phylogenies comprised of well-supported monophyletic clades that were subdivided into assemblages and subassemblages. A majority of clades contained receptors whose peptide hormone ligands are known from experimental studies in *D. melanogaster* or other species, which in turn provided evidence for the likely ligand of corresponding orthologs in mosquitoes. These 'characterized' GPCR clades and their known or predicted peptide hormone ligands are listed in Table 2 using the nomenclature of Vogel et al. (2013). Data on where different receptors are expressed in mosquitoes and drosophilids are difficult to summarize because of inconsistencies between studies in regard to the life stages and tissues that were examined. We therefore discuss such expression data, if available in mosquitoes, in Section 4.1.

Vogel et al. (2013) also identified several GPCR 'orphan' clades. In some cases placement in the phylogeny provides insights into predicted ligands. For example, subassemblage 1a of the Class A GPCRs contains two orphan clades (OA1 and OA2) that are sister to the bursicon and glycoprotein A2/B5 receptors. This makes them strong candidates for binding relaxin/ILPs such as ILP6 in *Ae. aegypti* and ILP8 in *D. melanogaster* (Brown et al., 2008). Notably, this prediction was recently supported experimentally for *D. melanogaster* ILP8, which deorphanizes OA1 (Garelli et al., 2015;

Table 2 Peptide Hormone GPCR Genes for Three Mosquito Species and *D. melanogaster*[a]

GPCR Class	Hormone	Ae. aegypti	An. gambiae	C. quinquefasciatus	D. melanogaster
A, Cluster 1					
1a	Bursicon	AAEL004777	AGAP008347	CPIJ011619	CG8930
1a	Glycoprotein A2/B5	AAEL004399	AGAP004035	CPIJ007712	CG7665
1a	Drosophila ILP8				CG31096
1b	CCAP R1	AAEL008652	AGAP001961	CPIJ006268	
1b	CCAP R2	AAEL008655	AGAP001962	CPIJ006269	CG33344
1b	Corazonin	AAEL011475	AGAP001558	CPIJ016679 CPIJ016678	CG10698
1b	ACP	AAEL009673	AGAP001532	CPIJ001199	
1b	AKH	AAEL011325	AGAP002156		CG11325
1b	Allatotropin	AAEL005310 AAEL011680		CPIJ014752	
2a	Sulfakinin R1	AAEL010207	AGAP001022	CPIJ005574	CG32540
2a	Sulfakinin R2			CPIJ016281 CPIJ016282	CG42301
2b	RYamide R1	AAEL008296	AGAP000351	CPIJ019394	CG5811
2b	RYamide R2	AAEL015418	AGAP000115	CPIJ018504	
2b	Tachykinin		AGAP002824	CPIJ014103 CPIJ007277	CG6515
2b	Natalisin	AAEL006947	AGAP001592		CG7887

2b	Kinin	AAEL011026 AAEL006636	AGAP010851	CPIJ012071	CG10626
2b	NPF	AAEL010626 NPYR2	AGAP004122–3	CPIJ006984 CPIJ018265	CG1147
2b	sNPF	AAEL013505 AAEL007924	AGAP012378	CPIJ13069	CG7395
2b	SIFamide	AAEL009698 AAEL005994	AGAP003335	CPIJ017622 CPIJ016970	CG10823
2d	CCH amide R1	AAEL016997	AGAP011452	CPIJ017639	CG30106
2d	CCH amide R2	AAEL017410	AGAP003631		CG14593
2d	ASTC R1	AAEL012920	AGAP010486	CPIJ011191	CG7285
2d	ASTC R2	AAEL012356	AGAP012268	CPIJ010469	CG13702
2d	ASTA R1	AAEL007169	AGAP003658	CPIJ016163	CG2872
2d	ASTA R2	AAEL006076–7	AGAP001773–4	CPIJ013095 CPIJ011118	CG10001
2e	Trissin	AAEL004732	AGAP008702	CPIJ001962–3	CG34381
2f	PK R1	AAEL017335 AAEL003747	AGAP003076	CPIJ011105–07	CG8784 CG8795D
2f	PK R2	AAEL012796 KC155994	AGAP000658	CPIJ014065 CPIJ019566 CPIJ015545	CG9918
2f	ETH R1	AAEL005803	AGAP002881	CPIJ003421	CG5911
2f	ETH R2	AAEL005804		CPIJ003422	

Continued

Table 2 Peptide Hormone GPCR Genes for Three Mosquito Species and *D. melanogaster*—cont'd

GPCR Class	Hormone	Ae. aegypti	An. gambiae	C. quinquefasciatus	D. melanogaster
A, Cluster 2					
	FMRFamide	AAEL015604 AAEL003378	AGAP001862	CPIJ007187	CG2114
	MIP	AAEL010313	AGAP012427	CPIJ007973	CG16752
	Myosuppressin	AAEL006283	AGAP005229	CPIJ000143	CG43745 CG8985
B					
	Pigment dispersing hormone	AAEL009024	AGAP003654	CPIJ009749	CG13758
	DH44 R1	AAEL008292	AGAP005464	CPIJ008822	CG12370
	DH44 R2	AAEL005894	AGAP005465	CPIJ008820	CG8422
	DH33	AAEL010043	AGAP009770	CPIJ014419	CG32843

[a]Gene entries for each species as in Table 1. Receptor classification per Vogel et al. (2013). The Class A GPCR phylogeny forms two clusters (1 and 2). Cluster 1 is further divided into subassemblages 1a, b, and 2a–f.

Vallejo et al., 2015). Other orphan clades provide no insights about their ligands but their relationship to other GPCRs in the phylogeny strongly suggest they bind peptide hormones. If correct, this suggests that mosquitoes and drosophilids produce several peptide hormones that have not been identified. Lastly, these data show that dipteran Class A and B GPCRs have experienced several instances of duplication or loss in particular species, which results in some characterized clades containing 'in group' orphans (Vogel et al., 2013). The functional significance of this is unclear although duplicated receptors may preferentially bind different isoforms of a given peptide hormone since differential processing can produce variants of the same ligand (Predel et al., 2010). Recent studies also provide some support for this suggestion (Liesch et al., 2013).

2.2.2 PKRs

Most PKRs are single subunit receptors that have an extracellular ligand binding domain, a single hydrophobic transmembrane-spanning domain, and an intracellular kinase domain that upon ligand binding autophosphorylates either tyrosine or serine/threonine residues (Hubbard and Miller, 2007; Lemmon and Schlessinger, 2010). Most PKRs form noncovalent dimers upon entry into the plasma membrane or ligand binding, although some, like the insulin receptor (IR) (see later), form a covalent heterotetramer through intra- and intersubunit disulphide bonds. PKRs with tyrosine phosphorylation activity are referred to as receptor tyrosine kinases (RTKs) which can activate one or more signalling pathways. PKRs with serine/threonine kinase domains are also capable of activating one or more signalling pathways.

PKR ligands in mammals include insulin, several growth factors, and certain cell surface molecules (Lemmon and Schlessinger, 2010). A homologue of the vertebrate IR, was cloned from *D. melanogaster* prior to sequencing of its genome, which was named the *Drosophila* IR (Fernandez-Almonacid and Rosen, 1987). Biochemical studies showed this receptor binds vertebrate insulin (Fernandez-Almonacid and Rosen, 1987; Sajid et al., 2011) while genetic studies provided evidence that several endogenous ILPs activate the insulin signalling pathway (Brogiolo et al., 2001). Genetic studies in *D. melanogaster* also identified an RTK as the receptor for PTTH (Rewitz et al., 2009). In mosquitoes, experiments conducted in *Ae. aegypti* show that its IR binds endogenous ILP3 with high affinity which activates the canonical insulin signalling pathway (Brown et al., 2008; Dhara et al., 2013; Wen et al., 2010). In contrast, no functional studies have been conducted in mosquitoes on the PTTH receptor (see Section 4.1).

Results from Vogel et al. (2013) assemble the mosquito and drosophilid PKR genes into two branches: 12 well-supported clades consisting of RTKs and 5 clades containing PKRs with conserved serine–threonine kinase domains that are related to mammalian transforming growth factor beta (TGF-β) receptors. Unlike GPCRs, this analysis also shows that dipteran PKRs have undergone few lineage-specific gains or losses, which suggest the same genes are likely present in most or all Diptera. Mosquito and drosophilid IRs form a single clade that is closely related to a clade that contains homologues of human anaplastic lymphoma RTK (ALK) and leucocyte RTK plus a second orphan clade (OR1) lost from *D. melanogaster*. The ALK-like receptor has been shown to have functional roles in development, ethanol sensitivity, and learning in *D. melaogaster* but its ligand is unknown (Gouzi et al., 2011; Lasek et al., 2011; Lorén et al., 2001). The OR1 receptor from *Ae. aegypti* was recently deorphanized by showing that it binds OEH (Vogel et al., 2015; see Section 4.1). Many of the remaining clades whose ligands are growth factors have been studied in *D. melanogaster* but have not been examined in mosquitoes. The RTK genes with known or predicted peptide hormone ligands in mosquitoes and *D. melanogaster* are summarized in Table 3.

2.2.3 RGCs

RGCs are conserved homodimeric membrane proteins which have an extracellular ligand binding domain, a single pass transmembrane domain, and an intracellular catalytic domain with adenylyl or guanylyl cyclase activities that are sensitive to intracellular calcium levels (Potter, 2011). RGCs also usually form dimers prior to or after ligand binding. Studies in mammals were the first to show that RGCs bind peptide hormones called brain and atrial natriuretic peptides (Potter, 2011). In insects, experiments with the oriental fruitfly, *Bactrocera dorsalis*, identified an RGC as the receptor for eclosion hormone (EH) (Chang et al., 2009), while a second RGC was identified as the receptor for neuropeptide-like peptide 1 (NPLP1) in *D. melanogaster* (Overend et al., 2012). Phylogenetic analysis of mosquito and drosophilid RGCs generated one clade containing predicted EH receptors, a second containing NPLP receptors, and four orphan clades (OGC1–4) (Vogel et al., 2013) (Table 3). One of these orphan clades appears orthologous to the vertebrate RGCs that bind natriuretic peptides, while a

Table 3 Peptide Hormone RTK and RGC Genes for Three Mosquito Species and *D. melanogaster*[a]

Receptor	Hormone	Ae. aegypti	An. gambiae	C. quinquefasciatus	D. melanogaster
RTK	ILPs	AAEL002317	AGAP012424	CPIJ006878	CG18402
RTK	ILPs	AAEL002317	AGAP012424	CPIJ006878	CG18402
RTK	PTTH	AAEL002404 AAEL010923	AGAP005763	CPIJ001844 CPIJ000058	CG1389
RGC	NPLPs	AAEL008390	AGAP012163	CPIJ010083	CG42636 CG42637
RGC	EH	AAEL008387	AGAP012161	CPIJ010082	CG10738

[a]Gene entries for each species as in Table 1.

second appears orthologous to vertebrate retinal RGCs. The other two orphan clades have no vertebrate ortholog.

3. SOURCES, PROCESSING, AND RELEASE OF PEPTIDE HORMONES IN MOSQUITOES

In mosquitoes, most insights into where different peptide hormones are produced derive from studies of adult females. This bias mirrors the literature at-large where most of what we know about mosquito genetics, biochemistry, physiology, and behaviour also come from studies of adults, which is the life stage of interest from the perspective of vector biology and disease transmission. A different situation, however, exists in a number of other insects including *D. melanogaster* where the source of many peptide hormones have been carefully mapped in larvae but distribution in adults is less detailed (Nässel and Winther, 2010). Studies on peptide hormone production and function in Lepidoptera also overall focus on larvae because of the historic interest in these insects as models for the study of moulting (Nijhout, 1998; Smith and Rybczynski, 2012; Truman and Riddiford, 2002). Thus, while most peptide hormone genes in mosquitoes are conserved with other insects, differences in functional priorities prevent some comparisons from being made in regard to where they are produced. The emphasis on adults has also affected which peptide hormones have been most studied in mosquitoes.

3.1 Peptide Hormone Producing Cells

Immunocytochemistry and peptidomics provide information on cells in mosquitoes where different peptide hormones have been detected. We list the species for which such data are available in Table 1 and illustrate these sites in Fig. 1. For the central nervous system several hormones are detected in the brain, specific neurosecretory cells, and ganglia of the ventral nerve chord (Brown and Cao, 2001; Brown and Lea, 1988; Brown et al., 2008; Cao and Brown, 2001; Estévez-Lao et al., 2013; Hellmich et al., 2014; Hernández-Martínez et al., 2005; Kaufmann and Brown, 2006; Kaufmann et al., 2009; Krieger et al., 2004; Marquez et al., 2011; Meola and Lea, 1972; Meola et al., 1998; Moffett and Moffett, 2005; Predel et al., 2010; Riehle et al., 2006; Siju et al., 2013; Stracker et al., 2002). The corpus cardiacum (CC) and perivisceral organs iteratively present in abdominal segments are neurohemal organs that store and release several different peptide hormones in mosquitoes (Brown and Cao, 2001; Brown and

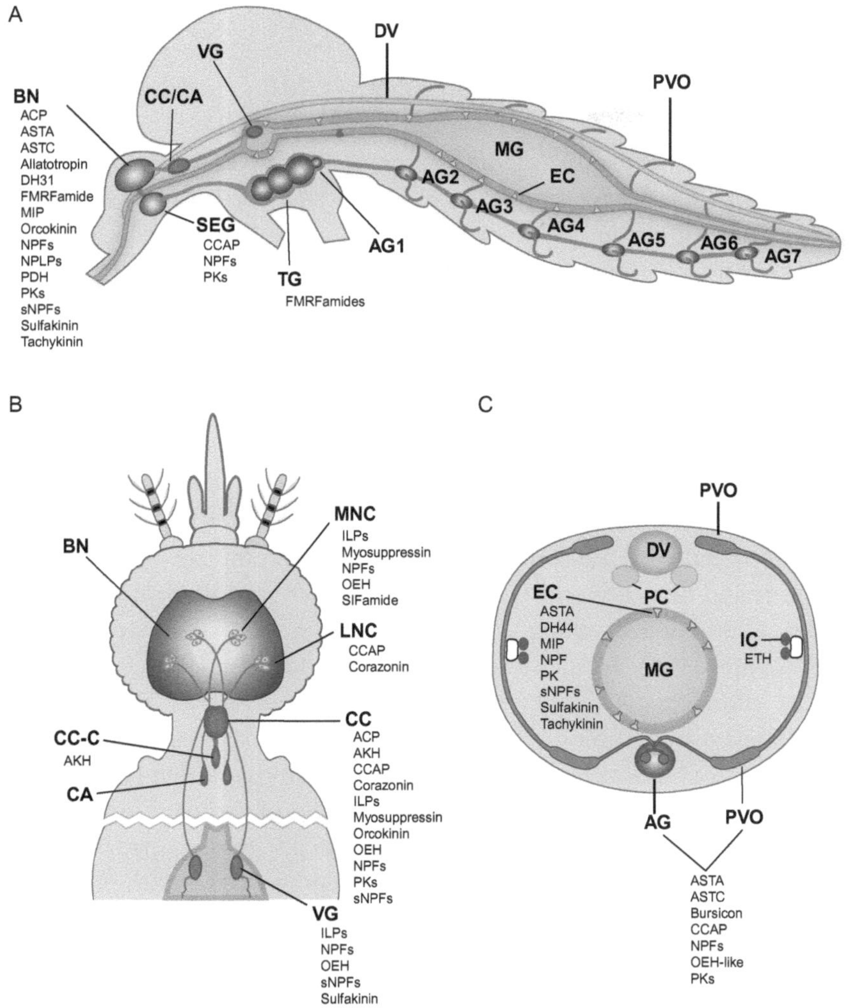

Fig. 1 Schematic illustrating the sources of different peptide hormones in adult female mosquitoes as determined by immunocytochemistry or peptidomics. Sagittal view of the whole body (A), dorsal view of the head and thoracic regions (B), and a cross section through the third abdominal segment (C). Select regions of the nervous system, organs and cells are labelled in bold. Peptide hormones in Table 1 are listed alphabetically below detected sites. Abbreviations: *AG1–7*, abdominal ganglia 1 through 7; *BN*, brain; *CA*, corpora allata; *CC*, corpus cardiacum; *CC-C*, corpus cardiacum intrinsic cells; *DV*, dorsal vessel; *EC*, gut endocrine cells; *IC*, Inka cells; *LNC*, lateral neurosecretory cells; *MG*, midgut; *MNC*, medial neurosecretory cells; *PC*, pericardia; *PVO*, perivisceral organs; *SEG*, subesophegial ganglion; *VG*, ventricular ganglion. Peptide hormone abbreviations are listed in Table 1. Most of these data derive from studies of adults with the exception being for ETH, which is shown in (C) although data derive from studies of larvae (see text). (See the colour plate.)

Lea, 1988; Cao and Brown, 2001; Estévez-Lao et al., 2013; Hernández-Martínez et al., 2005; Honegger et al., 2011; Kaufmann and Brown, 2006; Kaufmann et al., 2009; Marquez et al., 2011; Predel et al., 2010; Riehle et al., 2006; Veenstra and Costes, 1999). Studies conducted primarily in *D. melanogaster* show that the midgut epithelium contains intestinal stem cells, which self-renew and differentiate into two other cell types: enterocytes, that produce digestive enzymes and mediate nutrient uptake, and endocrine cells (Jiang and Edgar, 2012). In *Ae. aegypti*, midgut endocrine cells are known to produce and/or release several peptide hormones (Brown et al., 1986; Hernández-Martínez et al., 2005; Marquez et al., 2011; Moffett and Moffett, 2005; Predel et al., 2010; Stracker et al., 2002; Veenstra et al., 1995) (Table 1, Fig. 1).

3.2 Processing Enzymes and Hormone Secretion

Microscopy studies show that peptide hormones accumulate in the body of endocrine and neurosecretory cells where these are packaged into secretory vesicles as propeptides along with processing enzymes (Brown and Lea, 1988; Raabe, 1989). Studies of mammalian peptidergic cells indicate these include endopeptidases, which are primarily convertases that process propeptides, and other modifying enzymes that can alter the amino- and carboxyl termini of processed peptides in ways that are important for function (Hook et al., 2008; Isaac et al., 2009). Amidation of carboxyl termini also confers resistance to peptidases, which potentially extends half-life after release into circulation. Functional studies of convertases and the enzymes responsible for carboxy-terminal amidation have been conducted in *D. melanogaster* (Jiang et al., 2000; Reiher et al., 2011; Taghert and Veenstra, 2003; Wegener et al., 2011). Genes encoding homologous processing enzymes have been identified in mosquitoes but no functional studies have been conducted (Akbari et al., 2013; Matthews et al., 2016; Riehle et al., 2002).

Peptide hormones are released from endocrine and neurosecretory cells in response to external and internal stimuli. For neurosecretory cells in the brain, this release occurs primarily from axons extending to the CC and along the gut (Brown and Cao, 2001; Cao and Brown, 2001; Raabe, 1989). In contrast, neurosecretory cells located in other ganglia or tissues release peptide hormones in several other locations throughout the body (Raabe, 1989). Midgut endocrine cells release peptides into the space between epithelial cells for diffusion to neighbouring cells and the hemolymph (Brown et al., 1986, 1988). Several enzymes that degrade peptide hormones in circulation or after binding to target cells have been identified in mammals (Isaac et al., 2009). While homologues of these enzymes are known in both mosquitoes and *D. melanogaster* (Riehle et al., 2002; Taghert and Veenstra, 2003), little information is available regarding their expression or function. The only exception to this is in the context of data showing that some kinin agonists are resistant to proteolysis in mosquitoes, which suggests they could be useful as control agents (Taneja-Bageshwar et al., 2009).

Some peptide hormones are also likely degradated by endosomal enzymes after receptor binding and internalization. One example of this is insulin degrading enzymes (IDE, insulinases) that are conserved between mammals and insects including *An. gambiae* and *D. melanogaster* (Duckworth et al., 1989; Riehle et al., 2002). A recent study showed that an IDE is globally expressed in *D. melanogaster* throughout development, is present in both the membrane and cytoplasm of cells, and affects ILP levels and signalling (Galagovsky et al., 2014). Studies from mammals also suggest IDEs may have important roles in intracellular degradation of other peptides (Fernández-Gamba et al., 2009).

4. PEPTIDE HORMONE FUNCTIONS IN MOSQUITOES

Many studies have been conducted on the functions of peptide hormones in insects, but this literature is also large and difficult to summarize across species that have been studied. As a result, most reviews focus on either: (1) processes that peptide hormones have roles in regulating like metabolism (Baker and Thummel, 2007; Broughton and Partridge, 2009), moulting and metamorphosis (Smith and Rybczynski, 2012; White and Ewer, 2014), and water balance (Dow and Davies, 2006) or (2) particular types of peptide hormones such as FMRFamides (Nichols, 2003), ILPs (Antonova et al., 2012), or PRXamides (Jurenka, 2015). A few recent summaries discuss peptide hormone functions more broadly but they primarily emphasize nonmosquito species, which results in a number of studies from the mosquito literature being uncited (Hauser et al., 2008; Nässel and Winther, 2010; Verlinden et al., 2014). Our own assessment identifies functional data in one or more mosquito species for about a third of the entries in Table 1. Below, we first summarize these findings in alphabetical order with the absence of a given peptide from Table 1 indicating that no functional studies to our knowledge have been conducted in mosquitoes. We also provide a succinct summary of comparative information, so that the reader can put into firmer context findings from mosquitoes. We then follow this information with some concluding remarks.

4.1 Peptide Hormones with Reported Activities

4.1.1 Adipokinetic Hormone (AKH) and AKH/Corazonin Peptide (ACP)

Most insects including *D. melanogaster* have one AKH gene that is processed into an N-terminally pyroglutamate blocked peptide of 8–10 amino acids (Gäde, 2009). Functional studies outside of mosquitoes report that AKHs mobilize stored lipid and/or carbohydrate from the fat body into the hemolymph; making this hormone a functional analogue of mammalian glucagon (Van der Host, 2003). Studies from *D. melanogaster* focus primarily on AKH function in concert with ILPs in regulating metabolism (Baker and Thummel, 2007; Nässel and Winther, 2010) although one study also reports that AKH is cardiostimulatory (Noyes et al., 1995). As noted earlier, mosquitoes have AKH and ACP genes, which give rise to two peptides that bind class A GPCRs in subassemblage 1b. These receptors are also related to the receptors for CCAP and allatotropin (Belmont et al., 2006; Caers et al., 2012; Mugumbate et al., 2011, 2013; Vogel et al., 2013). Functional experiments indicate that AKH mobilizes stored carbohydrate but not lipid in *An. gambiae* while ACP has no effect on either lipid or carbohydrate mobilization (Kaufmann and Brown, 2008).

4.1.2 Allatostatin A and C (ASTA and C)

ASTs are structurally variable peptides from distinctly different genes (Veenstra, 2008). The first AST was identified from the cockroach *Diploptera punctata* as an allatostatin (Woodhead et al., 1989). However, studies report activities for ASTs in other insects that do not include the inhibition of juvenile hormone (JH). Thus, the name AST for these peptides is largely historical. We thus include allatostatin B (ASTB) with the myoinhibitory peptides (MIP) on the basis of sequence similarity (Table 1). In mosquitoes and drosophilids, ASTA and C bind related class A GPCRs that cluster together into subassemblage 2c along with the receptor for CCHamide (Mayoral et al., 2010; Vogel et al., 2013). ASTA and C have both been shown to regulate JH biosynthesis in *Ae. aegypti* (Li et al., 2004, 2006), while one report suggests a role for ASTs in gut motility (Onken et al., 2004). The role of ASTs in regulation of JH is comprehensively discussed in chapter "The Role of Juvenile Hormone in Mosquito Development and Reproduction" by Zhu and Noriega.

4.1.3 Allatotropin

Allatotropins are 14 amino acid peptides that stimulate JH biosynthesis in insects from several orders. As previously noted, however, drosophilids lack orthologous genes for both allatotropin and its receptor (Vogel et al., 2013). Functional studies identify the Class A, subassemblage 1a GPCR that binds allatotropin in *Ae. aegypti* while also demonstrating that allatotropin stimulates JH biosynthesis (Hernández-Martínez et al., 2007; Li et al., 2003; Nouzova et al., 2012). These results are further discussed in chapter "The Role of Juvenile Hormone in Mosquito Development and Reproduction" by Zhu and Noriega. Vogel et al. (2013) identified a closely related paralogue of the allatotropin receptor in *Ae. aegypti* but its ligand is unknown.

4.1.4 Bursicon and Partner of Bursicon

Bursicon has long been known in the literature as a factor required for tanning of insect cuticle after moulting (Nijhout, 1998). However, it was only recently identified in *D. melanogaster* as a 30 kDa heterodimer that is produced from products of the bursicon and partner of bursicon genes (Luo et al., 2005; Mendive et al., 2005). The Class A GPCR that binds bursicon in *D. melanogaster* clusters in subassemblage 1a with single orthologs in all mosquitoes (Vogel et al., 2013). This subassemblage also contains receptors that bind biogenic amines, steroid hormones, and certain large peptide ligands. As such, subassemblage 1a receptors are referred to as derived-peptide GPCRs (Vogel et al., 2013). The adult bias of the literature largely underlies the paucity of data on bursicon function in mosquitoes although one study does show that bursicon-producing neurosecretory cells undergo apoptosis in *An. gambiae* after adult ecdysis (Honegger et al., 2011).

4.1.5 Corazonin

An 11 amino acid peptide structurally similar to AKHs was isolated and named corazonin based on its activation of heart contractions in the cockroach *Periplaneta americana* (Veenstra, 1989). Characterization of the corazonin gene in mosquitoes and other insects revealed this similarity extended to their prepropeptides and led to the identification of a gene encoding another peptide named ACP (Hansen et al., 2010). These prepropeptides show structural similarity to the gonadotropin-releasing hormones (GnRH) in mammals. AKH, corazonin, and ACP bind different but closely related Class A GPCRs in *An. gambiae* that belong to subassemblage 1b (Belmont et al., 2006; Hansen et al., 2010; Vogel et al., 2013). These GPCRs also share homology with the GnRH GPCRs

(Hauser and Grimmelikhuijzen, 2014). The function and expression of ACP and its receptor are unknown in insects, whereas corazonin has a range of reported activities including being cardioactive (Hauser and Grimmelikhuijzen, 2014). However, studies in *An. gambiae* indicate no significant effect on dorsal vessel contraction after knockdown of this factor and its receptor by RNA interference (RNAi) (Hillyer et al., 2012).

4.1.6 Crustacean Cardioactive Peptide (CCAP)

CCAPs are cyclic, amidated nine amino acid peptides that were first identified from the crab *Carcinus maenas* but are widely conserved regulators of dorsal vessel contraction in insects and other arthropods (Stangier et al., 1999; White and Ewer, 2014). CCAP has also been implicated in regulating ecdysis (White and Ewer, 2014). The Class A, subassemblage 1b GPCR that binds CCAP was first identified in *D. melanogaster* and later in *An. gambiae* (Belmont et al., 2006). Expression studies in adult *An. gambiae* establish that CCAP localizes to brain neurosecretory cells with axons to the CC, and cells in abdominal ganglia with axons to the heart (Estévez-Lao et al., 2013) (Fig. 1). CCAP injected into adult female *An. gambiae* stimulates heart contractions, whereas RNAi knockdown of CCAP had the opposite effect (Chen and Hillyer, 2013; Estévez-Lao et al., 2013). Both water and sucrose deprivation reduce heart contraction rate but these effects occur independently of CCAP activity (Ellison et al., 2015).

4.1.7 Diuretic Hormone 31 (DH31, Calcitonin-Like)

This hormone was first identified from the cockroach *Di. punctata* as a 31 amino acid peptide with a C-terminal GXPamide that shares structural features and diuretic activity with vertebrate calcitonins (Furuya et al., 2000). DH31 homologues are now known from several other insects (Zandawala, 2012). Early studies showing that a peptide hormone regulates diuresis in mosquitoes after blood feeding (see earlier) was identified as a DH31 homologue in *An. gambiae* (Coast et al., 2005). Identification of the DH31 receptor in *D. melanogaster* as a Class B GPCR was followed by identification of its ortholog in *Ae. aegypti*, which is expressed as a gradient along the Malpighian tubules (Kwon et al., 2012). Expression in the hindgut further led to the discovery that DH31 exhibits myotropic activity (Kwon and Pietrantonio, 2013). These data are further elaborated upon in chapter "Renal Excretory Processes in Mosquitoes" by Piermarini.

4.1.8 Diuretic Hormone 44 (Corticotropin-Releasing Factor-Like)

A 41 amino acid peptide with diuretic activity that was initially identified from the moth *Manduca sexta* with its receptor (Kataoka et al., 1989; Reagan, 1994). DH44 sequences exhibit similarities to peptide hormones in the vertebrate corticotropin-releasing factor (CRF) family that bind Class B GPCRs (Balment and Lovejoy, 1999; Cardoso et al., 2014). A DH44 homologue was isolated from whole body extracts of *Culex salinarius* that increased cAMP levels in *M. sexta* Malpighian tubules and stimulated diuresis in *Ae. aegypti* tubules (Clark et al., 1998). Subsequently, two Class B GPCRs that bind DH44 were identified from *D. melanogaster* while one was identified from *Ae. aegypti* (Hector et al., 2009; Jagge and Pietrantonio, 2008). More information on DH44 is provided in chapter "Renal Excretory Processes in Mosquitoes" by Piermarini.

4.1.9 Ecdysis Triggering Hormone (ETH)

Studies conducted in *M. sexta* identified ETH as a 26 amino acid peptide that ended at its C-terminus with the sequence PRMamide (Zitnan et al., 1996). ETH genes encoding usually two ETH peptides have since been identified from insects in most orders including Diptera (Jurenka, 2015; Zitnan et al., 2003). As noted earlier, ETH peptides are conservatively produced in Inka (=epitracheal) cells (Fig. 1). ETH genes and the mature peptides they encode exhibit distinct features. However, the C-terminal PRXamide motif of ETHs is shared with another group of peptide hormones we refer to below as pyrokinins (see Section 4.1.15). ETHs also bind Class A GPCRs that are closely related to the pyrokinin receptors (Vogel et al., 2013). In *D. melanogaster* the ETH receptor gene is alternatively spliced to produce receptors that preferentially bind the two mature ETHs that are processed from prepro-ETH (Iversen et al., 2002; Park et al., 2002). Mosquito ETH genes also produce two ETH isoforms but no studies have examined receptor binding. Functional data support a conserved role for ETH in triggering ecdysis in *Ae. aegypti* during larval and pupal moulting (Dai and Adams, 2009). A recent study also implicates ETH in the timing of JH synthesis (Areiza et al., 2014).

4.1.10 FMRFamides

These neuropeptides are 7–10 amino acids in length, share C-terminal RFamide sequences, and have broadly conserved myotropic functions. The literature initially treated peptides ending in RF/Pamide including insect FMRFamides, sulfakinins, myosuppressins, NPF, and sNPF as related 'FMRFamide' or 'FaRPamide' peptides. These peptides also all bind Class A GPCRs. Subsequent studies, however, showed that FMRFamides, sulfakinins, myosuppressins, NPFs, and sNPFs derive from distinctly different genes, which suggests their similar C-termini reflects evolutionary convergence driven by currently unknown factors. The receptors for mosquito and drosophilid FMRFamides all reside in a single cluster 2 clade (Vogel et al., 2013). The myotropic action of FMRFamides on the heart and gut of *D. melanogaster* as well as their expression in the nervous system are well studied (Nichols, 2003). The first functional studies in mosquitoes used an FMRFamide antiserum to identify FMRFamide cells that were immunopositive in the nervous system and midgut of *Ae. aegypti* (Brown and Lea, 1988; Brown et al., 1986; Moffett and Moffett, 2005). These data though provide unclear results regarding which of these cells actually produce and/or store FMRFamides because the antisera used recognizes only the RFamide epitope. Studies in *An. gambiae* show that a single dose of FMRFamide stimulates larval heart contractions (Duttlinger et al., 2003). Hillyer et al. (2014) profiled FMRFamide gene expression in *An. gambiae* and predicted prepro-FMRFamide produces eight FMRFamide peptides, which is consistent with previously discussed peptidomic data showing that prepro-FMRFamide from *Ae. aegypti* produces nine peptides (Predel et al., 2010). Functional experiments indicate that two FMRFamide peptides increase heart contractions in *An. gambiae* at low doses but reduce contractions at high doses (Hillyer et al., 2014). A direction for study that would aid in interpreting the physiological importance of FMRFamides in mosquito physiology would be to clearly identify the cell sources for these peptides and release dynamics.

4.1.11 Insulin-Like Peptides (ILPs)

All ILPs are 6–8 kDa, share a common structural motif called the insulin fold, and are processed from precursors with similar domain structure (Pre, B, C, A) (De Meyts et al., 2009). ILPs are also broadly conserved among metazoans and are the most studied peptide hormones because of their important regulatory roles in metabolism, growth, and development. In mammals, insulin is the only ILP that binds with high affinity to the IR; an RTK, which activates the insulin signalling pathway (De Meyts et al., 2009). Mammals also produce two other types of ILPs: (1) insulin-like growth factors (IGFs) that preferentially bind another RTK, the IGF receptor (IGFR), and activate MAPK signalling and (2) relaxins that bind GPCRs and activate several signalling pathways (Bani, 1997; McDonald et al., 1989). Insulin primarily regulates metabolic responses, IGFs primarily have functions in growth, while relaxins are implicated in several activities.

This background is important because mosquitoes and other insects produce multiple ILPs but usually have only one IR/IGFR homologue which as previously noted is named the IR. Several but not all ILP family members are functionally redundant in *D. melanogaster* but it is currently unknown whether all ILP family members are capable of binding and activating the IR (Grönke et al., 2010; Zhang et al., 2009). That ILP8 requires a GPCR for function suggests it is relaxin-like and may not interact with the IR (Garelli et al., 2015; see earlier). Factors released from the fat body plus the peptide hormone CCHamide2 from gut endocrine cells are both implicated in regulating ILP release from brain medial neurosecretory cells in *D. melanogaster* (Sano et al., 2015).

Studies in mosquitoes have mostly been conducted in *Ae. aegypti* where five ILP genes (1, 3, 4, 7, and 8) are expressed in brain medial neurosecretory cells while three (2, 5, and 6) are detected in other tissues including the midgut and fat body (Riehle et al., 2006). Each prepro-ILP exhibits features consistent with processing like vertebrate insulin/relaxins except ILP6, which has features that suggest it is processed like a vertebrate IGF (Brown et al., 2008; Riehle et al., 2006). Most functional data focus on ILP3 because it is most similar to vertebrate insulin, binds with high affinity to the *Ae. aegypti* IR, and activates the insulin signalling pathway (Brown et al., 2008; Dhara et al., 2013; Wen et al., 2010).

ILP3 exhibits vertebrate insulin-like activity by reducing blood sugar (trehalose) in adult female *Ae. aegypti* while elevating carbohydrate and lipid stores in the fat body (Brown et al., 2008). ILP3 also has several functions in female reproduction. As noted earlier, the mosquito endocrinology literature began with characterization of EDNH, which was renamed OEH. In the process of identifying OEH, data suggested factors smaller than OEH were released from the brain after blood feeding that had similar activity (Brown et al., 1998; Matsumoto et al., 1989a). Subsequent experiments demonstrated this factor was ILP3 which is released with OEH from medial neurosecretory cells (Fig. 1). The factors that regulate ILP and OEH release after blood feeding are unknown but data strongly support that ILP3 directly stimulates ovary follicle cells to produce ecdysone by binding the IR (Brown et al., 2008; Riehle et al., 2002; Wen et al., 2010). ILP3 also stimulates late-phase trypsin expression by the midgut, which is required for blood meal digestion, and proliferation of hemocytes (Castillo et al., 2011;

Gulia-Nuss et al., 2011). Other factors involved in regulation of egg formation are discussed in chapter "Regulation of Reproductive Processes in Female Mosquitoes" by Roy et al.

Functional studies of ILPs in other mosquitoes are restricted two lines of investigation. First, data support a role for insulin signalling in arrest of ovary development and lipid storage in overwintering diapause by *C. pipiens* (Sim and Denlinger, 2009). Second, ingestion of vertebrate insulin in a blood meal activates insulin signalling and synthesis of reactive oxygen species in the midgut of anopheline mosquitoes (Kang et al., 2008).

4.1.12 Kinin

The first kinin was identified from the cockroach *Rhyparobia* (formerly *Leucophaea*) *maderae* (Holman et al., 1986), which was followed by identification of structurally similar peptides from other insects (Hayes et al., 1997; Hewes and Taghert, 2001). The single kinin gene in *Ae. aegypti* encodes a prepropeptide that is processed into three isoforms but the kinin gene from *D. melanogaster* is processed into only one (Hewes and Taghert, 2001; Veenstra et al., 1997b). In turn, *Ae. aegypti* encodes two class A GPCRs that bind kinins while *D. melanogaster* encodes one (Hewes and Taghert, 2001; Pietrantonio et al., 2005; Taneja-Bageshwar et al., 2009; Vogel et al., 2013). Immunostaining localizes the kinin receptor to the membrane of endocrine cells in the posterior midgut, Malpighian tubules, and the rectal pads of female *Ae. aegypti* (Kersch and Pietrantonio, 2011). Vogel et al. (2013) assigns these receptors to subassemblage 2b along with the receptors that bind RYamide, tachykinin, NPF, sNPF, and SIFamide. Functional studies in *Ae. aegypti* provide strong support for the role of kinins in diuresis (Beyenbach and Piermarini, 2010; Kersch and Pietrantonio, 2011), which is further discussed in chapter "Renal Excretory Processes in Mosquitoes" by Piermarini.

4.1.13 Neuropeptide F (NPF)

NPFs are 28–40 amino acid peptides that share a consensus C-terminal RxRFamide motif (Nässel and Wegener, 2011). They are broadly conserved among invertebrates and also share homology with vertebrate neuropeptide Y (NPY) and pancreatic polypeptides that regulate appetite, digestion, reproduction, stress, and other activities (Holzer et al., 2012). The first insect NPF was identified from *D. melanogaster* on the basis of its RFamide immunoreactivity (Brown et al., 1999), which was followed by cloning of an NPF gene from *Ae. aegypti* (Stanek et al., 2002). Drosophilids and mosquitoes both encode a single NPF gene while mature NPFs bind

Class A GPCRs in subassemblage 2b that are most closely related to the receptors that bind sNPFs (Nässel and Wegener, 2011; Vogel et al., 2013). This finding is interesting because NPFs and sNPFs are structurally distinct and derive from unrelated genes (see below). Functional studies in *D. melanogaster* indicate NPF has similar activities to NPY in mammals (Nässel and Wegener, 2011). In contrast, little is known about NPF function in mosquitoes. Expression studies detect NPF mRNA in the brain and midgut of *Ae. aegypti* while binding studies show that mature NPF from *Ae. aegypti* and *An. gambiae* both bind the *An. gambiae* NPF receptor (Garczynski et al., 2005; Stanek et al., 2002). More recently, studies of five Class A, subassemblage 2 GPCRs from *Ae. aegypti* identified two that bind *Ae. aegypti* NPF (Liesch et al., 2013). The only reported function for NPF in mosquitoes is the inhibition of anterior midgut peristalsis in larval stage *Ae. aegypti* (Onken et al., 2004), but it is highly likely these peptide hormones have other, more functionally significant activities, given the greater literature.

4.1.14 Ovary Ecdysteroidogenic Hormone (OEH)

OEH is a relatively large 9–13 kDa hormone that as previously noted was identified from *Ae. aegypti* (Brown et al., 1998). It is structurally related to neuroparsins which are peptides of unclear function that were first isolated from locusts (Badisco et al, 2007). OEH is released with ILPs from medial neurosecretory cells with both stimulating ovary follicle cells to produce ecdysone (Fig. 1). Functional redundancy with ILPs together with the absence of an identified receptor obscured the mode of action of OEH for many years. This changed with identification of the OEH receptor as an RTK that is closely related to the IR (Vogel et al., 2015). Results also show that OEH directly induces ecdysone production by activating PI3K/Akt signalling (Dhara et al., 2013; Vogel et al., 2015). Thus, redundancy is due to OEH and ILPs binding different receptors on follicle cells that appear to both activate insulin pathway signalling. In contrast, OEH does not activate late-phase trypsin expression or haemocyte proliferation because OEH receptor expression is restricted to ovary follicle cells (Vogel et al., 2015). Some mosquito species have evolved from blood feeding relatives to produce eggs without blood feeding (autogeny). Recent studies show that facultatively autogenous *Georgecraigius (= Aedes) atropalpus* releases OEH and ILP from medial neurosecretory cells shortly after adult emergence but OEH is the primary regulator of ecdysone synthesis by the ovaries (Gulia-Nuss et al., 2012). OEH alone also stimulates limited egg development in nonblood fed *Ae. aegypti* (Gulia-Nuss et al., 2015).

These findings suggest the evolution of autogeny is due in part to the shift from blood meal dependent to blood meal-independent release of OEH and ILPs from the nervous system.

4.1.15 Pyrokinins (PKs)

Peptide hormones we name PKs are relatively small and usually end with the C-terminal motif PRL/Vamide. This motif is similar to that of ETHs discussed earlier, which is why these peptides are often referred to collectively as PRXamides (Jurenka, 2015). The PRXamide literature is very difficult to follow for most individuals. One reason is because of the range of activities PRXamide peptides exhibit. The other is confusing nomenclature. PRXamides are named for either the biological activity or structural features exhibited by the first peptides that were identified. Thus, cardioacceleratory peptide (CAPA) was the name given for a PRXamide purified from *M. sexta*, leucopyrokinin (PK), and perviscerokinin (PVK) for myotropic PRXamides from cockroaches (*R. maderae* and *P. americana*), diapause hormone for PRXamides from *Bombyx mori* and *Helicoverpa zea* (DH-2), ETHs for PRXamides from *M. sexta*, pheromone biosynthesis activating peptide (PBAN) for a PRXamide from *H. zea*, and diuretic hormone (DH-1) for a PRXamide from *P. americana* (see Jurenka, 2015 for primary citations).

Peptide hormones are often named this way. It also works well for ETHs, which derive from a single gene and exhibit conservation in sequence and function. In contrast, the other PRXamides often exhibit cross-reactivity or activities that differ from given names when studied in different species. They also derive from two distinct genes. The first of these was named *pban* in *H. zea* because it encoded the corresponding peptide. However, *pban* from *H. zea* also contains four other PRXamides: DH-2 and three others referred to as PKs (Jurenka, 2015). Orthologs of this gene are known in many insects including *D. melanogaster* where it is named *hugin*. *Hugin* though encodes only two PRXamides named PK and hugin λ (see Nässel and Winther, 2010). The second gene is *capability* (*capa*), which was first identified in *D. melanogaster* with orthologs also now known from other insects. This gene usually contains three PRXamides: two named CAPA or CAPA-PVK peptides and a third named DH-1 or PK depending on publication (see Jurenka, 2015; Nässel and Winther, 2010). These names derive from sequence similarities with the originally identified CAPA, PVK, or PKs but again functional data often differ when studied in different species.

Given this background, we discuss ETHs as distinct peptide hormones (see earlier) because their structure and function support this. In contrast, the functional literature does not currently support giving different names to the peptides encoded by the *pban/hugin* or *capa* genes. We therefore name these genes in mosquitoes *pk1* and *pk2*. In *Ae. aegypti pk1* encodes four peptides we name PK1a–d while *pk2* encodes three we name PK2a–c. The Class A GPCRs in mosquitoes and drosophilids that bind peptides encoded by the above genes form one clade in subassemblage 2F while the receptors that bind ETHs form a sister group (Choi et al., 2013; Dai and Adams, 2009; Olsen et al., 2007; Vogel et al., 2013). Peptidomic and immunocytochemistry methods detect PKs in the CNS, periviseral organs, CC, and midgut of adult mosquitoes (Hellmich et al., 2014; Predel et al., 2010). The only functional data in mosquitoes show that PK2a from *Ae. aegypti* inhibits fluid secretion in larval Malpighian tubules at femtomolar concentration but stimulates diuresis at higher concentrations (Ionescu and Donini, 2012). By previous nomenclature PK2a would be referred to as CAPA1 (Jurenka, 2015), which is inconsistent with its diuretic activity.

4.1.16 Short Neuropeptide F (sNPF)

sNPFs are peptides of variable length (6–19 amino acids) that end with the consensus C-terminal motif PxLRLRFamide in insects (Nässel and Wegener, 2011). All sequenced insect and other invertebrate genomes contain one or more sNPF genes but orthologs are absent from vertebrates. Among holometabolous insects like mosquitoes, prepropeptides often contain multiple sNPF sequences, whereas hemimetabolous insects and other invertebrates tend to produce prepropeptides that contain only one sNPF. The naming of NPFs and sNPFs derives from early studies in the literature where multiple peptides with related C-termini were identified. Later studies showed that NPFs and sNPFs derive from different genes and bind distinct but related class A, GPCRs that for Diptera reside in subassemblage 2b (Nässel and Wegener, 2011; Vogel et al., 2013). We earlier noted that the sNPF gene has duplicated in *Ae. aegypti* but not in *An. gambiae* or *C. quinquefasciatus*. *Ae. aegypti* sNPF1 encodes what was historically named sNPF while sNPF2 contains three copies of what was originally named head peptide (Matsumoto et al., 1989b; Stracker et al., 2002).

A diversity of neurons in the CNS and chemosensory structures like the antennae express sNPF in *D. melanogaster* while functional data implicate sNPFs in regulating feeding, growth, and ILP production by medial neurosecretory cells (Nässel and Winther, 2010). The literature likewise suggests sNPFs are broadly expressed in the adult mosquito CNS (Garczynski et al., 2007; Matsumoto et al., 1989b; Predel et al., 2010; Siju et al., 2013; Stracker et al., 2002; Veenstra, 1999) (Fig. 1), while a recent study reports male accessory glands in *Ae. aegypti* produce and transfer sNPF2 to females during mating (Naccarati et al., 2012). Early functional experiments in *Ae. aegypti* showed that sNPF2 titer increases in the hemolymph of adult females after consuming a blood meal, which coincides with females exhibiting no host-seeking behaviour (Brown et al., 1994). Nonblood fed females in contrast are highly responsive to vertebrate hosts but injection of sNPF2 inhibited host seeking for up to 5 h which suggested a role for sNPF2 in regulating host-seeking behaviour (Brown et al., 1994). More recent studies confirm that *Ae. aegypti* NPF and sNPFs bind different receptors but also detected overlap in receptor binding between sNPF1 and sNPF2 (Liesch et al., 2013). Bioassays identified roles for both NPF and sNPFs in host-seeking behaviour by nonblood females, yet curiously mutagenesis of the GPCR that binds sNPFs has no effect on host seeking, sugar, and blood-feeding behaviour, egg laying, or locomotion (Liesch et al., 2013). Naccarati et al. (2012) tested whether sNPF2 regulates female mating receptivity in *Ae. aegypti* because a structurally unrelated factor named sex peptide in male accessory gland secretions of *D. melanogaster* exhibits this activity. However, sNPF2 had no effect on female mating receptivity. Lastly, one study reports that sNPF1 and two both inhibit peristalsis of the *Ae. aegypti* larval anterior midgut (Onken et al., 2004).

4.2 Concluding Remarks

Significant progress has been made in identifying peptide hormone genes and receptors in mosquitoes. The functional literature is also strong in a few areas of study that focus on adult females. These include the roles of peptide hormones in regulating egg formation, JH biosynthesis, and diuresis, which are also subjects of separate chapters in this volume. In contrast, mapping the cell sources of different peptides hormones across life stages remains weak as is the literature on functional activities during immature development. For example, the endocrine regulation of moulting and metamorphosis in *D. melanogaster* implicates PTTH in regulating ecdysone production by prothoracic glands (PGs) and insulin signalling in nutrition-dependent growth (Nijhout et al., 2014; Rewitz et al., 2009). No data, however, support that PTTH or ILPs directly activate ecdysone production (Niwa and Niwa, 2014). Given the literature on many other insects, it is somewhat ironic that surprisingly little is known about the regulation of moulting in mosquitoes. An early study showed that larval thoracic and abdominal wall tissues produce ecdysone yet also reported that PGs do not

(Jenkins et al., 1992). These findings together with the *Drosophila* literature indicate our understanding of peptide hormone functions in moulting are far from complete for Diptera. The orphan GPCR, RTK, and RGCs in mosquitoes also suggests some peptide hormones remain to be identified and characterized.

ACKNOWLEDGEMENTS

Research from the authors that is cited here was supported by the National Institutes of Health (R01AI1033108, R01AI106892, and F32GM109750) and Georgia Agricultural Experiment Station. We thank J.A. Johnson for her assistance with Fig. 1.

REFERENCES

Adams, M.D., Celniker, S.E., Holt, R.A., Evans, C.A., Gocayne, J.D., Amanatides, P.G., Scherer, S.E., Li, P.W., Hoskins, R.A., Galle, R.F., George, R.A., et al., 2000. The genome sequence of *Drosophila melanogaster*. Science 287, 2185–2195.

Akbari, O.S., Antoshechkin, I., Amrhein, H., Williams, B., Diloreto, R., Sandler, J., Hay, B.A., 2013. The developmental transcriptome of the mosquito *Aedes aegypti*, an invasive species and major arbovirus vector. G3 (Bethesda) 3, 1493–1509.

Antonova, Y., Arik, A.J., Moore, W., Riehle, M.R., Brown, M.R., 2012. Insulin-like peptides: structure, signaling, and function. In: Gilbert, L.I. (Ed.), Insect Endocrinology. Elsevier, New York, pp. 63–92.

Areiza, M., Nouzova, M., Rivera-Perez, C., Noriega, F.G., 2014. Ecdysis triggering hormone ensures proper timing of juvenile hormone biosynthesis in pharate adult mosquitoes. Insect Biochem. Mol. Biol. 54, 98–105.

Arensburger, P., Megy, K., Waterhouse, R.M., Abrudan, J., Amedeo, P., Antelo, B., Bartholomay, L., Bidwell, S., Caler, E., Camara, F., Campbell, C.L., Campbell, K.S., et al., 2010. Sequencing of *Culex quinquefasciatus* establishes a platform for mosquito comparative genomics. Science 330, 86–88.

Arsic, D., Guerin, P.M., 2008. Nutrient content of diet affects the signaling activity of the insulin/target of rapamycin/p70 S6 kinase pathway in the African malaria mosquito *Anopheles gambiae*. J. Insect Physiol. 54, 1226–1235.

Attwood, T.K., Findlay, J.B., 1994. Fingerprinting G-protein-coupled receptors. Protein Eng. 7, 195–203.

Badisco, L., Claeys, I., van Loy, T., van Hiel, M., Franssens, C.V., Simonet, G., Vanden Broeck, J., 2007. Neuroparsins, a family of conserved arthropod neuropeptides. Gen. Comp. Endocrinol. 153, 64–71.

Baker, K.D., Thummel, C.S., 2007. Diabetic larvae and obese flies-emerging studies of metabolism in *Drosophila*. Cell Metab. 6, 257–266.

Baker, D.A., Nolan, T., Fishcer, B., Pinder, A., Crisanti, A., Russell, S., 2011. A comprehensive gene expression atlas of sex- and tissue-specificity in the malaria vector, *Anopheles gambiae*. BMC Genomics 12, 296.

Balment, R.J., Lovejoy, D.A., 1999. Evolution and physiology of corticotropin-releasing factor (CRF) family of neuropeptides in vertebrates. Gen. Comp. Endocrinol. 115, 1–22.

Bani, D., 1997. Relaxin: a pleiotrophic hormone. Gen. Pharmacol. 28, 13–22.

Belmont, M., Cazzamali, G., Williamson, M., Hauser, F., Grimmelikhuijzen, C.J., 2006. Identification of four evolutionarily related G protein-coupled receptors from the malaria mosquito *Anopheles gambiae*. Biochem. Biophys. Res. Commun. 344, 160–165.

Beyenbach, K.W., Piermarini, P.M., 2010. Transcellular and paracellular pathways of transepithelial fluid secretion in Malphighian (renal) tubules of the yellow fever mosquito *Aedes aegypti*. Acta Physiol. 202, 387–407.

Biarc, J., Chalkley, R.J., Burlingame, A.L., Bradshaw, R.A., 2011. Receptor tyrosine kinase signaling—a proteomic perspective. Adv. Enzym. Regul. 51, 293–305.

Bockhaert, J., Pin, J.P., 1999. Molecular tinkering of G protein-coupled receptors: an evolutionary success. EMBO J. 18, 1723–1729.

Borovsky, D., Carlson, D.A., Griffin, P.R., Shabanowitz, J., Hunt, D.F., 1990. Mosquito oostatic factor: a novel decapeptide modulating trypsin-like enzyme biosynthesis in the midgut. FASEB J. 4, 3015–3020.

Brody, T., Cravchik, A., 2000. *Drosophila melanogaster* G protein-coupled receptors. J. Cell Biol. 150, F83–F88.

Brogiolo, W., Stocker, H., Ikeya, T., Rintelen, F., Fernandez, R., Hafen, E., 2001. An evolutionarily conserved function of the *Drosophila* insulin receptor and insulin-like peptides in growth control. Curr. Biol. 11, 213–221.

Broughton, S., Partridge, L., 2009. Insulin/IGF-like signaling, the central nervous system, and aging. Biochem. J. 15, 1–12.

Brown, M.R., Cao, C., 2001. Distribution of ovary ecdysteroidogenic hormone I in the nervous system and gut of mosquitoes. J. Insect Sci. 1, 3.

Brown, M.R., Lea, A.O., 1988. FMRFamide- and adipokinetic hormone-like immunoreactivity in the nervous system of the mosquito, *Aedes aegypti*. J. Comp. Neurol. 270, 606–614.

Brown, M.R., Crim, J.W., Lea, A.O., 1986. FMRFamide- and pancreatic polypeptide-like immunoreactivity of endocrine cells in the midgut of a mosquito. Tissue Cell 18, 419–428.

Brown, M.R., Raikhel, A.S., Lea, A.O., 1988. Ultrastructure of midgut endocrine cells in the adult mosquito, *Aedes aegypti*. Tissue Cell 17, 709–721.

Brown, M.R., Klowden, M.J., Crim, J.W., Young, L., Shrouder, L.A., Lea, A.O., 1994. Endogenous regulation of mosquito host-seeking behavior by a neuropeptide. J. Insect Physiol. 40, 399–406.

Brown, M.R., Graf, R., Swiderek, K.M., Fendley, D., Stracker, T.M., Champagne, D.E., Lea, A.O., 1998. Identification of a steroidogenic neurohormone in female mosquitoes. J. Biol. Chem. 273, 3967–3971.

Brown, M.R., Crim, J.W., Arata, R.C., Cai, H.N., Chun, C., Shen, P., 1999. Identification of a *Drosophila* brain-gut peptide related to the neuropeptide Y family. Peptides 20, 1035–1042.

Brown, M.R., Clark, K.D., Gulia, M., Zhao, Z., Garczynski, S.F., Crim, J.W., Sulderman, R.J., Strand, M.R., 2008. An insulin-like peptide regulates egg maturation and metabolism in the mosquito *Aedes aegypti*. Proc. Natl. Acad. Sci. U. S. A. 105, 5716–5721.

Caers, J., Peeters, L., Janssen, T., De Haes, W., Gade, G., Schoofs, L., 2012. Structure-activity studies of *Drosophila* adipokinetic (AKH) by a cellular expression system of dipteran AKH receptors. Gen. Comp. Endocrinol 77, 332–337.

Cao, C., Brown, M.R., 2001. Localization of an insulin-like peptide in brains of two flies. Cell Tissue Res. 304, 317–321.

Cardoso, J.C., Félix, R.C., Bergqvist, C.A., Larhammar, D., 2014. New insights into the evolution of vertebrate CRH (corticotropin-releasing hormone) and invertebrate DH44 (diuretic hormone 44) receptors in metazoans. Gen. Comp. Endocrinol. 209, 162–170.

Castillo, J., Brown, M.R., Strand, M.R., 2011. Blood feeding and insulin-like peptide 3 stimulate proliferation of hemocytes in the mosquito *Aedes aegypti*. PLoS Pathog. 7, e1002274.

Chang, J.C., Yang, R.B., Adams, M.E., Lu, K.H., 2009. Receptor guanylyl cyclases in Inka cells targeted by eclosion hormone. Proc. Natl. Acad. Sci. U. S. A. 106, 13371–13376.

Chen, W., Hillyer, J.F., 2013. FlyNap (triethylamine) increases the heart rate of mosquitoes and eliminates the cardioacceleratory effect of the neuropeptide CCAP. PLoS One 16, e70414.

Choi, M.Y., Estep, A., Sanscrainte, N., Becnel, J., Vander Meer, R.K., 2013. Identification and expression of PBAN/diapause hormone and GPCRs from *Aedes aegypti*. Mol. Cell. Endocrinol. 375, 113–120.

Clark, T.M., Hayes, T.K., Beyenbach, K.W., 1998. Dose-dependent effects of CRF-like diuretic peptide on transcellular and paracellular transport pathways. Am. J. Physiol. Renal Physiol. 274, F834–F840.

Clements, A.N., 1956. Hormonal control of ovary development in mosquitoes. J. Exp. Biol. 33, 211–223.

Coast, G.M., Garside, C.S., Webster, S.G., Schegg, K.M., Schooley, D.A., 2005. Mosquito natriuretic peptide identified as a calcitonin-like diuretic hormone in *Anopheles gambiae* (Giles). J. Exp. Biol. 208, 3281–3291.

Dai, L., Adams, M.E., 2009. Ecdysis triggering hormone signaling in the yellow fever mosquito *Aedes aegypti*. Gen. Comp. Endocrinol. 162, 43–51.

De Meyts, P., Gauguin, L., Svendsen, A.M., Sarhan, M., Knudsen, L., Nøhr, J., Kiselyov, V.V., 2009. Structural basis of allosteric ligand–receptor interactions in the insulin/relaxin peptide family: implications for other receptor tyrosine kinases and G-protein-coupled receptors. Ann. N. Y. Acad. Sci. 1160, 45–53.

Detinova, T.S., 1945. On the influence of glands of internal secretion upon the ripening of the gonads and the imaginal diapause of *Anopheles maculipennis*. Zool. Zhurnal 34, 291–298.

Dhara, A., Eum, J.H., Robertson, A., Gulia-Nuss, M., Vogel, K.J., Clark, K.D., Graf, R., Brown, M.R., Strand, M.R., 2013. Ovary ecdysteroidogenic hormone functions independently of the insulin receptor in the yellow fever mosquito, *Aedes aegypti*. Insect Biochem. Mol. Biol. 43, 1100–1108.

Dow, J.A., Davies, S.A., 2006. The Malphighian tubule: rapid insights from post-genomic biology. J. Insect Physiol. 52, 365–378.

Duckworth, W.C., Garcia, J.V., Liepnieks, J.J., Hamel, F.G., Hermodson, M.A., Frank, B.H., Rosner, M.R., 1989. *Drosophila* insulin degrading enzyme and rat skeletal muscle insulin protease cleave insulin at similar sites. Biochemistry 28, 2471–2477.

Duttlinger, A., Mispelon, M., Nichols, R., 2003. The structure of the FMRFamide receptor and activity of the cardioexcitatory neuropeptide are conserved in mosquito. Neuropeptides 37, 120–126.

Ellison, H.E., Estévez-Lao, T.Y., Murphree, C.S., Hillyer, J.F., 2015. Deprivation of both sucrose and water reduces the mosquito heart contraction rate while increasing the expression of nitric oxide synthase. J. Insect Physiol. 74, 1–9.

Estévez-Lao, T.Y., Boyce, D.S., Honegger, H.W., Hillyer, J.F., 2013. Cardioacceleratory function of the neurohormone CCAP in the mosquito *Anopheles gambiae*. J. Exp. Biol. 216, 601–613.

Fernandez-Almonacid, R., Rosen, O.M., 1987. Structure and ligand specificity of the *Drosophila melanogaster* insulin receptor. Mol. Cell. Biol. 7, 2718–2727.

Fernández-Gamba, A., Leal, M.C., Morelli, L., Castaño, E.M., 2009. Insulin-degrading enzyme: structure–function relationship and its possible roles in health and disease. Curr. Pharm. Des. 15, 3644–3655.

Fricker, L.D., 2012. Neuropeptides and Other Bioactive Peptides: From Discovery to Function. Morgan & Claypool Life Sciences, San Rafael, CA, 107 pp.

Furuya, K., Milchak, R.J., Schegg, K.M., Zhang, J., Tobe, S.S., Coast, G.M., Schooley, D.A., 2000. Cockroach diuretic hormone: characterization of a calcitonin-like peptide in insects. Proc. Natl. Acad. Sci. U. S. A. 97, 6469–6474.

Gäde, G., 2009. Peptides of the adipokinetic hormone/red pigment-concentrating hormone family: a new take on biodiversity. Ann. N. Y. Acad. Sci. 1163, 125–136.
Gäde, G., Hoffmann, K.H., Spring, J.H., 1997. Hormonal regulation in insects: facts, gaps, and future directions. Physiol. Rev. 77, 963–1032.
Galagovsky, D., Katz, M.J., Acevedo, J.M., Sorianello, E., Glavic, A., Wappner, P., 2014. The *Drosophila* insulin-degrading enzyme restricts growth by modulating the PI3K pathway in a cell-autonomous manner. Mol. Biol. Cell 25, 916–924.
Garczynski, S.F., Crim, J.W., Brown, M.R., 2005. Characterization of neuropeptide F and its receptor from the African malaria mosquito, *Anopheles gambiae*. Peptides 26, 99–107.
Garczynski, S.F., Crim, J.W., Brown, M.R., 2007. Characterization and expression of the short neuropeptide F receptor in the African malaria mosquito, *Anopheles gambiae*. Peptides 28, 109–118.
Garelli, A., Heredia, F., Casimiro, A.P., Macedo, A., Nunes, C., Garcez, M., Dias, A.R., Volonte, Y.A., Uhlmann, T., Caparros, E., Koyama, T., Gontijo, A.M., 2015. Dilp8 requires the neuronal relaxin receptor Lgr3 to couple growth to developmental timing. Nat. Commun. 6, 8732.
Gillett, J.D., 1956. Initiation and promotion of ovarian development in the mosquito *Aedes* (*Stegomyia*) *aegypti* (Linnaeus). Ann. Trop. Med. Parasitol. 50, 375–380.
Gouzi, J.Y., Moressis, A., Walker, J.A., Apostolopoulou, A.A., Palmer, R.H., Bernards, A., Skoulakis, E.M., 2011. The receptor tyrosine kinase Alk controls neurofibromin functions in *Drosophila* growth and learning. PLoS Genet. 7, e1002281.
Graf, R., Lea, A.O., Briegel, H., 1998. A temporal profile of the endocrine control of trypsin synthesis in the yellow fever mosquito, *Aedes aegypti*. J. Insect Physiol. 44, 451–454.
Grönke, S., Clarke, D.-F., Broughton, S., Andrews, T.D., Partridge, L., 2010. Molecular evolution and functional characterization of *Drosophila* insulin-like peptides. PLoS Genet. 6, e1000857.
Gulia-Nuss, M., Robertson, A.E., Brown, M.R., Strand, M.R., 2011. Insulin-like peptides and the target of rapamycin pathway coordinately regulate blood digestion and egg maturation in the mosquito *Aedes aegypti*. PLoS One 6, e20401.
Gulia-Nuss, M., Eum, J.H., Strand, M.R., Brown, M.R., 2012. Ovary ecdysteroidogenic hormone activates egg maturation in the mosquito *Georgecraigius atropalpus* after adult eclosion or a blood meal. J. Exp. Biol. 215, 3758–3767.
Gulia-Nuss, M., Elliot, A., Brown, M.R., Strand, M.R., 2015. Multiple factors contribute to anautogenous reproduction by the mosquito *Aedes aegypti*. J. Insect Physiol. 82, 8–16.
Hagedorn, H.H., O'Connor, J., Fuchs, M.S., Sage, B., Schlaeger, D.A., Bohm, M., 1975. The ovary as a source of alpha-ecdysone in an adult mosquito. Proc. Natl. Acad. Sci. U. S. A. 72, 3255–3259.
Hagedorn, H.H., Shapiro, J.P., Hanaoka, K., 1979. Ovarian ecdysone secretion is controled by a brain hormone in the adult mosquito. Nature 282, 92–94.
Hansen, K.K., Stafflinger, E., Schneider, M., Hauser, F., Cazzamali, G., Williamson, M., Kollmann, M., Schachtner, J., Grimmelikhuijzen, C.J., 2010. Discovery of a novel insect neuropeptide signaling system closely related to the insect adipokinetic hormone and corazonin hormonal systems. J. Biol. Chem. 285, 10736–10747.
Hauser, F., Grimmelikhuijzen, C.J., 2014. Evolution of the AKH/corazonin/ACP/GnRH receptor superfamily and their ligands in the Protostomia. Gen. Comp. Endocrinol. 209, 35–49.
Hauser, F., Nothacker, H.P., Grimmelikhuijzen, C.J., 1997. Molecular cloning, genomic organization, and developmental regulation of a novel receptor from *Drosophila melanogaster* structurally related to members of the thyroid-stimulating hormone, follicle-stimulating hormone, luteinizing hormone/choriogonadotropin receptor family from mammals. J. Biol. Chem. 272, 1002–1010.

Hauser, F., Cazzamali, G., Williamson, M., Blenau, W., Grimmelikhuijzen, C.J.P., 2006a. A review of neurohormone GPCRs present in the fruitfly *Drosophila melanogaster* and the honey bee *Apis mellifera*. Prog. Neurobiol. 80, 1–19.
Hauser, F., Williamson, M., Cazzamali, G., Grimmelikhuijzen, C.J., 2006b. Identifying neuropeptide and protein hormone receptors in *Drosophila melanogaster* by exploiting genomic data. Brief. Funct. Genomic. Proteomic. 4, 321–330.
Hauser, F., Cazzamali, G., Williamson, M., Park, Y., Li, B., Tanaka, Y., Predel, R., Neupert, S., Schachtner, J., Verleyen, P., Grimmelikhuijzen, C.J.P., 2008. A genome-wide inventory of neurohormone GPCRs in the red flour beetle *Tribolium castaneum*. Front. Neuroendocrinol. 29, 142–165.
Hauser, F., Neupert, S., Williamson, M., Predel, R., Tanaka, Y., Grimmelikhuijzen, C.J.P., 2010. Genomics and peptidomics of neuropeptides and protein hormones present in the parasitic wasp *Nasonia vitripennis*. J. Proteome Res. 9, 5296–5310.
Hayes, T.K., Strey, A., Belk, S., Holman, G.M., Nachman, R.J., Petzel, D., Readio, J., Meola, R., Pannabecker, T., Beyenbach, K., 1997. Biochemical characterization of mosquito kinin and related receptors. Ann. N. Y. Acad. Sci. 814, 342–345.
Hector, C.E., Bretz, C.A., Zhao, Y., Johnson, E.C., 2009. Functional differences between two CRF-related diuretic hormone receptors in *Drosophila*. J. Exp. Biol. 212, 3142–3147.
Hellmich, E., Nusawardani, T., Bartholomay, L., Jurenka, R., 2014. Pyrokinin/PBAN-like peptides in the central nervous system of mosquitoes. Cell Tissue Res. 356, 39–47.
Hernández-Martínez, S., Li, Y., Lanz-Mendoza, H., Rodríguez, M.H., Noriega, F.G., 2005. Immunostaining for allatotropin and allatostatin-A and -C in the mosquitoes *Aedes aegypti* and *Anopheles albimanus*. Cell Tissue Res. 321, 105–113.
Hernández-Martínez, S., Mayoral, J.G., Li, Y., Noriega, F.G., 2007. Role of juvenile hormone and allatotropin on nutrient allocation, ovarian development and survivorship in mosquitoes. J. Insect Physiol. 53, 230–234.
Hewes, R.S., Taghert, P.H., 2001. Neuropeptides and neuropeptide receptors in the *Drosophila melanogaster* genome. Genome Res. 11, 1126–1142.
Hill, C.A., Fox, A.N., Pitts, R.J., Kent, L.B., Tan, P.L., Chrystal, M.A., Cravchik, A., Collins, F.H., Robertson, H.M., Zwiebel, L.J., 2002. G protein coupled receptors in *Anopheles gambiae*. Science 298, 176–178.
Hillyer, J.F., Estevez-Lao, T.Y., Funkhouser, L.J., Aluoch, V.A., 2012. *Anopheles gambiae* corazonin: gene structure, expression and effect on mosquito heart physiology. Insect Mol. Biol. 21, 343–355.
Hillyer, J.F., Estevez-Lao, T.Y., de la Parte, L.E., 2014. Myotropic effects of FMRFamide containing peptides on the heart of the mosquito *Anopheles gambiae*. Dev. Comp. Immunol. 202, 15–25.
Holman, G.M., Cook, B.J., Nachman, R.J., 1986. Primary structure and synthesis of a blocked myotropic neuropeptide isolated from the cockroach, *Leucophaea maderae*. Comp. Biochem. Physiol. C 85, 219–224.
Holt, R.A., Subramanian, G.M., Halpern, A., Sutton, G.G., Charlab, R., Nusskern, D.R., Wincker, P., Clark, A.G., Ribeiro, J.M.C., Wides, R., Salzberg, S.L., et al., 2002. The genome sequence of the malaria mosquito *Anopheles gambiae*. Science 298, 129–135.
Holzer, P., Reichmann, F., Farzi, A., 2012. Neuropeptide Y, peptide YY and pancreatic polypeptide in the gut-brain axis. Neuropeptides 46, 261–274.
Honegger, H.W., Estevez-Lao, T.Y., Hillyer, J.F., 2011. Bursicon-expressing neurons undergo apoptosis after adult ecdysis in the mosquito *Anopheles gambiae*. J. Insect Physiol. 57, 1017–1022.
Hook, V., Funkelstein, L., Lu, D., Bark, S., Wegrzyn, J., Hwang, S.R., 2008. Proteases for processing proneuropeptides into peptide neurotransmitters and hormones. Annu. Rev. Pharmacol. Toxicol. 48, 393–423.

Hubbard, S.R., Miller, W.T., 2007. Receptor tyrosine kinases: mechanisms of activation and signaling. Curr. Opin. Cell Biol. 19, 117–123.
Hummon, A.B., Richmond, T.A., Verleyen, P., Baggerman, G., Huybrechts, J., Ewing, M.A., Vierstraete, E., Rodriguez-Zas, S.L., Schoofs, L., Robinson, G.E., Sweedler, J.V., 2006. From the genome to the proteome: Uncovering peptides in the *Apis* brain. Science 314, 647–649.
Ionescu, A., Donini, A., 2012. AedesCAPA-PVK-1 displays diuretic and dose dependent antidiuretic potential in the larval mosquito *Aedes aegypti* (Liverpool). J. Insect Physiol. 58, 1299–1306.
Isaac, R.E., Bland, N.D., Shirras, A.D., 2009. Neuropeptidases and the metabolic inactivation of insect neuropeptides. Gen. Comp. Endocrinol. 162, 8–17.
Iversen, A., Cazzamali, G., Williamson, M., Hauser, F., Grimmelikhuijzen, C.J.P., 2002. Molecular identification of the first insect ecdysis triggering hormone receptors. Biochem. Biophys. Res. Commun. 299, 924–931.
Jagge, C.L., Pietrantonio, P.V., 2008. Diuretic hormone 44 receptor in Malpighian tubules of the mosquito *Aedes aegypti*: evidence for transcriptional regulation paralleling urination. Insect Mol. Biol. 17, 413–426.
Jenkins, S.P., Brown, M.R., Lea, A.O., 1992. Inactive prothoracic glands in larvae and pupae of *Aedes aegypti*: Ecdysteroid release by tissues in the thorax and abdomen. Insect Biochem. Mol. Biol. 22, 553–559.
Jiang, H., Edgar, B.A., 2012. Intestinal stem cell function in *Drosophila* and mice. Curr. Opin. Genet. Dev. 22, 354–360.
Jiang, N., Kolhekarm, A.S., Jacobs, P.S., Mains, R.E., Eipper, B.A., Taghert, P.H., 2000. PHM is required for normal developmental transitions and for biosynthesis of secretory peptides in *Drosophila*. Dev. Biol. 226, 118–136.
Jiang, X., Peery, A., Hall, A.B., Sharma, A., Chen, X.G., Waterhouse, R.M., Komissarov, A., Riehle, M.M., Shouche, Y., Sharakhova, M.V., Lawson, D., et al., 2014. Genome analysis of a major urban malaria vector mosquito, *Anopheles stephensi*. Genome Biol. 15, 459.
Johnson, E.C., Bohn, L.M., Barak, L.S., Birse, R.T., Nässel, D.R., Caron, M.G., Taghert, P.H., 2003. Identification of *Drosophila* neuropeptide receptors by G protein-coupled receptors-beta-arrestin2 interactions. J. Biol. Chem. 278, 52172–52178.
Jurenka, R., 2015. The PRXamide neuropeptide signalling system: conserved in animals. Adv. Insect Physiol. 49, 123–169.
Kang, M.A., Mott, T.M., Tapley, E.C., Lewis, E.E., Luckhart, S., 2008. Insulin regulates aging and oxidative stress in *Anopheles stephensi*. J. Exp. Biol. 211, 741–748.
Kataoka, H., Troetschler, R.G., Li, J.P., Kramer, S.J., Carney, R.L., Schooley, D.A., 1989. Isolation and identification of a diuretic hormone from the tobacco hornworm, *Manduca sexta*. Proc. Natl. Acad. Sci. U.S.A. 86, 2976–2980.
Kaufmann, C., Brown, M.R., 2006. Adipokinetic hormones in the African malaria mosquito, *Anopheles gambiae*: identification and expression of genes for two peptides and a putative receptor. Insect Biochem. Mol. Biol. 36, 466–481.
Kaufmann, C., Brown, M.R., 2008. Regulation of carbohydrate metabolism and flight performance by a hypertrehalosaemic hormone in the mosquito *Anopheles gambiae*. J. Insect Physiol. 54, 367–377.
Kaufmann, C., Merzendorfer, H., Gade, G., 2009. The adipokinetic hormone system in Culicinae (Diptera: Culicidae): Molecular identification and characterization of two adipokinetic hormone (AKH) precursors from *Aedes aegypti* and *Culex pipiens* and two putative AKH receptor variants from *Ae. aegypti*. Insect Biochem. Mol. Biol. 39, 770–781.
Kersch, C.N., Pietrantonio, P.V., 2011. Mosquito *Aedes aegypti* (L.) leucokinin receptor is critical for in vivo fluid excretion post blood feeding. FEBS Lett. 585, 3507–3512.

Klowden, M.J., Lea, A.O., 1979. Humoral inhibition of host-seeking in *Aedes aegypti* during oocyte maturation. J. Insect Physiol. 25, 231–235.
Krieger, M.J.B., Jahan, N., Riehle, M.A., Cao, C., Brown, M.R., 2004. Molecular characterization of insulin-like peptide genes and their expression in the African malaria mosquito, *Anopheles gambiae*. Insect Mol. Biol. 13, 305–315.
Kwon, H., Pietrantonio, P.V., 2013. Calcitonin receptor 1 (AedaeGPCRCAL1) hindgut expression and direct role in myotropic action in females of the mosquito *Aedes aegypti* (L.). Insect Biochem.Mol. Biol. 43, 588–593.
Kwon, H., Lu, H.L., Longnecker, M.T., Pietrantonio, P.V., 2012. Role in diuresis of a calcitonin receptor (GPRCAL1) expressed in a distal-proximal gradient in renal organs of the mosquito *Aedes aegypti* (L.). PLos One 7, e50374.
Lasek, A.W., Lim, J., Kliethermes, C.L., Berger, K.H., Joslyn, G., Brush, G., Xue, L., Robertson, M., Moore, M.S., Vranizan, K., Morris, S.W., et al., 2011. An evolutionary conserved role for anaplastic lymphoma kinase in behavioral responses to ethanol. PLoS One 6, e22636.
Lea, A.O., 1967. The medial neurosecretory cells and egg maturation in mosquitoes. J. Insect Physiol. 13, 419–429.
Lea, A.O., 1972. Regulation of egg maturation in the mosquito by the neurosecretory system: the role of the corpus cardiacum. Gen. Comp. Endocrinol. 3 (Suppl.), 602–608.
Lea, A.O., Van Handel, E., 1970. Suppression of glycogen synthesis in the mosquito by a hormone from the medial neurosecretory cells. J. Insect Physiol. 16, 319–321.
Lemmon, M.A., Schlessinger, J., 2010. Cell signaling by receptor tyrosine kinases. Cell 141, 1117–1134.
Li, Y., Unnithan, G.C., Veenstra, J.A., Feyereisen, R., Noriega, F.G., 2003. Stimulation of JH biosynthesis by the corpora allata of adult female *Aedes aegypti* in vitro: effect of farnesoic acid and *Aedes* allatotropin. J. Exp. Biol. 206, 1825–1832.
Li, Y.P., Hernandez-Martinez, S., Noriega, F.G., 2004. Inhibition of juvenile hormone biosynthesis in mosquitoes: effect of allatostatic head factors, PISCF- and YXFGL-amide-allatostatins. Reg. Peptides 118, 175–182.
Li, Y., Hernandez-Martinez, S., Fernandez, F., Mayoral, J.G., Topalis, P., Priestap, H., Perez, M., Navare, A., Noriega, F.G., 2006. Biochemical, molecular, and functional characterization of PISCF-allatostatin, a regulator of juvenile hormone biosynthesis in the mosquito *Aedes aegypti*. J. Biol. Chem. 281, 34048–34055.
Li, B., Predel, R., Neupert, S., Hauser, F., Tanaka, Y., Cazzamali, G., Williamson, M., Arakane, Y., Verleyen, P., Schoofs, L., Schachtner, J., Grimmelikhuijzen, C.J.P., Park, Y., 2008. Genomics, transcriptomics, and peptidomics of neuropeptides and protein hormones in the red flour beetle *Tribolium castaneum*. Genome Res. 18, 113–122.
Liesch, J., Bellani, L.L., Vosshall, L.B., 2013. Functional and genetic characterization of neuropeptide Y-like receptors in *Aedes aegypti*. PLoS Negl. Trop. Dis. 7, e2486.
Lin, Y., Hamblin, M.T., Edwards, M.J., Barillas-Mury, C., Kanost, M.R., Knipple, D.C., Wolfner, M.F., Hagedorn, H.H., 1993. Structure, expression, and hormonal control of genes from the mosquito, *Aedes aegypti*, which encode proteins similar to the vitelline membrane proteins of *Drosophila melanogaster*. Dev. Biol. 155, 558–568.
Lorén, C.E., Scully, A., Grabbe, C., Edeen, P.T., Thomas, J., McKeown, M., Hunter, T., Palmer, R.H., 2001. Identification and characterization of DAlk: a novel *Drosophila melanogaster* RTK which drives ERK activation in vivo. Genes Cells 6, 531–544.
Luo, C.W., Dewey, E.M., Sudo, S., Ewer, J., Hsu, S.Y., Honegger, H.W., Hsueh, A.J.W., 2005. Bursicon, the insect cuticle-hardening hormone, is a heterodimeric cystine knot protein that activates G protein-coupled receptor LGR2. Proc. Natl. Acad. Sci. U. S. A. 102, 2820–2825.

Maddrell, S.H.P., Phillips, J.E., 1978. Induction of sulphate transport and hormonal control of fluid secretion by Malpighian tubules of larvae of the mosquito *Aedes taeniorhynchus*. J. Exp. Biol. 72, 181–202.
Marquez, A.G., Pietri, J.E., Smithers, H.M., Nuss, A., Antonova, Y., Drexler, A.L., Riehle, M.A., Brown, M.R., Luckhart, S., 2011. Insulin-like peptides in the mosquito *Anopheles stephensi*: identification and expression in response to diet and infection with *Plasmodium falciparum*. Gen. Comp. Endocrinol. 173, 303–312.
Matsumoto, S., Brown, M.R., Suzuki, A., Lea, A.O., 1989a. Isolation and characterization of ovarian ecdysteroidogenic hormones from the mosquito, *Aedes aegypti*. Insect Biochem. Mol. Biol. 19, 651–656.
Matsumoto, S., Brown, M.R., Crim, J.W., Vigna, S.R., Lea, A.O., 1989b. Isolation and primary structure of neuropeptides from the mosquito, *Aedes aegypti*, immunoreactive to FMRFamide antiserum. Insect Biochem. 19, 277–283.
Matthews, B.J., McBride, C.S., DeGennaro, M., Despo, O., Vosshall, L.B., 2016. The neurotransciptome of the *Aedes aegypti* mosquito. BMC Genomics 17, 32.
Mayoral, J.G., Nouzova, M., Brockhoff, A., Goodwin, M., Hernandez-Martinez, S., Richter, D., Meyerhof, W., Noriega, F.G., 2010. Allatostatin-C receptors in mosquitoes. Peptides 31, 442–450.
McDonald, N., Murray-Rust, J., Blundell, T., 1989. Structure–function relationships of growth factors and their receptors. Br. Med. Bull. 45, 554–569.
Mendive, F.M., Van Loy, T., Claeysen, S., Poels, J., Williamson, M., Hauser, F., Grimmelikhuijzen, C.J.P., Vassart, G., Broeck, J.V., 2005. *Drosophila* molting neurohormone bursicon is a heterodimer and the natural agonist of the orphan receptor DLGR2. FEBS Lett. 579, 2171–2176.
Meola, S.M., Lea, A.O., 1972. The ultrastructure of the corpus cardiacum of *Aedes sollicitans* and the histology of the cerebral neurosecretory system of mosquitoes. Gen. Comp. Endocrinol. 18, 210–234.
Meola, S.M., Clottens, F.L., Holman, G.M., Nachman, R.J., Nichols, R., Schoofs, L., Wright, M.S., Olson, J.K., Hayes, T.K., Pendleton, M.W., 1998. Isolation and immunocytochemical characterization of three tachykinin-related peptides from the mosquito, *Culex salinarius*. Neurochem. Res. 23, 189–202.
Moffett, S.B., Moffett, D.F., 2005. Comparison of immunoreactivity to serotonin, FMRFamide and SCPb in the gut and visceral nervous system of larvae, pupae and adults of the yellow fever mosquito *Aedes aegypti*. J. Insect Sci. 5, 20.
Mugumbate, G., Jackson, G.E., Van Der Spoel, D., 2011. Open conformation of adipokinetic hormone receptor from the malaria mosquito facilitates hormone binding. Peptides 32, 553–559.
Mugumbate, G., Jackson, G.E., van der Spoel, D., Kövér, K.E., Szilágyi, L., 2013. *Anopheles gambiae*, Anoga-HrTH hormone, free and bound structure—a nuclear magnetic resonance experiment. Peptides 41, 94–100.
Naccarati, C., Audsley, N., Keen, J.N., Kim, J.H., Howell, G.J., Kim, Y.J., Isaac, R.E., 2012. The host-seeking inhibitory peptide, Aea-HP-1, is made in the male accessory gland and transferred to the female during copulation. Peptides 34, 150–157.
Nässel, D.R., Wegener, C., 2011. A comparative review of short and long neuropeptide F signaling in invertebrates: any similarities to vertebrate neuropeptide Y signaling? Peptides 32, 1335–1355.
Nässel, D.R., Winther, A.M.E., 2010. *Drosophila* neuropeptides in regulation of physiology and behavior. Prog. Neurobiol. 92, 42–104.
Neafsey, D.E., Waterhouse, R.M., Abai, M.R., Aganezov, S.S., Alekseyev, M.A., Allen, J.E., Amon, J., Arcà, B., Arensburger, P., Artemov, G., Assour, L.A., et al., 2015. Mosquito genomics. Highly evolvable malaria vectors: the genomes of 16 Anopheles mosquitoes. Science 347, 1258522.

Nene, V., Wortman, J.R., Lawson, D., Haas, B., Kodira, C., Tu, Z.J., Loftus, B., Xi, Z.Y., Megy, K., Grabherr, M., Ren, Q.H., et al., 2007. Genome sequence of *Aedes aegypti*, a major arbovirus vector. Science 316, 1718–1723.

Nichols, R., 2003. Signaling pathways and physiological functions of *Drosophila melanogaster* FMRFamide-related peptides. Annu. Rev. Entomol. 48, 485–503.

Nijhout, H.F., 1998. Insect Hormones. Princeton University Press, Princeton, NJ, 260 pp.

Nijhout, H.F., Carrow, G.M., 1978. Diuresis after a bloodmeal in female *Anopheles freeborni*. J. Insect Physiol. 24, 293–298.

Nijhout, H.F., Riddiford, L.M., Mirth, C., Shingleton, A.W., Suzuki, Y., Callier, V., 2014. The developmental control of size in insects. Wiley Interdiscip. Rev. Dev. Biol. 3, 113–134.

Niwa, R., Niwa, Y.S., 2014. Enzymes for ecdysteroid biosynthesis: their biological functions in insects and beyond. Biosci. Biotechnol. Biochem. 78, 1283–1292.

Nouzova, M., Brockhoff, A., Mayoral, J.G., Goodwin, M., Meyerhof, W., Noriega, F.G., 2012. Functional characterization of an allatotropin receptor expressed in the corpora allata of mosquitoes. Peptides 34, 201–208.

Noyes, B.E., Katz, F.N., Schaffer, M.H., 1995. Identification and expression of the *Drosophila* adipokinetic hormone gene. Mol. Cell. Endocrinol. 109, 133–141.

Olsen, S.S., Cazzamali, G., Williamson, M., Grimmelikhuijzen, C.J.P., Hauser, F., 2007. Identification of one capa and two pyrokinin receptors from the malaria mosquito *Anopheles gambiae*. Biochem. Biophys. Res. Commun. 362, 245–251.

Onken, H., Moffett, S.B., Moffett, D.F., 2004. The anterior stomach of larval mosquitoes (*Aedes aegypti*): effects of neuropeptides on transepithelial ion transport and muscular motility. J. Exp. Biol. 207, 3731–3739.

Overend, G., Cabrero, P., Guo, A.X., Sebastian, S., Cundall, M., Armstrong, H., Mertens, I., Schoofs, L., Dow, J.A.T., Davies, S.A., 2012. The receptor guanylate cyclase Gyc76C and a peptide ligand, NPLP1-VQQ, modulate the innate immune IMD pathway in response to salt stress. Peptides 34, 209–218.

Park, Y., Filippov, V., Gill, S.S., Adams, M.E., 2002. Deletion of the ecdysis-triggering hormone gene leads to lethal ecdysis deficiency. Development 129, 493–503.

Pietrantonio, P.V., Jagge, C., Taneja-Bageshwar, S., Nachman, R.J., Barhoumi, R., 2005. The mosquito *Aedes aegypti* (L.) leucokinin receptor is a multiligand receptor for the three *Aedes* kinins. Insect Mol. Biol. 14, 55–67.

Pietri, J.E., Pietri, E.J., Potts, R., Riehle, M.A., Luckhart, S., 2015. *Plasmodium falciparum* suppresses the host immune response by inducing the synthesis of insulin-like peptides (ILPs) in the mosquito *Anopheles stephensi*. Dev. Comp. Immunol. 53, 134–144.

Potter, L.R., 2011. Regulation and therapeutic targeting of peptide-activated receptor guanylyl cyclases. Pharmacol. Ther. 130, 71–82.

Predel, R., Neupert, S., Garczynski, S.F., Crim, J.W., Brown, M.R., Russell, W.K., Kahnt, J., Russell, D.H., Nachman, R.J., 2010. Neuropeptidomics of the mosquito *Aedes aegypti*. J. Proteome Res. 9, 2006–2015.

Raabe, M., 1989. Recent Developments in Insect Neurohormones. Plenum Press, New York. 503 pp.

Radford, J.C., Terhzaz, S., Cabrero, P., Davies, S.A., Dow, J.A.T., 2004. Functional characterisation of the *Anopheles* leucokinins and their cognate G-protein coupled receptor. J. Exp. Biol. 207, 4573–4586.

Reagan, J.D., 1994. Expression cloning of an insect diuretic hormone receptor. A member of the calcitonin/secretin receptor family. J. Biol. Chem. 269, 9–12.

Reiher, W., Shirras, C., Kahnt, J., Baumeister, S., Isaac, R.E., Wegener, C., 2011. Peptidomics and peptide hormone processing in the *Drosophila* midgut. J. Proteome Res. 10, 1881–1892.

Rewitz, K.F., Yamanaka, N., Gilbert, L.I., O'Connor, M.B., 2009. The insect neuropeptide PTTH activates Receptor Tyrosine Kinase Torso to initiate metamorphosis. Science 326, 1403–1405.

Riehle, M.A., Garczynski, S.F., Crim, J.W., Hill, C.A., Brown, M.R., 2002. Neuropeptides and peptide hormones in *Anopheles gambiae*. Science 298, 172–175.

Riehle, M.A., Fan, Y., Cao, C., Brown, M.R., 2006. Molecular characterization of insulin-like peptides in the yellow fever mosquito, *Aedes aegypti*: expression, cellular localization, and phylogeny. Peptides 27, 2547–2560.

Robertson, H.M., Navik, J.A., Walden, K.K., Honegger, H.W., 2007. The bursicon gene in mosquitoes: an unusual example of mRNA trans-splicing. Genetics 176, 1351–1353.

Roller, L., Yamanaka, N., Watanabe, K., Daubnerova, I., Zitnan, D., Kataoka, H., Tanaka, Y., 2008. The unique evolution of neuropeptide genes in the silkworm *Bombyx mori*. Insect Biochem. Mol. Biol. 38, 1147–1157.

Sajid, W., Kulahin, N., Schluckebier, G., Ribel, U., Henderson, H.R., Tatar, M., Hansen, B.F., Svendsen, A.M., Kiselyov, V.V., Norgaard, P., Wahlund, P.O., Brandt, J., Kohanski, R.A., Andersen, A.S., De Meyts, P., 2011. Structural and biological properties of the *Drosophila* insulin-like peptide 5 show evolutionary conservation. J. Biol. Chem. 286, 661–673.

Sano, H., Nakamura, A., Texada, M.J., Truman, J.W., Ishimoto, H., Kamilkouchi, A., Nibu, Y., Kume, K., Ida, T., Kojima, M., 2015. The nutrient-responsive hormone CCHamide-2 controls growth by regulating insulin-like peptides in the brain of *Drosophila melanogaster*. PLoS Genet. 11, e1005209.

Scholler, S., Belmont, M., Cazzamali, G., Hauser, F., Williamson, M., Grimmelikhuijzen, C.J.P., 2005. Molecular identification of a myosuppressin receptor from the malaria mosquito *Anopheles gambiae*. Biochem. Biophys. Res. Commun. 327, 29–34.

Siju, K.P., Reifenrath, A., Scheiblich, H., Neupert, S., Predel, R., Hansson, B.S., Schachtner, J., Ignell, R., 2013. Neuropeptides in the antennal lobe of the yellow fever mosquito, *Aedes aegypti*. J. Comp. Neurol. 522, 592–608.

Sim, C., Denlinger, D.L., 2009. A shut-down in expression of an insulin-like peptide, ILP-1, halts ovarian maturation during the overwintering diapause of the mosquito *Culex pipiens*. Insect Mol. Biol. 18, 325–332.

Smith, W., Rybczynski, R., 2012. Prothoracicotropic hormone. In: Gilbert, L.I. (Ed.), Insect Endocrinology. Elsevier, New York, pp. 1–62.

Southey, B.R., Sweedler, J.V., Rodriguez-Zas, S.L., 2008. Prediction of neuropeptide cleavage sites in insects. Bioinformatics 24, 815–825.

Stanek, D.M., Pohl, J., Crim, J.W., Brown, M.R., 2002. Neuropeptide F and its expression in the yellow fever mosquito, *Aedes aegypti*. Peptides 23, 1367–1378.

Stangier, J., Hilbich, C., Beyreuther, K., Keller, R., 1999. Unusual cardioactive peptide (CCAP) from pericardial organs of the shore crab *Carcinus maenas*. Proc. Natl. Acad. Sci. U. S. A. 84, 575–579.

Stracker, T.H., Thompson, S., Grossman, G.L., Riehle, M.A., Brown, M.R., 2002. Characterization of the AeaHP gene and its expression in the mosquito *Aedes aegypti* (Diptera: Culicidae). J. Med. Entomol. 39, 331–342.

Szapiro, G., Barbour, B., 2009. Parasynaptic signalling by fast neurotransmitters: the cerebellar cortex. Neuroscience 162, 644–655.

Taghert, P.H., Veenstra, J.A., 2003. *Drosophila* neuropeptide signaling. Adv. Genet. 49, 1–65.

Taneja-Bageshwar, S., Strey, A., Isaac, R.E., Coast, G.M., Zubrzak, P., Pietrantonio, P.V., Nachman, R.J., 2009. Biostable agonists that match or exceed activity of native insect kinins on recombinant arthropod GPCRs. Gen. Comp. Endocrinol. 162, 122–128.

Truman, J.W., Riddiford, L.M., 2002. Endocrine insights into the evolution of metamorphosis in insects. Annu. Rev. Entomol. 47, 467–500.
Vallejo, D.M., Juarez-Carrewno, S., Bolivar, J., Morante, J., Dominguez, M., 2015. A brain circuit that syntchronizes growth and maturation revealed through Dilp8 binding to Lgr3. Science 350, aac6767.
Van der Host, D.J., 2003. Insect adipokinetic hormones: release and integration of flight energy metabolism. Comp. Biochem. Physiol. B Biochem. Mol. Biol. 136, 217–226.
Veenstra, J.A., 1989. Isolation and structure of corazonin, a cardioactive peptide in the American cockroach. FEBS Lett. 250, 231–234.
Veenstra, J.A., 1994. Isolation and identification of three leucokinins from the mosquito *Aedes aegypti*. Biochem. Biophys. Res. Commun. 202, 715–719.
Veenstra, J.A., 1999. Isolation and identification of three RFamide-immunoreactive peptides from the mosquito, *Aedes aegypti*. Peptides 20, 31–38.
Veenstra, J.A., 2008. Allatostatin C and its paralog allatostatin double C: the arthropod somatostatins. Insect Biochem. Mol. Biol. 39, 161–170.
Veenstra, J.A., Costes, L., 1999. Isolation and identification of a peptide and its cDNA from the mosquito *Aedes aegypti* related to *Manduca sexta* allatotropin. Peptides 20, 1145–1151.
Veenstra, J.A., Lau, G.W., Agricola, H.J., Petzel, D.H., 1995. Immunohistological localization of regulatory peptides in the midgut of the female mosquito *Aedes aegypti*. Histochem. Cell Biol. 104, 337–347.
Veenstra, J.A., Noriega, F.G., Graf, R., Feyereisen, R., 1997a. Identification of three allatostatins and their cDNA from the mosquito *Aedes aegypti*. Peptides 18, 937–942.
Veenstra, J.A., Pattillo, J.M., Petzel, D.H., 1997b. A single cDNA encodes all three *Aedes* leucokinins, which stimulate both fluid secretion by the Malpighian tubules and hindgut contractions. J. Biol. Chem. 272, 10402–10407.
Veenstra, J.A., Rombauts, S., Grbic, M., 2012. In silico cloning of genes encoding neuropeptides, neurohormones and their putative G-protein coupled receptors in a spider mite. Insect Biochem. Mol. Biol. 42, 277–295.
Verlinden, H., Vleugels, R., Zels, S., Dillen, S., Lenaerts, C., Crabbe, K., Spit, J., Broeck, J.V., 2014. Receptors for neuronal or endocrine signalling molecules as potential targets for the control of insect pests. Adv. Insect Physiol. 46, 167–303.
Vogel, K.J., Brown, M.R., Strand, M.R., 2013. Phylogenetic investigation of peptide hormone and growth factor receptors in five dipteran genomes. Front. Endocrinol. 4, 193.
Vogel, K.J., Brown, M.R., Strand, M.R., 2015. Ovary ecdysteroidogenic hormone requires a receptor tyrosine kinase to activate egg formation in the mosquito *Aedes aegypti*. Proc. Natl. Acad. Sci. U. S. A. 112, 5057–5062.
Waterhouse, R.M., 2015. A maturing understanding of the composition of the insect gene repertoire. Curr. Opin. Insect Sci. 7, 15–23.
Wegener, C., Herbert, H., Kahnt, J., Bender, M., Rhea, J.M., 2011. Deficiency of prohormone convertase dPC2 (AMONTILLADO) results in impaired production of bioactive neuropeptide hormones in *Drosophila*. J. Neurochem. 118, 581–595.
Wen, Z., Gulia, M., Clark, K.D., Dhara, A., Crim, J.W., Strand, M.R., Brown, M.R., 2010. Two insulin-like peptide family members from the mosquito *Aedes aegypti* exhibit differential biological and receptor binding activities. Mol. Cell. Endocrinol. 328, 47–55.
White, B.H., Ewer, J., 2014. Neural and hormonal control of postecdysial behaviors in insects. Annu. Rev. Entomol. 59, 363–381.
Wiegmann, B.M., Yeates, D.K., Thorne, J.L., Kishino, H., 2003. Time flies, a new molecular time-scale for brachyceran fly evolution without a clock. Syst. Biol. 52, 745–756.
Woodhead, A.P., Stay, B., Seidel, S.L., Khan, M.A., Tobe, S.S., 1989. Primary structure of four allatostains: neuropeptide inhibitors of juvenile hormone synthesis. Proc. Natl. Acad. Sci. U. S. A. 86, 5997–6001.

Zandawala, M., 2012. Calcitonin-like diuretic hormones in insects. Insect Biochem. Mol. Biol. 42, 816–825.

Zhang, Q., Denlinger, D.L., 2011. Molecular structure of the prothoracicotropic hormone gene in the northern house mosquito, *Culex pipiens*, and its expression analysis in association with diapause and blood feeding. Insect Mol. Biol. 20, 201–213.

Zhang, H., Liu, J., Li, C.R., Momen, B., Kohanski, R.A., Pick, L., 2009. Deletion of *Drosophila* insulin-like peptides causes growth defects and metabolic abnormalities. Proc. Natl. Acad. Sci. U. S. A. 106, 19617–19622.

Zitnan, D., Kingan, T.G., Hermesman, J.L., Adams, M.E., 1996. Identification of ecdysis-triggering hormone from an epitracheal endocrine system. Science 271, 88–91.

Zitnan, D., Spalovska, Z.I., Takac, P., Park, Y., Adams, M.E., 2003. Conservation of ecdysis-triggering hormone signalling in insects. J. Exp. Biol. 206, 1275–1289.

CHAPTER SEVEN

Functions of Small RNAs in Mosquitoes

M. Hussain, K. Etebari, S. Asgari
Australian Infectious Disease Research Centre, School of Biological Sciences, The University of Queensland, Brisbane, QLD, Australia

Contents

Abstract

The role of small noncoding RNAs in regulating gene expression and as a consequence their connection with most cellular pathways is becoming increasingly recognized. In insects, the three major small noncoding RNAs, microRNAs, PIWI-interacting RNAs and short interfering RNAs, are involved in a range of biological processes such as development, behaviour, dimorphism, immunity, host–pathogen interaction, antiviral responses and genome protection. In this chapter, we will review and discuss the latest findings pertaining the functions of these small RNAs in mosquito biology.

1. INTRODUCTION

Small noncoding RNAs (sncRNAs, ~20–30 nucleotides) are ubiquitous in eukaryotes and produced mostly through the RNA interference (RNAi) pathway to regulate gene expression at the transcriptional or post-transcriptional levels. Notably, the genes involved in this pathway are highly conserved indicating the appearance of the RNAi pathway early in the evolution of eukaryotes. sncRNAs contribute to silencing the expression of both endogenous as well as exogenous genes (of parasitic or pathogenic

Advances in Insect Physiology, Volume 51
ISSN 0065-2806
http://dx.doi.org/10.1016/bs.aiip.2016.04.001

origin) and destruction of replicative molecules (of viruses). In particular, RNAi is considered as a very potent antiviral defence in invertebrates and plants (Ding, 2010). The discovery of the RNAi pathway has opened up numerous avenues in research and potentials in pest and disease control relevant to agriculture and vector-borne diseases. In particular, our understanding of these pathways in insects has increased substantially in the last decade. Here, we review and discuss the latest advancements in the functions of sncRNAs in mosquito biology and their interaction with the pathogens and parasites that they carry/vector.

2. THE RNAi PATHWAYS AND THE BIOGENESIS OF sncRNAs

The RNAi pathway is involved in the production of three main types of sncRNAs, short interfering RNAs (siRNA), microRNAs (miRNAs) and PIWI-interacting RNAs (piRNAs). The main difference between the three types is the substrates used for their biogenesis (Fig. 1; see later). Comparative genomics studies have shown that the key components involved in the production of these sncRNAs are conserved in mosquitoes and contain NF-κB-related and Broad complex transcription factor binding sites, although species-specific differences exist in genes involved in the RNAi pathway (Campbell et al., 2008b). Evidence suggests rapid intraspecific evolution of these genes in mosquitoes (Bernhardt et al., 2012).

For *siRNA* production, double-stranded RNA (dsRNA) molecules are used as substrates for the production of siRNAs (~20–22 nt) (Fig. 1A). The dsRNA substrate could be produced inside the cell (endogenous) or introduced into the cell (exogenous) by transfection of synthetic dsRNA or by viruses (see later). The key enzyme involved in the biogenesis of siRNAs in insects is Dicer-2 (Galiana-Arnoux et al., 2006). The enzyme senses dsRNA in association with R2D2 (TRBP in human) and cleaves it into siRNA duplexes, which induce the formation of the RNA-induced silencing complex (RISC) (Blair and Olson, 2015). One of the main components of the RISC complex is the Argonaut 2 (Ago2) protein and R2D2 plays an important role in proper sorting of siRNAs into Ago2 (Okamura et al., 2011). In the complex, one of the strands is degraded and the leading strand guides the complex to complementary sequences in target mRNAs (or viral sequences) leading to their degradation. The complementarity of siRNA with the target sequences is 100%. While in plants and some other eukaryotes the RNAi response is amplified by an RNA-dependent RNA

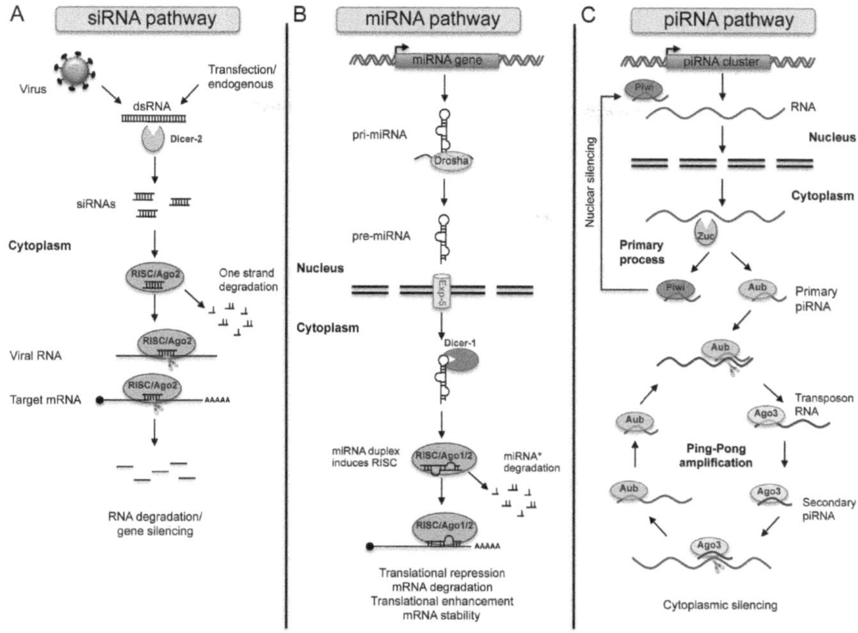

Fig. 1 RNAi pathways in insects; mostly based on *D. melanogaster* as a model. (A) In the *siRNA pathway*, the exogenous (with viral origin or by artificial transfection) or endogenous dsRNA is sensed by Dicer-2 and cleaved into 20–21 nt siRNAs that trigger the formation of the RISC complex in which Ago2 is involved. One of the strands is degraded and the RISC-guide strand binds to target RNA (viral or endogenous mRNA targeted for silencing) by complete complementarity leading to its cleavage. (B) In the *miRNA pathway*, a miRNA gene is transcribed into primary miRNA (pri-miRNA), which contains one or more stem-loop structures. The stem-loop is cleaved by the ribonuclease Drosha in the nucleus to form the precursor miRNA (pre-miRNA), which is transported into the cytoplasm by Exportin-5 (Exp-5). Subsequently, the hairpin head is removed by Dicer-1 and the resulting miRNA duplex is loaded into Ago1 or Ago2 as part of the RISC complex, in which one of the strands (miRNA*) is degraded. The mature miRNA (~22 nt) guides the RISC complex to mRNA targets by sequence complementarity (usually with mismatches). (C) In the *piRNA pathway*, piRNA clusters that mainly consist of repetitive and transposable DNA elements are transcribed and transported into the cytoplasm, where they appear to be cleaved by Zucchini (Zuc) into 24–27 nt piRNAs. These primary piRNAs are loaded onto Piwi and Aub proteins. Those associated with Piwi are translocated into the nucleus where they target the primary piRNA transcripts. On the other hand, the primary piRNAs associated with Aub target transposon transcripts in the cytoplasm. The cleavage products are transferred onto Ago3 and further processed into mature secondary piRNAs that target more transposon transcripts with the cleavage products loaded into Aub, trimmed into mature secondary piRNAs and enter in the cycle of Ping-Pong amplification. Therefore, the secondary piRNAs are involved in cytoplasmic silencing of transposon transcripts. (See the colour plate.)

polymerase (RdRP), a homolog of this gene does not exist in the genomes of dipteran insects examined (Zong et al., 2009). In fact, experimental evidence also suggests the absence of an RdRP-type amplification of RNAi in mosquitoes (Adelman et al., 2008; Hoa et al., 2003).

The substrate for the production of *miRNAs* is the primary miRNA (pri-miRNA) transcript that contains one or more stem-loop structures, often with mismatches in the stem (Fig. 1B) (Bartel, 2009). Pri-miRNA could be a coding or noncoding transcript that is produced by RNA polymerase II with the 5′-cap and a polyA tail in the nucleus. In the canonical pathway of miRNA biogenesis, the stem-loop is cleaved from the base by Drosha, a ribonuclease, in association with Pasha (DGCR8 in mammals) leading to the production of the precursor miRNA (pre-miRNA), which is a hairpin stem-loop of about 70 nt. pre-miRNA is then transported into the cytoplasm by Exportin-5, where the hairpin head is cleaved by Dicer-1 leading to the production of miRNA/miRNA* duplex which is loaded into Ago1/Ago2 (mostly Ago1) protein forming the RISC complex. In the complex, one of the strands may get degraded (known as miRNA* strand) or alternatively loaded into another RISC complex. In noncanonical pathways, miRNA could be produced by bypassing the cleavage by Drosha (eg, mirtrons produced through intron splicing) (Okamura et al., 2007) or bypassing Dicer-1 (eg, cleavage by Ago-2) (Yang and Lai, 2011). The mature miRNA (either from 5p or 3p arm) guides the RISC complex to its target, which is mostly mRNA, leading to its degradation, repression of translation or in some instances its increased stability (Hussain and Asgari, 2014). Unlike siRNAs, the complementarity of miRNAs and their target sequences are not usually 100% with some bulges and mismatches. While the overall sequence complementarity is important, complementarity of nucleotides 2–8 from the 5′-end (the seed region) of the miRNA with target sequences is central in the majority of instances (Lewis et al., 2005). Accordingly, most bioinformatics tools consider seed region complementarity as an important criterion for scanning targets.

The *piRNA* pathway is quite different and more complex as compared with the other two sncRNA pathways (Fig. 1C). In the primary processing pathway, long ssRNA precursors (also 5′-capped and polyadenylated) are produced from the so-called piRNA clusters from which a large number of piRNAs are eventually produced. This transcript is transported into the cytoplasm and localized to Yb bodies/nuage where they are cleaved and trimmed into 24–30 nt small RNAs by Zucchini and other unknown proteins. These are known as the primary piRNAs, which display a strong preference for a 5′-uridine (reviewed in Ishizu et al., 2012). The mature piRNAs are subsequently loaded into a PIWI protein (Piwi or Aubergine, Aub). Those loaded into the Piwi protein are transported into the nucleus where they interact with their target sequences in the genome leading to transcriptional gene silencing by deposition of silent H3K9me3 marks on the chromatin (Chambeyron and Seitz, 2014). The loaded primary piRNAs into Aub (with a strong bias for uracil in their position 1 from the 5′-end) can lead to the production of secondary piRNAs through the Ping-Pong cycle by mediating the cleavage of target RNAs (from transposons) by an endonuclease (Gunawardane et al., 2007). The antisense cleavage products are then loaded into Ago3 (another PIWI family protein) for further processing. Subsequently, these Ago3 loaded mature secondary piRNAs lead to the cleavage of more secondary piRNAs, which become loaded into Aub. This cycle of piRNA-mediated cleavages leads to the amplification of the products, which mostly target transposable elements (TEs) and in some cases other target transcripts. A signature for the secondary piRNAs is residue adenosine in their 10th nucleotide from the 5′-end (Ishizu et al., 2012). Due to their production pathway, primary and secondary piRNAs are complementary to each other for the first 10 nts from the 5′-end.

Of note, most of our knowledge on the piRNA biogenesis in insects described above come from *D. melanogaster* and little is known about this pathway and the molecules involved in it in mosquitoes. For example, as compared with *D. melanogaster*, there is an expansion in the number of PIWI proteins (one Ago3 and six *Piwi* genes) in *Ae. aegypti* and none of those could be assigned to Aub (Arensburger et al., 2011). However, recent research has shed some light on the role of these proteins in mosquitoes (see Section 2.2).

The sncRNAs produced through the different pathways of RNAi described above play important roles in various aspects of mosquito biology, which are reviewed in more details in the following sections.

2.1 The siRNA Pathway: Roles and Practical Uses

2.1.1 Endo-siRNAs: Regulators of Gene Expression and Protectors of the Genome

Based on a number of studies in various insects, including mosquitoes, the sources of dsRNA for the production of endogenous siRNAs (endo-siRNAs) could be complementary sense–antisense overlapping RNA transcripts (*cis*-natural antisense transcripts, *cis*-NAT) that are convergently or divergently transcribed, perfect stems formed by fold-back transcripts (hairpin-derived, hp, siRNAs) or TEs by hybridization of the terminal inverted repeats (Biryukova and Ye, 2015; Okamura and lai, 2008).

In *Anopheles gambiae*, abundant endo-siRNAs from *cis*-NAT transcripts of 21 nt in size were found in small RNA libraries from the gonads of the adult mosquitoes of both sexes, although more evident in the adult testes (Castellano et al., 2015). In contrast to *D. melanogaster*, only a small number of endo-siRNAs are produced from TEs in the larval or adult gonads of *An. gambiae*. Based on this observation, the authors suggested that endo-siRNAs might not be significant in TE control in the mosquito. However, in another study in which small RNA profiles from the whole *An. gambiae* mosquito samples (mixed somatic and germline tissues) were analysed, the majority of endo-siRNAs were produced from TEs (mostly DNA transposons as compared to retrotransposons) (Biryukova and Ye, 2015). Similar observation was made in the analysis of 21 nt siRNAs from *Aedes aegypti* whole mosquitoes (Arensburger et al., 2011). From these studies, it appears that endo-siRNAs may play a role in interaction with TEs and complement the function of piRNAs (see later) in suppressing TEs. While this function of endo-siRNAs has not been experimentally shown in mosquitoes, in *D. melanogaster*, loss/suppression of the siRNA pathway leads to increases in the expression of transposon mRNAs (Chung et al., 2008; Ghildiyal et al., 2008).

Similarly, there is no information about the function of hp-siRNAs in mosquitoes, but in *D. melanogaster* they regulate transcript levels of their targets. For example, hpRNA1 downregulates its target ATP synthase-β in testis, which is important in spermatogenesis and male fertility (Wen et al., 2015). In addition, endo-siRNAs provide more robustness for flies since loss of endo-siRNA production results in higher susceptibility to infection (see later) and less stress tolerance (Lim et al., 2011; Lucchetta et al., 2009). Production of endo-siRNAs could also be induced by infection. For instance it was shown that in *Ae. aegypti* a novel class of endo-siRNAs were produced upon Sindbis virus (SINV) infection (Adelman et al., 2012). These endo-siRNAs mapped to the entire length of the target transcripts and were derived from the antisense strands crossing exon–exon junctions. However, the function of these endo-siRNAs in mosquito–virus interaction is unknown.

2.1.2 Exo-siRNAs: Gene Silencing and Antiviral Defence

Exogenous dsRNAs introduced into eukaryotic cells induce a strong RNAi response by production of exo-siRNAs that destroy target sequences.

Most importantly, dsRNAs derived from viruses induce a potent antiviral response in host cells by production of exo-siRNAs that target viral sequences suppressing virus replication. These dsRNAs could be the viral genomic RNA (for dsRNA viruses), replicative intermediates produced during replication of single-stranded RNA (ssRNA) viruses, secondary structures in RNA genomes or bidirectional overlapping viral transcripts in the case of dsDNA viruses (Bronkhorst and van Rij, 2014). The regulation of virus replication by the exo-siRNA pathway is important for the establishment of nonpathogenic, but transmissive levels of arboviruses, which could otherwise compromise the survival of mosquitoes (Myles et al., 2008). For example, the RNAi response in the midgut of *Ae. aegypti* is crucial for providing an escape barrier to reduce dissemination of SINV to secondary tissues (Khoo et al., 2010). However, RNAi may play a bigger role in nonmidgut tissues depending on the mosquito; for instance, in *An. gambiae* when infected with O'nyong-nyong virus (Carissimo et al., 2015). The response may also spread from cell to cell inhibiting replication of incoming viruses, hence reducing virus dissemination (Attarzadeh-Yazdi et al., 2009). Furthermore, RNAi acts as a selection pressure to promote virus diversification as shown in the case of West Nile virus (WNV) (Brackney et al., 2009).

Several studies in different mosquitoes or their derived cell lines have shown that silencing genes involved in siRNA production, such as *Dicer-2*, *R2D2* and *Ago2*, results in significantly higher virus replication indicating the importance of the pathway in antiviral responses in mosquitoes (Campbell et al., 2008a; Cirimotich et al., 2009; Keene et al., 2004; Sanchez-Vargas et al., 2009; Schnettler et al., 2013; Waldock et al., 2012). This was also evident in the *Aedes albopictus* cell line C6/36, that is RNAi incompetent due to a defective Dicer-2 enzyme, in which arboviruses replicate to higher titres, and there are hardly viral siRNAs (vsiRNAs) produced in the cells (Brackney et al., 2010; Scott et al., 2010); although the higher virus production in these cells could also be partly due to other factors that may affect viral entry and processing of proteins. In addition, it was found that resistance to dengue virus (DENV) in field-collected and laboratory strains of *Ae. aegypti* was related to Dicer-2 genotype in the mosquitos (Bernhardt et al., 2012; Lambrechts et al., 2013). On the other hand, when a suppressor of RNAi (eg, B2 from Flock House virus (FHV)) is overexpressed by genetic modification of the viral genome, the number of vsiRNAs significantly declines resulting in more efficient replication of the viruses (Cirimotich et al., 2009), or in genetically modified mosquitoes that express B2 there was more DENV and SINV replication (Khoo et al., 2013). Considering that mosquitoes are important vectors of arboviruses, the significance of this pathway is very high both in terms of understanding regulation of virus replication in mosquitoes but also possible applications of the pathway to inhibit replication/transmission of arboviruses.

Next-generation sequencing has advanced our understanding of exo-siRNA production and their role in RNAi response. Several mosquito–arbovirus studies have taken advantage of the technology, in which large numbers of vsiRNAs have been identified. These include *Ae. aegypti* infected with SINV (Campbell et al., 2008a; Myles et al., 2008) and DENV (Schirtzinger et al., 2015), *Ae. albopictus*-derived U4.4 cells infected with Semliki Forest virus (SFV) (Siu et al., 2011) and Rift Valley fever virus (Léger et al., 2013). vsiRNAs are typically 21 nt that map to the entire length of the viral genome, although there are usually cold (no or few mapped siRNAs) as well as hot spots (high levels of mapped siRNAs) where small RNAs may map to both + and − strands or may show some bias towards either strands depending on the virus or host. vsiRNAs appear to come mostly from intermediate replicative dsRNAs rather than secondary structures (Siu et al., 2011). Interestingly, vsiRNAs from the hot spots tend to be less efficient in inducing the host cell RNAi response as compared with those from the cold spots. This has been shown in mosquito cells (U4.4 and Aag2) infected with SFV (Siu et al., 2011) and *D. melanogaster* S2 cells latently infected with FHV (Flynt et al., 2008). In the case of FHV, it was shown that the abundant vsiRNAs are not loaded efficiently into Ago2 (Flynt et al., 2008). It has been suggested that, in mosquitoes, the highly abundant vsiRNAs, which are not functional, may act as decoys to evade the host RNAi response to allow sufficient replication of the virus (Siu et al., 2011). On the other hand, vsiRNAs from the cold spots may balance and regulate replication of the virus so that it does not jeopardize the vector's survival.

Temperature, as an important environmental factor, may affect the efficiency of the RNAi response in mosquitoes to arboviruses. In *Ae. aegypti*, mosquitoes reared at 18°C exhibited lower RNAi response as compared to those reared at 28°C when transgenic sensor strains of the mosquito were tested (Adelman et al., 2013). As a consequence, mosquitoes reared at the cooler temperature had higher levels of infection when challenged with chikungunya (CHIKV) and yellow fever (YFV) viruses. This inefficiency in the RNAi response at the lower temperature was associated with the downstream events in the pathway, because vsiRNAs were produced normally in the mosquitoes.

2.1.3 Viral Suppressors of RNAi

To repress the host RNAi response, many plant and insect viruses encode viral suppressors of RNAi (VSRs) that inhibit the Dicer's or RISC's function or interfere with the loading of siRNAs into the RISC (Ding, 2010). While a defined and potent VSR has not yet been described from arboviruses, a number of studies have demonstrated VSR activities of a number of arboviruses. For example, a couple of studies showed that NS4B (Kakumani et al., 2013) and NS3 proteins from different serotypes of DENV suppress RNAi in human cells (Kakumani et al., 2015). These observations have not been confirmed in mosquitos. Further, the subgenomic flavivirus RNA (sfRNA) of WNV and DENV, produced by exoribonuclease XRN1 activity on genomic RNA, which mainly consists of the 3′-UTR of the viral genome, was shown to suppress the RNAi response in mosquito and mammalian cells (Schnettler et al., 2012). The VSR activity of sfRNA was further confirmed in *Culex quinquefasciatus* mosquitoes using the wild-type and an sfRNA-defective mutant of WNV, where knockdown of an endogenous chymotrypsin (CHYMO) was suppressed by the wild-type virus but not by the mutant virus (Moon et al., 2015). While the mechanism of RNAi suppression by sfRNA in mosquitoes remains unknown, it was shown in the same study that sfRNA from the virus interacts with Dicer and Ago2 proteins in human 293T cells using immunoprecipitation. An insect-specific entomobirnavirus, Culex Y virus (CYV), was shown to suppress the RNAi response in a *Culex tarsalis* cell line (CT) (van Cleef et al., 2014). In this case, the viral protein 3 (VP3) was shown to be the VSR, which was able to functionally replace the FHV B2 VSR. The VSR activity appears to be due to inhibition of Dicer-2 dicing of dsRNA by VP3, which has both dsRNA and siRNA binding activity.

2.1.4 Utilization of RNAi in Mosquitoes

RNAi has provided a powerful means for cell and organismal biology studies by specific silencing of target genes. Gene silencing using RNAi is particularly important when implementation of knockout mutants is not possible in model or nonmodel organisms. Mosquito research has also tremendously benefited from RNAi-mediated gene silencing. For this, target-specific long dsRNAs or siRNAs are synthesized *in vitro* and transfected into cells or injected into mosquitoes. Also, high throughput RNAi screening could be used to study host factors involved in replication of viruses. Although this has not been done in mosquitoes yet, studies of this type in *D. melanogaster* and human cells have resulted in the identification of insect factors involved in WNV (Yasunaga et al., 2014) and SINV (Panda et al., 2013) replication, and human factors in DENV replication (Barrows et al., 2014).

Pioneer studies in the 1990s showed that replication of arboviruses could be inhibited in mosquitoes or their cell lines by engineered SINV or mosquitoes themselves that produce dsRNAs specific to arbovirus sequences (Adelman et al., 2001; Gaines et al., 1996; Olson et al., 1996; Powers et al., 1994). This opened up avenues for exploitation of the RNAi response for arbovirus control by affecting the vector competence of mosquito vectors. In addition, RNAi has more recently been explored as an approach to control mosquito populations by targeting mosquito genes that affect reproduction (see later).

Suppression of arbovirus replication has been achieved in genetically modified (GM) mosquitoes expressing inverted repeat (IR) RNA (complementary to the target gene) or short-hairpin RNA (shRNA) to trigger the host RNAi response. For example, expression of an IR-RNA in *Ae. aegypti* midgut cells targeting the pre-membrane (prM) protein-coding region of DENV2 significantly inhibited replication of the virus in the mosquitoes (Franz et al., 2006). These mosquitoes were only refractory to different DENV2 genotypes but not other serotypes of the virus. However, the possibility of selecting regions that could inhibit all the four DENV serotypes exists (Korrapati et al., 2012). Similarly, expression of an IR-RNA (to DENV2 prM) in the salivary glands of *Ae. aegypti* mosquitoes using the *30K* gene driver led to significant reductions in DENV2 load in the gland and saliva (Mathur et al., 2010).

Stability of resistance to virus infection in GM mosquitoes is an issue of concern. In a laboratory setup, it was shown that while *Ae. aegypti* mosquitoes expressing an IR-RNA in the midgut triggers RNAi and suppresses DENV2 infection for 13 generations, this resistance diminished and was eventually lost by generation 17 (Franz et al., 2009). The loss of resistance was due to lack of expression of the IR-RNA, but not loss of the sequence from the genome. However, an identical line with two insertion sites was generated afterwards and shown to be stable and refractory to DENV2 for more than 33 generations (Franz et al., 2014). The homozygous DENV2 refractory line imposed minimal fitness costs on the GM mosquitoes providing proof of concept that the technology can be employed for the release of DENV incompetent mosquitoes in the field.

Sterile-insect technique (SIT) is a well-established approach in pest control, which involves the release of sterile males into the population that mate with wild female insects but no viable progeny is produced (Alphey et al., 2010). SIT can be achieved by radiation, chemical treatment or genetic modification of insects. However, there are disadvantages associated with each of those; for example, radiation or chemical treatment may affect the competitiveness of the treated insects and in the case of GM mosquitoes there are regulatory issues with their release in the wild (Helinski and Knols, 2008; Proverbs, 1969). Based on the fact that gene silencing has been successful by oral feeding of dsRNA to mosquitoes (Singh et al., 2013; Zhang et al., 2010), there have recently been developments in using RNAi for SIT by feeding dsRNA to mosquitoes. Whyard et al. fed *Escherichia coli* producing dsRNAs to *Ae. aegypti* larvae specific to a number of testis genes and a female sex determination gene (*doublesex*) to induce sterility in males and inhibition of development in females (Whyard et al., 2015). As a result, males were produced with significantly reduced fertility but still highly competitive. Feeding the ds-*doublesex* also led to the production of highly male-biased mosquito populations alleviating the need for sex-sorting insects before their release. The outcomes were very promising for insect-specific SIT, and this approach would get around concerns related to regulatory and public issues associated with GM mosquitoes. Using this technique, sterile males can be mass produced and released into the field or perhaps with costs associated with dsRNA production declining, they can be applied in certain focal areas that are highly populated by mosquitoes.

2.2 The piRNA Pathway: More Than Transposon Regulators

To date, from over 3000 mosquito species, the whole-genome sequences of only a few species are available. Among those are genome sequences of five medically important mosquitoes, *Ae. aegypti*, *Ae. albopictus*, *Anopheles darlingi*, *An. gambiae* and *C. quinquefasciatus* (Arensburger et al., 2010; Chen et al., 2015; Holt et al., 2002; Marinotti et al., 2013; Nene et al., 2007). Interestingly, large variations in their genome sizes exist, such as *Ae. albopictus* with 1967 Mb, *Ae. aegypti* 1376 Mb, *C. quinquefasciatus* 540 Mb, *An. gambiae* 278 Mb and *An. darlingi* with 174 Mb genome. These genome size differences could reflect various proportions of coding and noncoding transcripts as well as repetitive elements in these species.

In 2006, different research groups discovered piRNAs (25–32 nt) independently as a germline-specific, small RNA class from the mouse testes (Aravin et al., 2006; Girard et al., 2006; Lau et al., 2006). piRNA clusters are essential for detecting danger and their role is therefore highly conserved within animals (Izumi and Tomari, 2014; Siomi et al., 2011). piRNA clusters in the genome contain partial and complete transposon sequences, and they act as the memory banks for detecting TEs (Leslie, 2013). Therefore, they are at the forefront of the defence against transposons in germ cells. As described in Section 2 earlier, piRNA biogenesis is not yet fully understood, although they do not originate from hairpin-shaped precursors (different from miRNA) or dsRNAs (different from siRNA) and are Dicer independent. Of note, despite conservation of the piRNA pathway across organisms, the sequence identities of these small RNAs are weakly conserved (Lau et al., 2006). It is likely that sequences with similar length to piRNA could also be produced through other ncRNA pathways, degradation of mRNA or even background noise from next-generation sequencing platforms, which may affect the accuracy of piRNA discovery. Therefore, the lack of clear secondary structure and species conservation make the discovery and prediction of piRNAs especially challenging (Zhang et al., 2011).

Around 20% of the *Ae. aegypti* genome is occupied by piRNA clusters which produce 84% of piRNAs in this species (Arensburger et al., 2011). Low proportions of the predicted piRNA candidates in *Ae. aegypti* (19%) and *An. gambiae* (~40%) are transposon derived, while in *D. melanogaster* more than 50% of piRNAs originate from TEs (Arensburger et al., 2011; George et al., 2015). The majority of these TE linked piRNAs in *Ae. aegypti* map to long terminal repeat (LTR) retrotransposons. In the African malaria vector, *An. gambiae*, around 11% of the piRNA population map to protein-coding gene transcripts, which is noticeably higher than similar piRNA population (~6%) in *D. melanogaster* (George et al., 2015). Based on Gene Ontology (GO) analysis, genes involved in germline and embryonic development of *An. gambiae* generate a substantial portion of genic piRNAs (George et al., 2015). This suggests that piRNAs in mosquitoes may have other functions apart from TE silencing and probably involve in the epigenetic regulation of reproductive processes. Epigenetic activities and the ability to regulate gene expression or translation, similar to miRNAs, have also recently been reported as functions of piRNAs (Ross et al., 2014; Siddiqi and Matushansky, 2012).

Traditionally, piRNAs were associated with silencing TEs in germline cells, but a few studies, including some on insects, have described other possible functionalities in somatic cells for this group of ncRNAs (Arensburger et al., 2011; Kiuchi et al., 2014; Malone et al., 2009; Ross et al., 2014; Yan et al., 2011). A recent piRNA re-discovery analysis revealed that more than

55% of *D. melanogaster* somatic tissues' piRNA clusters (head and imaginal discs) overlapped with their previously reported ovarian piRNA clusters (Yan et al., 2011).

While the majority of piRNAs are transcribed from discrete genomic loci or particular intergenic repetitive elements on the genome known as piRNA clusters (Aravin et al., 2007), some studies have shown that a smaller portion of piRNAs match to a few protein-coding genes or their 3′-UTR (Arensburger et al., 2011; Robine et al., 2009). In *An. gambiae*, the majority of genic piRNAs do not exhibit the Ping-Pong sequence signatures and are 29 nt in length, while TE-associated piRNAs are almost all 26–27 nt in length, and most likely produced by a "Ping-Pong" dependent mechanism (Biryukova and Ye, 2015). In another study on the same mosquito, Castellano et al. (2015) also observed a similar result that many TEs and the 3′-UTR of some endogenous genes produce an abundant peak of 29 nt small RNAs with piRNA-like characteristics. They showed that the abundant 29 nt small RNAs are not exclusively genic piRNAs or expressed from the 3′-UTR of endogenous protein-coding genes. Interestingly, in all tissues, they express from the antisense strand of TEs such as *GYPSY12* and *COPIA3* (Castellano et al., 2015).

2.2.1 piRNAs and Host–Pathogen Interactions

piRNAs are understudied in host–pathogen interactions and their role in immunity is not clear; however, recent investigations have raised the possibility that antiviral activity might be linked to the piRNA pathway and PIWI proteins in mosquitoes (Chotkowski et al., 2008; Mayoral et al., 2014; Miesen et al., 2015, 2016; Schnettler et al., 2013; Scott et al., 2010; Vodovar et al., 2012; Wu et al., 2010). piRNA-like virus-derived small RNAs have been identified in mosquitoes or cell lines infected with arboviruses such as DENV (Hess et al., 2011), WNV, the bunyavirus La Crosse virus (LACV), SINV (Brackney et al., 2010), CHIKV (Morazzani et al., 2012), SFV (Schnettler et al., 2013) and RVFV (Léger et al., 2013). Based on the investigations cited above, the small RNAs in the size of piRNAs produced in DENV and WNV infected cells appear to lack the Ping-Pong signature in contrast to the ones from CHIKV, SFV, SINV, LACV and RVFV. Interestingly, unlike *Ae. albopictus* RNAi-deficient C6/36 cells, in *Culex pipiens* mosquitoes infected with WNV no evidence of viral piRNAs was found (Fros et al., 2015). Knockdown of the piRNA pathway proteins in *Ae. aegypti* Aag2 cells infected with SFV showed that among the proteins only depletion of Piwi4 increased the replication of this mosquito-borne virus (Schnettler et al., 2013). While Piwi4 is not essential for virus-specific piRNA-like production as shown for SFV (Schnettler et al., 2013) and SINV (Miesen et al., 2015), it might be used in targeting the virus, at least in the case of SFV (Schnettler et al., 2013). In C6/36 cells infected with SINV, RVFV or LACV, the majority of virus-derived small RNAs were of 27 nt with piRNA properties (Brackney et al., 2010; Léger et al., 2013; Vodovar et al., 2012). Therefore, the piRNA pathway may compensate the absence of the exo-siRNA pathway as an antiviral response. The population of virus-derived small RNAs may also change depending on the state of infection. For example, in Aag2 cells infected with RVFV, during the acute infection, the majority of the virus-derived small RNAs were 21 nt (siRNA size), whereas during the persistent infection it was mainly 24–27 nt (piRNA size) (Léger et al., 2013). Similar to the role of some miRNAs (Wu et al., 2011), piRNAs could also potentially play a role in the establishment of persistent infection.

Vodovar et al. (2012) clearly showed that arboviruses use similar approaches as transposons to activate the piRNA and the siRNA pathways in mosquitoes. They concluded that the usual characteristics of the parasitic nucleic acids, rather than their origin, might only be distinguishable by the RNAi pathways (Vodovar et al., 2012). Recently, it has been shown that distinct sets of PIWI proteins (Piwi5 and Ago3) are required for piRNA biogenesis from arboviral (SINV) precursors compared with endogenous transposon precursors in *Ae. aegypti* (Miesen et al., 2015). Similarly, depletion of Piwi5 and Ago3 in Aag2 cells led to undetectable levels of DENV-2 piRNAs (Miesen et al., 2016). Interestingly, knockdown of the siRNA or the piRNA pathway had no effect on replication of DENV-2 in Aag2 cells. In another study, the critical role of PIWI-related small RNA regulatory pathway proteins for the production of piRNA-like viral small RNAs, with an antiviral activity, in DENV-infected mosquitoes was identified (Hess et al., 2011). Therefore, it appears that viral nucleic acids are necessary to determine their recognition for processing into piRNA-like molecules. However, our knowledge about host piRNA response to these types of viral infections is very limited.

In *Ae. aegypti* Aag2 cells, differential expression of a number of piRNAs in response to *Wolbachia* infection was observed (Mayoral et al., 2014). It was speculated that changes in *PIWI* genes transcription level or interruption of PIWI proteins trafficking in the presence of *Wolbachia* could be responsible for piRNA profile modifications. Consistent with this assumption, an overall increase in the transcription levels of Ago1, 2 and 3 in *Wolbachia*-infected

Aag2 cells was recorded, which may also explain the presence of more small RNA reads in those samples (Mayoral et al., 2014). Further, another recent study indicated that *Wolbachia* blocks the distribution of Ago1 to the nucleus, which controls small RNA trafficking from the cytoplasm to the nucleus (Hussain et al., 2013a). The results suggested that microbial challenges, such as *Wolbachia* infection, could also modify the expression of endogenous piRNAs. Blood feeding also induced the expression of Ago3, Piwi and Aubergine in *Anopheles stephensi* (Maclas et al., 2014). While the exact role of piRNAs in mosquito-pathogen interaction remains largely unknown, evidence gathered so far strongly suggest their involvement in responses to infection.

2.3 The miRNA Pathway and Its Discovery in Mosquitoes

Since the first paper appeared in 2007 on identification of miRNAs in *Anopheles* mosquitoes, several novel and conserved miRNAs have been found in other species as well (Li et al., 2009; Mead and Tu, 2008; Winter et al., 2007). Previously, 55 miRNAs were identified in *An. gambiae* and *An. stephensi* through direct cloning and bioinformatics approaches (Mead and Tu, 2008). Later, in a separate study aimed at different developmental stages of *An. gambiae*, a total of 117 miRNAs were found through small RNA sequencing including 14 novel ones (Hu et al., 2015). In *Ae. aegypti* embryo and midgut, 86 miRNAs were identified through 454 sequencing (Li et al., 2009). Further attempts exceeded the number of miRNAs to 120 in *Ae. aegypti* (Hu et al., 2015). In another study, 77 miRNAs were found in blood fed *C. quinquefasciatus* mosquitoes and 65 miRNAs in *Ae. albopictus* C7/10 cell line (Skalsky et al., 2010). Recent RNA sequencing conducted in *Ae. albopictus* samples collected at diverse developmental stages identified 119 miRNA genes producing 215 miRNAs (Gu et al., 2013). In these studies, mainly RNA sequencing coupled with bioinformatics was used to generate the miRNA profiles and still large numbers of them have not been validated with Northern blot analysis or miRNA-specific RT-PCR.

In all the mosquito miRNA discovery studies, most of the miRNAs have been found conserved in diverse mosquito species and other insect species, while a few are species specific. Among all mosquito miRNAs, at least 25 appear to be mosquito specific when compared with four *Drosophila* genomes (*D. melanogaster*, *D. ananassae*, *D. pseudoobscura*, *D. grimshawi*) (Gu et al., 2013). These mosquito-specific miRNAs may provide insights into the molecular basis of mosquito-specific adaptations and biology. Several studies have shown global changes in the expression levels of mosquito miRNAs in response to viral, *Plasmodium* or bacterial infections, and blood feeding and developmental changes; however, the function and target genes for a very few of them have experimentally been validated.

2.3.1 Functional Significance of Mosquito miRNAs

During the last 2–3 years, several review articles have covered the role of miRNAs in insect biology (Asgari, 2013, 2014, 2015; Hussain and Asgari, 2014; Lucas and Raikhel, 2013; Lucas et al., 2013, 2015b). Here, we discuss published work on functional aspects of some of the mosquito miRNAs experimentally shown to be involved in various biological processes such as blood digestion, egg development, metamorphosis and defence against pathogens (summarized in Fig. 2).

2.3.1.1 miRNAs in Blood Digestion, Reproductive Development and Lifespan

2.3.1.1.1 miR-1174, the Gut-Specific miRNA, Controls Blood Digestion Recently, miR-1174 and miR-1175 were reported to be upregulated in the midguts of *Ae. aegypti* and *An. gambiae* post blood meal (PBM) (Liu et al., 2014). An earlier study also showed upregulation of

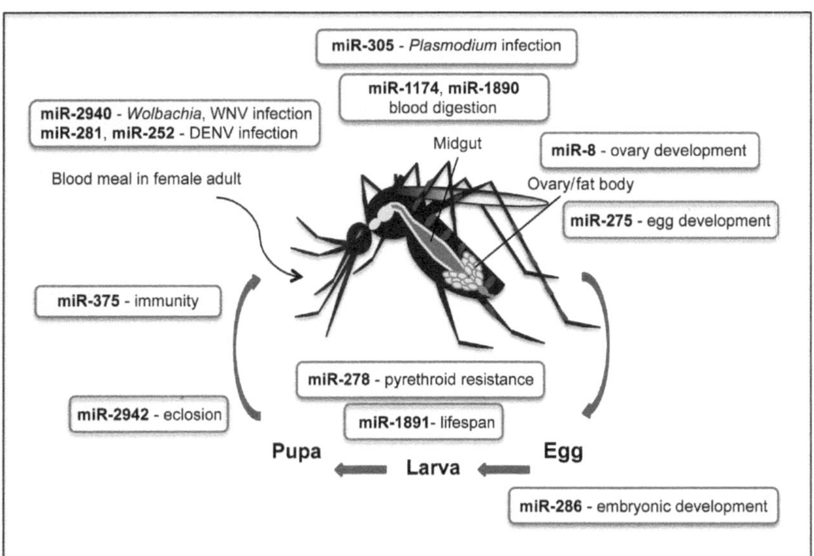

Fig. 2 A summary of miRNAs involved in various processes in mosquito biology. Refer to text for details. (See the colour plate.)

the miR-1174/miR-1175 cluster PBM in *An. gambiae* (Winter et al., 2007). Both of these miRNAs are located at the same locus with a gap of only 83 nucleotides between them, and hence their expression is co-regulated. *In situ* hybridization showed that miR-1174 is localized in the posterior midgut suggesting its potential role in blood digestion (Liu et al., 2014). In miR-1174 depleted *Ae. aegypti* or *An. gambiae*, bloated abdomen phenotype (BAP) was observed that made them unable to fly and feed on blood. Other noticeable changes in these mosquitoes were severe defects in blood feeding, less sugar absorption and fluid excretion that finally affected egg development. Interestingly, inhibition of miR-1175 did not produce such phenotypes in mosquitoes, thus making miR-1174 knockdown effect very specific and authentic. In this report, the mechanism of miR-1174 depletion effect on phenotype was further dissected through working on its possible target gene(s). It was shown through *in vitro* as well as *in vivo* analyses that miR-1174 specifically targets *serine hydroxymethyltransferase* (*SHMT*) in its 3′-UTR and downregulates its transcript levels (Liu et al., 2014). Moreover, RNAi silencing of *SHMT* in mosquitoes led to similar phenotypes as observed in miR-1174 depleted ones.

2.3.1.1.2 miR-1890 and Blood Digestion, Regulating a Serine Protease In *Ae. aegypti*, similar to other haematophagous mosquitoes, there is an established link between vertebrate blood meal and egg development. Serine proteases (SP) are of prime importance in this whole process of blood digestion (Isoe et al., 2009). There are two major types of SPs in blood fed midguts: trypsin-like and chymotrypsin-like SPs (reviewed in Isoe et al., 2009). Interestingly, transcripts of the *early phase trypsin* (*AaET*) gene are present in the midgut of *Ae. aegypti* before blood feeding, which are then translated into protein following blood feeding (Isoe et al., 2009). The expression of *AaET* gene is controlled by Juvenile hormone (JH) (Noriega et al., 1997). JHA15, a chymotrypsin-like SP, is also present in the midgut before blood feeding similar to AaET and its expression is also controlled by JH, but unlike AeET, JHA15 protein can be detected before and after blood feeding (Bian et al., 2008). In female mosquitoes, JHA15 has high expression levels after eclosion and drops significantly following blood feeding (Lucas et al., 2015c). This differential expression of JHA15 led researchers to look if miRNA could be involved in its regulation. Target sites in the 3′-UTR of *JHA15* were found for several miRNAs, but because of specificity and greater complementarity miR-1890 was selected for further analysis (Lucas et al., 2015c). In PBM mosquitoes, miR-1890 level was highest at 24 h and drastically decreased at 36 h, which showed an opposite trend to JHA15 levels. The specific interaction of miR-1890 with JHA15 was confirmed in vitro and further, upregulation of JHA15 in miR-1890 depleted mosquitoes was observed at 24 h PBM. Inhibition of miR-1890 in mosquitoes also led to abnormal blood digestion that was later rescued after RNAi silencing of *JHA15* in another experiment. Expression level of miR-1890 in PBM mosquitoes on itself was an interesting question, which was examined in the same study. The investigators found that 20-hydroxyecdysone (20E), but not insulin, regulates miR-1890 expression most likely via binding of E74B protein (one of the genes of 20E genetic hierarchy) to the promoter region of the miR-1890 gene. This study further demonstrated the importance of miRNAs in mosquito blood digestion.

2.3.1.1.3 miR-8 in Regulation of Reproductive Development miR-8 has been one of the well-studied miRNAs associated with animal development and disease. In *Drosophila* and cockroach, miR-8 is involved in neuromuscular development and proper motor coordination (Karres et al., 2007; Rubio et al., 2013). Other studies have shown that miR-8 regulates body size through its interaction with the target U-shaped, and also regulates innate immune homeostasis and notch signalling in *Drosophila* (Choi and Hyun, 2012; Etebari and Asgari, 2013; Jin et al., 2012; Lee and Hyun, 2014). In *Ae. aegypti*, miR-8 was found highly induced in the fat body cells following blood feeding (Bryant et al., 2010), which led researchers to examine its involvement in mosquito physiology by blocking it in mosquitoes. Microinjection of miR-8 antagomir in female mosquitoes resulted in abnormal vitellogenic ovaries during the first gonadotrophic cycle (Lucas et al., 2015a). The average length of primary follicle in the ovary of PBM mosquitoes was 200–210 μm that reduced to 154.7 μm in miR-8 depleted mosquitoes with significant effect on fecundity. A miR-8 depleted mosquito could only lay 36.3 eggs per female on average as compared to 92.8 in a control mosquito. In the same study, a *Drosophila* orthologue of *secreted wingless-interacting molecule* "*swim*" gene, associated with long-range wingless signalling, was also identified in *Ae. aegypti* as a target for miR-8. The specific/direct interaction of aaeSwim with miR-8 was confirmed in vitro (luciferase system) as well as in vivo (vg-Gal4/UAS system). Further, the investigators found that miR-8 depletion did not affect expression of vitellogenin and lipophorin (LP) at the transcript or protein level, but disrupted secretion of the proteins by the fat body and reduced lipid accumulation in developing oocytes (Lucas et al., 2015a). Interestingly, miR-8 depletion did not affect blood digestion, previtellogenic ovary development and host-seeking behaviours of female mosquitoes.

2.3.1.1.4 miR-275 is Essential for Blood Digestion and Egg Development In mosquitoes, blood feeding activates vitellogenesis, which is central to egg maturation (Attardo et al., 2005). Since the fat body is the main site of production of yolk protein precursors, involvement of miRNAs in vitellogenesis in the fat body of *Ae. aegypti* was examined by exploring differential expression of a number of conserved miRNAs in the tissue during vitellogenesis (Bryant et al., 2010). Along with two other miRNAs (miR-8 and miR-14), miR-275 was one of the miRNAs found highly induced during the process. Further analysis concluded that miR-275 expression was upregulated under the control of 20-hydroxyecdysone (20E) and amino acid/target of rapamycin (AA/TOR) pathways (Bryant et al., 2010). Promising functional changes were observed in miR-275 depleted mosquitoes such as problems in intake and digestion of blood, and inhibition of egg development. In addition, anatomical abnormalities, such as incomplete peritrophic matrix, muscle weakening in the anterior midgut and inefficient apoptosis of nurse cells were also observed in these mosquitoes. Although miR-275 inhibition in female mosquitoes produced drastic disruptions in blood digestion, fluid excretion and egg development, no target gene(s) for miR-275 was determined to find the basic mechanism for this miRNA's functioning (Bryant et al., 2010).

2.3.1.1.5 miR-286 in Embryogenesis In *Ae. albopictus*, miR-286b was found to be embryo-specific with a very high copy number of 18703 that reduces drastically with development to larval (9 copies) or pupal (1 copy) stages (Puthiyakunnon et al., 2013). The same study highlighted its role in hatching of embryos in terms of both time and number. Inhibition of miR-286 resulted in a decrease in the number of embryos hatched as well as delay in the process of hatching. Interestingly, in *Drosophila*, miR-286 is one of the members of the miR-309~6 cluster that encodes a polycistronic transcript to make eight miRNAs (Biemar et al., 2005).

2.3.1.1.6 miR-2942 Controls Eclosion miR-2942 was found to be important during the development of *Ae. albopictus* (Puthiyakunnon et al., 2013). This miRNA is upregulated in larvae with a huge increase in copy number from 23 in embryo to 3044 specifically in larvae, and it is downregulated at the pupal stage. A significant reduction in the eclosion rate at the larval stage was observed in miR-2942 depleted *Ae. albopictus* by using microinjection of the miRNA inhibitor. No target gene for miR-2942 has been confirmed, therefore, whether this miRNA has a direct or indirect role in eclosion is yet to be investigated.

2.3.1.1.7 miR-1891 in Lifespan Extreme increase in the abundance of miR-1891 in adults of *Ae. albopictus* suggested its possible importance in lifespan (Puthiyakunnon et al., 2013). Copy number for miR-1891 increased from six in embryo to 7658 in male and 12518 in female mosquitoes. Depletion of miR-1891 in both male and female *Ae. albopictus* resulted in marked decrease in lifespan. In males, the mean lifespan of 16.5 days decreased to 11.33 days, while in females from 24.75 to 13.35 days in miR-1891 inhibitor injected mosquitoes. Inhibition of the miRNA also affected fecundity as it was significantly reduced in females.

2.3.1.2 miRNAs in the Regulation of Defence Responses to Foreign Invaders and Chemicals

2.3.1.2.1 miR-305 Helps Plasmodium Infection The malarial mosquito, *An. gambiae* exhibits defence against the *Plasmodium* parasite via Toll and IMD pathways (Clayton et al., 2014). *Plasmodium* entry into the midgut elicits transcriptional changes of immune genes in mosquitoes providing the basis for defence. An interesting study showed involvement of a cellular miRNA in susceptibility of *An. gambiae* to the *Plasmodium* parasite (Dennison et al., 2015). Initially, the investigators observed a general effect on miRNA levels by knockdown of Dicer-1 and Ago1 through RNAi in mosquitoes prior to *Plasmodium* infection that resulted in resistance against *Plasmodium falciparum*. Two miRNAs, miR-989 and miR-305, were highly upregulated upon *P. falciparum* infection in the midgut. Further, the investigators found that miR-305, but not miR-989, helps in establishing *P. falciparum* infection, as its inhibition by using the inhibitor of miR-305 enhanced resistance, while supplying the miR-305 mimic increased susceptibility. Notably, the role of miR-305 was also observed in controlling the midgut microbiota because in *An. gambiae*, anti-*Plasmodium* and antibacterial defences are commonly regulated by the IMD pathway (Dennison et al., 2015). The microbiota load in the midgut of a single mosquito decreased by fivefold after miR-305 depletion, whereas application of the mimic of miR-305 led to an increase in the midgut microbiota. Dennison et al. identified the anti-*Plasmodium* effector genes *APL1C* and immune genes, *PGRP-LB* and *PGRP-LC*, as potential targets of miR-305 through in silico prediction, which await experimental validation.

A couple of other research publications have shown changes in *Anopheles* species miRNA expression in response to *Plasmodium* infection that were distinct from the ones changed upon blood feeding. Jain et al. found relatively less number of upregulated miRNAs in *An. stephensi* as compared to blood feeding, signifying general downregulation of miRNAs upon *Plasmodium* infection (Jain et al., 2014). Another study has shown induced expression of miR-317 and miR-2940 in response to *Plasmodium* infection in *An. gambiae* (Biryukova and Ye, 2015). Keeping in view the global threat of malaria, further research on miRNA interaction with *Plasmodium* infection is of paramount importance. Recent investigations on miRNAs, piRNAs and other small RNAs produced in the germline of *An. gambiae* concluded that large numbers of miRNAs are upregulated in the gonads during gametogenesis (Castellano et al., 2015). The promising ones were miR-2944a-2, miR-286b, miR-989 and miR-13061b. Among these, miR-989 has already been shown to express in the ovaries of *An. stephensi* and *Ae. aegypti*, and is required for oogenesis in *Drosophila* (Kugler et al., 2013; Mead and Tu, 2008).

2.3.1.2.2 miR-375 in Immunity miR-375 is highly induced upon blood feeding in *Ae. aegypti* and regulates immune-related genes, *cactus* and *REL1* (Hussain et al., 2013b). Moreover, the miR-375 mimic in Aag2 cell line was found to enhance DENV infection via upregulation of cactus, which inhibits activation of the NF-κB transcription factor. In a separate study, silencing *cactus* led to the attenuation of DENV-2 infection in *Ae. aegypti* mosquitoes (Xi et al., 2008). Recently, another study showed significant decreases in CHIKV titre in Aag2 and BHK-21 cells transfected with the inhibitor of miR-375, which supported its association with virus infection (Maharaj et al., 2015). miR-375 was also found to be upregulated during metamorphosis transition of pupa to adult in female *An. stephensi*, but not in males (Jain et al., 2014).

2.3.1.2.3 miR-2940 Associates with Wolbachia and Arbovirus Infection On investigating the connection of mosquito miRNAs with the endosymbiotic bacterium, *Wolbachia w*Melpop strain, transinfected into *Ae. aegypti*, miR-2940-5p was found important in maintaining *Wolbachia* density (Hussain et al., 2011). miR-2940-5p is a mosquito-specific miRNA and its abundance is greatly induced in *w*Melpop-infected mosquitoes. In three separate studies, functional analyses of *Ae. aegypti* target genes regulated by miR-2940-5p were reported (Hussain et al., 2011; Zhang et al., 2013, 2014). First, it was found that miR-2940-5p targets the 3′-UTR region of *metalloprotease m41 ftsh* (*MetP*) transcripts and enhances its transcript levels (Hussain et al., 2011). Interestingly, either inhibition of miR-2940-5p or RNAi silencing of *MetP* in both *Wolbachia*-infected cells and adult mosquitoes led to a significant reduction in *Wolbachia* density (Hussain et al., 2011). Second, miR-2940-5p was found to downregulate the expression of *the DNA (cytosine-5) methyltransferase* (*AaDnmt2*) (Zhang et al., 2013). Ectopic expression of AaDnmt2 in mosquito cells led to reduced *Wolbachia* density, but promoted replication of DENV-2. Third, miR-2940-5p regulation of *arginine methyltransferase 3* (*AeArgM3*) had an effect on *Wolbachia* density, but not on DENV-2 replication (Zhang et al., 2014). Silencing of *AeArgM3* led to significant reduction in *Wolbachia* density. In addition to miR-2940-5p, miR-12 was shown to be involved in *Wolbachia* maintenance by regulating *DNA replication licensing* (*MCM6*) and *monocarboxylate transporter* (*MCT1*) genes (Osei-Amo et al., 2012). Inhibition of miR-12 led to reduced *Wolbachia* density.

Another study showed that in response to WNV infection, miR-2940-5p is downregulated in *Ae. albopictus* C6/36 cells (Slonchak et al., 2014). miR-2940-5p was found to be required for efficient WNV replication by regulating MetP. In addition, two other papers also showed downregulation of miR-2940-5p and miR-2940-3p in CHIKV and DENV-infected *Ae. albopictus* cells (Liu et al., 2015; Shrinet et al., 2014). Both miR-2940-5p and miR-2940-3p are highly expressed miRNAs and possibly target several other genes that demands more investigation into their role in mosquito biology.

2.3.1.2.4 miR-981 Regulates Importin Translocation of Ago1 from the cytoplasm to the nucleus in *Ae. aegypti* cells and mosquitoes infected with *Wolbachia w*MelPop is reduced (Hussain et al., 2013a). RNA silencing of all the importin genes (involved in the cytoplasm to the nucleus trafficking) revealed that importin β-4 was one of the two importins that facilitates Ago1 entry to the nucleus. Interestingly, the expression level of importin β-4 protein was significantly lower in *w*MelPop-infected cells or mosquitoes as compared to uninfected ones. Downregulation of importin β-4 was found to be under the control of one of the *Wolbachia*-induced miRNAs, miR-981. A follow-up bioinformatics study revealed changes in the distribution of miRNAs in the cytoplasm and the nucleus of Aag2 cells upon *Wolbachia w*MelPop infection (Mayoral et al., 2014).

2.3.1.2.5 miR-278 in Pyrethroid Resistance For the chemical control of *Culex pipiens* mosquitoes, pyrethroid insecticides have been extensively used worldwide (Scott et al., 2015). Excessive use of the pyrethroid has led to the development of resistance, with resistance levels of up to 7000-fold found in the larvae of field-collected mosquitoes. Research conducted by different groups has concluded two major mechanisms of pyrethroid resistance in *C. pipiens*, including target site insensitivity (mutations in Vssc) and overexpression of P450 that is related to enhanced detoxification (Scott et al., 2015). In order to find out the role of miRNAs in pyrethroid resistance in *C. pipiens*, total miRNA profile was created through Illumina sequencing of deltamethrin-sensitive (DS) and deltamethrin-resistant (DR) strains (Hong et al., 2014). 28 miRNAs were differentially expressed in between the two strains. In this study, miR-71 level significantly declined in adult females of the DR-strain suggesting the potential involvement of miR-71 in deltamethrin resistance. Another study from the same group based on these sequencing data also highlighted the importance of miR-278-3p in pyrethroid resistance, which was upregulated in the DS strain (Lei et al., 2015). The investigators found a target of miR-278-3p in the 3′-UTR of one of the P450 genes, *CYP6AG11*. Initially, they confirmed the contrasting expression of both miRNA and the target genes in the two strains. In the DR-strain, the abundance of miR-278-3p was 4.2-fold lower, and that of *CYP6AG11* 13.5-fold higher as compared with the DS strain. They further conducted experiments to confirm the specific interaction of miR-278-3p with *CYP6AG11* through luciferase reporter as well as ectopic expression construct in C6/36 cells. Transfection of miR-278-3p mimic downregulated *CYP6AG11* transcript levels, thus increasing the susceptibility to deltamethrin in cells, while overexpression of *CYP6AG11* by transfection of mimics into C6/36 cells provided deltamethrin resistance (Lei et al., 2015). Similarly, microinjection of miR-278-3p mimic in mosquitoes downregulated *CYP6AG11* by 1.3-fold confirming miRNA–mRNA interaction in vivo; although this effect is quite marginal. In two different field populations of *C. pipiens* mosquitoes, higher abundance of miR-278-3p (2.1- and 1.7-fold) was found in the DS-strains as compared to their corresponding DR-strains, which led the authors to propose the idea of using this miRNA as a biomarker for detecting resistance to the insecticide.

2.3.1.2.6 Antagonistic and Protagonist miRNAs to DENV One of the best characterized examples of interaction of a cellular miRNA with a virus is the binding of liver specific miR-122 to the 5′-UTR of Hepatitis C virus, resulting in enhanced virus replication (Shimakami et al., 2012). A similar kind of work was published in *Ae. albopictus* in which a midgut specific miR-281 was found to be upregulated upon DENV infection (Zhou et al., 2014). Interestingly, miR-281 mimic and inhibitor significantly increased and decreased DENV gRNA levels, respectively, in C6/36 cells as well as mosquitoes. These results showed that miR-281 facilitates DENV replication and infection. The authors further found that miR-281 binds to the 5′-UTR of DENV and validated the specific interaction in C6/36 cells using an EGFP reporter vector system.

To date, miR-252 is the only mosquito miRNA reported to have a direct interaction with the viral genomic RNA within the ORF (Yan et al., 2014). In C6/36 cells, it was shown that miR-252's abundance increased threefold in response to DENV-2 infection. In this work, miR-252 was suggested to target and downregulate DENV E (envelope) protein resulting in negative impact on DENV replication. By utilizing miR-252 inhibitor and mimic, increase and decrease in DENV genomic RNA copies and E protein were found, respectively, suggesting this miRNA's role as antiviral. However, the authors did not examine the effect of miR-252 on other viral proteins and looked at *E protein* as an isolated gene. Considering that DENV genomic RNA has one open reading frame coding for a polyprotein that is subsequently processed into 10 structural and nonstructural proteins, if miR-252 targets the E protein-coding sequence, it should basically affect the whole reading frame (essentially the viral genome) either at the transcript or translational level. Recently, an interesting study revealed specific degradation of miR-252 by a conserved exoribonuclease, Dis3 in *Drosophila* (Towler et al., 2015). Silencing of *Dis3* led to increase in miR-252 abundance and produced wingless phenotype as wing imaginal disc growth was disrupted. To find out the mechanism of miR-252 induction in DENV-infected cells, it might be worth examining if Dis3 level is changed in these cells.

3. CONCLUSIONS

In the 21st century, mosquitoes continue to be medically important in regards to transmission of diseases. While there have been successful and promising stories in vector and vector-borne disease control, they still pose great risks to human and animal health worldwide necessitating continuous efforts in devising novel approaches in mosquito and disease control. This can only be achieved by sustained basic and applied research on the biology of mosquito vectors and their interaction with the pathogens they carry. Small noncoding RNAs have become an important component of research in biological studies due to their key roles in gene regulation and involvement in various biological processes. In addition, sncRNAs are used as powerful tools in the study of gene functions (eg, for gene silencing), but also potentially in vector and disease control. For example, application of RNAi in sterile insect technique has proven promising for suppressing vector populations of dengue virus, leading to a reduction in the transmission of the virus, and potentially Zika virus where they coincide. A number of miRNAs have been shown to play important roles in mosquito egg development, blood digestion, eclosion and host–pathogen interactions, which could potentially be utilized in disrupting mosquitoes' life cycle or replication of pathogens. Further, piRNAs and miRNAs are suggested to be involved in establishing persistent virus infection in mosquitoes; therefore, they could be good targets for intervention as a persistent viral infection of mosquitoes is crucial for successful transmission of arboviruses to their mammalian hosts. While the release of genetically modified mosquitoes remains an obstacle in translation of these studies, it should not discourage basic research on better understanding the role of sncRNAs in mosquito biology and proof-of-concept investigations to demonstrate their potential for exploitation in vector and disease control.

ACKNOWLEDGEMENTS

The authors would like to thank Dr Esther Schnettler from the University of Glasgow for her comments and critical reading of the chapter. Work in Asgari's laboratory has been supported by the Australian Research Council (DP110102112, DP150101782, DE120101512) and National Health and Medical Research (APP1062983, APP1027110).

REFERENCES

Adelman, Z.N., Blair, C.D., Carlson, J.O., Beaty, B.J., Olson, K.E., 2001. Sindbis virus-induced silencing of dengue viruses in mosquitoes. Insect Mol. Biol. 10, 265–273.

Adelman, Z.N., Anderson, M.A., Morazzani, E.M., Myles, K.M., 2008. A transgenic sensor strain for monitoring the RNAi pathway in the yellow fever mosquito, *Aedes aegypti*. Insect Biochem. Mol. Biol. 38, 705–713.

Adelman, Z., Anderson, M., Liu, M., Zhang, L., Myles, K., 2012. Sindbis virus induces the production of a novel class of endogenous siRNAs in *Aedes aegypti* mosquitoes. Insect Mol. Biol. 21, 357–368.

Adelman, Z.N., Anderson, M.A., Wiley, M.R., Murreddu, M.G., Samuel, G.H., Morazzani, E.M., Myles, K.M., 2013. Cooler temperatures destabilize RNA interference and increase susceptibility of disease vector mosquitoes to viral infection. PLoS Negl. Trop. Dis. 30, e2239.

Alphey, L., Benedict, M., Bellini, R., Clark, G., Dame, D., Service, M., Dobson, S., 2010. Sterile-insect methods for control of mosquito-borne diseases: an analysis. Vector Borne Zoonotic Dis. 10, 295–311.

Aravin, A., Gaidatzis, D., Pfeffer, S., Lagos-Quintana, M., Landgraf, P., Iovino, N., Morris, P., Brownstein, M.J., Kuramochi-Miyagawa, S., Nakano, T., Chien, M.C., Russo, J.J., Ju, J.Y., Sheridan, R., Sander, C., Zavolan, M., Tuschl, T., 2006. A novel class of small RNAs bind to MILI protein in mouse testes. Nature 442, 203–207.

Aravin, A.A., Sachidanandam, R., Girard, A., Fejes-Toth, K., Hannon, G.J., 2007. Developmentally regulated piRNA clusters implicate MILI in transposon control. Science 316, 744–747.

Arensburger, P., Megy, K., Waterhouse, R.M., Abrudan, J., Amedeo, P., Antelo, B., Bartholomay, L., Bidwell, S., Caler, E., Camara, F., Campbell, C.L., Campbell, K.S., Casola, C., Castro, M.T., Chandramouliswaran, I., Chapman, S.a.B., Christley, S., Costas, J., Eisenstadt, E., Feshotte, C., Fraser-Liggett, C., Guigo, R., Haas, B., Hammond, M., Hansson, B.S., Hemingway, J., Hill, S., Howarth, C., Ignell, R., Kennedy, R.C., Kodira, C.D., Lobo, N.F., Mao, C., Mayhew, G., Michel, K., Mori, A., Liu, N., Naveira, H., Nene, V., Nguyen, N., Pearson, M.D., Pritham, E.J., Puiu, D., Qi, Y., Ranson, H., Ribeiro, J.M.C., Roberston, H.M., Severson, D.W., Shumway, M., Stanke, M., Strausberg, R., Sun, C., Sutton, G., Tu, Z., Tubio, J.M.C., Unger, M.F., Vanlandingham, D.L., Vilella, A.J., White, O., White, J.R., Wondji, C.S., Wortman, J., Zdobnov, E.M., Birren, B., Christensen, B.M., Collins, F.H., Cornel, A., Dimopoulos, G., Hannick, L.I., Higgs, S., Lanzaro, G.C., Lawson, D., Lee, N.H., Muskavitch, M., Raikhel, A.S., Atkinson, P.W., 2010. Sequencing of *Culex quinquefasciatus* establishes a platform for mosquito comparative genomics. Science 330, 86–88.

Arensburger, P., Hice, R.H., Wright, J.A., Craig, N.L., Atkinson, P.W., 2011. The mosquito *Aedes aegypti* has a large genome size and high transposable element load but contains a low proportion of transposon-specific piRNAs. BMC Genomics 12, 606.

Asgari, S., 2013. MicroRNA functions in insects. Insect Biochem. Mol. Biol. 43, 388–397.

Asgari, S., 2014. Role of microRNAs in arbovirus/vector interactions. Viruses 6, 3514–3534.

Asgari, S., 2015. Regulatory role of cellular and viral microRNAs in insect-virus interactions. Curr. Opin. Insect Sci. 8, 104–110.

Attardo, G., Hansen, I., Raikhel, A., 2005. Nutritional regulation of vitellogenesis in mosquitoes: implications for anautogeny. Insect Biochem. Mol. Biol. 35, 661–675.

Attarzadeh-Yazdi, G., Fragkoudis, R., Chi, Y., Siu, R., Ulper, L., Barry, G., Rodriguez-Andres, J., Nash, A., Bouloy, M., Merits, A., Fazakerley, J., Kohl, A., 2009. Cell-to-cell spread of the RNA interference response suppresses Semliki Forest virus (SFV) infection of mosquito cell cultures and cannot be antagonized by SFV. J. Virol. 83, 5735–5748.

Barrows, N., Jamison, S., Bradrick, S., Le Sommer, C., Kim, S., Pearson, J., Ma, G.-B., 2014. Functional genomics approach for the identification of human host factors supporting dengue viral propagation. Methods Mol. Biol. 1138, 285–299.

Bartel, D.P., 2009. MicroRNAs: target recognition and regulatory functions. Cell 23, 215–233.

Bernhardt, S., Simmons, M., Olson, K., Beaty, B., Blair, C., Black, W., 2012. Rapid intraspecific evolution of miRNA and siRNA genes in the mosquito *Aedes aegypti*. PLoS One 7, e44198.

Bian, G., Raikhel, A.S., Zhu, J., 2008. Characterization of a juvenile hormone-regulated chymotrypsin-like serine protease gene in *Aedes aegypti* mosquito. Insect Biochem. Mol. Biol. 38, 190–200.

Biemar, F., Zinzen, R., Ronshaugen, M., Sementchenko, V., Manak, J., Levine, M., 2005. Spatial regulation of microRNA gene expression in the *Drosophila* embryo. Proc. Natl. Acad. Sci. U.S.A. 102, 15907–15911.

Biryukova, I., Ye, T., 2015. Endogenous siRNAs and piRNAs derived from transposable elements and genes in the malaria vector mosquito *Anopheles gambiae*. BMC Genomics 16, 278.

Blair, C., Olson, K., 2015. The role of RNA interference (RNAi) in arbovirus-vector interactions. Viruses 7, 820–843.

Brackney, D., Beane, J., Ebel, G., 2009. RNAi targeting of West Nile virus in mosquito midguts promotes virus diversification. PLoS Pathog. 5, e1000502.

Brackney, D.E., Scott, J.C., Sagawa, F., Woodward, J.E., Miller, N.A., Schilkey, F.D., Mudge, J., Wilusz, J., Olson, K.E., Blair, C.D., Ebel, G.D., 2010. C6/36 *Aedes albopictus* cells have a dysfunctional antiviral RNA interference response. PLoS Negl. Trop. Dis. 4, e856.

Bronkhorst, A., Van Rij, R., 2014. The long and short of antiviral defense: small RNA-based immunity in insects. Curr. Opin. Virol. 7, 19–28.

Bryant, B., Macdonad, W., Raikhel, A.S., 2010. microRNA miR-275 is indispensable for blood digestion and egg development in the mosquito *Aedes aegypti*. Proc. Natl. Acad. Sci. U.S.A. 107, 22391–22398.

Campbell, C., Keene, K., Brackney, D., Olson, K., Blair, C., Wilusz, J., Foy, B., 2008a. *Aedes aegypti* uses RNA interference in defense against Sindbis virus infection. BMC Microbiol. 8, 47.

Campbell, C.L., Black Iv, W.C., Hess, A.M., Foy, B.D., 2008b. Comparative genomics of small RNA regulatory pathway components in vector mosquitoes. BMC Genomics 9, 425.

Carissimo, G., Pondeville, E., Mcfarlane, M., Dietrich, I., Mitri, C., Bischoff, E., Antoniewski, C., Bourgouin, C., Failloux, A., Kohl, A., Vernick, K., 2015. Antiviral immunity of *Anopheles gambiae* is highly compartmentalized, with distinct roles for RNA interference and gut microbiota. Proc. Natl. Acad. Sci. U.S.A. 112, E176–E185.

Castellano, L., Rizzi, E., Krell, J., Di Cristina, M., Galizi, R., Mori, A., Tam, J., De Bellis, G., Stebbing, J., Crisanti, A., Nolan, T., 2015. The germline of the malaria mosquito produces abundant miRNAs, endo-siRNAs, piRNAs and 29-nt small RNAs. BMC Genomics 16, 100.

Chambeyron, S., Seitz, H., 2014. Insect small non-coding RNA involved in epigenetic regulations. Curr. Opin. Insect Sci. 1, 1–9.

Chen, X.-G., Jiang, X., Gu, J., Xu, M., Wu, Y., Deng, Y., Zhang, C., Bonizzoni, M., Dermauw, W., Vontas, J., Armbruster, P., Huang, X., Yang, Y., Zhang, H., He, W., Peng, H., Liu, Y., Wu, K., Chen, J., Lirakis, M., Topalis, P., Van Leeuwen, T., Hall, A.B., Jiang, X., Thorpe, C., Mueller, R.L., Sun, C., Waterhouse, R.M., Yan, G., Tu, Z.J., Fang, X., James, A.A., 2015. Genome sequence of the Asian Tiger mosquito, *Aedes albopictus*, reveals insights into its biology, genetics, and evolution. Proc. Natl. Acad. Sci. U.S.A. 112, E5907–E5915.

Choi, I.K., Hyun, S., 2012. Conserved microRNA miR-8 in fat body regulates innate immune homeostasis in *Drosophila*. Dev. Comp. Immunol. 37, 50–54.

Chotkowski, H.L., Ciota, A.T., Jia, Y., Puig-Basagoiti, F., Kramer, L.D., Shi, P.Y., Glaser, R.L., 2008. West Nile virus infection of *Drosophila melanogaster* induces a protective RNAi response. Virology 377, 197–206.

Chung, W.J., Okamura, K., Martin, R., Lai, E.C., 2008. Endogenous RNA interference provides a somatic defense against *Drosophila* transposons. Curr. Biol. 18, 795–802.

Cirimotich, C.M., Scott, J.C., Phillips, A.T., Geiss, B.J., Olson, K.E., 2009. Suppression of RNA interference increases alphavirus replication and virus-associated mortality in *Aedes aegypti* mosquitoes. BMC Microbiol. 9, 49.

Clayton, A.M., Dong, Y., Dimopoulos, G., 2014. The Anopheles innate immune system in the defense against malaria infection. J. Innate Immun. 6, 169–181.

Dennison, N.J., Benmarzouk-Hidalgo, O.J., Dimopoulos, G., 2015. MicroRNA-regulation of *Anopheles gambiae* immunity to *Plasmodium falciparum* infection and midgut microbiota. Dev. Comp. Immunol. 49, 170–178.

Ding, S.-W., 2010. RNA-based antiviral immunity. Nat. Rev. Immunol. 10, 632–644.

Etebari, K., Asgari, S., 2013. Conserved microRNA miR-8 blocks activation of the Toll pathway by up-regulating Serpin 27 transcripts. RNA Biol. 10, 1356–1364.

Flynt, A., Liu, N., Martin, R., Lai, E., 2008. Dicing of viral replication intermediates during silencing of latent *Drosophila* viruses. Proc. Natl. Acad. Sci. U.S.A. 106, 5270–5275.

Franz, A., Sanchez-Vargas, I., Adelman, Z., Blair, C., Beaty, B., James, A., Olson, K., 2006. Engineering RNA interference-based resistance to dengue virus type 2 in genetically modified *Aedes aegypti*. Proc. Natl. Acad. Sci. U.S.A. 103, 4198–4203.

Franz, A., Sanchez-Vargas, I., Piper, J., Smith, M., Khoo, C., James, A., Olson, K., 2009. Stability and loss of a virus resistance phenotype over time in transgenic mosquitoes harbouring an antiviral effector gene. Insect Mol. Biol. 18, 661–672.

Franz, A., Sanchez-Vargas, I., Raban, R., Black, W., James, A., Olson, K., 2014. Fitness impact and stability of a transgene conferring resistance to dengue-2 virus following introgression into a genetically diverse *Aedes aegypti* strain. PLoS Negl. Trop. Dis. 8, e2833.

Fros, J., Miesen, P., Vogels, C., Gaibani, P., Sambri, V., Martina, B., Koenraadt, C., Van Rij, R.P., Vlak, J.M., Takken, W., Pijlman, G.P., 2015. Comparative Usutu and West Nile virus transmission potential by local *Culex pipiens* mosquitoes in north-western Europe. One Health 1, 31–36.

Gaines, P.J., Olson, K.E., Higgs, S., Powers, A.M., Beaty, B.J., Blair, C.D., 1996. Pathogen-derived resistance to dengue type 2 virus in mosquito cells by expression of the pre-membrane coding region of the viral genome. J. Virol. 70, 2132–2137.

Galiana-Arnoux, D., Dostert, C., Schneemann, A., Hoffman, J.A., Imler, J.-L., 2006. Essential function in vivo for Dicer-2 in host defense against RNA viruses in *Drosophila*. Nat. Immunol. 7, 590–597.

George, P., Jensen, S., Pogorelcnik, R., Lee, J., Xing, Y., Brasset, E., Vaury, C., Sharakhov, I.V., 2015. Increased production of piRNAs from euchromatic clusters and genes in *Anopheles gambiae* compared with *Drosophila melanogaster*. Epigenetics Chromatin. 8, 50.

Ghildiyal, M., Seitz, H., Horwich, M., Li, C., Du, T., Lee, S., Xu, J., Kittler, E., Zapp, M., Weng, Z., Zamore, P., 2008. Endogenous siRNAs derived from transposons and mRNAs in *Drosophila* somatic cells. Science 320, 1077–1081.

Girard, A., Sachidanandam, R., Hannon, G.J., Carmell, M.A., 2006. A germline-specific class of small RNAs binds mammalian Piwi proteins. Nature 442, 199–202.

Gu, J., Hu, W., Wu, J., Zheng, P., Chen, M., James, A., Chen, X., Tu, Z., 2013. miRNA genes of an invasive vector mosquito, *Aedes albopictus*. PLoS One 8, e67638.

Gunawardane, L., Saito, K., Nishida, K., Miyoshi, K., Kawamura, Y., Nagami, T., Siomi, H., Siomi, M.C., 2007. A slicer-mediated mechanism for repeat-associated siRNA 5′ end formation in *Drosophila*. Science 315, 1587–1590.

Helinski, M., Knols, B., 2008. Mating competitiveness of male *Anopheles arabiensis* mosquitoes irradiated with a partially or fully sterilizing dose in small and large laboratory cages. J. Med. Entomol. 45, 698–705.

Hess, A.M., Prasad, A.N., Ptitsyn, A., Ebel, G.D., Olson, E.N., Barbacioru, C., Monighetti, C., Campbell, C.L., 2011. Small RNA profiling of Dengue virus-mosquito interactions implicates the PIWI RNA pathway in anti-viral defense. BMC Microbiol. 11, 45.

Hoa, N., Keene, K., Olson, K., Zheng, L., 2003. Characterization of RNA interference in an *Anopheles gambiae* cell line. Insect Biochem. Mol. Biol. 33, 949–957.

Holt, R.A., Subramanian, G.M., Halpern, A., Sutton, G.G., Charlab, R., Nusskern, D.R., Wincker, P., Clark, A.G., Ribeiro, J.M., Wides, R., Salzberg, S.L., Loftus, B., Yandell, M., Majoros, W.H., Rusch, D.B., Lai, Z., Kraft, C.L., Abril, J.F., Anthouard, V., Arensburger, P., Atkinson, P.W., Baden, H., De, B.V., Baldwin, D., Benes, V., Biedler, J., Blass, C., Bolanos, R., Boscus, D., Barnstead, M., Cai, S., Center, A., Chaturverdi, K., Christophides, G.K., Chrystal, M.A., Clamp, M., Cravchik, A., Curwen, V., Dana, A., Delcher, A., Dew, I., Evans, C.A., Flanigan, M., Grundschober-Freimoser, A., Friedli, L., Gu, Z., Guan, P., Guigo, R., Hillenmeyer, M.E., Hladun, S.L., Hogan, J.R., Hong, Y.S., Hoover, J., Jaillon, O., Ke, Z., Kodira, C., Kokoza, E., Koutsos, A., Letunic, I., Levitsky, A., Liang, Y., Lin, J.J., Lobo, N.F., Lopez, J.R., Malek, J.A., Mcintosh, T.C., Meister, S., Miller, J., Mobarry, C., Mongin, E., Murphy, S.D., O'brochta, D.A., Pfannkoch, C., Qi, R., Regier, M.A., Remington, K., Shao, H., Sharakhova, M.V., Sitter, C.D., Shetty, J., Smith, T.J., Strong, R., Sun, J., Thomasova, D., Ton, L.Q., Topalis, P., Tu, Z., Unger, M.F., Walenz, B., Wang, A., Wang, J., Wang, M., Wang, X., Woodford, K.J., Wortman, J.R., Wu, M., Yao, A., Zdobnov, E.M., Zhang, H., Zhao, Q., et al., 2002. The genome sequence of the malaria mosquito *Anopheles gambiae*. Science 298, 129–149.

Hong, S., Guo, Q., Wang, W., Hu, S., Fang, F., Lv, Y., Yu, J., Zou, F., Lei, Z., Ma, K., Ma, L., Zhou, D., Sun, Y., Zhang, D., Shen, B., Zhu, C., 2014. Identification of differentially expressed microRNAs in *Culex pipiens* and their potential roles in pyrethroid resistance. Insect Biochem. Mol. Biol. 55, 39–50.

Hu, W., Criscione, F., Liang, S., Tu, Z., 2015. MicroRNAs of two medically important mosquito species: *Aedes aegypti* and *Anopheles stephensi*. Insect Mol. Biol. 24, 240–252.

Hussain, M., Asgari, S., 2014. A glance on the role of miRNAs in insect life. In: Hoffmann, K.H. (Ed.), Insect Molecular Biology and Ecology. CRC Press, Boca Raton, USA.

Hussain, M., Frentiu, F.D., Moreira, L.A., O'neill, S.L., Asgari, S., 2011. *Wolbachia* utilizes host microRNAs to manipulate host gene expression and facilitate colonization of the dengue vector *Aedes aegypti*. Proc. Natl. Acad. Sci. U.S.A. 108, 9250–9255.

Hussain, M., O'neill, S.L., Asgari, S., 2013a. *Wolbachia* interferes with the intracellular distribution of Argonaute 1 in the dengue vector *Aedes aegypti* by manipulating the host microRNAs. RNA Biol. 10, 1868–1875.

Hussain, M., Walker, T., O'neill, S., Asgari, S., 2013b. Blood meal induced microRNA regulates development and immune associated genes in the Dengue mosquito vector, *Aedes aegypti*. Insect Biochem. Mol. Biol. 43, 146–152.

Ishizu, H., Siomi, H., Siomi, M., 2012. Biology of PIWI-interacting RNAs: new insights into biogenesis and function inside and outside of germlines. Genes Dev. 26, 2361–2373.

Isoe, J., Rascón, A.A., Kunz, S., Miesfeld, R.L., 2009. Molecular genetic analysis of midgut serine proteases in *Aedes aegypti* mosquitoes. Insect Biochem. Mol. Biol. 39, 903.

Izumi, N., Tomari, Y., 2014. Diversity of the piRNA pathway for nonself silencing: worm-specific piRNA biogenesis factors. Genes Dev. 28, 665–671.

Jain, S., Rana, V., Shrinet, J., Sharma, A., Tridibes, A., Sunil, S., Bhatnagar, R., 2014. Blood feeding and *Plasmodium* infection alters the miRNome of *Anopheles stephensi*. PLoS One 9, e98402.

Jin, H., Kim, V.N., Hyun, S., 2012. Conserved microRNA miR-8 controls body size in response to steroid signaling in *Drosophila*. Genes Dev. 26, 1427–1432.

Kakumani, P., Ponia, S., Rk, S., Sood, V., Chinnappan, M., Banerjea, A., Medigeshi, G., Malhotra, P., Mukherjee, S., Bhatnagar, R., 2013. Role of RNA interference (RNAi) in dengue virus replication and identification of NS4B as an RNAi suppressor. J. Virol. 87, 8870–8883.

Kakumani, P., Rajgokul, K., Ponia, S., Kaur, I., Mahanty, S., Medigeshi, G., Banerjea, A., Chopra, A., Malhotra, P., Mukherjee, S., Bhatnagar, R., 2015. Dengue NS3, an RNAi suppressor, modulates the human miRNA pathways through its interacting partner. Biochem. J. 471, 89–99.

Karres, J.S., Hilgers, V., Carrera, I., Treisman, J., Cohen, S.M., 2007. The conserved microRNA miR-8 tunes atrophin levels to prevent neurodegeneration in *Drosophila*. Cell 131, 136–145.

Keene, K., Foy, B., Sanchez-Vargas, I., Beaty, B., Blair, C., Olson, K., 2004. RNA interference acts as a natural antiviral response to O'nyong-nyong virus (Alphavirus; Togaviridae) infection of *Anopheles gambiae*. Proc. Natl. Acad. Sci. U.S.A. 101, 17240–17245.

Khoo, C., Piper, J., Sanchez-Vargas, I., Olson, K., Franz, A., 2010. The RNA interference pathway affects midgut infection- and escape barriers for Sindbis virus in *Aedes aegypti*. BMC Microbiol. 10, 130.

Khoo, C., Doty, J., Heersink, M., Olson, K., Franz, A., 2013. Transgene-mediated suppression of the RNA interference pathway in *Aedes aegypti* interferes with gene silencing and enhances Sindbis virus and dengue virus type 2 replication. Insect Mol. Biol. 22, 104–114.

Kiuchi, T., Koga, H., Kawamoto, M., Shoji, K., Sakai, H., Arai, Y., Ishihara, G., Kawaoka, S., Sugano, S., Shimada, T., Suzuki, Y., Suzuki, M., Katsuma, S., 2014. A single female-specific piRNA is the primary determiner of sex in the silkworm. Nature 509, 633–636.

Korrapati, A., Swaminathan, G., Gingh, A., Khanna, N., Swaminathan, S., 2012. Adenovirus delivered short hairpin RNA targeting a conserved site in the 5′ non-translated region inhibits all four serotypes of dengue viruses. PLoS Negl. Trop. Dis. 6, e1735.

Kugler, J.-M., Chen, Y.-W., Weng, R., Cohen, S.M., 2013. miR-989 is required for border cell migration in the *Drosophila* ovary. PLoS One 8, e67075.

Lambrechts, L., Quillery, E., Noël, V., Richardson, J., Jarman, R., Scott, T., Chevillon, C., 2013. Specificity of resistance to dengue virus isolates is associated with genotypes of the mosquito antiviral gene Dicer-2. Proc. Biol. Sci. 280, 20122437.

Lau, N.C., Seto, A.G., Kim, J., Kuramochi-Miyagawa, S., Nakano, T., Bartel, D.P., Kingston, R.E., 2006. Characterization of the piRNA complex from rat testes. Science 313, 363–367.

Lee, G., Hyun, S., 2014. Multiple targets of the microRNA miR-8 contribute to immune homeostasis in *Drosophila*. Dev. Comp. Immunol. 45, 245–251.

Léger, P., Lara, E., Jagla, B., Sismeiro, O., Mansuroglu, Z., Coppée, J., Bonnefoy, E., Bouloy, M., 2013. Dicer-2- and Piwi-mediated RNA interference in Rift Valley fever virus-infected mosquito cells. J. Virol. 87, 1631–1648.

Lei, Z., Lv, Y., Wang, W., Guo, Q., Zou, F., Hu, S., Fang, F., Tian, M., Liu, B., Liu, X., Ma, K., Ma, L., Zhou, D., Zhang, D., Sun, Y., Shen, B., Zhu, C., 2015. MiR-278-3p regulates pyrethroid resistance in *Culex pipiens* pallens. Parasitol. Res. 114, 699–706.

Leslie, M., 2013. The immune system's compact genomic counterpart. Science 339, 25–27.

Lewis, B.P., Burge, C.B., Bartel, D.P., 2005. Conserved seed pairing, often flanked by adenosines, indicates that thousands of human genes are microRNA targets. Cell 120, 15–20.

Li, S., Mead, E.A., Liang, L., Tu, Z., 2009. Direct sequencing and expression analysis of a large number of miRNAs in *Aedes aegypti* and a multi-species survey of novel mosquito miRNAs. BMC Genomics 10, 581.

Lim, D., Oh, C., Lee, L., Hong, J., Noh, S., Hwang, S., Kim, S., Han, S., Lee, Y., 2011. The endogenous siRNA pathway in *Drosophila* impacts stress resistance and lifespan by regulating metabolic homeostasis. FEBS Lett. 585, 3079–3085.

Liu, S., Lucas, K., Roy, S., Ha, J., Raikhel, A., 2014. Mosquito-specific microRNA-1174 targets serine hydroxymethyltransferase to control key functions in the gut. Proc. Natl. Acad. Sci. U.S.A. 111, 14460–14465.

Liu, Y., Zhou, Y., Wu, J., Zheng, P., Li, Y., Zheng, X., Puthiyakunnon, S., Tu, Z., Chen, X.-G., 2015. The expression profile of *Aedes albopictus* miRNAs is altered by dengue virus serotype-2 infection. Cell Biosci. 5, 16.

Lucas, K., Raikhel, A.S., 2013. Insect microRNAs: biogenesis, expression profiling and biological functions. Insect Biochem. Mol. Biol. 43, 24–38.

Lucas, K., Myles, K., Raikhel, A., 2013. Small RNAs: a new frontier in mosquito biology. Trends Parasitol. 29, 295–303.

Lucas, K., Roy, S., Ha, J., Gervaise, A., Kokoza, V., Raikhel, A., 2015a. MicroRNA-8 targets the Wingless signaling pathway in the female mosquito fat body to regulate reproductive processes. Proc. Natl. Acad. Sci. U.S.A. 112, 1440–1445.

Lucas, K.J., Zhao, B., Liu, S., Raikhel, A.S., 2015b. Regulation of physiological processes by microRNAs in insects. Curr. Opin. Insect Sci. 11, 1–7.

Lucas, K.J., Zhao, B., Roy, S., Gervaise, A.L., Raikhel, A.S., 2015c. Mosquito-specific microRNA-1890 targets the juvenile hormone-regulated serine protease JHA15 in the female mosquito gut. RNA Biol. 12, 1383–1390.

Lucchetta, E., Carthew, R., Ismagilov, R., 2009. The endo-siRNA pathway is essential for robust development of the *Drosophila* embryo. PLoS One 4, e7576.

Maclas, V., Coleman, J., Bonizzoni, M., James, A., 2014. piRNA pathway gene expression in the malaria vector mosquito *Anopheles stephensi*. Insect Mol. Biol. 23, 579–586.

Maharaj, P., Widen, S., Huang, J., Wood, T., Thangamani, S., 2015. Discovery of mosquito saliva microRNAs during CHIKV infection. PLoS Negl. Trop. Dis. 9, e0003386.

Malone, C.D., Brennecke, J., Dus, M., Stark, A., Mccombie, W.R., Sachidanandam, R., Hannon, G.J., 2009. Specialized piRNA pathways act in germline and somatic tissues of the *Drosophila* ovary. Cell 137, 522–535.

Marinotti, O., Cerqueira, G., De Almeida, L., Ferro, M., Loreto, E., Zaha, A., Teixeira, S., Wespiser, A., Almeida, E.S.A., Schlindwein, A., Pacheco, A., Silva, A., Graveley, B., Walenz, B., Lima, B.A., Ribeiro, C., Nunes-Silva, C., De Carvalho, C., Soares, C., De Menezes, C., Matiolli, C., Caffrey, D., Araújo, D., De Oliveira, D., Golenbock, D., Grisard, E., Fantinatti-Garboggini, F., De Carvalho, F., Barcellos, F., Prosdocimi, F., May, G., Azevedo, J.G., Guimarães, G., Goldman, G., Padilha, I., Batista, J.S., Ferro, J., Ribeiro, J., Fietto, J., Dabbas, K., Cerdeira, L., Agnez-Lima, L., Brocchi, M., De Carvalho, M., Teixeira, M.M., Diniz, M.M.M., Goldman, M., Cruz Schneider, M., Felipe, M., Hungria, M., Nicolás, M., Pereira, M., Montes, M., Cantão, M., Vincentz, M., Rafael, M., Silverman, N., Stoco, P., Souza, R., Vicentini, R., Gazzinelli, R., Neves, R.O., Silva, R., Astolfi-Filho, S., Maciel, T., Urményi, T., Tadei, W., Camargo, E., De Vasconcelos, A., 2013. The genome of *Anopheles darlingi*, the main neotropical malaria vector. Nucleic Acids Res. 41, 7387–7400.

Mathur, G., Sánchez-Vargas, I., Alvarez, D., Olson, K., Marinotti, O., James, A.A., 2010. Transgene-mediated suppression of dengue viruses in the salivary glands of the yellow fever mosquito, *Aedes aegypti*. Insect Mol. Biol. 19, 753–763.

Mayoral, J.G., Etebari, K., Hussain, M., Khromykh, A.A., Asgari, S., 2014. *Wolbachia* infection modifies the profile, shuttling and structure of microRNAs in a mosquito cell line. PLoS One 9, e96107.

Mead, E.A., Tu, Z., 2008. Cloning, characterization, and expression of microRNAs from the Asian malaria mosquito, *Anopheles stephensi*. BMC Genomics 9, 244.

Miesen, P., Girardi, E., Van Rij, R.P., 2015. Distinct sets of PIWI proteins produce arbovirus and transposon-derived piRNAs in *Aedes aegypti* mosquito cells. Nucleic Acids Res. 43, 6545–6556.

Miesen, P., Ivens, A., Buck, A., Van Rij, R., 2016. Small RNA profiling in dengue virus 2-infected *Aedes* mosquito cells reveals viral piRNAs and novel host miRNAs. PLoS Negl. Trop. Dis. 10, e0004452.

Moon, S., Dodd, B., Brackney, D., Wilusz, C., Ebel, G., Wilusz, J., 2015. Flavivirus sfRNA suppresses antiviral RNA interference in cultured cells and mosquitoes and directly interacts with the RNAi machinery. Virology 485, 322–329.

Morazzani, E., Wiley, M., Murreddu, M., Adelman, Z., Myles, K., 2012. Production of virus-derived ping-pong-dependent piRNA-like small RNAs in the mosquito soma. PLoS Pathog. 8, e1002470.

Myles, K., Wiley, M., Morazzani, E., Adelman, Z., 2008. Alphavirus-derived small RNAs modulate pathogenesis in disease vector mosquitoes. Proc. Natl. Acad. Sci. U.S.A. 105, 19938–19943.

Nene, V., Wortman, J., Lawson, D., Haas, B., Kodira, C., Tu, Z., Loftus, B., Xi, Z., Megy, K., Grabherr, M., Ren, Q., Zdobnov, E., Lobo, N., Campbell, K., Brown, S., Bonaldo, M., Zhu, J., Sinkins, S., Hogenkamp, D., Amedeo, P., Arensburger, P., Atkinson, P., Bidwell, S., Biedler, J., Birney, E., Bruggner, R., Costas, J., Coy, M., Crabtree, J., Crawford, M., Debruyn, B., Decaprio, D., Eiglmeier, K., Eisenstadt, E., El-Dorry, H., Gelbart, W., Gomes, S., Hammond, M., Hannick, L., Hogan, J., Holmes, M., Jaffe, D., Johnston, J., Kennedy, R., Koo, H., Kravitz, S., Kriventseva, E., Kulp, D., Labutti, K., Lee, E., Li, S., Lovin, D., Mao, C., Mauceli, E., Menck, C., Miller, J., Montgomery, P., Mori, A., Nascimento, A., Naveira, H., Nusbaum, C., O'leary, S., Orvis, J., Pertea, M., Quesneville, H., Reidenbach, K., Rogers, Y., Roth, C., Schneider, J., Schatz, M., Shumway, M., Stanke, M., Stinson, E., Tubio, J., Vanzee, J., Verjovski-Almeida, S., Werner, D., White, O., Wyder, S., Zeng, Q., Zhao, Q., Zhao, Y., Hill, C., Raikhel, A., Soares, M., Knudson, D., Lee, N., Galagan, J., Salzberg, S., Paulsen, I., Dimopoulos, G., Collins, F., Birren, B., Fraser-Liggett, C., Severson, D., 2007. Genome sequence of *Aedes aegypti*, a major arbovirus vector. Science 316, 1718–1723.

Noriega, F.G., Shah, D.K., Wells, M.A., 1997. Juvenile hormone controls early trypsin gene transcription in the midgut of *Aedes aegypti*. Insect Mol. Biol. 6, 63–66.

Okamura, K., Lai, E., 2008. Endogenous small interfering RNAs in animals. Nat. Rev. Mol. Cell Biol. 9, 673–678.

Okamura, K., Hagen, J.W., Duan, H., Tyler, D.M., Lai, E.C., 2007. The mirtron pathway generates microRNA-class regulatory RNAs in *Drosophila*. Cell 130, 89–100.

Okamura, K., Robine, N., Liu, Y., Liu, Q., Lai, E., 2011. R2D2 organizes small regulatory RNA pathways in *Drosophila*. Mol. Cell. Biol. 31, 884–896.

Olson, K.E., Higgs, S., Gaines, P.J., Powers, A.M., Davis, B.S., Kamrud, K.I., Carlson, J.O., Blair, C.D., Beaty, B.J., 1996. Genetically engineered resistance to dengue-2 virus transmission in mosquitoes. Science 272, 884–886.

Osei-Amo, S., Hussain, M., O'neill, S.L., Asgari, S., 2012. *Wolbachia*-induced aae-miR-12 miRNA negatively regulates the expression of MCT1 and MCM6 genes in *Wolbachia*-infected mosquito cell line. PLoS One 7, e50049.

Panda, D., Rose, P., Hanna, S., Gold, B., Hopkins, K., Lyde, R., Marks, M., Cherry, S., 2013. Genome-wide RNAi screen identifies SEC61A and VCP as conserved regulators of Sindbis virus entry. Cell Rep. 5, 1737–1748.

Powers, A.M., Olson, K.E., Higgs, S., Carlson, J.O., Beaty, B.J., 1994. Intracellular immunization of mosquito cells to LaCrosse virus using a recombinant Sindbis virus vector. Virus Res. 32, 57–67.

Proverbs, M., 1969. Induced sterilization and control of insects. Annu. Rev. Entomol. 14, 81–102.

Puthiyakunnon, S., Yao, Y., Li, Y., Gu, J., Peng, H., Chen, X., 2013. Functional characterization of three microRNAs of the Asian tiger mosquito, *Aedes albopictus*. Parasit. Vectors 6, 230.

Robine, N., Lau, N.C., Balla, S., Jin, Z., Okamura, K., Kuramochi-Miyagawa, S., Blower, M.D., Lai, E.C., 2009. A broadly conserved pathway generates 3′ UTR-directed primary piRNAs. Curr. Biol. 19, 2066–2076.

Ross, R.J., Weiner, M.M., Lin, H., 2014. PIWI proteins and PIWI-interacting RNAs in the soma. Nature 505, 353–359.

Rubio, M., Montañez, R., Perez, L., Milan, M., Belles, X., 2013. Regulation of atrophin by both strands of the mir-8 precursor. Insect Biochem. Mol. Biol. 43, 1009–1014.

Sanchez-Vargas, I., Scott, J., Poole-Smith, B., Franz, A., Barbosa-Solomieu, V., Wilusz, J., Olson, K., Blair, C., 2009. Dengue virus type 2 infections of *Aedes aegypti* are modulated by the mosquito's RNA interference pathway. PLoS Pathog. 5, e1000299.

Schirtzinger, E., Andrade, C., Devitt, N., Ramaraj, T., Jacobi, J., Schilkey, F., Hanley, K., 2015. Repertoire of virus-derived small RNAs produced by mosquito and mammalian cells in response to dengue virus infection. Virology 476, 54–60.

Schnettler, E., Sterken, M.G., Leung, J.Y., Metz, S.W., Geertsema, C., Goldbach, R.W., Vlak, J.M., Kohl, A., Khromykh, A.A., Pijlman, G.P., 2012. Noncoding flavivirus RNA displays RNA interference suppressor activity in insect and mammalian cells. J. Virol. 86, 13486–13500.

Schnettler, E., Donald, C.L., Human, S., Watson, M., Siu, R.W., Mcfarlane, M., Fazakerley, J.K., Kohl, A., Fragkoudis, R., 2013. Knockdown of piRNA pathway proteins results in enhanced Semliki Forest virus production in mosquito cells. J. Gen. Virol. 94, 1680–1689.

Scott, J.C., Brackney, D.E., Campbell, C.L., Bondu-Hawkins, V., Hjelle, B., Ebel, G.D., Olson, K.E., Blair, C.D., 2010. Comparison of Dengue virus type 2-specific small RNAs from RNA interference-competent and -incompetent mosquito cells. PLoS Negl. Trop. Dis. 4, e848.

Scott, J.G., Yoshimizu, M.H., Kasai, S., 2015. Pyrethroid resistance in *Culex* pipiens mosquitoes. Pestic. Biochem. Physiol. 120, 68–76.

Shimakami, T., Yamane, D., Jangra, R.K., Kempf, B.J., Spaniel, C., Barton, D.J., Lemon, S.M., 2012. Stabilization of hepatitis C virus RNA by an Ago2-miR-122 complex. Proc. Natl. Acad. Sci. U.S.A. 109, 941–946.

Shrinet, J., Jain, S., Jain, J., Bhatnagar, R., Sunil, S., 2014. Next generation sequencing reveals regulation of distinct *Aedes* microRNAs during chikungunya virus development. PLoS Negl. Trop. Dis. 8, e2616.

Siddiqi, S., Matushansky, I., 2012. Piwis and piwi-interacting RNAs in the epigenetics of cancer. J. Cell. Biochem. 113, 373–380.

Singh, A.D., Wong, S., Ryan, C.P., Whyard, S., 2013. Oral delivery of double-stranded RNA in larvae of the yellow fever mosquito *Aedes aegypti*: implications for pest mosquito control. J. Insect Sci. 13, 69.

Siomi, M.C., Sato, K., Pezic, D., Aravin, A.A., 2011. PIWI-interacting small RNAs: the vanguard of genome defence. Nat. Rev. Mol. Cell Biol. 12, 246–258.

Siu, R., Fragkoudis, R., Simmonds, P., Donald, C., Chase-Topping, M., Barry, G., Attarzadeh-Yazdi, G., Rodriguez-Andres, J., Nash, A., Merits, A., Fazakerley, J., Kohl, A., 2011. Antiviral RNA interference responses induced by Semliki Forest virus infection of mosquito cells: characterization, origin, and frequency-dependent functions of virus-derived small interfering RNAs. J. Virol. 85, 2907–2917.

Skalsky, R., Vanlandingham, D.L., Scholle, F., Higgs, S., Cullen, B.R., 2010. Identification of microRNAs expressed in two mosquito vectors, *Aedes albopictus* and *Culex quinquefasciatus*. BMC Genomics 11, 119.

Slonchak, A., Hussain, M., Torres Morales, S., Asgari, S., Khromykh, A.A., 2014. Expression of mosquito microRNA aae-miR-2940-5p is down-regulated in response to West Nile virus infection to restrict viral replication. J. Virol. 88, 8457–8467.

Towler, B.P., Jones, C.I., Viegas, S.C., Apura, P., Waldron, J.A., Smalley, S.K., Arraiano, C.M., Newbury, S.F., 2015. The 3′-5′ exoribonuclease Dis3 regulates the expression of specific microRNAs in *Drosophila* wing imaginal discs. RNA Biol. 12, 728–741.

Van Cleef, K., Van Mierlo, J., Miesen, P., Overheul, G., Fros, J., Schuster, S., Marklewitz, M., Pijlman, G., Junglen, S., Van Rij, R., 2014. Mosquito and *Drosophila* entomobirnaviruses suppress dsRNA- and siRNA-induced RNAi. Nucleic Acids Res. 42, 8732–8744.

Vodovar, N., Bronkhorst, A., Van Cleef, K., Miesen, P., Blanc, H., Van Rij, R., Saleh, M.C., 2012. Arbovirus-derived piRNAs exhibit a ping-pong signature in mosquito cells. PLoS One 7, e30861.

Waldock, J., Olson, K., Christophides, G., 2012. *Anopheles gambiae* antiviral immune response to systemic O'nyong-nyong infection. PLoS Negl. Trop. Dis. 6, E1565.

Wen, J., Duan, H., Bejarano, F., Okamura, K., Fabian, L., Brill, J., Bortolamiol-Becet, D., Martin, R., Ruby, J., Lai, E., 2015. Adaptive regulation of testis gene expression and control of male fertility by the *Drosophila* hairpin RNA pathway. Mol. Cell 57, 165–178.

Whyard, S., Erdelyan, C., Partridge, A., Singh, A., Beebe, N., Capina, R., 2015. Silencing the buzz: a new approach to population suppression of mosquitoes by feeding larvae double-stranded RNAs. Parasit. Vectors 8, 96.

Winter, F., Edaye, S., Hüttenhofer, A., Brunel, C., 2007. *Anopheles gambiae* miRNAs as actors of defence reaction against *Plasmodium* invasion. Nucleic Acids Res. 35, 6953–6962.

Wu, Q., Luo, Y., Lu, R., Lau, N., Lai, E.C., Li, W.-X., Ding, S.-W., 2010. Virus discovery by deep sequencing and assembly of virus-derived small silencing RNAs. Proc. Natl. Acad. Sci. U.S.A. 107, 1606–1611.

Wu, Y.-L., Wu, C.P., Liu, C.Y.Y., Hsu, P.W.-C., Wu, E.C., Chao, Y.C., 2011. A non-coding RNA of insect HzNV-1 virus establishes latent viral infection through micro-RNA. Sci. Rep. 1, 60.

Xi, Z., Ramirez, J.L., Dimopoulos, G., 2008. The *Aedes aegypti* Toll pathway controls dengue virus infection. PLoS Pathog. 4, e1000098.

Yan, Z., Hu, H.Y., Jiang, X., Maierhofer, V., Neb, E., He, L., Hu, Y., Hu, H., Li, N., Chen, W., Khaitovich, P., 2011. Widespread expression of piRNA-like molecules in somatic tissues. Nucleic Acids Res. 39, 6596–6607.

Yan, H., Zhou, Y., Liu, Y., Deng, Y., Chen, X., 2014. miR-252 of the Asian tiger mosquito *Aedes albopictus* regulates dengue virus replication by suppressing the expression of the dengue virus envelope protein. J. Med. Virol. 86, 1428–1436.

Yang, J.-S., Lai, E.C., 2011. Alternative miRNA biogenesis pathways and the interpretation of core miRNA pathway mutants. Mol. Cell 43, 892–903.

Yasunaga, A., Hanna, S., Li, J., Cho, H., Rose, P., Spiridigliozzi, A., Gold, B., Diamond, M., Cherry, S., 2014. Genome-wide RNAi screen identifies broadly-acting host factors that inhibit arbovirus infection. PLoS Pathog. 10, e1003914.

Zhang, X., Zhang, J., Zhu, K., 2010. Chitosan/double-stranded RNA nanoparticlemediated RNA interference to silence chitin synthase genes through larval feeding in the African malaria mosquito (*Anopheles gambiae*). Insect Mol. Biol. 19, 683–693.

Zhang, Y., Wang, X., Kang, L., 2011. A k-mer scheme to predict piRNAs and characterize locust piRNAs. Bioinformatics 27, 771–776.

Zhang, G., Hussain, M., O'neill, S.L., Asgari, S., 2013. *Wolbachia* uses a host microRNA to regulate transcripts of a methyltransferase contributing to dengue virus inhibition in *Aedes aegypti*. Proc. Natl. Acad. Sci. U.S.A. 110, 10276–10281.

Zhang, G., Hussain, M., Asgari, S., 2014. Regulation of arginine methyltransferase 3 by a *Wolbachia*-induced microRNA in *Aedes aegypti* and its effect on *Wolbachia* and dengue virus replication. Insect Biochem. Mol. Biol. 53, 81–88.

Zhou, Y., Liu, Y., Yan, H., Li, Y., Zhang, H., Xu, J., Puthiyakunnon, S., Chen, X., 2014. miR-281, an abundant midgut-specific miRNA of the vector mosquito *Aedes albopictus* enhances dengue virus replication. Parasit. Vectors 7, 488.

Zong, J., Yao, X., Yin, J., Zhang, D., Ma, H., 2009. Evolution of the RNA-dependent RNA polymerase (RdRP) genes: duplications and possible losses before and after the divergence of major eukaryotic groups. Gene 447, 29–39.

CHAPTER EIGHT

The Complement System of Malaria Vector Mosquitoes

M. Povelones*, M.A. Osta†, G.K. Christophides‡
*School of Veterinary Medicine, University of Pennsylvania, Philadelphia, PA, United States
†American University of Beirut, Beirut, Lebanon
‡Imperial College London, London, United Kingdom

Contents

Abstract

Mosquitoes are vectors for numerous human and animal diseases, including malaria that is caused by protozoan *Plasmodium* parasites and is responsible for over half a million deaths annually. The immune system of the major African malaria vector *Anopheles gambiae* provides a robust barrier to infection by *Plasmodium*. A system analogous to vertebrate complement is responsible for the majority of parasite losses. Work over the past decade has characterized several components of the mosquito complement pathway, provided a tentative scheme of the hierarchical events leading to its activation and identified multiple effector mechanisms that are differentially activated depending on the pathogen type. Although most parasites are killed by complement reactions, some escape and continue transmission. In this review, we synthesize these

Advances in Insect Physiology, Volume 51
ISSN 0065-2806
http://dx.doi.org/10.1016/bs.aiip.2016.06.001

discoveries and examine parasite evasion tactics in an evolutionary context. The impact of the complement pathway in shaping the global malaria transmission is discussed together with future research challenges and opportunities for disease control.

1. INTRODUCTION

Within 24–36 h inside the mosquito midgut, the malaria parasite completes complex and finely regulated developmental processes that include gametogenesis, sexual reproduction, zygote differentiation into a motile ookinete, traversal of both the peritrophic matrix and the midgut epithelium and transformation to oocyst. These processes are associated with various mosquito responses including behavioural, hormonal and homeostatic (Gendrin et al., 2015; Mendes et al., 2011; Upton et al., 2015). Importantly, parasites encounter an orchestrated sequence of robust immune responses that take place within two mosquito compartments: the midgut lumen, and the midgut epithelium and its associated basal labyrinth that is an extension of the haemocoel and is thus awash in haemolymph (Clayton et al., 2014). These mosquito immune responses lead to elimination of most parasites and a strong parasite population bottleneck, often leading to premature termination of transmission (Fig. 1). Nonetheless, a small number of parasites manage to escape these reactions and develop to oocysts by evading the mosquito immune system.

In the mosquito midgut lumen, immune responses largely involve the antibacterial NF-κB signalling pathway Imd. After a bloodmeal, midgut bacterial levels increase dramatically resulting in a corresponding increase in bacterial elicitors of the immune system, like peptidoglycan (Kumar et al., 2010), ultimately causing enhanced Imd signalling once the tolerance levels are breached (Dong et al., 2009; Meister et al., 2009). Target genes of Imd signalling can lead to parasite losses in the midgut lumen (Fig. 2). Interestingly, direct parasite recognition by the Imd pathway is also shown, but the mechanisms involved are yet unknown (Garver et al., 2012).

When ookinetes invade the midgut epithelium, activation of the c-Jun N-terminal kinase (JNK) pathway triggers a nitration reaction with two important outcomes: apoptosis of the invaded cells and marking of ookinetes for destruction by the mosquito complement system (Garver et al., 2013). Killing by mosquito complement is the most critical of all immune reactions against parasites, as well as other microbes invading the haemolymph, and is the focus of this review. Systematic research in the past 15 years has revealed

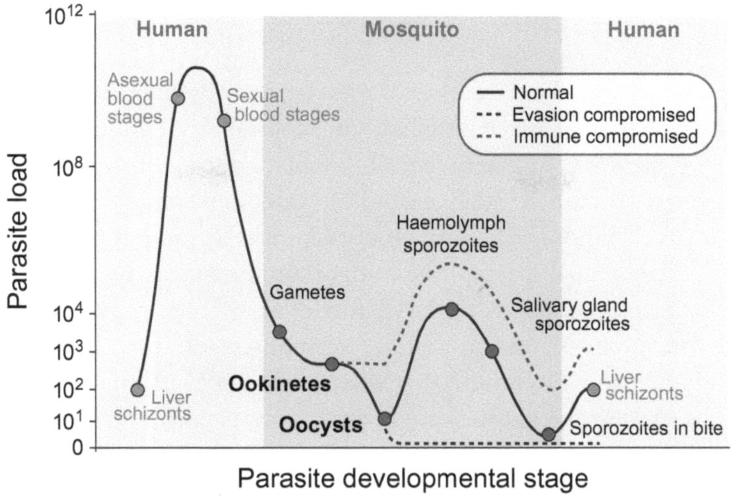

Fig. 1 *Plasmodium* population bottleneck in the *A. gambiae* midgut. Malaria parasites are confronted by robust reactions of the mosquito immune system in the mosquito midgut, especially during the ookinete-to-oocyst developmental transition. In laboratory conditions, of approximately 1000 ookinetes, the average number that survive through this stage and develop to oocysts is about 20 (*black solid line*). The killed ookinetes are eliminated by reactions of the mosquito complement pathway. Inactivation of this pathway increases the oocyst load between 5- and 50-fold with respective increases of the number of sporozoites (*red dashed line*). Conditions that disrupt the ability of parasites to evade the complement pathway completely abolishes transmission (*blue dashed line*). The numbers indicated are approximate values based on Vogel (2010). (See the colour plate.)

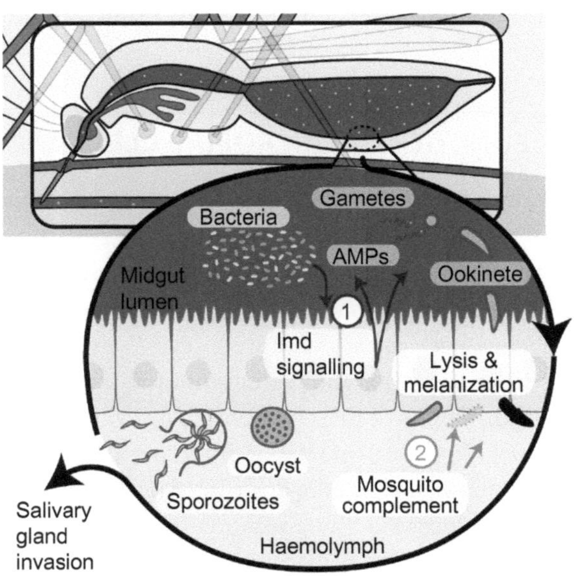

Fig. 2 *Plasmodium* life cycle in the *A. gambiae.* Schematic diagram of mosquito stages of the *Plasmodium* life cycle highlighting the site of parasite killing for the Imd pathway (1; *blue lines*) and the mosquito complement system as ookinetes enter the haemolymph (2; *orange lines*). (See the colour plate.)

a complex pathway that resembles the vertebrate complement largely conceptually rather than mechanistically or evolutionary. Mosquito complement is a network of haemolymph-circulating proteins, including the Complement Factor 3 (C3) homolog thioester-containing protein 1 (TEP1), which binds to the surface of ookinetes as soon as they appear in the basal subepithelial space of the midgut, orchestrating their lysis or melanization (Blandin et al., 2004; Povelones et al., 2013). These reactions are assisted by actin-mediated cellular responses of the invaded cells (Schlegelmilch and Vlachou, 2013; Shiao et al., 2006). A question that tantalized researchers for many years is how some parasites manage to evade such a robust immune reaction. A partial answer came from the identification of Pfs47, a surface molecule of *Plasmodium falciparum* gametocytes and ookinetes. Some natural variants of Pfs47 do not activate JNK signalling in the invaded midgut cells. In this case, ookinetes are not nitrated and thus escape subsequent attack by mosquito complement (Molina-Cruz et al., 2013; Ramphul et al., 2015). Ookinetes expressing other Pfs47 variants, or those in which the gene has been deleted, activate JNK signalling, nitration and complement-mediated attack.

Here, we review our current understanding of the mosquito complement system, focusing on its responses against malaria parasites but also discussing its responses to other microbes, including bacteria and fungi, that have been used extensively in dissecting this pathway. Future research challenges to further unravel this key immune pathway and opportunities for its use in disease control are also presented.

2. TEP1 FUNCTION IN MICROBIAL CLEARANCE

The mosquito complement pathway is summarized in Fig. 3. The hallmark of the mosquito complement pathway is TEP1, a homolog of vertebrate C3. It was discovered by targeted homology searches and initially shown to bind to bacteria via its reactive thioester motif mediating their phagocytosis in *Anopheles gambiae* cultured cells (Levashina et al., 2001). Later, it was shown to be important in adult mosquitoes for bacterial phagocytosis (Moita et al., 2005). TEP1 is structurally similar to vertebrate C3 (Baxter et al., 2007). However, one major difference is that TEP1, like other insect TEPs, lacks an anaphylatoxin domain. Another major difference is that after being secreted into the haemolymph as a full-length protein (TEP1-F), TEP1 undergoes a maturation cleavage to form $TEP1_{cut}$ (Le et al., 2012). Given that the protease cleavage site is likely to adopt

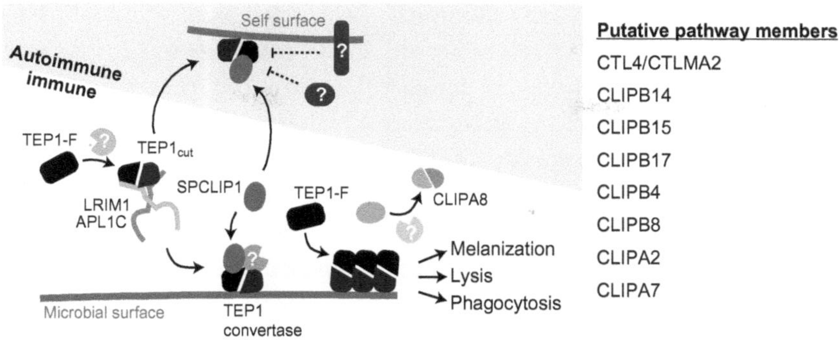

Fig. 3 Schematic model of *A. gambiae* complement pathway. *Black arrows* indicate known interactions or proteolytic reactions. Hypothesized components or interactions are indicated with a *question mark* or *dashed line*, respectively. Activating events are indicated with *arrows*. Inhibitory events are shown with a *flat line*. Mosquito complement components that are not yet placed within this model but whose function is known to depend on TEP1 (CTL4/CTLM2, CLIPA2), or highly expected to depend on TEP1 based on their RNAi phenotypes (CLIPB14, CLIPB15, CLIPB17, CLIPB4, CLIPB8 and CLIPA7) are listed in the space on the *right*. (See the colour plate.)

several extended conformations, it is hypothesized that it might be targeted by diverse proteases (Baxter et al., 2007). The identity of the protease(s) that perform this cleavage is not known. Candidate proteases belonging to the family of CLIP domain serine proteases have been silenced individually without affecting the amount of $TEP1_{cut}$ formation. This suggests that either CLIPs function redundantly for TEP1 processing or another class of proteases performs this function. A protease capable of processing TEP1-F exists in the haemolymph as it was shown that injection of TEP1-F into mosquitoes led to the rapid appearance of $TEP1_{cut}$ (Fraiture et al., 2009). It is also possible that TEP1-F is processed to $TEP1_{cut}$ prior to being secreted by TEP1-producing cells. Identification of the TEP1 maturing protease(s) is an important frontier for understanding the function of mosquito complement.

Following its maturation cleavage, $TEP1_{cut}$ instantly binds to a complex of two leucine-rich repeat proteins, LRIM1 and APL1C, which appear to act as scaffold keeping $TEP1_{cut}$ in an active state until it is delivered to the microbial surface and preventing autoimmune reaction of its thioester motif with the haemolymph and surrounding tissues (Fraiture et al., 2009; Povelones et al., 2009). One intriguing observation is that western blots of proteins eluted from the surfaces of bacteria and probed with anti-TEP1 antibody always revealed one major band corresponding to $TEP1_{cut}$, and in some instances an additional faint full-length TEP1-F band (Levashina et al., 2001; Povelones et al., 2013; Yassine et al., 2014). Since TEP1 was shown to have an active thioester motif (Levashina et al., 2001) one would expect to obtain several higher molecular weight bands containing covalent complexes of TEP1 and bacterial surface molecules. Indeed, high-molecular-weight complexes were observed when proteins eluted from bacteria following incubation with mammalian serum were probed with anti-C3b antibody, indicating the covalent binding of C3b to surface structures (Barrio et al., 1991; Vogel et al., 1997). Even C3b multimers corresponding to C3b nucleophilic attack on other C3b molecules have been detected in these assays (Barrio et al., 1991; Berends et al., 2015). This questions whether the thioester motif in TEP1 is indeed involved in covalent linkage to microbial molecules as shown for mammalian C3b and C4b. The fact that methylamine treatment reduced $TEP1_{cut}$ interaction with bacterial surfaces may suggest so (Levashina et al., 2001); however, the concentration of methylamine used to hydrolyse the thioester bond is sufficient to disrupt the covalent linkage between LRIM1 and APL1C (M. Povelones, unpublished observation). The methylamine results may also suggest that hydrolysis of the thioester motif is a prerequisite for $TEP1_{cut}$ dissociation from LRIM1/APL1C complex and localization to bacterial surfaces in an unusual mechanism that remains to be elucidated.

TEP1 is a major anti-*Plasmodium* factor. TEP1 silencing increases the numbers of *Plasmodium berghei* and (Blandin et al., 2004) *P. falciparum* (NF54 strain) oocysts in *A. gambiae* (Dong et al., 2006a). TEP1 localizes to the surfaces of dead *P. berghei* ookinetes as evidenced by their irregular morphology, loss of nuclear stain and the presence of bubble-like projections from their surface. Additionally, since these parasites expressed cytoplasmic GFP, the GFP signal was lost following TEP1 binding suggesting that TEP1 is likely triggering parasite lysis (Blandin et al., 2004). How TEP1 mediates parasite lysis remains unknown. Despite the fact that TEP1 was shown to localize to *P. berghei* ookinetes by electron microscopy, there has been no evidence of pore formation analogous to that induced by the membrane attack complex of mammalian complement.

A multimeric molecule equivalent to the LRIM1/APL1C/$TEP1_{cut}$ complex has not been described to date in vertebrates. LRIM1 was the first to be discovered through a functional genetic screen for molecules that alter the infection intensity of the rodent parasite *P. berghei* that is often used as a laboratory model in such studies (Osta et al., 2004). Soon after, the *APL1* locus was identified as a candidate region that conferred resistance to both field *P. falciparum* isolates and *P. berghei* (Riehle et al., 2006). The locus contains three closely related genes, *APL1A*, *APL1B* and *APL1C* that have presumably undergone recent duplication (Povelones et al., 2009; Riehle et al., 2008; Waterhouse et al., 2010). Therefore, in the first functional studies of APL1, the dsRNA used for gene silencing targeted all three proteins (Fraiture et al., 2009; Povelones et al., 2009; Riehle et al., 2006). Later, the *A. gambiae* genome was manually annotated and it was appreciated that the *APL1* locus contained three recently duplicated genes. These updated gene models allowed the design of gene-specific dsRNAs targeting unique sequences in the UTRs for functional analysis of the individual roles of APL1A, APL1B and APL1C in mosquito immunity, as discussed later.

TEP1 is also required for the clearance of *Escherichia coli* and *Staphylococcus aureus* bacteria injected into the haemocoel (Moita et al., 2005). Enhancing TEP1 activity by silencing either of two negative regulators of the pathway (Kamareddine et al., 2016; Yassine et al., 2014), triggered the rapid clearance of *E. coli* from the haemolymph of *A. gambiae* mosquitoes, indicating that TEP1 is clearly involved in mosquito resistance to systemic bacterial infections. It is worth noting that in all these assays the numbers of bacteria injected into the haemocoel are likely to exceed by far those occurring in a natural infection, which may saturate several effector responses including phagocytosis. This is unlikely to occur in natural conditions were most likely few cells succeed initially in entering the haemolymph establishing an infection. It remains to be shown whether in the latter case, TEP1 role remains essential or becomes redundant. Interestingly, *TEP1* knockdown mosquitoes infected with the entomopathogenic fungus *Beauveria bassiana* harboured significantly higher numbers of circulating hyphal bodies relative to the control group (Yassine et al., 2012), also suggesting a role for complement in mosquito resistance to fungal infections. In this study, a natural mode of infection was adopted based on spraying mosquitoes with a suspension of fungal spores to mimic natural infections occurring by tarsal contact with fungal spores (Kamareddine et al., 2013). More recently, TEP1 was shown to bind and remove damaged sperm cells contributing to male fertility suggesting a broader role extending to tissue homeostasis (Pompon and Levashina, 2015).

3. *Plasmodium* EVASION OF MOSQUITO COMPLEMENT

Rodent malaria parasites are a valuable model and have helped uncover much of what is known about mosquito complement and immunity in general. Rodent parasites are relatively easy to work with in the laboratory and powerful tools have been established to genetically manipulate them. Transformation of rodent parasites, primarily *P. berghei*, has been particularly helpful in studying the early steps of mosquito infection and in the analysis of complement pathway components. However, the strong RNAi phenotypes exhibited by the key complement pathway members LRIM1, CTL4 and CTLMA2 using *P. berghei* infections were not observed when *P. falciparum* field isolates were first used to infect mosquitoes (Cohuet et al., 2006). Initially, this raised concerns as to the *Plasmodium* species-specificity of mosquito immune reactions and questioned the appropriateness of using *P. berghei* as a model malaria parasite. However, an alternative interpretation of these data is that the human parasites which coevolved with *A. gambiae* mosquitoes are able to evade mosquito complement while rodent parasites, which are not transmitted by *A. gambiae* cannot. In this context, the human parasites may utilize specific molecules that allow them to actively avoid complement-mediated killing. In support of this hypothesis, using a different combination of *P. falciparum* strain and *A. gambiae*, it was shown that the central mosquito complement component, TEP1, was strongly associated with protecting mosquitoes from infection by both rodent and human parasites (Dong et al., 2006a). Furthermore, silencing the proteins that form the core of the mosquito complement pathway (LRIM1, APL1C and SPCLIP1) in *A. gambiae* increased infection by *P. falciparum* (Dong et al., 2006a; Garver et al., 2012; Mitri et al., 2015; Molina-Cruz et al., 2012, 2013). Despite this, it remained unclear for years whether there are absolute differences in the way the mosquito immune system attacked different plasmodia.

While the molecular basis of complement evasion is not fully understood, recent genetic experiments have shed light on this important question. A survey of *P. falciparum* isolates from different geographical locations revealed striking differences in their ability to infect *A. gambiae*. A Brazilian isolate (7G8) was strongly attacked and melanized in the refractory L3–5 strain of mosquitoes, whereas an African *P. falciparum* isolate (3D7) was not (Molina-Cruz et al., 2013) and produced viable oocysts. Analysis of recombinant progeny of a cross of 7G8 and 3D7 was used to map parasite determinants of killing and melanization. Interestingly, a single locus, Pfs47, was implicated in a series of elegant mapping and functional analysis.

Pfs47 is a *Plasmodium* surface protein and has been an intense target of study in recent years due, in part, to its potential as a transmission-blocking vaccine target in the human host (van Schaijk et al., 2006). Expressed in female gametocytes, the surface protein is not required for the fertility of

P. falciparum (Molina-Cruz et al., 2013; Vaughan et al., 2012), unlike its homolog in *P. berghei* (van Dijk et al., 2010). Following gamete fusion, Pfs47 maintains its surface localization in ookinetes. Pfs47 KO strain parasites overwhelmingly succumb to the TEP1-dependent immune response, but are able to survive readily in TEP1 knockdown mosquitoes (Molina-Cruz et al., 2013). Further investigation has revealed that many parasite strains are closely associated with their respective vector strains by Pfs47 haplotype—a given haplotype protects the parasite only from its own vector's complement-mediated immune control, while leaving it vulnerable to that of another vector (Molina-Cruz et al., 2015).

The mechanism of action of Pfs47 is not yet clear; however, its disruption of the anopheline immune response is not limited to the complement cascade, with recent evidence suggesting effects on the JNK-mediated apoptotic cascade (Ramphul et al., 2015). While recent research indicates that polymorphisms in Pfs47 are not solely responsible for strain-specific immune response, it is clear that Pfs47 is crucially important to evasion of vector immune responses in at least some parasite strains (Eldering et al., 2016). Whether the interaction between Pfs47 and the *A. gambiae* complement system involves direct interaction between the surface protein and specific complement components remains a topic of active research.

4. COMPLEMENT FIXATION ON MICROBIAL SURFACES: SPONTANEOUS OR PRR MEDIATED?

TEP1 is not the only protein that exhibits effector functions against *Plasmodium* parasites. Several other proteins exhibited similar RNAi phenotypes as TEP1 among which are putative pattern recognition receptors (PRRs) including leucine-rich immune proteins (LRIMs), Gram-negative binding proteins (GNBPs), fibrinogen-related proteins (FREPs) and immunoglobulin superfamily proteins. LRIM1 was the first LRIM protein discovered to exhibit a similar RNAi phenotype to TEP1 characterized by a dramatic increase in *P. berghei* oocysts in the vector (Osta et al., 2004). Subsequently, another LRIM called APL1C was identified as an anti-*Plasmodium* factor (Riehle et al., 2006, 2008) and was later shown to form a complex with LRIM1 that stabilizes TEP1$_{cut}$ in the mosquito haemolymph (Fraiture et al., 2009; Povelones et al., 2009) as described in the previous section. So far these two leucine-rich proteins are the only putative PRRs that act upstream of TEP1. LRRD7 is another member of the LRIM family with anti-*P. berghei* and *P. falciparum* activities; however, it is still not clear whether it functions through TEP1 (Dong et al., 2006a). Systematic functional genetic analysis of the GNBP family in *A. gambiae* identified GNBPA2 as an anti-*P. falciparum* molecule, while GNBPB3 and GNBPB4 were implicated in defence against *P. berghei*. GNBP4 knockdown reduced the expression of several immune-related genes in bacteria-challenged cell lines including several AMPs suggesting that it activates NF-kB pathways in response to infection (Warr et al., 2008). Several members of the FREP gene family, specifically FBN8, FBN9 and FBN39, also exhibit anti-*Plasmodium* activities, with FBN39 regulating mosquito resistance to *P. falciparum*, whereas FBN8 and FBN9 provide resistance to both *P. falciparum* and to *P. berghei* (Dong et al., 2006a). AgDscam is a hypervariable receptor of the immunoglobulin-domains superfamily which was also implicated in mosquito resistance to *P. berghei* (Dong et al., 2006b) and *P. falciparum* (Dong et al., 2012) infections. Alternative splicing of Dscam can generate over 31,000 different splice forms with variable adhesive domains. The generation of Dscam splice variants seems to be pathogen species specific and is controlled by splice factors whose transcript levels are regulated by Imd and Toll pathways (Dong et al., 2012).

Whether these PRRs (with the exception of LRIM1 and APL1C) are acting upstream of TEP1 or involved in TEP1 recruitment to surfaces is still unknown. Despite the fact that GNBPB4 (Warr et al., 2008), Dscam (Dong et al., 2012) and FBN9 (Dong and Dimopoulos, 2009) were shown to localize to the surface of *Plasmodium* ookinetes, it is still not clear whether they interact with dead or live ookinetes and whether their recruitment to these surfaces precedes or follows TEP1 binding. Colocalization studies with TEP1 and gene silencing experiments coupled with quantitative measurements of TEP1 binding to ookinetes or other microbial surfaces will be needed to clarify these points. Hence, it remains an open question as to whether the complement system in mosquitoes resembles more the classical and lectin pathways in mammals from the perspective of requiring an upstream PRR to trigger the response or the alternative (spontaneous) pathway whereby TEP1 is spontaneously activated. Since LRIM1 and APL1C, the only upstream regulators of TEP1, do not seem to interact with microbial surfaces and that TEP1 is required for the efficient melanization of sephadex beads which exhibit no microbial patterns on their surface (Warr et al., 2006), favour the scenario of the spontaneous pathway. In this scenario, one would expect that TEP1 spontaneous activation is amplified on microbial (foreign) surfaces, whereas it is controlled in the fluid phase and on the surface of host cells by regulatory proteins that are yet to be determined.

5. MOSQUITO COMPLEMENT REGULATION

In mammals, complement amplification by the C3 convertases increases the deposition of C3b opsonin on microbial surfaces triggering their quick uptake by phagocytic cells. Studies in *Anopheles gambiae* point to a potential analogous amplification process in insect complement. It was shown that systemic bacterial infections trigger a quick reduction in TEP1-F levels in the mosquito haemolymph (Povelones et al., 2013). However, knockdown of the serine protease homolog *SPCLIP1* rescues this TEP1-F loss suggesting the presence of a putative convertase-like activity that may enhance TEP1-F conversion to active $TEP1_{cut}$ on microbial surfaces. This activity is thought to be triggered by the initial deposition of the LRIM1/APL1C loaded $TEP1_{cut}$ on microbial surfaces followed by the recruitment of SPCLIP1 (Povelones et al., 2013) and a putative catalytic enzyme that remain to be identified. Using an *E. coli* bioparticle surface extraction assay to monitor TEP1 levels on the surfaces of bacteria that were recovered from the mosquito haemolymph soon after injection, it was clearly shown that *SPCLIP1* kd reduced the amount of $TEP1_{cut}$ deposited on *E. coli* surfaces (Povelones et al., 2013). It is worth noting here that the source of $TEP1_{cut}$ that accumulates on microbial surfaces may also be the active form carried by the LRIM1/APL1C complex rather than TEP1-F itself. Further experiments are needed to clarify this point.

In contrast to SPCLIP1, CLIPA2, another serine protease homologue, seems to act as a negative regulator of mosquito complement amplification. *CLIPA2* knockdown enhanced the consumption of haemolymph TEP1-F following bacterial systemic infections. Further, *CLIPA2* knockdown mosquitoes exhibited increased resistance to infection (Yassine et al., 2014) in TEP1-dependent manner (Kamareddine et al., 2016). Silencing CLIPA2 did not influence TEP1 transcript levels (Kamareddine et al., 2016; Yassine et al., 2014) strongly suggesting that CLIPA2 function is restricted to the control of TEP1 activation dynamics. However, while *SPCLIP1* silencing reduced the amount of $TEP1_{cut}$ eluted from *E. coli* bioparticle surfaces, *CLIPA2* silencing, on the other hand, did not show an increase in the amount of $TEP1_{cut}$ eluted from these surfaces. This could be explained by enhanced bacterial clearance in the latter genotype which would reduce the amount of recovered bioparticles. SPCLIP1 and CLIPA2 are now considered as components of the complement pathway in mosquitoes in addition to the LRIM1/APL1C heterodimer and possibly others factors yet to be discovered. This conclusion was reached following the observation that silencing LRIM1 in naive mosquitoes triggered the concomitant loss of TEP1$_{cut}$, SPCLIP1 and CLIPA2 from the haemolymph indicating that these proteins are components of a functional protein module whose actual and putative constituents are summarized in Fig. 3. Interestingly, although the autoimmune attack of self-surfaces recapitulates some aspects of complement activation on a microbial surface, TEP1-F is not depleted and there are no signs of mosquito mortality or tissue damage or melanization (Povelones et al., 2013). This strongly implies the presence of regulatory factors that prevent complete TEP1 convertase formation on mosquito tissue (Fig. 3). Identifying regulatory factors is important since interfering with them would likely lead to increased mosquito mortality and may ultimately provide novel avenues for controlling mosquito populations.

In addition to serine protease homologues, complement activity seems to be regulated by other classes of proteins including C-type lectins (CTLs) and apolipophorins. Systematic functional genetic analysis of selected members of the CTL gene family identified two genes, CTL4 and CTLMA2, whose silencing triggered the massive melanization of *P. berghei* ookinetes in the mosquito midgut (Osta et al., 2004). Interestingly, the melanotic response triggered in these mosquito genotypes was found to be TEP1-dependent (Povelones et al., 2011), indicating that these CTLs act downstream of TEP1. CTL4 and CTLMA2 form a disulfide-linked heterodimer in the haemolymph which is also involved in resistance to *E. coli* infections (Schnitger et al., 2009). However, it remains unknown whether this latter role requires mosquito complement or instead reflects the direct antimicrobial potential of these molecules. In the small intestine of humans and mice, simple CTLs homologous to CTL4 and CTLMA2 were shown to be important components of the mucosal immune defence whereby they directly lyse Gram-positive bacteria (Cash et al., 2006). It is tempting to speculate that the CTL4-CTLMA2 dimer is a component of the lytic arm of complement, the absence of which switches the response to melanization. Studies are ongoing to highlight the mode of action of these lectins and their specific role in the context of mosquito complement.

Apolipophorins are complex assemblies of multifunctional molecules that are mainly involved in lipid transport. Initially, it was reported that silencing the *Apo-II/I* (Lipophorin) gene in *A. gambiae* mosquitoes reduced the numbers of *P. berghei* and *P. falciparum* oocysts that successfully developed in their midguts (Mendes et al., 2008). Microscopic analysis revealed that parasite death in these mosquitoes occurred at the ookinete stage.

Later, it was shown that ookinete killing in *Apo-II/I* knockdown mosquitoes required TEP1 activity (Rono et al., 2010). Recently, *Apo-II/I* knockdown mosquitoes were shown to exhibit increased resistance to systemic infections with *B. bassiana* and *E. coli*, also in a TEP1-dependent manner (Kamareddine et al., 2016). Interestingly, silencing *Apo-II/I* upregulated the expression of *TEP1* following systemic infections in a JNK pathway dependent manner (Kamareddine et al., 2016). Altogether, these studies suggest that mosquito Lipophorin plays an important role in regulating mosquito complement during systemic infections. Deciphering the molecular basis of this regulation will provide novel insight into the functional interplay between lipid metabolism and immune gene regulation.

6. EFFECTOR MECHANISMS MEDIATED BY MOSQUITO COMPLEMENT

6.1 Phagocytosis

The first evidence for an effector role for TEP1 came from phagocytosis assays performed in the mosquito haemocyte-like 5.1* cell line (Levashina et al., 2001). These assays revealed that bacteria preincubated in conditioned medium depleted of TEP1 (by RNAi) were less efficiently phagocytosed when presented to 5.1* cells. These observations were later confirmed using an in vivo phagocytosis assay based on time-lapse microscopy to follow the fate of fluorescently labelled *E. coli* and *S. aureus* bioparticles injected into the abdomens of mosquitoes silenced for specific genes by RNAi (Moita et al., 2005). This in vivo assay confirmed the role of TEP1 in the phagocytosis of *E. coli* and *S. aureus*. The authors also showed that TEP1 silencing increased the accumulation of live *E. coli* and *S. aureus* cells in adult mosquitoes (Moita et al., 2005), suggesting a significant contribution of TEP1 to bacterial clearance. It was not clear though whether this increase is due to the impaired lytic or phagocytic arm of TEP1. TEP1 binds to bacterial surfaces and is proposed to act as an opsonin (Levashina et al., 2001). However, the nature of the receptor that binds TEP1 opsonin remains unknown. It is also possible that this putative receptor may not bind TEP1 directly but another molecule recruited downstream of TEP1.

6.2 Lysis

While TEP1 was shown to bind bacterial surfaces and trigger their phagocytosis it remained unknown whether it triggered bacterial lysis as well, similar to mammalian complement. Also, if it triggered lysis what would be its relative contribution midst other mosquito factors that are well recognized to be involved in that process, namely the antimicrobial proteins and peptides. This question was addressed using a simple infection model based on injecting adult *A. gambiae* mosquitoes with *E. coli*-expressing cytoplasmic GFP and measuring the amount of GFP protein released into the haemolymph at 30 min after bacterial injection. Haemolymph samples were centrifuged to eliminate GFP carryover from live bacteria present in the haemolymph. In this assay, the intensity of the GFP signal in western blot analysis of haemolymph samples is expected to correlate with the extent of bacterial lysis that occurred in the haemolymph (Kamareddine et al., 2016). Using this approach, a barely detectable GFP signal was observed in the haemolymph of *TEP1* knockdown mosquitoes compared to that of control mosquitoes and those of other genotypes. These results indicate that TEP1 is indeed involved in bacterial lysis and that in the very early phases of bacterial systemic infection the main lytic activity in the haemolymph is attributed to TEP1. What remains to be shown is the relative contribution of the opsonic and lytic branches of TEP1 to bacterial defence. To be able to address this question, specific factors involved in each branch should be identified first to allow the targeted genetic disruption of each branch at a time. In humans, individuals with genetic deficiencies in components of the complement membrane attack complex that triggers cellular lysis seem to cope well with almost all bacterial infections but are more susceptible to infections with *Nisseria* (Pettigrew et al., 2009). This suggests that the lytic arm of complement may have evolved under selective pressure from few bacterial species particularly adept at escaping opsonization such as *Nisseria*. One particular difficulty in addressing this question in *A. gambiae* is the lack of information on the nature of bacterial pathogens encountered by these mosquitoes in their natural habitats.

6.3 Melanization

The first evidence for the role of TEP1 in the melanization response came from the observation that *TEP1* silencing completely abolished *P. berghei* ookinete melanization in the refractory *A. gambiae* strain L3–5 leading to normal oocyst development (Blandin et al., 2004). *CTL4* knockdown in susceptible *A. gambiae* strains triggered the massive melanization of *P. berghei* ookinetes resulting in a phenotype similar to that of L3–5 mosquitoes (Osta et al., 2004). Cosilencing *CTL4* and *TEP1* in the susceptible Ngousso strain abolished melanization leading to the normal development of oocysts, suggesting that TEP1 is required for parasite melanization irrespective of the mosquito genetic background (Povelones et al., 2011). However, these studies did not provide a definitive proof that links TEP1 to the initiation of the melanization response since melanization could have been triggered in response to the dead ookinetes killed by TEP1 rather than directly by TEP1 itself. Subsequent studies using bacterial infections of mosquitoes provided irrefutable evidence on the essential role of TEP1 in triggering melanization. In fact, *TEP1* knockdown abolished prophenoloxidase activity in the haemolymph in response to *E. coli* systemic infections (Povelones et al., 2013). CLIPA8 is an essential SPH in melanization which is cleaved soon after the injection of bacteria into the mosquito haemolymph (Schnitger et al., 2007). *TEP1* kd abolished also CLIPA8 cleavage suggesting that TEP1 is upstream of the melanization response. CLIPA2 is another SPH that negatively regulates TEP1 consumption during systemic infections. *CLIPA2* silencing enhanced TEP1 activity leading to an exaggerated phenoloxidase activity in the haemolymph following *E. coli* infections (Yassine et al., 2012). Furthermore, *TEP1* kd abolished the recruitment of prophenoloxidase to *B. bassiana* hyphae invading the mosquito haemocoel, hence inhibiting melanization (Yassine et al., 2012). Altogether, these studies indicate that the infection-induced melanization response in the mosquito is indeed an effector branch downstream of TEP1 in parallel with the lytic and opsonization branches. We have previously identified several catalytic CLIPs including CLIPB17, CLIPB8 and CLIPB4 that are involved in the melanization of *P. berghei* ookinetes through a systematic in vivo RNAi screen (Volz et al., 2006). The fact that TEP1 acts as a key trigger of the infection-induced melanization response suggests that these CLIPBs are additional components of the TEP1 pathway. How does TEP1 coordinate the activation of its downstream effector responses to deal with specific microbial insults? What are the molecular components of these effector responses and their hierarchical organization? What is the relative contribution of each of these responses to microbial clearance in different infection settings? These are all challenging questions that remain to be addressed in order to draw a comprehensive picture of mosquito complement.

7. STRUCTURE–FUNCTION STUDIES OF COMPLEMENT COMPONENTS

7.1 APL1C

The APL1C protein is the most extensively studied of the APL1 proteins. APL1C circulates in the mosquito haemolymph as a disulfide-linked heterodimer with LRIM1 (Fraiture et al., 2009; Povelones et al., 2009). This heterodimer formation appears to be important for the secretion or stability of APL1C. When LRIM1 is knocked down in *A. gambiae* mosquitoes, APL1C is almost depleted from the haemolymph (Fraiture et al., 2009; Povelones et al., 2009). Cell culture studies also supported the preference of a heterodimer. Transfection of insect cells lacking any endogenous LRIM family members with LRIM1 or APL1C alone will secrete monomers and homodimers of each protein into the cell culture medium. Cells cotransfected with both secrete each protein more efficiently and in a predominantly heterodimeric form, although monomers and homodimers are still present at low levels (Baxter et al., 2010; Povelones et al., 2011). A static crystal structure has been published for both LRIM1 and APL1C individually. Importantly, the structure of the LRIM1/APL1C heterodimer has been elucidated as well (Baxter et al., 2010). The three-dimensional structure reveals that the five cysteine residues that are characteristic of LRIM family members form two internal disulfide bonds (Baxter et al., 2010) that are important for the stability and secretion of the proteins (Baxter et al., 2010; Fraiture et al., 2009; Povelones et al., 2009, 2011). The final cysteine forms the disulfide bond between LRIM1 and APL1C. Mutagenesis of this residue completely abrogates the ability of LRIM1 and APL1C to form a covalent heterodimer (Baxter et al., 2010; Povelones et al., 2011). However, this mutation does not compromise the stability of LRIM1 or APL1C. Proteins lacking this cysteine residue are still capable of interacting with each other noncovalently and maintain their ability to enhance the secretion of their partner into cell culture medium (Baxter et al., 2010; Povelones et al., 2011).

7.2 APL1A

Detailed protein analysis has been performed for APL1A. Interestingly, several alleles of APL1A are found in *A. gambiae* mosquitoes colonized from Ngousso (Holm et al., 2012). These alleles contain natural variants of those studied in APL1C. An APL1A protein containing all five cysteine residues was efficiently secreted by transfected cells and formed high-molecular-weight complexes. Alleles of APL1A that truncate the protein, either removing or misaligning the bridging cysteine, produce proteins that are not secreted from cells and are found strongly aggregated in vesicles inside the transfected cells. The *A. gambiae* 4a3A cell line used for this analysis were previously shown to have very low levels of LRIM1 and APL1C (Povelones et al., 2009) and it remains unknown whether either of these potential partners could assist in their secretion.

7.3 APL1B

APL1B can also heterodimerize with LRIM1 (Williams et al., 2015). In addition, the APL1B protein was found to have the capacity to homodimerize. In contrast to LRIM1/APL1A and APL1B/APL1C heterodimers, homodimers of APL1B occur via a noncovalent interaction between their N-terminal domains that is independent of the cysteine rich motif (Williams et al., 2015).

7.4 LRIM1

Despite its ability to heterodimerize with all APL1 proteins in cell culture experiments, to date, only one band containing LRIM1 is detectable in the haemolymph of *A. gambiae* corresponding to the LRIM1/APL1C heterodimer (Povelones et al., 2009). This would suggest that if complexes form between LRIM1 and APL1A or APL1B they are secreted at levels below the limit of detection, are present in other tissues, or are retained within cells and not secreted into the haemolymph. It would be interesting to produce APL1 antibodies that cross react with all forms of APL1 to screen mosquito haemolymph following specific knockdown of APL1A, APL1B and APL1C.

8. CONCLUDING REMARKS

Mosquito complement protein TEP1 plays a key role in mosquito immune defence against *Plasmodium*, bacteria, and fungi. Several TEP1 alleles are present in mosquito populations suggesting that TEP1 is under selective pressure from pathogens prevailing in mosquito habitats, especially the aquatic larval habitats. Mosquitoes have many more TEP genes than *Drosophila*, which contains only six TEPs that do not seem to play significant roles in immunity. These observations raise interesting questions that remain to be addressed in order to understand in what context the role of mosquito complement is unique and nonredundant with other effector responses. Do TEPs play key roles in other mosquito species? Does their role extend to other distant blood-sucking insects? Do TEPs provide an advantage to blood-sucking insects? What classes of microbes specifically require TEPs to be cleared? Answers to these questions await rigorous functional genetic studies in different classes of insects.

REFERENCES

Barrio, E., Anton, L.C., Marques, G., Sanchez, A., Vivanco, F., 1991. Formation of covalently linked C3-C3 dimers on IgG immune aggregates. Eur. J. Immunol. 21, 343–349.

Baxter, R.H., et al., 2007. Structural basis for conserved complement factor-like function in the antimalarial protein TEP1. Proc. Natl. Acad. Sci. U.S.A. 104, 11615–11620.

Baxter, R.H., et al., 2010. A heterodimeric complex of the LRR proteins LRIM1 and APL1C regulates complement-like immunity in Anopheles gambiae. Proc. Natl. Acad. Sci. U.S.A. 107, 16817–16822.

Berends, E.T., et al., 2015. Molecular insights into the surface-specific arrangement of complement C5 convertase enzymes. BMC Biol. 13, 93.

Blandin, S., et al., 2004. Complement-like protein TEP1 is a determinant of vectorial capacity in the malaria vector Anopheles gambiae. Cell 116, 661–670.

Cash, H.L., Whitham, C.V., Behrendt, C.L., Hooper, L.V., 2006. Symbiotic bacteria direct expression of an intestinal bactericidal lectin. Science 313, 1126–1130.

Clayton, A.M., Dong, Y., Dimopoulos, G., 2014. The Anopheles innate immune system in the defense against malaria infection. J. Innate Immun. 6, 169–181.

Cohuet, A., et al., 2006. Anopheles and Plasmodium: from laboratory models to natural systems in the field. EMBO Rep. 7, 1285–1289.

Dong, Y., Dimopoulos, G., 2009. Anopheles fibrinogen-related proteins provide expanded pattern recognition capacity against bacteria and malaria parasites. J. Biol. Chem. 284, 9835–9844.

Dong, Y., et al., 2006a. Anopheles gambiae immune responses to human and rodent Plasmodium parasite species. PLoS Pathog. 2, e52.

Dong, Y., Taylor, H.E., Dimopoulos, G., 2006b. AgDscam, a hypervariable immunoglobulin domain-containing receptor of the Anopheles gambiae innate immune system. PLoS Biol. 4, e229.

Dong, Y., Manfredini, F., Dimopoulos, G., 2009. Implication of the mosquito midgut microbiota in the defense against malaria parasites. PLoS Pathog. 5, e1000423.

Dong, Y., Cirimotich, C.M., Pike, A., Chandra, R., Dimopoulos, G., 2012. Anopheles NF-kappaB-regulated splicing factors direct pathogen-specific repertoires of the hypervariable pattern recognition receptor AgDscam. Cell Host Microbe 12, 521–530.

Eldering, M., et al., 2016. Variation in susceptibility of African Plasmodium falciparum malaria parasites to TEP1 mediated killing in Anopheles gambiae mosquitoes. Sci. Rep. 6, 20440.

Fraiture, M., et al., 2009. Two mosquito LRR proteins function as complement control factors in the TEP1-mediated killing of Plasmodium. Cell Host Microbe 5, 273–284.

Garver, L.S., et al., 2012. Anopheles Imd pathway factors and effectors in infection intensity-dependent anti-Plasmodium action. PLoS Pathog. 8, e1002737.

Garver, L.S., de Almeida Oliveira, G., Barillas-Mury, C., 2013. The JNK pathway is a key mediator of Anopheles gambiae antiplasmodial immunity. PLoS Pathog. 9, e1003622.

Gendrin, M., et al., 2015. Antibiotics in ingested human blood affect the mosquito microbiota and capacity to transmit malaria. Nat. Commun. 6, 5921.

Holm, I., et al., 2012. Diverged alleles of the Anopheles gambiae leucine-rich repeat gene APL1A display distinct protective profiles against Plasmodium falciparum. PLoS One 7, e52684.

Kamareddine, L., Fan, Y., Osta, M.A., Keyhani, N.O., 2013. Expression of trypsin modulating oostatic factor (TMOF) in an entomopathogenic fungus increases its virulence towards Anopheles gambiae and reduces fecundity in the target mosquito. Parasites Vectors 6, 22.

Kamareddine, L., Nakhleh, J., Osta, M.A., 2016. Functional interaction between apolipophorins and complement regulate the mosquito immune response to systemic infections. J. Innate Immun. 8, 314–326.

Kumar, S., Molina-Cruz, A., Gupta, L., Rodrigues, J., Barillas-Mury, C., 2010. A peroxidase/dual oxidase system modulates midgut epithelial immunity in Anopheles gambiae. Science 327, 1644–1648.

Le, B.V., Williams, M., Logarajah, S., Baxter, R.H., 2012. Molecular basis for genetic resistance of Anopheles gambiae to Plasmodium: structural analysis of TEP1 susceptible and resistant alleles. PLoS Pathog. 8, e1002958.

Levashina, E.A., et al., 2001. Conserved role of a complement-like protein in phagocytosis revealed by dsRNA knockout in cultured cells of the mosquito, Anopheles gambiae. Cell 104, 709–718.

Meister, S., et al., 2009. Anopheles gambiae PGRPLC-mediated defense against bacteria modulates infections with malaria parasites. PLoS Pathog. 5, e1000542.

Mendes, A.M., et al., 2008. Conserved mosquito/parasite interactions affect development of Plasmodium falciparum in Africa. PLoS Pathog. 4, e1000069.

Mendes, A.M., et al., 2011. Infection intensity-dependent responses of Anopheles gambiae to the African malaria parasite Plasmodium falciparum. Infect. Immun. 79, 4708–4715.

Mitri, C., et al., 2015. An evolution-based screen for genetic differentiation between Anopheles Sister taxa enriches for detection of functional immune factors. PLoS Pathog. 11, e1005306.

Moita, L.F., et al., 2005. In vivo identification of novel regulators and conserved pathways of phagocytosis in A. gambiae. Immunity 23, 65–73.

Molina-Cruz, A., et al., 2012. Some strains of Plasmodium falciparum, a human malaria parasite, evade the complement-like system of Anopheles gambiae mosquitoes. Proc. Natl. Acad. Sci. U.S.A. 109, E1957–E1962.

Molina-Cruz, A., et al., 2013. The human malaria parasite Pfs47 gene mediates evasion of the mosquito immune system. Science 340, 984–987.

Molina-Cruz, A., et al., 2015. Plasmodium evasion of mosquito immunity and global malaria transmission: the lock-and-key theory. Proc. Natl. Acad. Sci. U.S.A. 112, 15178–15183.

Osta, M.A., Christophides, G.K., Kafatos, F.C., 2004. Effects of mosquito genes on Plasmodium development. Science 303, 2030–2032.

Pettigrew, H.D., Teuber, S.S., Gershwin, M.E., 2009. Clinical significance of complement deficiencies. Ann. N. Y. Acad. Sci. 1173, 108–123.

Pompon, J., Levashina, E.A., 2015. A new role of the mosquito complement-like cascade in male fertility in Anopheles gambiae. PLoS Biol. 13, e1002255.

Povelones, M., Waterhouse, R.M., Kafatos, F.C., Christophides, G.K., 2009. Leucine-rich repeat protein complex activates mosquito complement in defense against Plasmodium parasites. Science 324, 258–261.

Povelones, M., Upton, L.M., Sala, K.A., Christophides, G.K., 2011. Structure-function analysis of the Anopheles gambiae LRIM1/APL1C complex and its interaction with complement C3-like protein TEP1. PLoS Pathog. 7, e1002023.

Povelones, M., et al., 2013. The CLIP-domain serine protease homolog SPCLIP1 regulates complement recruitment to microbial surfaces in the malaria mosquito Anopheles gambiae. PLoS Pathog. 9, e1003623.

Ramphul, U.N., Garver, L.S., Molina-Cruz, A., Canepa, G.E., Barillas-Mury, C., 2015. Plasmodium falciparum evades mosquito immunity by disrupting JNK-mediated apoptosis of invaded midgut cells. Proc. Natl. Acad. Sci. U.S.A. 112, 1273–1280.

Riehle, M.M., et al., 2006. Natural malaria infection in Anopheles gambiae is regulated by a single genomic control region. Science 312, 577–579.

Riehle, M.M., et al., 2008. Anopheles gambiae APL1 is a family of variable LRR proteins required for Rel1-mediated protection from the malaria parasite, Plasmodium berghei. PLoS One 3, e3672.

Rono, M.K., Whitten, M.M., Oulad-Abdelghani, M., Levashina, E.A., Marois, E., 2010. The major yolk protein vitellogenin interferes with the anti-plasmodium response in the malaria mosquito Anopheles gambiae. PLoS Biol. 8, e1000434.

Schlegelmilch, T., Vlachou, D., 2013. Cell biological analysis of mosquito midgut invasion: the defensive role of the actin-based ookinete hood. Pathog. Glob. Health 107, 480–492.

Schnitger, A.K., Kafatos, F.C., Osta, M.A., 2007. The melanization reaction is not required for survival of Anopheles gambiae mosquitoes after bacterial infections. J. Biol. Chem. 282, 21884–21888.

Schnitger, A.K., Yassine, H., Kafatos, F.C., Osta, M.A., 2009. Two C-type lectins cooperate to defend Anopheles gambiae against Gram-negative bacteria. J. Biol. Chem. 284, 17616–17624.

Shiao, S.H., Whitten, M.M., Zachary, D., Hoffmann, J.A., Levashina, E.A., 2006. Fz2 and cdc42 mediate melanization and actin polymerization but are dispensable for Plasmodium killing in the mosquito midgut. PLoS Pathog. 2, e133.

Upton, L.M., Povelones, M., Christophides, G.K., 2015. Anopheles gambiae blood feeding initiates an anticipatory defense response to Plasmodium berghei. J. Innate Immun. 7, 74–86.

van Dijk, M.R., et al., 2010. Three members of the 6-cys protein family of Plasmodium play a role in gamete fertility. PLoS Pathog. 6, e1000853.

van Schaijk, B.C., et al., 2006. Pfs47, paralog of the male fertility factor Pfs48/45, is a female specific surface protein in Plasmodium falciparum. Mol. Biochem. Parasitol. 149, 216–222.

Vaughan, A.M., et al., 2012. A transgenic Plasmodium falciparum NF54 strain that expresses GFP-luciferase throughout the parasite life cycle. Mol. Biochem. Parasitol. 186, 143–147.

Vogel, G., 2010. The "do unto others" malaria vaccine. Science 328, 847–848.

Vogel, U., et al., 1997. Complement factor C3 deposition and serum resistance in isogenic capsule and lipooligosaccharide sialic acid mutants of serogroup B Neisseria meningitidis. Infect. Immun. 65, 4022–4029.

Volz, J., Muller, H.M., Zdanowicz, A., Kafatos, F.C., Osta, M.A., 2006. A genetic module regulates the melanization response of Anopheles to Plasmodium. Cell. Microbiol. 8, 1392–1405.

Warr, E., Lambrechts, L., Koella, J.C., Bourgouin, C., Dimopoulos, G., 2006. Anopheles gambiae immune responses to Sephadex beads: involvement of anti-Plasmodium factors in regulating melanization. Insect Biochem. Mol. Biol. 36, 769–778.

Warr, E., Das, S., Dong, Y., Dimopoulos, G., 2008. The Gram-negative bacteria-binding protein gene family: its role in the innate immune system of anopheles gambiae and in anti-Plasmodium defence. Insect Mol. Biol. 17, 39–51.

Waterhouse, R.M., Povelones, M., Christophides, G.K., 2010. Sequence-structure-function relations of the mosquito leucine-rich repeat immune proteins. BMC Genomics 11, 531.

Williams, M., Summers, B.J., Baxter, R.H., 2015. Biophysical analysis of anopheles gambiae leucine-rich repeat proteins APL1A1, APL1B [corrected] and APL1C and their interaction with LRIM1. PLoS One 10, e0118911.

Yassine, H., Kamareddine, L., Osta, M.A., 2012. The mosquito melanization response is implicated in defense against the entomopathogenic fungus Beauveria bassiana. PLoS Pathog. 8, e1003029.

Yassine, H., Kamareddine, L., Chamat, S., Christophides, G.K., Osta, M.A., 2014. A serine protease homolog negatively regulates TEP1 consumption in systemic infections of the malaria vector Anopheles gambiae. J Innate Immun 6, 806–818.

CHAPTER NINE

Influences of the Mosquito Microbiota on Vector Competence

S. van Tol, G. Dimopoulos
Bloomberg School of Public Health, Johns Hopkins University, Baltimore, MD, United States

Contents

Abstract

Mosquitoes have a diverse intestinal microbiota which plays essential roles in a variety of the insect's physiological functions. This microbiota can also influence vector competence for human pathogens in multiple ways, including immune system activation or suppression, the regulation of other physiological processes, direct inhibition of the pathogen, and competition with the pathogen for limited resources. The mosquito's intestinal microbiome is responsible for eliciting basal immune activity, which in addition to preventing overproliferation of bacteria, also acts against exogenous bacteria, fungi, parasites, and viruses. Some endogenous microorganisms directly impair pathogen infectivity or viability, independent of the mosquito, through mechanisms involving

Advances in Insect Physiology, Volume 51
ISSN 0065-2806
http://dx.doi.org/10.1016/bs.aiip.2016.04.006

reactive oxygen species and other more stable bacteria-/fungi-produced metabolites. Other microorganisms, particularly those with entomopathogenic activity, can influence pathogen transmission by shaping mosquito population dynamics. Exploration of the mosquito gut microbiome's antipathogen and entomopathogenic activities can provide essential knowledge for the development of novel disease-control approaches based on blocking pathogen infection in the mosquito and vertebrate host and/or shortening the mosquito's lifespan.

1. INTRODUCTION

Mosquitoes, mainly those of the *Aedes*, *Anopheles*, and *Culex* species, transmit pathogens relevant to human health. Each year, hundreds of millions of people become infected with a mosquito-borne pathogen (Gulland, 2012; Hay et al., 2010; Murray et al., 2013). In order for a mosquito to play a viable role in a pathogen's transmission cycle, whether for a protozoan parasite, helminth, or virus, the mosquito must be a competent vector. Vector competence is the ability of a susceptible mosquito to become infected with a pathogen, support the pathogen's replication and/or development, and transmit the pathogen to a susceptible vertebrate host, typically following infection of the salivary glands (Christofferson and Mores, 2011; Sardelis et al., 2001; Zhou et al., 2014). In fact, most mosquitoes are not competent vectors for a given pathogen, because of the presence of numerous barriers to the infection.

The midgut is the first tissue an ingested pathogen encounters. Barriers present in this tissue include the peritrophic matrix, digestive enzymes, immune effectors, blood meal components, and endogenous microorganisms, which together greatly reduce the number of viable pathogens (Abraham and Jacobs-Lorena, 2004; Billingsley and Rudin, 1992; Dong et al., 2009; Ghosh et al., 2011; Noden et al., 2011; Ramirez et al., 2012). This bottleneck environment is critical in allowing the mosquito to control infection by any pathogens that are able to invade the midgut epithelium. Once the pathogen enters the midgut epithelial cells, infection-induced responses, including innate immune pathway activation and other factors such as pathogen–host complementarity and resource competition with endogenous microorganisms, threaten the pathogen's ability to replicate, disseminate, or invade other tissues (Caragata et al., 2013; Lambrechts et al., 2012; Martin-Acebes et al., 2011; Molina-Cruz et al., 2013, 2015). In the later stages of infection, the immune response remains a critical determinant of competence; however, the early-stage responses regulate the magnitude of the responses needed to control infection later.

Thus, reciprocal interactions among the vector's endogenous microbiota, the pathogen, and the mosquito vector all serve to alter vector competence. The natural colonization of the mosquito midgut with bacteria and eukaryotic organisms potentially alters the mosquito's innate immune responses, either through increased basal expression of immune effectors or amplification of immune responses effective against both the pathogen and the vector's microbiota (Christophides et al., 2002; Ramphul et al., 2014; Rodrigues et al., 2010; Vlachou et al., 2005; Xi et al., 2008). Competition for limited resources that are critical for the life cycle of both the pathogen and endogenous microorganisms may also contribute to reduced vector competence (Caragata et al., 2013; Herren et al., 2014). In addition to indirect inhibition, components of the microbiota can produce factors that directly impair a pathogen's viability or infectivity (Bahia et al., 2014; Cirimotich et al., 2011; Ramirez et al., 2014b). Other members of the midgut microbiota can increase mosquito mortality, preventing any infected females from surviving long enough to transmit the pathogen to a susceptible host (Akhouayri et al., 2013; Bahia et al., 2014; Blanford et al., 2005; Gonzalez-Ceron et al., 2003; Ramirez et al., 2014b). The multifactorial role of the microbiota in regulating vector competence potentiates the introduction of microbiota-based disease-control strategies.

This chapter discusses the acquisition and composition of the mosquito's endogenous microbiota, outlines the mechanisms by which the microbiota modulates vector competence, and discusses the potential role of the microbiota in the control of vector-borne diseases.

2. ACQUISITION AND COMPOSITION OF THE MOSQUITO'S ENDOGENOUS MICROBIOTA

The mosquito's microbiota influences the life history of the mosquito, from development and reproductive capacity to nutrition and infection susceptibility (Bian et al., 2013; Blanford et al., 2005, 2012; Dobson et al., 2002; Dong et al., 2009; Gaio et al., 2011; Garza-Hernandez et al., 2013; Joshi et al., 2014; McMeniman et al., 2009; Minard et al., 2013; Nnakumusana, 1986; Ondiaka et al., 2015; Scholte et al., 2006; Walker et al., 2011). The microbiota includes the nonpathogenic and pathogenic microorganisms found in association with the mosquito. As with vertebrates, a variety of microorganisms naturally inhabit the mosquito (Coon et al., 2014; Duguma et al., 2015; Gimonneau et al., 2014). Although the midgut microbiota is the most investigated tissue because of its interaction with potential pathogens, other tissues (the fat body, salivary glands, and reproductive organs) are colonized as well (Akhouayri et al., 2013; Damiani et al., 2010; Favia et al., 2007; Gimonneau et al., 2014; Gusmao et al., 2010; Lindh et al., 2008; Ricci et al., 2011; Sharma et al., 2014; Wang et al., 2011). The reciprocal interactions of the midgut microbiota with the host and pathogen influence vector competence. Knowledge of how the microbiota is acquired and its composition offers insight into the mechanisms of pathogen interference and guidance in the design of future applications.

The mosquito's microbiota is acquired at several stages throughout the mosquito's lifespan. As larvae, mosquitoes are in an aquatic environment that hosts a diverse microbial community (Coon et al., 2014; Duguma et al., 2015; Gimonneau et al., 2014). The larvae filter-feed on organic matter, including bacteria and eukaryotic microorganisms. Ingestion of bacteria is essential for larval development (Chouaia et al., 2012; Coon et al., 2014; Mitraka et al., 2013; Wotton et al., 1997). In addition to ingestion of microorganisms naturally present in the water, female mosquitoes can transmit bacteria, fungi, or yeasts vertically (Akhouayri et al., 2013; Bian et al., 2013; Damiani et al., 2010; Favia et al., 2007; Gusmao et al., 2010; Hughes et al., 2014; Lindh et al., 2008; Mitraka et al., 2013; Ricci et al., 2011; Wang et al., 2011). This intergenerational transmission can occur through infection of the egg cytoplasm (Werren, 1997), colonization of the egg surface (Damiani et al., 2010; Favia et al., 2007), or inoculation of the breeding water at oviposition (Lindh et al., 2008). Any microorganisms transferred to the breeding water can then be ingested once the larvae hatch.

When the larvae undergo metamorphosis into pupae, the midgut epithelium turns over, and a peritrophic matrix surrounds the sloughed midgut (Moll et al., 2001; Moncayo et al., 2005). The pupal stage does not feed, so no additional microorganisms are introduced at this point in the life cycle (Moll et al., 2001; Moncayo et al., 2005). At the transition from pupa to adult, a second peritrophic matrix may form around the first, and both are egested, leaving the adult gut essentially sterile (Moll et al., 2001; Moncayo et al., 2005). At emergence, the adult ingests water from the breeding site, and some of the imbibed bacteria and fungi colonize the adult gut (Pumpuni et al., 1996). The efficiency of the microorganisms' persistence from larva to adult, independent of ingestion at the adult stage (termed trans-stadial transmission), is unclear (Duguma et al., 2015; Favia et al., 2007; Gimonneau et al., 2014; Lindh et al., 2008; Mitraka et al., 2013; Pumpuni et al., 1996; Ricci et al., 2011; Seye et al., 2009). Although ingestion of larval water seems to be the most efficient mode of adult colonization, microorganisms can also be acquired from nectar (Alvarez-Perez et al., 2012; Pumpuni et al., 1996) and, more rarely, via sexual transmission (Damiani et al., 2008; Favia et al., 2007).

Over the past several years, a number of studies have evaluated the mosquito's endogenous bacterial population. Through the use of both cultivable and noncultivable methods, such as metagenomic sequencing, the bacterial composition of a variety of mosquito vectors across the various life stages has been evaluated (Andrews et al., 2013; Apte-Deshpande et al., 2012; Boissiere et al., 2012; Chandel et al., 2013; Demaio et al., 1996; Dinparast Djadid et al., 2011; Duguma et al., 2015; Gimonneau et al., 2014; Ngwa et al., 2013; Osei-Poku et al., 2012; Pidiyar et al., 2004; Ramirez et al., 2012; Sharma et al., 2014; Terenius et al., 2008; Valiente Moro et al., 2013; Wang et al., 2011; Zouache et al., 2010). Despite the variability in bacterial composition among and within mosquito populations, several themes have emerged, including a "core microbiota", shared colonization patterns, and the presence of a dominant bacterial isolate.

Consistent identification of bacteria with particular characteristics hints at the existence of a core mosquito microbiota. In the midguts of *Aedes*, *Anopheles*, and *Culex* mosquitoes, Proteobacteria is the dominant phylum (Andrews et al., 2013; Apte-Deshpande et al., 2012; Boissiere et al., 2012; Chandel et al., 2013; Dinparast Djadid et al., 2011; Duguma et al., 2015; Gimonneau et al., 2014; Osei-Poku et al., 2012; Pidiyar et al., 2004; Ramirez et al., 2012; Sharma et al., 2014; Valiente Moro et al., 2013; Wang et al., 2011; Zouache et al., 2010). Other dominant phyla include Bacteriodetes, Firmicutes, and Actinobacteria; however, the contribution of these bacteria and less common phyla to the overall midgut microbiota is more variable across vector species (Boissiere et al., 2012; Chandel et al., 2013; Osei-Poku et al., 2012; Ramirez et al., 2012; Sharma et al., 2014; Valiente Moro et al., 2013; Wang et al., 2011; Zouache et al., 2010). Within the dominant phyla, several genera have been repeatedly identified, including *Acinetobacter*, *Enterobacter*, *Bacillus*, *Serratia*, *Pseudomonas*, *Pantoea*, and *Asaia* (Andrews et al., 2013; Boissiere et al., 2012; Chandel et al., 2013; Chouaia et al., 2010; De Freece et al., 2013; Dinparast Djadid et al., 2011; Gusmao et al., 2010; Minard et al., 2013; Osei-Poku et al., 2012; Pidiyar et al., 2004; Pumpuni et al., 1996; Ramirez et al., 2012; Straif et al., 1998; Terenius et al., 2008, 2012). Several of the bacteria are found across mosquito genera and in both field-caught and laboratory populations (Boissiere et al., 2012; Duguma et al., 2015; Rani et al., 2009; Steyn et al., 2015; Wang et al., 2011). These commonalities in composition suggest that mosquito colonization may be semiselective or may reflect an increased propensity for the mosquito to encounter these particular bacteria.

The composition studies also offer insight into the dynamics of microbiota acquisition. As filter feeders, the larvae ingest the microorganisms available in the aquatic environment. Several studies show that the diversity in the water exceeds the diversity found in the larval gut (Boissiere et al., 2012; Coon et al., 2014; Duguma et al., 2015; Gimonneau et al., 2014; Wotton et al., 1997; Zouache et al., 2010). The larval habitat is a primary contributor to variability among and within vector species (Chavshin et al., 2012; Demaio et al., 1996; Duguma et al., 2015; Gimonneau et al., 2014; Moll et al., 2001; Moncayo et al., 2005; Ngwa et al., 2013; Steyn et al., 2015; Tajedin et al., 2009; Wang et al., 2011).

When larvae develop into the pupae and adults, the number of unique bacteria identified decreases, consistent with the physical turnover of the midgut at emergence (Chavshin et al., 2012; Demaio et al., 1996; Duguma et al., 2015; Gimonneau et al., 2014; Moll et al., 2001; Moncayo et al., 2005; Ngwa et al., 2013; Steyn et al., 2015; Tajedin et al., 2009; Wang et al., 2011). Despite the proliferation of bacteria that occurs following a blood meal (Gusmão et al., 2010; Pumpuni et al., 1996), diversity decreases in the adult, and only a few bacterial genera remain associated with the adult gut (Terenius et al., 2012; Wang et al., 2011). The continued reduction in unique bacteria colonizing the gut at all life stages suggests that the midgut environment is semiselective, with only a particular subset of bacteria capable of persisting throughout the mosquito's lifespan. Although a great deal of research is needed to elucidate the mechanisms determining microbiota composition, recent experiments have shown that both host and bacterial genetics can influence the capacity of a bacterium to colonize the adult midgut (Bando et al., 2013; Pei et al., 2015; Stathopoulos et al., 2014). For example, *Enterobacter* requires *waaL*, a gene necessary for proper LPS structure, for efficient colonization in the mosquito (Pei et al., 2015).

The majority of the variability is related to which constituents are dominant and which are minor. Typically, only one or two bacteria are dominant, whereas many others are present in much smaller numbers (Andrews et al., 2013; Boissiere et al., 2012; Chandler et al., 2015; Chouaia et al., 2010; De Freece et al., 2013; Demaio et al., 1996; Dinparast Djadid et al., 2011; Minard et al., 2014; Ngwa et al., 2013; Osei-Poku et al., 2012; Pidiyar et al.,

2004; Ramirez et al., 2012; Rani et al., 2009; Sharma et al., 2014; Straif et al., 1998; Terenius et al., 2008). The operational taxonomic units (OTUs) defined by metagenomic sequencing methods reflect this observation. In a given sample, a single OTU contributes the majority of the reads, while unique, minor OTUs represent the bacteria present in low numbers (Boissiere et al., 2012; Ngwa et al., 2013; Wang et al., 2011). The dominant bacterium is typically a member of the "core microbiota", Gram-negative, and possessing general shared characteristics, such as catalase activity, osmotolerance, and the ability to colonize a microaerophilic environment (Alvarez-Perez et al., 2012; Boissiere et al., 2012; Chandel et al., 2013; Gimonneau et al., 2014; Ngwa et al., 2013; Wang et al., 2011). Several factors contribute to the variability with which a bacterium dominates the adult gut, including larval habitat (Boissiere et al., 2012; Chandel et al., 2013; Gimonneau et al., 2014), sex (Rani et al., 2009; Valiente Moro et al., 2013; Zouache et al., 2010), diet (Colman et al., 2012; Wang et al., 2011), and genetics (Bando et al., 2013; Pei et al., 2015; Stathopoulos et al., 2014).

Although less is known about the eukaryotic organisms associated with mosquitoes, similar patterns in colonization dynamics are observed. As observed for the bacteria, the population of eukaryotic microorganisms is more diverse in larvae than in adults, one isolate is dominant within an individual, and several genera are repeatedly identified (Chandler et al., 2015; Frants and Mertvetsova, 1986; Gusmao et al., 2010; Ignatova et al., 1996; Lara da Costa and Cunha de Oliveira, 1998; Pereira et al., 2009; Ricci et al., 2010; Steyn et al., 2015; Tajedin et al., 2009). *Wickerhamomyces*, *Aspergillus*, *Saccharomyces*, and *Penicillium* are among the most commonly isolated eukaryotic genera (Chandler et al., 2015; Frants and Mertvetsova, 1986; Gusmao et al., 2010; Ignatova et al., 1996; Lara da Costa and Cunha de Oliveira, 1998; Pereira et al., 2009; Ricci et al., 2010; Steyn et al., 2015; Tajedin et al., 2009). Additional studies are needed in the future to identify other members of the mosquito mycobiota and to dissect the factors contributing to colonization.

3. THE MOSQUITO INNATE IMMUNE SYSTEM AND INTERACTIONS WITH THE ENDOGENOUS MICROBIOTA

The endogenous microbiota plays a role in regulating the mosquito's innate immune system. The innate immune system is critical for mitigating pathogen-induced responses, and the effects of reciprocal microbiota–host interactions on immune signalling are important to dissect. The mosquito's innate immune system consists of three main pathways: the Toll, immunodeficiency (IMD), and Janus kinase/signal transducers and activators of transcription (JAK/STAT) pathways (Fragkoudis et al., 2009; Kingsolver et al., 2013; Levashina, 2004). Each of the three main pathways follows a similar general path: pathogen recognition receptors (PRRs) recognize pathogens via interactions with pathogen-associated molecular patterns (PAMPs); downstream of this pathogen recognition, an intracellular signal cascade is induced (Cirimotich et al., 2010). The intracellular cascade results in the nuclear translocation of transcription factors that regulate the expression of a variety of genes encoding PRRs, immune effectors, and proteins involved in immune regulation (Cirimotich et al., 2010).

Despite the crosstalk among these pathways, different pathways are activated preferentially in response to certain pathogens. The Toll pathway, analogous to the interleukin-1 receptor pathway in mammals (Barillas-Mury et al., 1996; Belvin et al., 1995), is most important in regulating responses to viruses (Dodson et al., 2014; Dong et al., 2012a; Sanders et al., 2005; Sim et al., 2013; Xi et al., 2008; Zambon et al., 2005), fungi (Bian et al., 2005; Dong et al., 2012a; Shin et al., 2005; Wang et al., 2015), Gram-positive bacteria (Barillas-Mury et al., 1996; Dimopoulos et al., 1997; Dong et al., 2009; Michel et al., 2001; Nicolas et al., 1998; Warr et al., 2008), and the rodent malaria parasite *Plasmodium berghei* (Fraiture et al., 2009; Frolet et al., 2006; Garver et al., 2009; Mitri et al., 2009; Ramirez et al., 2014a; Riehle et al., 2008). Melanization and phagocytosis, both of which are haemocyte mediated (Kim and Kim, 2005; Ligoxygakis et al., 2002; Povelones et al., 2013; Ramirez et al., 2014a; Wang et al., 2013; Zou et al., 2008), are the primary effector mechanisms promoted by the Toll pathway. The IMD pathway, analogous to the tumour necrosis pathway (Kleino et al., 2005), mainly responds to Gram-negative bacteria (Antonova et al., 2009; Bahia et al., 2014; Bian et al., 2010; Dong and Dimopoulos, 2009; Georgel et al., 2001; Hoa and Zheng, 2007; Kallio et al., 2005; Meister et al., 2009; Oliveira et al., 2011; Sluss et al., 1996; Stathopoulos et al., 2014) and the human malaria parasite *Plasmodium falciparum* (Antonova et al., 2009; Garver et al., 2009, 2012, 2013; Meister et al., 2009). Among the main effector mechanisms, IMD regulates nitration and lysis via Rel2, an NF-kB-like transcription factor (Antonova et al., 2009; Dong and Dimopoulos, 2009; Frolet et al., 2006; Garver et al., 2012, 2013; Lin et al., 2007; Luna et al., 2006; Meister et al., 2005; Mitri et al., 2009), and cellular responses such as cytoskeletal rearrangement and apoptosis through the Jun-terminal kinase branch (Georgel et al., 2001; Leulier et al., 2000; Oliveira et al., 2012; Ramphul et al., 2014; Silverman et al., 2003). Although the JAK/STAT pathway is the most important in mediating antiviral (Carissimo et al., 2014; Dong et al., 2012a; Dostert et al., 2005; Fragkoudis et al., 2008; Sim et al., 2013; Souza-Neto et al., 2009) and antifungal responses (Dong et al., 2012a), like the mammalian interferon pathway (Dupuis et al., 2003; Fu et al., 1992), this pathway also contributes to anti-*Plasmodium* (Bahia et al., 2011; Gupta et al., 2009) and antibacterial responses (Gupta et al., 2009; Oliveira et al., 2011). In the case of each of these pathways, exposure to bacteria results in the translocation of primary transcription factors, although not all pathways contribute equally to this process (Barillas-Mury et al., 1996; Gupta et al., 2009; Hoa and Zheng, 2007; Nicolas et al., 1998; Sluss et al., 1996).

The exogenous RNA interference (RNAi) pathway is involved in virus defence (Brackney et al., 2009; Campbell et al., 2008; Carissimo et al., 2014; Franz et al., 2006; Keene et al., 2004; McFarlane et al., 2014; Myles et al., 2008; Sanchez-Vargas et al., 2009; Sim et al., 2013). Although it is not directly involved in bacterial responses, this pathway intersects with the IMD and JAK/STAT pathways (Paradkar et al., 2012, 2014). In addition, the endogenous RNAi pathway is involved in the regulation of host mRNA expression, and therefore the expression of genes important to bacterial colonization (Dennison et al., 2015; Osei-Amo et al., 2012; Zhang et al., 2013).

The mosquito's innate immune pathways are important in regulating the load and composition of the microbiota. RNAi-mediated silencing of the Toll and IMD pathway NF-kB-like transcription factors, Rel1 and Rel2, respectively, results in the proliferation of the endogenous microbiota (Dong et al., 2009; Meister et al., 2009; Yassine et al., 2014). Also, silencing of the complement-like protein TEP1 allows the bacteria to replicate (Dong et al., 2009). Both Rel1 and Rel2 control the expression of *TEP1* expression (Fraiture et al., 2009; Frolet et al., 2006; Garver et al., 2012, 2013; Riehle et al., 2008). Multiple PRRs, including AgDscam and PGRP-LC, also control the endogenous microbiota (Dong and Dimopoulos, 2009; Dong et al., 2006b, 2009; Meister et al., 2009). In a study analysing small nucleotide polymorphisms associated with *Serratia marcescens* colonization, Stathopoulos et al. (2014) made the interesting observation that certain AgDscam homologs are important in controlling proliferation once ingested.

Conversely, the bacteria present in the mosquito also regulate the expression of immunity-related genes. Xi et al. (2008) have shown that *Aedes aegypti* mosquitoes treated with antibiotics experience reduced expression of antimicrobial peptides (AMPs) when compared to mosquitoes with an intact microbiota. AMPs, which are controlled by both Rel1 and Rel2, inhibit viruses, fungi, bacteria, and *Plasmodium* through the disruption of the plasma membrane without affecting the cells of the mosquito (Antonova et al., 2009; Barillas-Mury et al., 1996; Boutros et al., 2002; De Gregorio et al., 2002; Dong et al., 2012a; Hoa and Zheng, 2007; Lemaitre et al., 1996; Lin et al., 2007; Luna et al., 2006; Manfruelli et al., 1999; Meister et al., 2005; Shin et al., 2005; Xi et al., 2008). Other antibacterial effectors (eg, PRRs, including AgDscam and PGRP-LC) and immune regulator genes are known to be upregulated in bacteria-colonized mosquitoes (Dong et al., 2009). Also, genes involved in the melanization pathway, such as serine proteases and serine protease inhibitors, are differentially regulated to control the endogenous microbiota (Dong et al., 2009).

Microbiota–mosquito immune interactions are further exemplified by the antibacterial immune responses that are induced following the proliferation of the microbiota after a blood meal. When female mosquitoes take a blood meal, the endogenous microbiota proliferates (Gusmao et al., 2010; Pumpuni et al., 1996); ingestion of blood, and more specifically heme, reduces the production of reactive oxygen species (ROS) via reduction in the expression of the heme peroxidase, IMPer (Oliveira et al., 2011). This reduction in IMPer leads to a breakdown in the dual oxidase, DUOX, and the IMPer network that increases the permeability of this barrier to AMPs (Kumar et al., 2010; Oliveira et al., 2011). Although the peritrophic matrix, a physical barrier composed of chitin fibrils, chitin-binding domain proteins, and other components, forms around the blood bolus, some of the proliferating bacteria may be able to activate the mosquito's innate immune pathways, which are more sensitive to activation when ROS are downregulated (Ha et al., 2009; Kumar et al., 2010; Molina-Cruz et al., 2007; Oliveira et al., 2011).

Although little is known about the role of endogenous eukaryotic microorganisms in immune regulation, each of the pathways has been shown to be involved in the clearance of fungal pathogens (Bian et al., 2005; Dong et al., 2012a; Shin et al., 2005; Wang et al., 2015; Yassine et al., 2012). Presumably, the native eukaryotic organisms can regulate the innate immune pathways in a manner similar to the native bacteria; however, further research is required to investigate these relationships.

Overall, the mosquito's innate immune system is composed of three major pathways that are involved in the regulation of bacterial responses, and the endogenous microbiota influences the baseline expression of the mosquito's immune effectors (Fig. 1). This reciprocal interaction is important to consider when evaluating the role exerted by bacteria on vector competence through immune regulation.

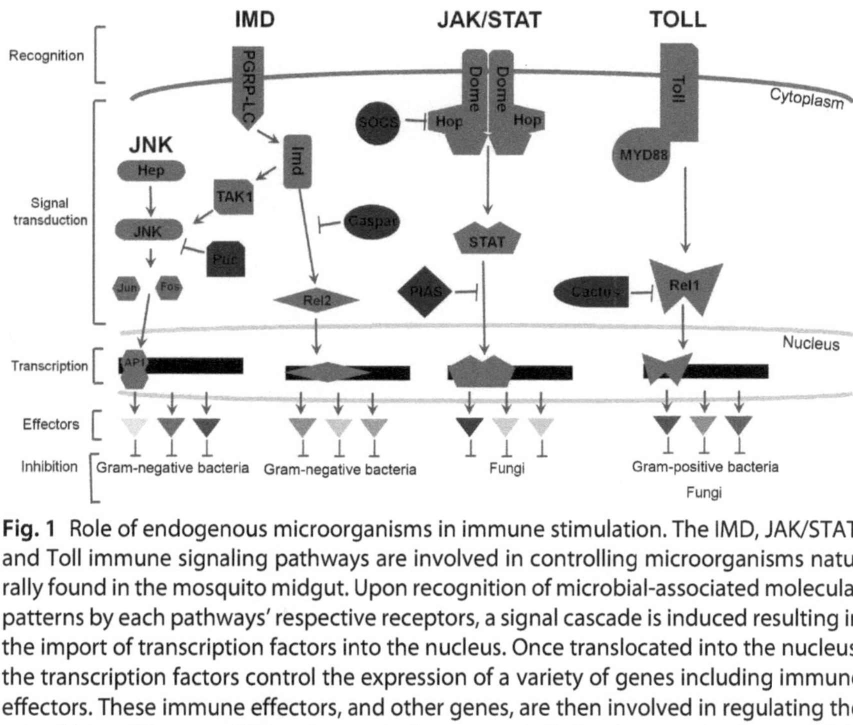

Fig. 1 Role of endogenous microorganisms in immune stimulation. The IMD, JAK/STAT, and Toll immune signaling pathways are involved in controlling microorganisms naturally found in the mosquito midgut. Upon recognition of microbial-associated molecular patterns by each pathways' respective receptors, a signal cascade is induced resulting in the import of transcription factors into the nucleus. Once translocated into the nucleus, the transcription factors control the expression of a variety of genes including immune effectors. These immune effectors, and other genes, are then involved in regulating the microbial proliferation. Although a particular category of microbe can stimulate multiple immune responses, one pathway is stimulated predominantly in association with a particular category of microorganisms (Gram-negative bacteria stimulate both the JNK and Imd branches of the IMD pathway; fungi stimulate both the JAK/STAT and Toll pathways; and Gram-positive bacteria stimulate the Toll pathway). Receptors, signal transduction proteins, and transcription factors of immune pathways shown in *blue*; negative regulators of pathways shown in *red*; DNA represented by the *black rectangles*. (See the colour plate.)

4. INFLUENCE OF THE MOSQUITO MICROBIOTA ON VECTOR COMPETENCE: *PLASMODIUM*

4.1 *Plasmodium*'s Life Cycle in the Mosquito

Protozoan parasites of the genus *Plasmodium* are the etiologic agents of malaria. *Anopheles* mosquitoes serve as the vector for the *Plasmodium* parasites that infect humans (Sinka et al., 2012). When the mosquito feeds on an infected host, blood cells containing the male and female gametocytes are ingested (Smith et al., 2014). Once inside the mosquito, the parasite undergoes sexual replication to form a motile ookinete (Smith et al., 2014). At the ookinete stage, the parasite must overcome barriers to infection, such as the peritrophic matrix, to invade the midgut columnar epithelial cells (Abraham and Jacobs-Lorena, 2004; Han, 2000). The invasion process induces cellular stress and nitric oxide synthase (Han, 2000; Kumar et al., 2010; Lim et al., 2005; Molina-Cruz et al., 2007, 2013; Smith et al., 2012). As a result of the damage and oxidative stress that occurs, invaded cells may undergo apoptosis (Han, 2000). The resulting damage may allow endogenous microorganisms to leak through and induce antibacterial responses, some of which overlap with and amplify the anti-*Plasmodium* responses (Christophides et al., 2002; Ramphul et al., 2014; Rodrigues et al., 2010; Vlachou et al., 2005). A subset of ookinetes that successfully invade the epithelial cells then traverse the invaded cells to reach the basal lamina, where they mature into oocysts (Smith et al., 2014). During cell traversal, the parasite is subject to additional immune responses, including melanization and other haemocyte-mediated effector responses (Aguilar et al., 2007; Baxter et al., 2007; Dong et al., 2006a, 2009; Fraiture et al., 2009; Frolet et al., 2006; Habtewold et al., 2008; Molina-Cruz et al., 2013; Oliveira et al., 2012; Povelones et al., 2009; Ramphul et al., 2014; Riehle et al., 2008; Rodrigues et al., 2010; Wang et al., 2013). The few parasites that survive through the oocyst stage asexually reproduce and release thousands of sporozoites (Smith et al., 2014). The sporozoites released into the haemocoel must evade additional immune challenges in order to invade the salivary glands (Bian et al., 2013; Habtewold et al., 2008). Once viable sporozoites reach the salivary glands and invade the salivary ducts, the mosquito is vector competent (Aly et al., 2009). The time from ingestion of gametocyte-contaminated blood meal to transmissible sporozoites is considered the extrinsic incubation period (EIP), typically around 2 weeks (Aly et al., 2009). Several biotic and abiotic factors contribute to a vector's competence for the malaria parasite, including the midgut microbiota (Beier et al., 1994; Dong et al., 2009; Gendrin et al., 2015; Noden et al., 2011; Sharma et al., 2013).

4.2 Bacteria and *Plasmodium*

The presence of bacteria in the midgut influences the vector competence of mosquitoes for transmitting *Plasmodium*. Depletion of the midgut bacteria with antibiotics increases the oocyst burden and the prevalence of multiple human and rodent malaria parasites in various multiple *Anopheles* species (Beier et al., 1994; Dong et al., 2009; Gendrin et al., 2015; Noden et al., 2011; Sharma et al., 2013). In addition, reintroduction of live bacteria, particularly Gram-negative constituents, decreases the *Plasmodium* infection intensity in a dose-dependent manner (Dong et al., 2009; Gonzalez-Ceron et al., 2003). These observations indicate that the midgut bacteria limit the parasite's ability to infect the mosquito. On closer inspection, the presence of bacteria appears to alter the mosquito's vector competence for the malaria parasites through altered immune signalling or direct interactions with the parasite.

4.2.1 Microbiota-Mediated Immune Activation

Endogenous bacteria mediate both the basal and *Plasmodium*-induced immune responses of the mosquito, which alter the parasite's ability to complete its life cycle within the mosquito. The Toll and IMD pathways are particularly important for *P. berghei* and *P. falciparum* responses, respectively (Fraiture et al., 2009; Frolet et al., 2006; Garver et al., 2009, 2012, 2013; Meister et al., 2009; Mitri et al., 2009; Ramirez et al., 2014a; Riehle et al., 2008). Immune genes involved in shared responses encode a variety of PRRs, effectors, and immune regulators, particularly those mediating melanization, lysis, and phagocytotic responses (An et al., 2010; Christophides et al., 2002; Dimopoulos et al., 1997, 2000, 2002; Eappen et al., 2013; Ramphul et al., 2014; Rodrigues et al., 2010; Vlachou et al., 2005). Overexpression of shared immune genes at the time of parasite exposure allows the mosquito to mount responses more quickly.

When the ookinete invades the midgut epithelium, the damage allows bacteria in the midgut to amplify the mosquito's immune responses that coincidently control *Plasmodium*. The peptidoglycan recognition protein, PGRP-LC, a PRR that acts as the transmembrane receptor of the IMD pathway, is involved in both anti-*Plasmodium* and antibacterial responses

(Dong et al., 2009; Meister et al., 2009). Meister et al. (2009) reported that PGRP-LC is important in regulating endogenous and exogenous bacteria as well as immunity to *Plasmodium*. Following RNAi-mediated silencing of PGRP-LC3, endogenous bacteria proliferate, and the mosquito becomes more susceptible to exogenous bacterial challenge and to *Plasmodium* (Meister et al., 2009). Treatment of mosquitoes with antibiotics prior to an infectious blood meal abrogates the role of PGRP-LC3-mediated protection against *Plasmodium* (Meister et al., 2009). The reliance on endogenous bacteria to induce the antiparasitic responses suggests that the bacteria activate PGRP-LC3 more strongly than does the ookinete, and genes activated downstream of PGRP-LC are effective in controlling both the bacteria and *Plasmodium*. Also, when mosquitoes are infected with *P. berghei* after PGRP-LC3 silencing, the melanization of the ookinetes increases, suggesting an increased reliance on Toll-mediated responses (Meister et al., 2009). Furthermore, the IMD pathway negatively regulates the melanization pathway (Frolet et al., 2006; Meister et al., 2009), suggesting that the presence of the endogenous bacteria can influence which effector response is preferentially activated following parasite invasion.

Other PRRs are likewise involved in bacteria–*Plasmodium* shared responses. Gram-negative binding proteins, particularly GNBPB4 (Warr et al., 2008); the fibronectin-related proteins, especially FBN9 (Dong and Dimopoulos, 2009; Dong et al., 2009); and immunoglobulin superfamily proteins, such as infection response with immunoglobulin domain protein (IRID)-6 (Garver et al., 2008) and the alternatively spliced hypervariable immunoglobulin domain-encoding gene, AgDscam (Dong et al., 2006b, 2012b), have been shown to be important in restricting both bacteria and *Plasmodium*.

Recent experiments also suggest that the presence of bacteria is critical for the induction of a "priming" response against *Plasmodium* in *Anopheles gambiae*. Rodrigues et al. (2010) have hypothesized that parasite invasion allows bacteria-mediated activation of a haemocyte differentiation factor that induces the differentiation of haemocytes into granulocytes. The increase in granulocytic haemocytes protects against either a subsequent *Plasmodium* or exogenous bacterial challenge (Rodrigues et al., 2010). Also, the "priming" effect results in the increased expression of *TEP1* and *LRIM1*, both of which are critical in haemocyte-mediated responses against bacteria and *Plasmodium* (Rodrigues et al., 2010). Treatment of mosquitoes with antibiotic prior to the initial *Plasmodium* exposure does not result in the production of the haemocyte differentiation factor, and protective levels of

TEP1 and LRIM1 are not induced (Ramirez et al., 2015; Rodrigues et al., 2010). Although these experiments further support the role of bacteria in the activation of anti-*Plasmodium* effectors, *An. albimanus* mosquitoes do not rely on the presence of bacteria to generate a "priming" response to *P. berghei* (Contreras-Garduno et al., 2015). These conflicting results concerning the involvement of the microbiota in regulating *Plasmodium* infection responses serve as a reminder that underlying species-specific differences in immunity play a critical role in controlling infection.

Reintroduction of Gram-negative bacterial isolates from field-collected mosquitoes can also alter the expression of genes that are important for *Plasmodium* defence. In addition to the general immune stimulation described earlier, exposure to *Wolbachia*, *S. marcescens*, *Enterobacter* spp. *Zambia* (*Esp_Z*), *Pantoea* sp., or *Pseudomonas putida* increases the expression of AMPs (Bahia et al., 2014; Cirimotich et al., 2011; Moreira et al., 2009; Ramirez et al., 2014b). The induction of AMPs may offer protection against *Plasmodium*; however, Gram-positive isolates, such as *Bacillus pumilus*, that do not enable a *Plasmodium*-protective phenotype also induce AMP expression (Cirimotich et al., 2011). Although the contribution of AMPs to *Plasmodium* defence is unclear, other immune factors that negatively affect *Plasmodium* development, including PGRP-LC, TEP1, Rel2, FBN9, and ROS, are upregulated (Bahia et al., 2014; Bian et al., 2013; Hughes et al., 2011b; Kambris et al., 2010; Moreira et al., 2009; Stathopoulos et al., 2014). Interestingly, natural variation within *S. marcescens* related to the *flhDC* plasmid influences the bacterium's ability to restrict *Plasmodium* infection, suggesting that the tripartite interactions can be quite specific (Bando et al., 2013).

Although the majority of overlapping *Plasmodium* and immune responses minimize *Plasmodium* infection, some bacteria induce responses that enhance vector competence. For example, CTL4 and CTLMA2 are overexpressed in Gram-negative or *P. berghei*-challenged mosquitoes (Osta et al., 2004; Povelones et al., 2013; Schnitger et al., 2009). The expression of these C-type lectins protects against *Escherichia coli* infection but enhances *P. berghei* infection through an inhibition of the melanization pathway (Osta et al., 2004; Povelones et al., 2013; Schnitger et al., 2009).

4.2.2 *Direct Inhibition*

In addition to indirectly inhibiting *Plasmodium* development, some bacteria interact directly with the parasite. Coincubation of the sporogonic stage of

Plasmodium with certain Gram-negative bacteria inhibits ookinete formation or damages the parasite in vitro (Bahia et al., 2014; Cirimotich et al., 2011; Ramirez et al., 2014b). Such mosquito-independent inhibition is observed for bacteria that commonly colonize *Anopheles* mosquitoes, including *Comamonas* sp., *Acinetobacter, Pseudomonas, Pantoea, Serratia, Enterobacter,* and *Elizabethkingia* (Bahia et al., 2014; Cirimotich et al., 2011; Ramirez et al., 2014b). The ability of bacteria that naturally associate with mosquitoes to impede *Plasmodium* competence may contribute to the natural variation in vector competence.

Bacteria directly impair *Plasmodium* development through species-specific mechanisms. The molecular mechanisms of parasite inhibition by bacteria with in vitro activity are beginning to be explored. Cirimotich et al. (2011) have demonstrated that an *Enterobacter* isolated from *Anopheles* in Zambia *Esp_Z* produces hydrogen peroxide. Introduction of antioxidants either in vitro or in vivo negates *Esp_Z*'s ability to block parasite development (Cirimotich et al., 2011). Other bacteria can secrete other protective factors that may be active against multiple parasite stages. For example, *S. marcescens* and *Chromobacterium* spp. *Panama* (*Csp_P*) inhibit both the asexual and sporogonic stages of *Plasmodium* when bacteria-free culture supernatant is tested (Bahia et al., 2014; Ramirez et al., 2014b). Further experiments are required to identify the molecules responsible for the inhibition and their mechanisms of action.

4.3 Eukaryotic Microorganisms and *Plasmodium*

Despite the minimal research that has been done thus far to investigate the interactions of endogenous yeast and fungi with the mosquito, a couple of studies have implicated entomopathogenic eukaryotic microorganisms in the impairment of *Plasmodium* transmissibility. Exposure to *Beauveria bassiana* or *Metarhizium anisopliae,* two entomopathogenic fungi, reduces sporozoite prevalence to zero (Blanford et al., 2005). The underlying fungi-induced responses that prevent parasite transmission are not yet characterized. Certain strains of a symbiotic yeast, *Wicherhamomyces anomalus,* produce a killer toxin that interacts with β-1,3-glucans, a fungal cell wall component also found in *Plasmodium,* damaging the ookinetes and limiting the number of parasites invading the mosquito's epithelium (Valzano et al., 2016). A large gap exists in our current knowledge of mycobiota–mosquito–*Plasmodium* tripartite interactions and the direct effects of eukaryotic microorganisms on *Plasmodium.*

5. INFLUENCE OF THE MOSQUITO MICROBIOTA ON VECTOR COMPETENCE: VIRUSES

Multiple mosquito species serve as vectors of viruses transmissible to humans, more broadly known as arboviruses. *Aedes* spp. are the primary vectors for multiple viruses, including dengue virus (DENV), chikungunya virus (CHIKV), zika virus (ZIKV), and yellow fever virus (YFV); *Culex* spp. are competent vectors of West Nile virus (WNV); and *Anopheles* spp. are vectors for O'nyong-nyong virus (ONNV) (Carissimo et al., 2014; Coffey et al., 2013; Keene et al., 2004). When the mosquito ingests an infectious blood meal, a population of viruses enters the insect, with only a relatively small proportion of virus particles being infectious to the mosquito (Coffey et al., 2013). Much like the situation in *Plasmodium* infection, the virus must navigate midgut infection barriers to enter midgut epithelial cells (Beaty and Marquardt, 1996). Once in the host cell, the virus must recruit the mosquito's cellular machinery and interfere with the mosquito's immune and cellular responses, such as apoptosis and autophagy, in order to replicate (Kanthong et al., 2010; Moy et al., 2014; Ocampo et al., 2013; Vaidyanathan and Scott, 2006; Wang et al., 2012a). Progeny virions produced in the midgut epithelium are then released from the basal lamina into the haemocoel, where they disseminate and infect other tissues (Beaty and Marquardt, 1996). When the virus infects other tissues, secondary replication occurs, and the virus is further amplified to a titre high enough to infect the salivary glands (Beaty and Marquardt, 1996). Once the salivary glands are infected, the virus replicates, and the progeny are shed into the saliva (Beaty and Marquardt, 1996). To act as a competent vector, the mosquito must support the virus' entire life cycle and survive long enough to transmit virus particles that are infectious to a susceptible vertebrate host (Beaty and Marquardt, 1996; Coffey et al., 2013). A plethora of biotic and abiotic factors influence either competence or the length of the EIP; these include immune responses (Avadhanula et al., 2009; Brackney et al., 2009; Campbell et al., 2008; Carissimo et al., 2014; Keene et al., 2004; McFarlane et al., 2014; Sanchez-Vargas et al., 2009; Sanders et al., 2005; Sim et al., 2013; Souza-Neto et al., 2009; Xi et al., 2008), temperature (Alto and Bettinardi, 2013; Dohm et al., 2002; Reisen et al., 2006; Richards et al., 2007; Turell, 1989), larval environment (Alto et al., 2008), and the mosquito's microbiota (Carissimo et al., 2014; Ramirez et al., 2012; Xi et al., 2008).

5.1 Bacteria and Viruses

The influence of the endogenous microbiota on vector competence for viruses is variable. The literature largely supports the conclusion that the presence of Gram-negative bacteria protects against arboviruses, mainly DENV (Ramirez et al., 2012, 2014b; Xi et al., 2008). Treatment of *A. aegypti* with antibiotics prior to a DENV-infected blood meal increases DENV titres (Xi et al., 2008), whereas introduction of Gram-negative bacteria, *Proteus* spp., *Paenibacillus*, or *Csp_P*, reduces DENV titres in vivo (Ramirez et al., 2012, 2014b). Despite this observation in the *Ae. aegypti*-DENV system, Carissimo et al. (2014) have shown that treatment of *An. gambiae* with antibiotics reduces susceptibility to ONNV, and reintroduction of bacteria increases ONNV midgut titres. The endogenous bacteria's influence on immune signalling, direct interaction with the virus, and competition for shared resources may alter vector competence.

5.1.1 Microbiota-Mediated Immune Activation

The presence of bacteria enhances the basal expression of AMPs that act as antibacterial and antiviral effector molecules (Dong et al., 2012b; Luplertlop et al., 2011; Ramirez et al., 2012; Xi et al., 2008). Xi et al. (2008) have shown that *Ae. aegypti* with an intact microbiota have higher AMP expression than do mosquitoes treated with antibiotics. AMPs, particularly cecropins, protect the mosquitoes from arboviral infection (Dong et al., 2012b; Luplertlop et al., 2011; Ramirez et al., 2012; Xi et al., 2008). Another AMP, lysozyme C, protects against DENV, but the protection only persists as long as the bacteria are present (Ramirez et al., 2012). When *Ae. aegypti* are exposed to *Csp_P*, several AMPs are upregulated, and this upregulation may contribute to the observed inhibition of DENV in vivo (Ramirez et al., 2014b).

Wolbachia, an endosymbiotic bacterium associated with many insect species, influences mosquito immune responses in a pathogen–vector-specific manner. *Ae. aegypti* do not naturally harbour *Wolbachia*, and transinfection of some strains of the bacterium, including *w*Mel, *w*MelPop, and *w*AlbB, decreases the titres for several viruses, and immune pathways tend to be upregulated (Frentiu et al., 2010, 2014; Kambris et al., 2009, 2010; Moreira et al., 2009; van den Hurk et al., 2012; Walker et al., 2011). In mosquito species naturally harbouring *Wolbachia*, such as *Aedes albopictus*, the bacterium's presence does not alter the replication of the same viruses (Kambris et al., 2010; Lu et al., 2012). Bian et al. (2010) have shown that transinfection of *Wolbachia* into *Ae. aegypti* induces the expression of *Rel1* and *Rel2*. Although several groups have demonstrated *Wolbachia*'s ability to enhance the expression of innate immune pathways, the relative importance of the upregulation of these genes in controlling viral replication and dissemination is unclear. Some studies have indicated that the *Wolbachia*-stimulated immune genes are required for protection, while others have shown that RNAi silencing of the upregulated immune genes does not alter *Wolbachia*'s inhibitory phenotype (Kambris et al., 2009, 2010; Moreira et al., 2009; Pan et al., 2011). More delicate experiments are needed to clarify the role of *Wolbachia*-stimulated immune genes in blocking arboviruses in *Ae. aegypti*. In the case of *Culex tarsalis*, another population naturally devoid of *Wolbachia*, transinfection with *w*AlbB suppresses the Toll pathway and enhances WNV infection (Dodson et al., 2014).

Another example of bacteria-mediated enhancement of competence involves the colonization of *Ae. aegypti* with *Serratia odorifera*. Titres for both CHIKV and DENV increase when *S. odorifera* is present (Apte-Deshpande et al., 2012, 2014). Although the mechanism is not fully characterized, the immune response seems to be suppressed as the result of the expression of a bacterial protein that interacts with the mosquito (Apte-Deshpande et al., 2014).

5.1.2 Direct Inhibition

In a few instances, exposure to bacteria inhibits arboviruses directly. Joyce et al. (2011) have coincubated bacterial isolates from *Ae. albopictus* with LaCrosse virus and shown that viral infection of mammalian cells decreases in response. The cell-free supernatant collected from *Csp_P*, grown as a biofilm, also inhibits DENV infection of baby hamster kidney cells (Ramirez et al., 2014b). The ability of *Csp_P* to interfere with virus infectivity without the bacterial cells being present indicates that a secreted factor may contribute to the in vivo anti-DENV response (Ramirez et al., 2014b). Although the experiments evaluated infection of mammalian cells, the bacteria can likely alter mosquito cell infectivity as well. Further studies are needed to identify the factor or factors responsible for the inhibitory phenotype, as well as to determine the molecular mechanisms involved and their contributions to in vivo activity.

5.1.3 Resource Competition

Bacteria can interfere with the arboviruses through competition for limited resources critical to viral replication. Bacterial symbionts, such as *Wolbachia* and *Spiroplasma*, require cholesterol and fatty acids to replicate (Caragata et al., 2013; Herren et al., 2014; Williamson et al., 1996), as do the enveloped viruses DENV and WNV (Fu et al., 2015; Heaton et al., 2010; Jupatanakul et al., 2014; Martin-Acebes et al., 2011; Perera et al., 2012). The ability of a DENV protein, NS3, to recruit the host's fatty acid synthase to sites of replication exemplifies the importance of fatty acid synthesis for DENV replication (Heaton et al., 2010). In addition, a number of lipid-related proteins function as agonists to DENV infection, including sterol carrier protein-2, myeloid differentiation 2-related lipid recognition protein, and Niemann Pick-Type CI protein (Fu et al., 2015; Jupatanakul et al., 2014). Although bacteria–virus resource competition has not been well explored in mosquito systems, an experiment conducted in *Drosophila* supports the hypothesis that resource competition can significantly affect viral replication and dissemination (Caragata et al., 2013). In the experiment, *w*MelPop or *w*Mel did not protect the insects against *Drosophila* C virus when they were fed on a cholesterol-supplemented diet (Caragata et al., 2013). In the future, the role of endogenous bacteria in lipid homeostasis could offer insight into the importance of resource competition in regulating arboviral replication and dissemination within mosquitoes.

5.2 Fungi and Viruses

The mycobiota also influences arbovirus infection. Although no experiments have evaluated the role of endogenous fungi in the modulation of mosquito-borne virus life cycles, a few experiments in *Ae. aegypti* using the entomopathogenic fungi *B. bassiana* and *M. anisopliae* have shown that infection with these fungi reduces DENV replication and dissemination (Dong et al., 2012a; Garza-Hernandez et al., 2013). Dong et al. (2012a) have indicated that *B. bassiana* likely limits DENV infection through activation of both the Toll and JAK/STAT pathways. The Toll pathway-regulated genes *Rel1*, *attacin*, *cecropin G*, *defensin A* and *C*, and *diptericin* and the JAK/STAT pathway genes *Dome* and *dengue virus restriction factors 1* and *2* are upregulated in *B. bassiana*-exposed mosquitoes, and all of these gene products can mediate an anti-DENV response (Dong et al., 2012a). The alteration of immune signalling following *B. bassiana* exposure suggests that other fungi can similarly influence vector competence. Fungi and yeasts are likely capable of producing factors that directly interact with viruses and can compete with viruses for host-derived resources; however, no evidence for such factors currently exists.

6. ENTOMOPATHOGENIC BACTERIA AND FUNGI: LIFESPAN ATTENUATION AND VECTOR COMPETENCE

Preventing female mosquitoes from surviving through the completion of the EIP interferes with vector competence. Several entomopathogenic bacteria and fungi prevent the transmission of mosquito-borne pathogens by killing the adult mosquitoes.

For several decades, the entomopathogenic bacteria *Bacillus thuringiensis* and *B. spaericus* have been used to control vector populations. Since the aquatic, immature life stages of the mosquito are the primary targets, these bacteria decrease the population density (Barbosa et al., 2007; Becker et al., 1995; Fernandez-Luna et al., 2010; Fillinger et al., 2003; Futami et al., 2011; Goldberg and Margalit, 1977; Joshi et al., 2014; Kahindi et al., 2008; Karch et al., 1992; Majambere et al., 2007; Mittal et al., 1993; Nartey et al., 2013; Pantuwatana and Sattabongkot, 1990; Paris et al., 2010; Romi et al., 1993; Singer, 1973; Skovmand and Sanogo, 1999). Reducing the vector population decreases the efficiency of baseline transmission, commonly referred to as the vectorial capacity (Geissbuhler et al., 2009; Karch et al., 1991; Kroeger et al., 1995; Kumar et al., 1994); however, the effects of larval exposure on the vector competence of surviving adults are understudied. One study has found that adults exposed to *B. sphaericus* as larvae have a reduced prevalence of oocysts, but the intensity of infection does not differ among *Plasmodium*-infected individuals (Young et al., 1990).

Several other bacteria possess mosquitocidal properties that are active against multiple life stages. Although the majority of identified bacteria possess mosquitocidal activity against larvae and pupae (Geetha and Manonmani, 2010; Geetha et al., 2010, 2011; Prabakaran et al., 2002; Ramirez et al., 2014b), several isolates also attenuate adult survival (Akhouayri et al., 2013; Bahia et al., 2014; Gonzalez-Ceron et al., 2003; Ramirez et al., 2014b). The *Chromobacterium* previously described as possessing in vivo anti-dengue and anti-*Plasmodium* activities kills both the adult and larval stages of *Anopheles* and *Aedes* (Ramirez et al., 2014b). Other bacteria commonly isolated from mosquitoes, including *S. marcescens* and *Elizabethkingia*, increase mortality in adults (Akhouayri et al., 2013; Bahia et al., 2014; Gonzalez-Ceron et al., 2003). Also, bacteria-colonized mosquitoes die more quickly following a blood meal than do antibiotic-treated mosquitoes (Dong et al., 2009). Transinfection of *Anopheles* and *Ae. aegypti* with some strains of *Wolbachia* has also been associated with enhanced mortality (Hughes et al., 2011a,c; Kambris et al., 2009). In the case of *An. gambiae* and *w*MelPop, mortality is greatly increased immediately following a blood meal (Hughes et al., 2011a). Increased adult mortality, especially following a blood meal, suggests that a proportion of the mosquitoes that take infectious blood meals will not survive to transmit the pathogen.

Many fungi also possess entomopathogenic properties. The fungal entomopathogens *B. bassiana* and *M. anisopliae* decrease vector competence through a variety of mechanisms, including immune signalling (Blanford et al., 2005; Dong et al., 2012a; Garza-Hernandez et al., 2013) and lifespan attenuation of adult females (Blanford et al., 2005, 2012; Bukhari et al., 2011; Clark et al., 1968; Farenhorst et al., 2009, 2010; Howard et al., 2010, 2011; Ondiaka et al., 2015; Scholte et al., 2006; Valero-Jimenez et al., 2014; Weiser, 1984). Like *B. thuringiensis* and *B. sphaericus*, *M. anisopliae* and *B. bassiana* also reduce vectorial capacity by reducing larval survival (Mohanty and Prakash, 2000, 2002; Muspratt, 1963; Nnakumusana, 1986; Prakash et al., 2010; Priyanka et al., 2001; Seye et al., 2009), fecundity (Garza-Hernandez et al., 2013; Nnakumusana, 1986; Scholte et al., 2006), and fertility (Nnakumusana, 1986). Other fungi with mosquitocidal properties have been identified, including *Culicinomyces, Fusarium, Coelomomyces, Chrysosporium, Trichophyton, Lencanicillium, Penicillium, Aspergillus, Saccharomomonas,* and *Isaria* (Blanford et al., 2012; Luz et al., 2010; Maketon et al., 2013; Mohanty and Prakash, 2000; Muspratt, 1963; Nnakumusana, 1986; Prakash et al., 2010; Priyanka et al., 2001; Seye et al., 2009; Singh and Prakash, 2011).

7. MICROBIOTA-BASED DISEASE CONTROL STRATEGIES

Exploitation of the microbiota's capacity to influence vector competence provides multiple outlets for the control of mosquito-transmitted pathogens. Strategies currently employed include the use of *Wolbachia*-infected *Ae. aegypti* to control DENV in Australia and entomopathogens to control vector population density. Future applications under development include the introduction of competence-limiting bacteria and fungi into mosquito populations.

7.1 Wolbachia

Wolbachia potentially influences vector competence through a variety of mechanisms, including cytoplasmic incompatibility, lifespan attenuation, resource competition, and immune gene regulation. Importantly, these competence-modulating properties are not mutually exclusive (Dobson et al., 2002; McMeniman et al., 2009; Walker et al., 2011). For example, the cytoplasmic incompatibility drives the replacement of a DENV-susceptible *Ae. aegypti* population with *w*Mel-*Ae. Aegypti*, which have been shown to be more refractory to DENV in Australia (Hoffmann et al., 2011). Since the *Ae. aegypti* populations are naturally *Wolbachia* naïve, the mosquitoes infected with *Wolbachia* outcompete uninfected individuals and rise to near-fixation in months (Hoffmann et al., 2011). Even a few years after the release of the *w*Mel-*Ae. aegypti*, the *Wolbachia*-infected population was still dominant and retained a refractory phenotype against three DENV serotypes (Frentiu et al., 2014).

Despite the success of *w*Mel-*Ae. aegypti* in reducing competence for DENV, multiple factors can eclipse the effectiveness of *Wolbachia* as an effective means of control for other arboviruses. As previously discussed, *Wolbachia* infection enhances WNV infection in *C. tarsalis*, although *Wolbachia* does not naturally associate with this vector (Dodson et al., 2014). In addition, naturally *Wolbachia*-infected species such as *Ae. albopictus* are not protected from DENV or CHIKV (Bourtzis et al., 2000; Frentiu et al., 2010; Kambris et al., 2010; Lu et al., 2012; Rances et al., 2012; Skelton et al., 2015), suggesting that as the vector, pathogen, and *Wolbachia* coevolve, the effectiveness of the competence impairment may dissipate. The apparent *Wolbachia*/virus/vector-specific interaction necessitates a thorough investigation of the mechanisms of virus interference prior to additional field introductions.

Interactions of *Wolbachia* with the native microbiota and *Plasmodium* species-specific effects further complicate the potential for *Wolbachia* introduction into *Anopheles* populations. Until recently, *Wolbachia* had not been detected in natural *Anopheles* populations (Baldini et al., 2014; Kittayapong et al., 2000), theoretically making *Anopheles* a good target population for *Wolbachia*-driven population replacement, since infection of *Anopheles* with *w*MelPop reduced *An. gambiae*'s competence for the human malaria parasite, *P. falciparum* (Bian et al., 2013; Hughes et al., 2011a; Kambris et al., 2010). Despite the inhibitory effects observed with *P. falciparum* and wMelPop, the effectiveness against *P. berghei* is *Wolbachia* strain dependent (Hughes et al., 2011c; Kambris et al., 2010), and against *P. yoelii* is temperature dependent (Murdock et al., 2014). Competition with *Asaia* for ovary colonization further complicates the introduction of the bacterium into *Anopheles* (Hughes et al., 2014). Hughes et al. (2014) have demonstrated that natural colonization of the mosquito's ovaries with *Asaia* spp. impedes vertical transmission of *Wolbachia*; eliminating *Asaia* from the ovaries enables *Wolbachia* to infect the ovaries and be transferred intergenerationally. The high prevalence of *Asaia* in *Anopheles* and other mosquito species may therefore argue against the implementation of *Wolbachia*-mediated *Plasmodium* control.

7.2 Entomopathogenic Bacteria and Fungi

Inclusion of entomopathogenic microorganisms in vector control programmes limits pathogen transmission through interference with multiple contributors to vectorial capacity, including vector competence. Entomopathogenic bacteria such as *B. thuringiensis* and *B. sphaericus* mainly limit vector density through efficient killing of larvae, thereby reducing the number of blood-feeding females (Geissbuhler et al., 2009; Kroeger et al., 1995; Kumar et al., 1994). Several field studies utilizing either of these mosquitocidal bacteria have reported reduced malaria transmission and risk of infection with *Plasmodium* (Geissbuhler et al., 2009; Kroeger et al., 1995; Kumar et al., 1994). In addition to larval killing, fecundity and fertility are decreased in adults exposed to entomopathogenic pathogens as larvae (Garza-Hernandez et al., 2013; Joshi et al., 2014; Nnakumusana, 1986; Scholte et al., 2006). The intergenerational effects on vector density contribute to the observed reduction in vectorial capacity. In addition, some entomopathogenic fungi decrease the propensity of females to blood-feed (Blanford et al., 2005, 2012; Scholte et al., 2006), further limiting the likelihood of pathogen transmission. Although adult killing does not alter the length of the EIP, lifespan attenuation can prevent pathogen-infected females from surviving long enough to infect a susceptible host (Akhouayri et al., 2013; Bahia et al., 2014; Blanford et al., 2005; Bukhari et al., 2011; Gonzalez-Ceron et al., 2003; Ramirez et al., 2014b). Also, multiple adulticidal fungi and bacteria interfere with immune signalling and directly interfere with pathogen development, such as *S. marcescens* (Bahia et al., 2014; Gonzalez-Ceron et al., 2003; Stathopoulos et al., 2014). The multifactorial role of entomopathogenic microorganisms in controlling pathogen transmission necessitates the development of effective introduction strategies, discussed later, and characterization of killing mechanisms for recently identified entomopathogens.

Despite their potential to cause significant impairment of mosquito-borne pathogens, resistance, nonspecific effects, and a lack of sustainability have limited the effectiveness of current entomopathogenic-based strategies. Resistance to entomopathogic fungi and bacteria is subject to slower selection than that to chemical insecticides; however, persistent use of these microorganisms may lead to a loss of efficacy over time (Goldman et al., 1986; Mittal et al., 1998; Paris et al., 2010, 2011). Also, the alleles associated with resistance to *B. thuringiensis* are selected against in the absence of bacterial exposure because of the incurred fitness costs (Paris et al., 2011). The entomopathogenic fungi *B. bassiana* and *M. anisopliae* kill insecticide-resistant mosquitoes efficiently, and fungi–insecticide combinations synergistically kill mosquitoes (Farenhorst et al., 2009, 2010; Howard et al., 2010, 2011). Another significant limitation is the short-term activity of the pathogens. Even after the formulations are altered to maximize activity, the biocontrol agents need to be reapplied on a frequent basis (Darbro and Thomas, 2009; Fillinger et al., 2003; Howard et al., 2010, 2011; Karch et al., 1991, 1992; Kroeger et al., 1995; Majambere et al., 2007; Mittal et al., 1993), often weekly, a requirement that may not be sustainable in resource-poor areas where mosquito-borne disease are more prevalent. In addition, *M. anisopliae* and *B. bassiana* can cause opportunistic infections in humans and are aeroallergens (Darbro and Thomas, 2009; de Garcia et al., 1997; Figueira et al., 2012). Future research should focus on designing more sustainable and safer methods for introducing entompathogens.

7.3 Introduction of Bacteria and Fungi into Adult Mosquito Populations to Reduce Vector Competence

Several strains of bacteria and fungi isolated from mosquitoes confer pathogen-inhibitory phenotypes, but their viability as control options requires effective introduction into mosquito populations. As previously described, microorganisms can alter competence by means of a variety of mechanisms, including direct inhibition of the pathogen, stimulation of protective immune responses, competition with the pathogen for limited resources, and lifespan attenuation. With the exception of *Wolbachia*, such competence-altering microorganisms have yet to be field-applied. Several options for introduction of the fungi and pathogens include baited sugar meals (Muller et al., 2010a,b; Qualls et al., 2014; Stewart et al., 2013) and gravid (egg-laying) traps (Barbosa et al., 2007; Futami et al., 2011; Lindh et al., 2008). Attractive sugar baits, currently used to introduce toxic chemicals such as boric acid, could be modified to introduce fungi and bacteria (Muller et al., 2010a,b; Qualls et al., 2014; Stewart et al., 2013). The sugar solutions attract mosquitoes semispecifically, although nontarget effects remain a concern. In addition to sugar baits, gravid traps could be treated with bacteria or fungi that the larvae will readily ingest

(Thiery et al., 1991). Both of these methods would be most efficient for the initial introduction of microorganisms that can be vertically transmitted, thus enabling propagation throughout the population.

Prior to the introduction of these bacteria or fungi into natural mosquito populations, potential off-target effects on other mosquito-borne pathogens require attention. For example, the introduction of a bacterium intended to control the competence in *Ae. aegypti* for DENV cannot necessarily enhance competence for another pathogen the vector is known to transmit, such as CHIKV. Also, the microorganism cannot increase the ability of another vector within geographic proximity to transmit a pathogen. These introduced bacteria and fungi must also be safe for humans and other nontarget animals.

7.4 Paratransgenesis

In paratransgenesis, a mosquito symbiont is genetically manipulated to express a factor that hinders vector competence. Effector gene expression through the microorganisms limits the cost of gene expression to the mosquito, and bacteria and fungi tend to be more malleable to genetic manipulation. Wang and Jacobs-Lorena (2013) have reviewed the premise of paratransgenesis and outline the five necessary factors to develop an effective paratransgenic system. First, the bacteria, fungi, or yeast used must associate with the target vector population (Wang and Jacobs-Lorena, 2013). For example, using a bacterium that is naturally vertically transmitted, such as *Asaia*, potentiates population-wide dissemination following initial introduction (Akhouayri et al., 2013; Favia et al., 2007; Lindh et al., 2008; Wang et al., 2011). Also, the microorganisms must grow efficiently in a laboratory setting and must be malleable genetically (Wang and Jacobs-Lorena, 2013). Examples of bacteria that are amenable to genetic manipulation, associate with the mosquito, and grow efficiently outside the mosquito include *Asaia* spp. and *Pantoea agglemorans* (Bisi and Lampe, 2011; Favia et al., 2007; Riehle et al., 2007; Straif et al., 1998; Wang et al., 2012b). Once genetically manipulated, the microorganism must retain fitness equivalent to that of the wild type and must colonize the mosquito efficiently (Wang and Jacobs-Lorena, 2013). The gene, when expressed more effectively, must inhibit the pathogen when secreted (Wang and Jacobs-Lorena, 2013). In several proof-of-concept experiments, the transgene did not effectively prevent pathogen invasion when expressed on the bacterium's surface, as compared to genes with secretion signals (Bisi and Lampe, 2011; Riehle et al., 2007; Wang et al., 2012b). Finally, an efficient means of environmental introduction and subsequent propagation is needed (Wang and Jacobs-Lorena, 2013). As previously discussed, attractive sugar baits or gravid traps could serve as media for the initial introduction of transgenic symbionts that are expected to be capable of intergenerational maintenance. The ability of the microorganism to be transmitted sexually would improve the efficacy of the paratransgenic system.

Although the paratransgenesis approach has not been implemented in the field for mosquito-borne pathogen control, several proof-of-concept experiments have validated this approach (Bisi and Lampe, 2011; Fang et al., 2011; Riehle et al., 2007; Wang et al., 2012b). Paratransgenesis has also been applied in *Rhodnius prolixus*, the vector of the Chagas disease parasite, using its genetically manipulated symbiont, *Rhodococcus rhodnii* (Durvasula et al., 1999; Fieck et al., 2010).

The effector molecules used can target either the vector or the pathogen. Examples of genes could include functional antibody fragments that either bind the pathogen- or mosquito-expressed receptors required for pathogen invasion (Bisi and Lampe, 2011; Durvasula et al., 1999; Fang et al., 2011; Ghosh et al., 2011). Also, analogues to pathogen motifs, such as the enolase–plasminogen interaction peptide that binds plasminogen, can be used to prevent invasion factors from interacting with the pathogen (Wang et al., 2012b). Host-derived genes, such as SM1 that binds to the salivary gland and midgut epithelium to impair the infectivity of *Plasmodium*, or immune effectors, like AMPs, can also be used (Bisi and Lampe, 2011; Fieck et al., 2010; Kim et al., 2004; Moreira et al., 2002; Possani et al., 1998). Introduction of multiple transgenic microorganisms expressing effectors that target a variety of steps in the pathogen's life cycle can increase the effectiveness (Fang et al., 2011).

The specificity paratransgenesis offers could minimize unwanted off-target effects; however, other limitations can impair the efficiency of this method. For example, the natural microbiota colonizing the reproductive organs can inhibit propagation of the introduced transgenic symbiont. Reciprocal inhibition of *Asaia* and *Wolbachia* in the colonization of the ovaries suggests that the natural microbiota can inhibit transmission in natural mosquito populations (Hughes et al., 2014; Rossi et al., 2015). Also, coevolution of the paratransgenic unit with a pathogen can lead to the evolution of resistance over time; the use of multiple effector transgenic systems would slow this process. The proof-of-concept experiments are encouraging, and semifield work would provide information needed to assess the efficacy of this approach.

8. CONCLUSION

The tripartite interaction among mosquito, pathogen, and microbiota influences the outcome of pathogen infection (summarized in Table 1). The microbiota employs a variety of mechanisms to modulate vector competence, including regulation of mosquito immunity, competition with the pathogen for limited resources, production of factors that directly inhibit the pathogen, and attenuation of the mosquito's lifespan. The endogenous microbiota can alter the baseline expression of genes important in early pathogen blocking or recognition of ingested pathogens, thus limiting initial infection. Also, the proliferation of the bacteria after a blood meal induces the expression of immune genes that overlap with pathogen responses, further limiting pathogen replication and development. Despite the fact that the majority of the evidence indicates a protective role for the microbiota in mosquito–pathogen interactions, the few instances of infection enhancement exemplify the complexity of these host–microbiota–pathogen interactions. The competition between members of the microbiota and pathogens for host-derived resources, such as cholesterol and fatty acids, can impair the replication and dissemination of arboviruses. Certain bacteria and fungi can also produce factors such as ROS or secondary metabolites that directly damage the pathogen or impair invasion. Entomopathogenic microorganisms, some of which naturally associate with mosquitoes, shorten the lifespan of the adults, decreasing the probability that an infected mosquito will live through the completion of the EIP.

The multifactorial role of the microbiota in shaping vector competence offers several opportunities to implement microbiota-mediated disease-control methods. Several options for intervention include the introduction of entomopathogenic bacteria and fungi, *Wolbachia*, or other microorganisms that impair vector competence naturally or paratransgenically. Currently applications are limited to *Wolbachia*–DENV in *Ae. aegypti* and the entomopathogenic bacteria *B. thuringiensis* and *B. sphaericus*. Because of the decline in the efficacy of insecticides and antimalarial treatments and the persistent lack of effective vaccines for mosquito-borne pathogens, the need for more effective vector-targeted control methods is increasing. In the future, the underlying mechanisms governing microbiota-mediated reductions in vector competence need further characterization, and the methods for effective introduction require evaluation. Overall, the microbiota plays a critical role in shaping vector competence, and the exploitation of the microbiota as a form of control can play a valuable role in decreasing the burden of mosquito-borne pathogens.

Table 1 Role of the Microbiota in Influencing Pathogen Transmission

	Category of Pathogen Transmission Interference	Mechanism of Action	Examples
Mosquito–pathogen interactions (vector competence)	Immune stimulation	Alters basal expression of immune effectors with dual activity against pathogens Activate immune pathways that overlap with antipathogen responses, leading to amplification upon pathogen invasion	*Plasmodium* ▪ Pathogen recognition • AgDscam • FBN9 • GNBPB4 • IRID-6 ▪ PGRP-LC effectors • Antimicrobial peptides • TEP1 ▪ Regulators • Rel1 • Rel2 • Serine proteases • Serine protease inhibitors Viruses ▪ Pathogen recognition • Dome ▪ Regulators • Rel1 • Rel2 ▪ Effectors • Antimicrobial peptides • Dengue virus restriction factors

Continued

Table 1 Role of the Microbiota in Influencing Pathogen Transmission—cont'd

	Category of Pathogen Transmission Interference	Mechanism of Action	Examples
	Direct inhibition	Microorganisms can produce factors that, independent of the mosquito, impair the viability or infectivity of a pathogen	*Plasmodium* ▪ Unidentified secreted factors (proteins or secondary metabolites) • *Chromobacterium* (*Csp_P*) • *Serratia marcescens* • *Wicherhamomyces anomalus* ▪ Reactive oxygen species production • *Enterobacter* (*Esp_Z*) Viruses ▪ Unidentified secreted factors (proteins or secondary metabolites) • *Chromobacterium* (*Csp_P*)-dengue virus
	Resource competition	Other microorganisms utilize limited resources which can delay or impair pathogen replication or growth	Viruses ▪ Cholesterol ▪ Fatty acids
Mosquito population dynamics (vector–vertebrate host interactions)	Vector density *Larvacidal activity*	Decrease in the number of mosquitoes that eclose as adults	▪ *Bacillus sphaericus* ▪ *B. thuringiensis* ▪ *Beauveria bassiana* ▪ *Chromobacterium* (*Csp_P*) ▪ *Metharhizium anisopliae*
	Fecundity and Fertility	Fewer offspring in the next generation	▪ *Beauveria bassiana* ▪ *Metharhizium anisopliae*

	Propensity to blood feed	Reduces the likelihood that an adult female will interact with a vertebrate host; either blocking pathogen acquisition or transmission	▪ *Beauveria bassiana*
	Daily survival	Decreases probability of surviving long enough to complete the extrinsic incubation period/ transmit a pathogen if infected Decreases the likelihood of encountering an infected host	▪ *Beauveria bassiana* ▪ *Chromobacterium* (*Csp_P*) ▪ *Elizabethkingia* ▪ *Metharhizium anisopliae* ▪ *Serratia marcescens*

Bacteria and eukaryotic organisms naturally associated with mosquitoes influence their capacity to transmit pathogens, the interactions listed above are further described in the text.

REFERENCES

Abraham, E., Jacobs-Lorena, M., 2004. Mosquito midgut barriers to malaria parasite development. Insect Biochem. Mol. Biol. 34, 667–671.

Aguilar, R., Das, S., Dong, Y., Dimopoulos, G., 2007. Continuous exposure to Plasmodium results in decreased susceptibility and transcriptomic divergence of the Anopheles gambiae immune system. BMC Genomics 8, 451.

Akhouayri, I., Habtewold, T., Christophides, G., 2013. Melanotic pathology and vertical transmission of the gut commensal Elizabethkingia meningoseptica in the major malaria vector Anopheles gambiae. PLoS One 8, e77619.

Alto, B., Bettinardi, D., 2013. Temperature and dengue virus infection in mosquitoes: independent effects on the immature and adult stages. Am. J. Trop. Med. Hyg. 88, 497–505.

Alto, B., Lounibos, L., Mores, C., Reiskind, M., 2008. Larval competition alters susceptibility of adult Aedes mosquitoes to dengue infection. Proc. R. Soc. B Biol. Sci. 275, 463–471.

Alvarez-Perez, S., Herrera, C., Vega, C., 2012. Zooming-in on floral nectar: a first exploration of nectar-associated bacteria in wild plant communities. FEMS Microbiol. Ecol. 80, 591–602.

Aly, A., Vaughan, A., Kappe, S., 2009. Malaria parasite development in the mosquito and infection of the mammalian host. Annu. Rev. Microbiol. 63, 195–221.

An, C., Budd, A., Kanost, M., Michel, K., 2010. Characterization of a regulatory unit that controls melanization and affects longevity of mosquitoes. Cell. Mol. Life Sci. 68, 1929–1939.

Andrews, E., Xu, G., Rich, S., 2013. Microbial communities within field-collected Culiseta melanura and Coquillettidia perturbans. Med. Vet. Entomol. 28, 125–132.

Antonova, Y., Alvarez, K., Kim, Y., Kokoza, V., Raikhel, A., 2009. The role of NF-kB factor REL2 in the Aedes aegypti immune response. Insect Biochem. Mol. Biol. 39, 303–314.

Apte-Deshpande, A., Paingankar, M., Gokhale, M., Deobagkar, D., 2012. Serratia odorifera a midgut inhabitant of Aedes aegypti mosquito enhances its susceptibility to dengue-2 virus. PLoS One 7, e40401.

Apte-Deshpande, A., Paingankar, M., Gokhale, M., Deobagkar, D., 2014. Serratia odorifera mediated enhancement in susceptibility of Aedes aegypti for chikungunya virus. Indian J. Med. Res. 139, 762–768.

Avadhanula, V., Weasner, B., Hardy, G., Kumar, J., Hardy, R., 2009. A novel system for the launch of alphavirus RNA synthesis reveals a role for the Imd pathway in arthropod antiviral response. PLoS Pathog. 5, e1000582.

Bahia, A., Kubota, M., Tempone, A., Araujo, H., Guedes, B., Orfano, A., Tadei, W., Rios-Velasquez, C., Han, Y., Secundino, N., Barillas-Mury, C., Pimenta, P., Traub-Cseko, Y., 2011. The JAK-STAT pathway controls Plasmodium vivax load in early stages of Anopheles aquasalis infection. PLoS Negl. Trop. Dis. 5, e1317.

Bahia, A., Dong, Y., Blumberg, B., Mlambo, G., Tripathi, A., BenMarzouk-Hidalgo, O., Chandra, R., Dimopoulos, G., 2014. Exploring Anopheles gut bacteria for Plasmodium blocking activity. Environ. Microbiol. 16, 2980–2994.

Baldini, F., Segata, N., Pompon, J., Marcenac, P., Robert Shaw, W., Dabire, R., Diabate, A., Levashina, E., Catteruccia, F., 2014. Evidence of natural Wolbachia infections in field populations of Anopheles gambiae. Nat. Commun. 5, 3985.

Bando, H., Okado, K., Guelbeogo, W., Badolo, A., Aonuma, H., Nelson, B., Fukumoto, S., Xuan, X., Sagnon, N., Kanuka, H., 2013. Intra-specific diversity of Serratia marcescens in Anopheles mosquito midgut defines Plasmodium transmission capacity. Sci. Rep. 3, 1641.

Barbosa, R., Souto, A., Eiras, A., Regis, L., 2007. Laboratory and field evaluation of an oviposition trap for Culex quinquefasciatus(Diptera: Culicidae). Mem. Inst. Oswaldo Cruz 102, 523–529.

Barillas-Mury, C., Charlesworth, A., Gross, I., Richman, A., Hoffmann, J., Kafatos, F., 1996. Immune factor Gambif1, a new rel family member from the human malaria vector, Anopheles gambiae. EMBO J. 15, 4691–4701.

Baxter, R., Chang, C., Chelliah, Y., Blandin, S., Levashina, E., Deisenhofer, J., 2007. Structural basis for conserved complement factor-like function in the antimalarial protein TEP1. Proc. Natl. Acad. Sci. U.S.A. 104, 11615–11620.

Beaty, B., Marquardt, W., 1996. The Biology of Disease Vectors. University Press of Colorado, Niwot, CO, pp. 57–60.

Becker, N., Zgomba, M., Petric, D., Beck, M., Ludwig, M., 1995. Role of larval cadavers in recycling processes of Bacillus sphaericus. J. Am. Mosq. Control Assoc. 11, 329–334.

Beier, M., Pumpuni, C., Beier, J., Davis, J., 1994. Effects of para-aminobenzoic acid, insulin, and gentamicin on Plasmodium falciparum development in Anopheline mosquitoes (Diptera: Culicidae). J. Med. Entomol. 31, 561–565.

Belvin, M., Jin, Y., Anderson, K., 1995. Cactus protein degradation mediates Drosophila dorsal-ventral signaling. Genes Dev. 9, 783–793.

Bian, G., Shin, S., Cheon, H., Kokoza, V., Raikhel, A., 2005. Transgenic alteration of Toll immune pathway in the female mosquito Aedes aegypti. Proc. Natl. Acad. Sci. U.S.A. 102, 13568–13573.

Bian, G., Xu, Y., Lu, P., Xie, Y., Xi, Z., 2010. The endosymbiotic bacterium Wolbachia induces resistance to dengue virus in Aedes aegypti. PLoS Pathog. 6, e1000833.

Bian, G., Joshi, D., Dong, Y., Lu, P., Zhou, G., Pan, X., Xu, Y., Dimopoulos, G., Xi, Z., 2013. Wolbachia invades Anopheles stephensi populations and induces refractoriness to Plasmodium infection. Science 340, 748–751.

Billingsley, P., Rudin, W., 1992. The role of the mosquito peritrophic membrane in bloodmeal digestion and infectivity of Plasmodium species. J. Parasitol. 78, 430.

Bisi, D., Lampe, D., 2011. Secretion of anti-plasmodium effector proteins from a natural Pantoea agglomerans isolate by using PelB and HlyA secretion signals. Appl. Environ. Microbiol. 77, 4669–4675.

Blanford, S., Chan, B., Jenkins, N., Sim, D., Turner, R., Read, A., Thomas, M., 2005. Fungal pathogen reduces potential for malaria transmission. Science 308, 1638–1641.

Blanford, S., Jenkins, N., Read, A., Thomas, M., 2012. Evaluating the lethal and pre-lethal effects of a range of fungi against adult Anopheles stephensi mosquitoes. Malar. J. 11, 365.

Boissiere, A., Tchioffo, M., Bachar, D., Abate, L., Marie, A., Nsango, S., Shahbazkia, H., Awono-Ambene, P., Levashina, E., Christen, R., Morlais, I., 2012. Midgut microbiota of the malaria mosquito vector Anopheles gambiae and interactions with Plasmodium falciparum infection. PLoS Pathog. 8, e1002742.

Bourtzis, K., Pettigrew, M., O'Neill, S., 2000. Wolbachia neither induces nor suppresses transcripts encoding antimicrobial peptides. Insect Mol. Biol. 9, 635–639.

Boutros, M., Agaisse, H., Perrimon, N., 2002. Sequential activation of signaling pathways during innate immune responses in Drosophila. Dev. Cell 3, 711–722.

Brackney, D., Beane, J., Ebel, G., 2009. RNAi targeting of West Nile virus in mosquito midguts promotes virus diversification. PLoS Pathog. 5, e1000502.

Bukhari, T., Takken, W., Koenraadt, C., 2011. Development of Metarhizium anisopliae and Beauveria bassiana formulations for control of malaria mosquito larvae. Parasit. Vectors 4, 23.

Campbell, C., Keene, K., Brackney, D., Olson, K., Blair, C., Wilusz, J., Foy, B., 2008. Aedes aegypti uses RNA interference in defense against Sindbis virus infection. BMC Microbiol. 8, 47.

Caragata, E., Rances, E., Hedges, L., Gofton, A., Johnson, K., O'Neill, S., McGraw, E., 2013. Dietary cholesterol modulates pathogen blocking by Wolbachia. PLoS Pathog. 9, e1003459.

Carissimo, G., Pondeville, E., McFarlane, M., Dietrich, I., Mitri, C., Bischoff, E., Antoniewski, C., Bourgouin, C., Failloux, A., Kohl, A., Vernick, K., 2014. Antiviral immunity of Anopheles gambiae is highly compartmentalized, with distinct roles for RNA interference and gut microbiota. Proc. Natl. Acad. Sci. U.S.A. 112, E176–E185.

Chandel, K., Mendki, M., Parikh, R., Kulkarni, G., Tikar, S., Sukumaran, D., Prakash, S., Parashar, B., Shouche, Y., Veer, V., 2013. Midgut microbial community of Culex quinquefasciatus mosquito populations from India. PLoS One 8, e80453.

Chandler, J., Liu, R., Bennett, S., 2015. RNA shotgun metagenomic sequencing of northern California (USA) mosquitoes uncovers viruses, bacteria, and fungi. Front. Microbiol. 6, 185.

Chavshin, A., Oshaghi, M., Vatandoost, H., Pourmand, M., Raeisi, A., Enayati, A., Mardani, N., Ghoorchian, S., 2012. Identification of bacterial microflora in the midgut of the larvae and adult of wild caught Anopheles stephensi: a step toward finding suitable paratransgenesis candidates. Acta Trop. 121, 129–134.

Chouaia, B., Rossi, P., Montagna, M., Ricci, I., Crotti, E., Damiani, C., Epis, S., Faye, I., Sagnon, N., Alma, A., Favia, G., Daffonchio, D., Bandi, C., 2010. Molecular evidence for multiple infections as revealed by typing of Asaia bacterial symbionts of four mosquito species. Appl. Environ. Microbiol. 76, 7444–7450.

Chouaia, B., Rossi, P., Epis, S., Mosca, M., Ricci, I., Damiani, C., Ulissi, U., Crotti, E., Daffonchio, D., Bandi, C., Favia, G., 2012. Delayed larval development in Anopheles mosquitoes deprived of Asaia bacterial symbionts. BMC Microbiol. 12, S2.

Christofferson, R., Mores, C., 2011. Estimating the magnitude and direction of altered Arbovirus transmission due to viral phenotype. PLoS One 6, e16298.

Christophides, G., Zdobno, E., Barillas-Mury, C., Birney, E., Blandin, S., Blass, C., Brey, P., Collins, F., Danielli, A., Dimopoulos, G., Hetru, C., Hoa, N., Hoffmann, J., et al., 2002. Immunity-related genes and gene families in Anopheles gambiae. Science 298, 159–165.

Cirimotich, C., Dong, Y., Garver, L., Sim, S., Dimopoulos, G., 2010. Mosquito immune defenses against Plasmodium infection. Dev. Comp. Immunol. 34, 387–395.

Cirimotich, C., Dong, Y., Clayton, A., Sandiford, S., Souza-Neto, J., Mulenga, M., Dimopoulos, G., 2011. Natural microbe-mediated refractoriness to Plasmodium infection in Anopheles gambiae. Science 332, 855–858.

Clark, T., Kellen, W., Fukuda, T., Lindegren, J., 1968. Field and laboratory studies on the pathogenicity of the fungus Beauveria bassiana to three genera of mosquitoes. J. Invertebr. Pathol. 11, 1–7.

Coffey, L., Forrester, N., Tsetsarkin, K., Vasilakis, N., Weaver, S., 2013. Factors shaping the adaptive landscape for arboviruses: implications for the emergence of disease. Future Microbiol. 8, 155–176.

Colman, D., Toolson, E., Takacs-Vesbach, C., 2012. Do diet and taxonomy influence insect gut bacterial communities? Mol. Ecol. 21, 5124–5137.

Contreras-Garduno, J., Rodriguez, M., Hernandez-Martinez, S., Martinez-Barnetche, J., Alvarado-Delgado, A., Izquierdo, J., Herrera-Ortiz, A., Moreno-Garcia, M., Velazquez-Meza, M., Valverde, V., Argotte-Ramos, R., Rodriguez, M., Lanz-Mendoza, H., 2015. Plasmodium berghei induced priming in Anopheles albimanus independently of bacterial co-infection. Dev. Comp. Immunol. 52, 172–181.

Coon, K., Vogel, K., Brown, M., Strand, M., 2014. Mosquitoes rely on their gut microbiota for development. Mol. Ecol. 23, 2727–2739.

Damiani, C., Ricci, I., Crotti, E., Rossi, P., Rizzi, A., Scuppa, P., Esposito, F., Bandi, C., Daffonchio, D., Favia, G., 2008. Paternal transmission of symbiotic bacteria in malaria vectors. Curr. Biol. 18, R1087–R1088.

Damiani, C., Ricci, I., Crotti, E., Rossi, P., Rizzi, A., Scuppa, P., Capone, A., Ulissi, U., Epis, S., Genchi, M., Sagnon, N., Faye, I., Kang, A., Chouaia, B., Whitehorn, C., Moussa, G., Mandrioli, M., Esposito, F., Sacchi, L., Bandi, C., Daffonchio, D., Favia, G., 2010. Mosquito-bacteria symbiosis: the case of Anopheles gambiae and Asaia. Microb. Ecol. 60, 644–654.

Darbro, J., Thomas, M., 2009. Spore persistence and likely aeroallergenicity of entomopathogenic fungi used for mosquito control. Am. J. Trop. Med. Hyg. 80 (6), 992–997.

De Freece, C., Damiani, C., Valzano, M., D'amelio, S., Cappelli, A., Ricci, I., Favia, G., 2013. Detection and isolation of the α-proteobacterium Asaia in Culex mosquitoes. Med. Vet. Entomol. 28, 438–442.

de Garcia, M., Arboleda, M., Barraquer, F., Grose, E., 1997. Fungal keratitis caused by Metarhizium anisopliae var. anisopliae. Med. Mycol. 35, 361–363.

De Gregorio, E., Spellman, P., Tzou, P., Rubin, G., Lemaitre, B., 2002. The Toll and Imd pathways are the major regulators of the immune response in Drosophila. EMBO J. 21, 2568–2579.

Demaio, J., Pumpuni, C., Kent, M., Beier, J., 1996. The midgut bacterial flora of wild Aedes triseriatus, Culex pipiens, and Psorophora columbiae mosquitoes. Am. J. Trop. Med. Hyg. 54, 219–223.

Dennison, N., BenMarzouk-Hidalgo, O., Dimopoulos, G., 2015. MicroRNA-regulation of Anopheles gambiae immunity to Plasmodium falciparum infection and midgut microbiota. Dev. Comp. Immunol. 49, 170–178.

Dimopoulos, G., Richman, A., Muller, H., Kafatos, F., 1997. Molecular immune responses of the mosquito Anopheles gambiae to bacteria and malaria parasites. Proc. Natl. Acad. Sci. U.S.A. 94, 11508–11513.

Dimopoulos, G., Casavant, T., Chang, S., Scheetz, T., Roberts, C., Donohue, M., Schultz, J., Benes, V., Bork, P., Ansorge, W., Soares, M., Kafatos, F., 2000. Anopheles gambiae pilot gene discovery project: identification of mosquito innate immunity genes from expressed sequence tags generated from immune-competent cell lines. Proc. Natl. Acad. Sci. U.S.A. 97, 6619–6624.

Dimopoulos, G., Christophides, G., Meister, S., Schultz, J., White, K., Barillas-Mury, C., Kafatos, F., 2002. Genome expression analysis of Anopheles gambiae: responses to injury, bacterial challenge, and malaria infection. Proc. Natl. Acad. Sci. U.S.A. 99, 8814–8819.

Dinparast Djadid, N., Jazayeri, H., Raz, A., Favia, G., Ricci, I., Zakeri, S., 2011. Identification of the midgut microbiota of An. stephensi and An. maculipennis for their application as a paratransgenic tool against malaria. PLoS One 6, e28484.

Dobson, S., Fox, C., Jiggins, F., 2002. The effect of Wolbachia-induced cytoplasmic incompatibility on host population size in natural and manipulated systems. Proc. R. Soc. B Biol. Sci. 269, 437–445.

Dodson, B., Hughes, G., Paul, O., Matacchiero, A., Kramer, L., Rasgon, J., 2014. Wolbachia enhances West Nile virus (WNV) infection in the mosquito Culex tarsalis. PLoS Negl. Trop. Dis. 8, e2965.

Dohm, D., O'Guinn, M., Turell, M., 2002. Effect of environmental temperature on the ability of Culex pipiens (Diptera: Culicidae) to transmit West Nile virus. J. Med. Entomol. 39, 221–225.

Dong, Y., Dimopoulos, G., 2009. Anopheles fibrinogen-related proteins provide expanded pattern recognition capacity against bacteria and malaria parasites. J. Biol. Chem. 284, 9835–9844.

Dong, Y., Aguilar, R., Xi, Z., Warr, E., Mongin, E., Dimopoulos, G., 2006a. Anopheles gambiae immune responses to human and rodent Plasmodium parasite species. PLoS Pathog. 2, e52.

Dong, Y., Taylor, H., Dimopoulos, G., 2006b. AgDscam, a hypervariable immunoglobulin domain-containing receptor of the Anopheles gambiae innate immune system. PLoS Biol. 4, e229.

Dong, Y., Manfredini, F., Dimopoulos, G., 2009. Implication of the mosquito midgut microbiota in the defense against malaria parasites. PLoS Pathog. 5, e1000423.

Dong, Y., Cirimotich, C., Pike, A., Chandra, R., Dimopoulos, G., 2012a. Anopheles NF-kB-regulated splicing factors direct pathogen-specific repertoires of the hypervariable pattern recognition receptor AgDscam. Cell Host Microbe 12, 521–530.

Dong, Y., Morton, J., Ramirez, J., Souza-Neto, J., Dimopoulos, G., 2012b. The entomopathogenic fungus Beauveria bassiana activate toll and JAK-STAT pathway-controlled effector genes and anti-dengue activity in Aedes aegypti. Insect Biochem. Mol. Biol. 42, 126–132.

Dostert, C., Jouanguy, E., Irving, P., Troxler, L., Galiana-Arnoux, D., Hetru, C., Hoffmann, J., Imler, J., 2005. The Jak-STAT signaling pathway is required but not sufficient for the antiviral response of drosophila. Nat. Immunol. 6, 946–953.

Duguma, D., Hall, M., Rugman-Jones, P., Stouthamer, R., Terenius, O., Neufeld, J., Walton, W., 2015. Developmental succession of the microbiome of Culex mosquitoes. BMC Microbiol. 15, 140.

Dupuis, S., Jouanguy, E., Al-Hajjar, S., Fieschi, C., Al-Mohsen, I., Al-Jumaah, S., Yang, K., Chapgier, A., Eidenschenk, C., Eid, P., Ghonaium, A., Tufenkeji, H., Frayha, H., Al-Gazlan, S., Al-Rayes, H., Schreiber, R., Gresser, I., Casanova, J., 2003. Impaired response to interferon-alpha/beta and lethal viral disease in human STAT1 deficiency. Nat. Genet. 33, 388–391.

Durvasula, R., Gumbs, A., Panackal, A., Kruglov, O., Taneja, J., Kang, A., Cordon-Rosales, C., Richards, F., Whitham, R., Beard, C., 1999. Expression of a functional antibody fragment in the gut of Rhodnius prolixus via transgenic bacterial symbiont Rhodococcus rhodnii. Med. Vet. Entomol. 13, 115–119.

Eappen, A., Smith, R., Jacobs-Lorena, M., 2013. Enterobacter-activated mosquito immune responses to Plasmodium involve activation of SRPN6 in Anopheles stephensi. PLoS One 8, e62937.

Fang, W., Vega-Rodriguez, J., Ghosh, A., Jacobs-Lorena, M., Kang, A., St. Leger, R., 2011. Development of transgenic fungi that kill human malaria parasites in mosquitoes. Science 331, 1074–1077.

Farenhorst, M., Mouatcho, J., Kikankie, C., Brooke, B., Hunt, R., Thomas, M., Koekemoer, L., Knols, B., Coetzee, M., 2009. Fungal infection counters insecticide resistance in African malaria mosquitoes. Proc. Natl. Acad. Sci. U.S.A. 106, 17443–17447.

Farenhorst, M., Knols, B., Thomas, M., Howard, A., Takken, W., Rowland, M., N'Guessan, R., 2010. Synergy in efficacy of fungal entomopathogens and permethrin against West African insecticide-resistant Anopheles gambiae mosquitoes. PLoS One 5, e12081.

Favia, G., Ricci, I., Damiani, C., Raddadi, N., Crotti, E., Marzorati, M., Rizzi, A., Urso, R., Brusetti, L., Borin, S., Mora, D., Scuppa, P., Pasqualini, L., Clementi, E., Genchi, M., Corona, S., Negri, I., Grandi, G., Alma, A., Kramer, L., Esposito, F., Bandi, C., Sacchi, L., Daffonchio, D., 2007. Bacteria of the genus Asaia stably associate with Anopheles stephensi, an Asian malarial mosquito vector. Proc. Natl. Acad. Sci. U.S.A. 104, 9047–9051.

Fernandez-Luna, M., Tabashnik, B., Lanz-Mendoza, H., Bravo, A., Soberon, M., Miranda-Rios, J., 2010. Single concentration tests show synergism among Bacillus thuringiensis subsp. israelensis toxins against the malaria vector mosquito Anopheles albimanus. J. Invertebr. Pathol. 104, 231–233.

Fieck, A., Hurwitz, I., Kang, A., Durvasula, R., 2010. Trypanosoma cruzi: synergistic cytotoxicity of multiple amphipathic anti-microbial peptides to T. cruzi and potential bacterial hosts. Exp. Parasitol. 125, 342–347.

Figueira, L., Pinheiro, D., Moreira, R., Pinto, E., Simoes, J., Camisa, E., Torrao, L., Palmares, J., Falco-Reis, F., 2012. Beauveria bassiana keratitis in bullous keratopathy: antifungal sensitivity testing and management. Eur. J. Opthalmol. 22, 814–818.

Fillinger, U., Knols, B., Becker, N., 2003. Efficacy and efficiency of new Bacillus thuringiensis var. israelensis and Bacillus sphaericus formulations against Afrotropical anophelines in Western Kenya. Trop. Med. Int. Health 8, 37–47.

Fragkoudis, R., Chi, Y., Siu, R., Barry, G., Attarzadeh-Yazdi, G., Merits, A., Nash, A., Fazakerley, J., Kohl, A., 2008. Semliki Forest virus strongly reduces mosquito host defence signaling. Insect Mol. Biol. 17, 647–656.

Fragkoudis, R., Attarzadeh-Yazdi, G., Nash, A., Fazakerley, J., Kohl, A., 2009. Advances in dissecting mosquito innate immune responses to arbovirus infection. J. Gen. Virol. 90, 2061–2072.

Fraiture, M., Baxter, R., Steinert, S., Chelliah, Y., Frolet, C., Quispe-Tintaya, W., Hoffmann, J., Blandin, S., Levashina, E., 2009. Two mosquito LRR proteins function as complement control factors in the TEP1-mediated killing of Plasmodium. Cell Host Microbe 5, 273–284.

Frants, T., Mertvetsova, O., 1986. Yeast associations with mosquitoes of the genus Aedes Mg. (Diptera, Culicidae) in the Tom-Ob river region. Nauchnye Doki Vyss Shkoly Biol Nauki 4, 94–98.

Franz, A., Sanchez-Vargas, I., Adelman, Z., Blair, C., Beaty, B., James, A., Olson, K., 2006. Engineering RNA interference-based resistance to dengue virus type 2 in genetically modified Aedes aegypti. Proc. Natl. Acad. Sci. U.S.A. 103, 4198–4203.

Frentiu, F., Robinson, J., Young, P., McGraw, E., O'Neill, S., 2010. Wolbachia-mediated resistance to dengue virus infection and death at the cellular level. PLoS One 5, e13398.

Frentiu, F., Zakir, T., Walker, T., Popovici, J., Pyke, A., van den Hurk, A., McGraw, E., O'Neill, S., 2014. Limited dengue virus replication in field-collected Aedes aegypti mosquitoes infected with Wolbachia. PLoS Negl. Trop. Dis. 8, e2688.

Frolet, C., Thoma, M., Blandin, S., Hoffmann, J., Levashina, E., 2006. Boosting NF-kB-dependent basal immunity of Anopheles gambiae aborts development of Plasmodium berghei. Immunity 25, 677–685.

Fu, X., Schindler, C., Improta, T., Aebersold, R., Darnell, J., 1992. The proteins of ISGF-3, the interferon alpha-induced transcriptional activator, define a gene family involved in signal transduction. Proc. Natl. Acad. Sci. U.S.A. 89, 7840–7843.

Fu, Q., Inankur, B., Yin, J., Striker, R., Lan, Q., 2015. Sterol carrier protein 2, a critical host factor for dengue virus infection, alters the cholesterol distribution in mosquito Aag2 cells. J. Med. Entomol. 52, 1124–1134.

Futami, K., Kongere, J., Mwania, M., Lutiali, P., Njenga, S., Minakawa, N., 2011. Effects of Bacillus thuringiensis israelensis on Anopheles arabiensis. J. Am. Mosq. Control Assoc. 27, 81–83.

Gaio, A., Gusmão, D., Santos, A., Berbert-Molina, M., Pimenta, P., Lemos, F., 2011. Contribution of midgut bacteria to blood digestion and egg production in Aedes aegypti (Diptera: Culicidae) (L.). Parasit. Vectors 4, 105.

Garver, L., Xi, Z., Dimopoulos, G., 2008. Immunoglobulin superfamily members play an important role in the mosquito immune system. Dev. Comp. Immunol. 32, 519–531.

Garver, L., Dong, Y., Dimopoulos, G., 2009. Caspar controls resistance to Plasmodium falciparum in diverse Anopheline species. PLoS Pathog. 5, e1000335.

Garver, L., Bahia, A., Das, S., Souza-Neto, J., Shiao, J., Dong, Y., Dimopoulos, G., 2012. Anopheles Imd pathway factors and effectors in infection intensity-dependent anti-Plasmodium action. PLoS Pathog. 8, e1002737.

Garver, L., de Almeida Oliveira, G., Barillas-Mury, C., 2013. The JNK pathway is a key mediator of Anopheles gambiae antiplasmodial immunity. PLoS Pathog. 9, e1003622.

Garza-Hernandez, J., Rodriguez-Perez, M., Salazar, M., Russell, T., Adeleke, M., de Luna-Santillana, E., Reyes-Villanueva, F., 2013. Vectorial capacity of Aedes aegypti for dengue virus type 2 is reduced with co-infection of Metarhizium anisopliae. PLoS Negl. Trop. Dis. 7, e2013.

Geetha, I., Manonmani, A., 2010. Surfactin: a novel mosquitocidal biosurfactant produced by Bacillus subtilis ssp. subtilis (VCRC B471) and influence of abiotic factors on its pupicidal efficacy. Lett. Appl. Microbiol. 51, 406–412.

Geetha, I., Manonmani, A., Paily, K., 2010. Identification and characterization of a mosquito pupicidal metabolite of a Bacillus subtilis subsp. subtilis strain. Appl. Microbiol. Biotechnol. 86, 1737–1744.

Geetha, I., Manonmani, A., Prabakaran, G., 2011. Bacillus amyloliquefaciens: a mosquitocidal bacterium from mangrove forests of Andaman & Nicobar islands, India. Acta Trop. 120, 155–159.

Geissbuhler, Y., Kannady, K., Chaki, P., Emidi, B., Govella, N., Mayagaya, V., Kiama, M., Mtasiwa, D., Mshinda, H., Lindsay, S., Tanner, M., Fillinger, U., de Castro, M., Killeen, G., 2009. Microbial larvicide application by a large-scale, community-based program reduces malaria infection prevalence in urban Dar Es Salaam, Tanzania. PLoS One 4, e5107.

Gendrin, M., Rodgers, F., Yerbanga, R., Ouedraogo, J., Basanez, M., Cohuet, A., Christophides, G., 2015. Antibiotics in ingested human blood affect the mosquito microbiota and capacity to transmit malaria. Nat. Commun. 6, 5921.

Georgel, P., Naitza, S., Kappler, C., Ferrandon, D., Zachary, D., Swimmer, C., Kopczynski, C., Duyk, G., Reichhart, J., Hoffmann, J., 2001. Drosophila immune deficiency (IMD) is a death domain protein that activates antibacterial defense and can promote apoptosis. Dev. Cell 1, 503–514.

Ghosh, A., Coppens, I., Gardsvoll, H., Ploug, M., Jacobs-Lorena, M., 2011. Plasmodium ookinetes coopt mammalian plasminogen to invade the mosquito midgut. Proc. Natl. Acad. Sci. U.S.A. 108, 17153–17158.

Gimonneau, G., Tchioffo, M., Abate, L., Boissiere, A., Awono-Ambene, P., Nsango, S., Christen, R., Morlais, I., 2014. Composition of Anopheles coluzzii and Anopheles gambiae microbiota from larval to adult stages. Infect. Genet. Evol. 28, 715–724.

Goldberg, L., Margalit, J., 1977. A bacterial spore demonstrating rapid larvicidal activity against Anopheles sergentii, Uranotaenia unguiculata, Culex univitattus, Aedes aegypti, and Culex pipiens. Mosq. News 37, 355–358.

Goldman, I., Arnold, J., Carlton, B., 1986. Selection for resistance to Bacillus thuringiensis subspecies israelensis in field and laboratory populations of the mosquito Aedes aegypti. J. Invertebr. Pathol. 47, 317–324.

Gonzalez-Ceron, L., Santillan, F., Rodriguez, M., Mendez, D., Hernandez-Avila, J., 2003. Bacteria in midguts of field-collected Anopheles albimanus block Plasmodium vivax sporogonic development. J. Med. Entomol. 40, 371–374.

Gulland, A., 2012. Death toll from malaria is double the WHO estimate, study finds. BMJ 344, e895.

Gupta, L., Molina-Cruz, A., Kumar, S., Rodrigues, J., Dixit, R., Zamora, R., Barillas-Mury, C., 2009. The STAT pathway mediates late-phase immunity against Plasmodium in the mosquito Anopheles gambiae. Cell Host Microbe 5, 498–507.

Gusmao, D., Santos, A., Marini, D., Bacci, M., Berbert-Molina, M., Lemos, F., 2010. Culture-dependent and culture-independent characterization of microorganisms associated with Aedes aegypti (Diptera: Culicidae) (L.) and dynamics of bacterial colonization in the midgut. Acta Trop. 115, 275–281.

Ha, E., Lee, K., Seo, Y., Kim, S., Lim, J., Oh, B., Kim, J., Lee, W., 2009. Coordination of multiple dual oxidase-regulatory pathways in responses to commensal and infectious microbes in drosophila gut. Nat. Immunol. 10, 949–957.

Habtewold, T., Povelones, M., Blagborough, A., Christophides, G., 2008. Transmission blocking immunity in the malaria non-vector mosquito Anopheles quadriannulatus Species A. PLoS Pathog. 4, e1000070.

Han, Y., 2000. Molecular interactions between Anopheles stephensi midgut cells and Plasmodium berghei: the time bomb theory of ookinete invasion of mosquitoes. EMBO J. 19, 6030–6040.

Hay, S., Okiro, E., Gething, P., Patil, A., Tatem, A., Guerra, C., Snow, R., 2010. Estimating the global clinical burden of Plasmodium falciparum malaria in 2007. PLoS Med. 7, e1000290.

Heaton, N., Perera, R., Berger, K., Khadka, S., LaCount, D., Kuhn, R., Randall, G., 2010. Dengue virus nonstructural protein 3 redistributes fatty acid synthase to sites of viral replication and increases cellular fatty acid synthesis. Proc. Natl. Acad. Sci. U.S.A. 107, 17345–17350.

Herren, J., Paredes, J., Schupfer, F., Arafah, K., Bulet, P., Lemaitre, B., 2014. Insect endosymbiont proliferation is limited by lipid availability. Elife 3, e02964.

Hoa, N., Zheng, L., 2007. Functional characterization of the NF-kB transcription factor gene REL2 from Anopheles gambiae. Insect Sci. 14, 175–184.

Hoffmann, A., Montgomery, B., Popovici, J., Iturbe-Ormaetxe, I., Johnson, P., Muzzi, F., Greenfield, M., Durkan, M., Leong, Y., Dong, Y., Cook, H., Axford, J., Callahan, A., Kenny, N., Omodei, C., McGraw, E., Ryan, P., Ritchie, S., Turelli, M., O'Neill, S., 2011. Successful establishment of Wolbachia in Aedes populations to suppress dengue transmission. Nature 476, 454–457.

Howard, A., Koenraadt, C., Farenhorst, M., Knols, B., Takken, W., 2010. Pyrethroid resistance in Anopheles gambiae leads to increased susceptibility to the entomopathogenic fungi Metarhizium anisopliae and Beauveria bassiana. Malar. J. 9, 168.

Howard, A., N'Guessan, R., Koenraadt, C., Asidi, A., Farenhorst, M., Akogbeto, M., Knols, B., Takken, W., 2011. First report of the infection of insecticide-resistant malaria vector mosquitoes with an entomopathogenic fungus under field conditions. Malar. J. 10, 24.

Hughes, G., Koga, R., Xue, P., Fukatsu, T., Rasgon, J., 2011a. Wolbachia infections are virulent and inhibit the human malaria parasite Plasmodium falciparum in Anopheles gambiae. PLoS Pathog. 7, e1002043.

Hughes, G., Ren, X., Ramirez, J., Sakamoto, J., Bailey, J., Jedlicka, A., Rasgon, J., 2011b. Wolbachia infections in Anopheles gambiae cells: transcriptomic characterization of a novel host-symbiont interaction. PLoS Pathog. 7, e1001296.

Hughes, G., Vega-Rodriguez, J., Xue, P., Rasgon, J., 2011c. Wolbachia strain wAlbB enhances infection by the rodent malaria parasite Plasmodium berghei in Anopheles gambiae mosquitoes. Appl. Environ. Microbiol. 78, 1491–1495.

Hughes, G., Dodson, B., Johnson, R., Murdock, C., Tsujimoto, H., Suzuki, Y., Patt, A., Cui, L., Nossa, C., Barry, R., Sakamoto, J., Hornett, E., Rasgon, J., 2014. Native microbiome impedes vertical transmission of Wolbachia in Anopheles mosquitoes. Proc. Natl. Acad. Sci. U.S.A. 111, 12498–12503.

Ignatova, E., Nagornaia, S., Povazhnaia, T., Ianishevskaia, G., 1996. The yeast flora of blood-sucking mosquitoes. Microbiol. Z. 58, 12–15.

Joshi, D., McFadden, M., Bevins, D., Zhang, F., Xi, Z., 2014. Wolbachia strain wAlbB confers both fitness costs and benefit on Anopheles stephensi. Parasit. Vectors 7, 336.

Joyce, J., Nogueira, J., Bales, A., Pittman, K., Anderson, J., 2011. Interactions between La Crosse virus and bacteria isolated from the digestive tract of Aedes albopictus (Diptera: Culicidae). J. Med. Entomol. 48, 389–394.

Jupatanakul, N., Sim, S., Dimopoulos, G., 2014. Aedes aegypti ML and Niemann-Pick type C family members are agonists of dengue virus infection. Dev. Comp. Immunol. 43, 1–9.

Kahindi, S., Midega, J., Mwangangi, J., Kibe, L., Nzovu, J., Luethy, P., Githure, J., Mbogo, C., 2008. Efficacy of Vectobac DT and Culinexcombi against mosquito larvae in unused swimming pools in Malindi, Kenya. J. Am. Mosq. Control Assoc. 24, 538–542.

Kallio, J., Leinonen, A., Ulvila, J., Valanne, S., Ezekowitz, R., Ramet, M., 2005. Functional analysis of immune response genes in Drosophila identifies JNK pathway as a regulator of antimicrobial peptide gene expression in S2 cells. Microbes Infect. 7, 811–819.

Kambris, Z., Cook, P., Phuc, H., Sinkins, S., 2009. Immune activation by life-shortening Wolbachia and reduced filarial competence in mosquitoes. Science 326, 134–136.

Kambris, Z., Blagborough, A., Pinto, S., Blagrove, M., Godfray, H., Sinden, R., Sinkins, S., 2010. Wolbachia stimulates immune gene expression and inhibits Plasmodium development in Anopheles gambiae. PLoS Pathog. 6, e1001143.

Kanthong, N., Laosutthipong, C., Flegel, T., 2010. Response to dengue virus infections altered by cytokine-like substances from mosquito cell cultures. BMC Microbiol. 10, 290.

Karch, S., Manzambi, Z., Salaun, J., 1991. Field trials with VectolexTM (Bacillus sphaericus) and VectobacTM (Bacillus thuringiensis H-14) against Anopheles gambiae and Culex quinquefasciatus breeding in Zaire. J. Am. Mosq. Control Assoc. 7, 176–179.

Karch, S., Asidi, N., Manzambi, Z., Salaun, J., 1992. Efficacy of Bacillus sphaericus against the malaria vector Anopheles gambiae and other mosquitoes in swamps and rice fields in Zaire. J. Am. Mosq. Control Assoc. 8, 376–380.

Keene, K., Foy, B., Sanchez-Vargas, I., Beaty, B., Blair, C., Olson, K., 2004. RNA interference acts as a natural antiviral response to O'nyong-nyong virus (Alphavirus; Togaviridae) infection of Anopheles gambiae. Proc. Natl. Acad. Sci. U.S.A. 101, 17240–17245.

Kim, T., Kim, Y., 2005. Overview of innate immunity in Drosophila. J. Biochem. Mol. Biol. 38, 121–127.

Kim, W., Koo, H., Richman, A., Seeley, D., Vizioli, J., Klocko, A., O'brochta, D., 2004. Ectopic expression of a Cecropin transgene in the human malaria vector mosquito Anopheles gambiae (Diptera: Culicidae): effects on susceptibility to Plasmodium. J. Med. Entomol. 41, 447–455.

Kingsolver, M., Huang, Z., Hardy, R., 2013. Insect antiviral innate immunity: pathways, effectors, and connections. J. Mol. Biol. 425, 4921–4936.

Kittayapong, P., Baisley, K., Baimai, V., O'Neill, S., 2000. Distribution and diversity of Wolbachia infections in Southeast Asian mosquitoes (Diptera: Culicidae). J. Med. Entomol. 37, 340–345.

Kleino, A., Valanne, S., Ulvila, J., Kallio, J., Myllymaki, H., Enwald, H., Stoven, S., Poidevin, M., Ueda, R., Hultmark, D., Lemaitre, B., Rämet, M., 2005. Inhibitor of apoptosis 2 and TAK1-binding protein are components of the Drosophila Imd pathway. EMBO J. 24, 3423–3434.

Kroeger, A., Horstick, O., Riedl, C., Kaiser, A., Becker, N., 1995. The potential for malaria control with the biological larvicide Bacillus thuringiensis israelensis (Bti) in Peru and Ecuador. Acta Trop. 60, 47–57.

Kumar, A., Sharma, V., Sumodan, P., Thavaselvam, D., Kamat, R., 1994. Malaria control utilizing Bacillus sphaericus against Anopheles stephensi in Panaji, Goa. J. Am. Mosq. Control Assoc. 10, 534–539.

Kumar, S., Molina-Cruz, A., Gupta, L., Rodrigues, J., Barillas-Mury, C., 2010. A peroxidase/dual oxidase system modulates midgut epithelial immunity in Anopheles gambiae. Science 327, 1644–1648.

Lambrechts, L., Quillery, E., Noel, V., Richardson, J., Jarman, R., Scott, T., Chevillon, C., 2012. Specificity of resistance to dengue virus isolates is associated with genotypes of the mosquito antiviral gene Dicer-2. Proc. R. Soc. B Biol. Sci. 280, 20122437.
Lara da Costa, G., Cunha de Oliveira, P., 1998. Penicillium species in mosquitoes from two Brazilian regions. J. Basic Microbiol. 38, 343–347.
Lemaitre, B., Nicolas, E., Michaut, L., Reichhart, J., Hoffmann, J., 1996. The dorsoventral regulatory gene cassette spatzle/Toll/cactus controls the potent antifungal response in Drosophila adults. Cell 86, 973–983.
Leulier, F., Rodriguez, A., Khush, R., Abrams, J., Lemaitre, B., 2000. The Drosophila caspase Dredd is required to resist Gram-negative bacterial infection. EMBO Rep. 1, 353–358.
Levashina, E., 2004. Immune responses in Anopheles gambiae. Insect Biochem. Mol. Biol. 34, 673–678.
Ligoxygakis, P., Pelte, N., Ji, C., Duvic, B., Belvin, M., Jiang, H., Hoffmann, J., Reichhart, J., 2002. A serpin mutant links Toll activation to melanization in the host defence of Drosophila. EMBO J. 21, 6330–6337.
Lim, J., Gowda, D., Krishnegowda, G., Luckhart, S., 2005. Induction of nitric oxide synthase in Anopheles stephensi by Plasmodium falciparum: mechanism of signaling and the role of parasite glycosylphosphatidylinositols. Infect. Immun. 73, 2778–2789.
Lin, H., Zhang, L., Luna, C., Hoa, N., Zheng, L., 2007. A splice variant of PGRP-LC required for expression of antimicrobial peptides in Anopheles gambiae. Insect Sci. 14, 185–192.
Lindh, J., Borg-Karlson, A., Faye, I., 2008. Transstadial and horizontal transfer of bacteria within a colony of Anopheles gambiae (Diptera: Culicidae) and oviposition response to bacteria-containing water. Acta Trop. 107, 242–250.
Lu, P., Bian, G., Pan, X., Xi, Z., 2012. Wolbachia induces density-dependent inhibition to dengue virus in mosquito cells. PLoS Negl. Trop. Dis. 6, e1754.
Luna, C., Hoa, N., Lin, H., Zhang, L., Nguyen, H., Kanzok, S., Zheng, L., 2006. Expression of immune responsive genes in cell lines from two different Anopheline species. Insect Mol. Biol. 15, 721–729.
Luplertlop, N., Surasombatpattana, P., Patramool, S., Dumas, E., Wasinpiyamongkol, L., Saune, L., Hamel, R., Bernard, E., Sereno, D., Thomas, F., Piquemal, D., Yssel, H., Briant, L., Misse, D., 2011. Induction of a peptide with activity against a broad spectrum of pathogens in the Aedes aegypti salivary gland, following infection with dengue virus. PLoS Pathog. 7, e1001252.
Luz, C., Mnyone, L., Sangusangu, R., Lyimo, I., Rocha, L., Humber, R., Russell, T., 2010. A new resting trap to sample fungus-infected mosquitoes, and the pathogenicity of Lecanicillium muscarium to culicid adults. Acta Trop. 116, 105–107.
Majambere, S., Lindsay, S., Green, C., Kandeh, B., Fillinger, U., 2007. Microbial larvicides for malaria control in The Gambia. Malar. J. 6, 76.
Maketon, M., Amnuaykanjanasin, A., Kaysorngup, A., 2013. A rapid knockdown effect of Penicillium citrinum for control of the mosquito Culex quinquefasciatus in Thailand. World J. Microbiol. Biotechnol. 30, 727–736.
Manfruelli, P., Reichhart, J., Steward, R., Hoffmann, J., Lemaitre, B., 1999. A mosaic analysis in Drosophila fat body cells of the control of antimicrobial peptide genes by the Rel proteins Dorsal and DIF. EMBO J. 18, 3380–3391.
Martin-Acebes, M., Blazquez, A., Jimenez de Oya, N., Escribano-Romero, E., Saiz, J., 2011. West Nile virus replication requires fatty acid synthesis but is independent on phosphatidylinositol-4-phosphate lipids. PLoS One 6, e24970.
McFarlane, M., Arias-Goeta, C., Martin, E., O'Hara, Z., Lulla, A., Mousson, L., Rainey, S., Misbah, S., Schnettler, E., Donald, C., Merits, A., Kohl, A., Failloux, A., 2014. Characterization of Aedes aegypti innate-immune pathways that limit Chikungunya virus replication. PLoS Negl. Trop. Dis. 8, e2994.

McMeniman, C., Lane, R., Cass, B., Fong, A., Sidhu, M., Wang, Y., O'Neill, S., 2009. Stable introduction of a life-shortening Wolbachia infection into the mosquito Aedes aegypti. Science 323, 141–144.

Meister, S., Kanzok, S., Zheng, X., Luna, C., Li, T., Hoa, N., Clayton, J., White, K., Kafatos, F., Christophides, G., Zheng, L., 2005. Immune signaling pathways regulating bacterial and malaria parasite infection of the mosquito Anopheles gambiae. Proc. Natl. Acad. Sci. U.S.A. 102, 11420–11425.

Meister, S., Agianian, B., Turlure, F., Relogio, A., Morlais, I., Kafatos, F., Christophides, G., 2009. Anopheles gambiae PGRPLC-mediated defense against bacteria modulates infections with malaria parasites. PLoS Pathog. 5, e1000542.

Michel, T., Reichhart, J., Hoffmann, J., Royet, J., 2001. Drosophila Toll is activated by Gram-positive bacteria through a circulating peptidoglycan recognition protein. Nature 414, 756–759.

Minard, G., Mavingui, P., Moro, C., 2013. Diversity and function of bacterial microbiota in the mosquito holobiont. Parasit. Vectors 6, 146.

Minard, G., Tran, F., Dubost, A., Tran-Van, V., Mavingui, P., Valiente Moro, C., 2014. Pyrosequencing 16S rRNA genes of bacteria associated with wild tiger mosquito Aedes albopictus: a pilot study. Front. Cell. Infect. Microbiol. 4, 59.

Mitraka, E., Stathopoulos, S., Siden-Kiamos, I., Christophides, G., Louis, C., 2013. Asaia accelerates larval development of Anopheles gambiae. Pathog. Glob. Health 107, 305–311.

Mitri, C., Jacques, J., Thiery, I., Riehle, M., Xu, J., Bischoff, E., Morlais, I., Nsango, S., Vernick, K., Bourgouin, C., 2009. Fine pathogen discrimination within the APL1 gene family protects Anopheles gambiae against human and rodent malaria species. PLoS Pathog. 5, e1000576.

Mittal, P., Adak, T., Batra, C., Sharma, V., 1993. Laboratory and field evaluation of Spherix, a formulation of Bacillus sphaericus (B-101) to control breeding of Anopheles stephensi and Culex quinquefasciatus. Indian J. Malariol. 30, 81–89.

Mittal, P., Adak, T., Sharma, V., 1998. Variations in the response to Bacillus sphaericus toxins in different strains of Anopheles stephensi Liston. Indian J. Malariol. 35, 178–184.

Mohanty, S., Prakash, S., 2000. Laboratory evaluation of Trichophyton ajelloi, a fungal pathogen of Anopheles stephensi and Culex quinquefasciatus. J. Am. Mosq. Control Assoc. 16, 254–257.

Molina-Cruz, A., DeJong, R., Charles, B., Gupta, L., Kumar, S., Jaramillo-Gutierrez, G., Barillas-Mury, C., 2007. Reactive oxygen species modulate Anopheles gambiae immunity against bacteria and Plasmodium. J. Biol. Chem. 283, 3217–3223.

Molina-Cruz, A., Garver, L., Alabaster, A., Bangiolo, L., Haile, A., Winikor, J., Ortega, C., van Schaijk, B., Sauerwein, R., Taylor-Salmon, E., Barillas-Mury, C., 2013. The human malaria parasite Pfs47 gene mediates evasion of the mosquito immune system. Science 340, 984–987.

Molina-Cruz, A., Canepa, G., Kamath, N., Pavlovic, N., Mu, J., Ramphul, U., Ramirez, J., Barillas-Mury, C., 2015. Plasmodium evasion of mosquito immunity and global malaria transmission: the lock-and-key theory. Proc. Natl. Acad. Sci. U.S.A. 112, 15178–15183.

Moll, R., Romoser, W., Modrakowski, M., Moncayo, A., Lerdthusnee, K., 2001. Meconial peritrophic membranes and the fate of midgut bacteria during mosquito (Diptera: Culicidae) metamorphosis. J. Med. Entomol. 38, 29–32.

Moncayo, A., Lerdthusnee, K., Leon, R., Robich, R., Romoser, W., 2005. Meconial peritrophic matrix structure, formation, and meconial degeneration in mosquito pupae/pharate adults: histological and ultrastructural aspects. J. Med. Entomol. 42, 939–944.

Moreira, L., Ito, J., Ghosh, A., Devenport, M., Zieler, H., Abraham, E., Crisanti, A., Nolan, T., Catteruccia, F., Jacobs-Lorena, M., 2002. Bee venom phospholipase inhibits malaria parasite development in transgenic mosquitoes. J. Biol. Chem. 277, 40839–40843.

Moreira, L., Iturbe-Ormaetxe, I., Jeffery, J., Lu, G., Pyke, A., Hedges, L., Rocha, B., Hall-Mendelin, S., Day, A., Riegler, M., Hugo, L., Johnson, K., Kay, B., McGraw, E., van den Hurk, A., Ryan, P., O'Neill, S., 2009. A Wolbachia symbiont in Aedes aegypti limits infection with dengue, Chikungunya, and Plasmodium. Cell 139, 1268–1278.

Moy, R., Gold, B., Molleston, J., Schad, V., Yanger, K., Salzano, M., Yagi, Y., Fitzgerald, K., Stanger, B., Soldan, S., Cherry, S., 2014. Antiviral autophagy restricts Rift Valley fever virus infection and is conserved from flies to mammals. Immunity 40, 51–65.

Muller, G., Beier, J., Traore, S., Toure, M., Traore, M., Bah, S., Doumbia, S., Schlein, Y., 2010a. Successful field trial of attractive toxic sugar bait (ATSB) plant-spraying methods against malaria vectors in the Anopheles gambiae complex in Mali, West Africa. Malar. J. 9, 210.

Muller, G., Junnila, A., Schlein, Y., 2010b. Effective control of adult Culex pipiens by spraying an attractive toxic sugar bait solution in the vegetation near larval habitats. J. Med. Entomol. 47, 63–66.

Murdock, C., Blanford, S., Hughes, G., Rasgon, J., Thomas, M., 2014. Temperature alters Plasmodium blocking by Wolbachia. Sci. Rep. 4, 3932.

Murray, N., Quam, M., Wilder-Smith, A., 2013. Epidemiology of dengue: past, present and future prospects. Clin. Epidemiol. 5, 299–309.

Muspratt, J., 1963. Destruction of the larvae of Anopheles gambiae Giles by a Coelomomyces fungus. Bull. World Health Organ. 29, 81–86.

Myles, K., Wiley, M., Morazzani, E., Adelman, Z., 2008. Alphavirus-derived small RNAs modulate pathogenesis in disease vector mosquitoes. Proc. Natl. Acad. Sci. U.S.A. 105, 19938–19943.

Nartey, R., Owusu-Dabo, E., Kruppa, T., Baffour-Awuah, S., Annan, A., Oppong, S., Becker, N., Obiri-Danso, K., 2013. Use of Bacillus thuringiensis var israelensis as a viable option in an Integrated Malaria Vector Control Programme in the Kumasi Metropolis, Ghana. Parasit. Vectors 6, 116.

Ngwa, C., Glockner, V., Abdelmoshen, U., Scheuermayer, M., Fischer, R., Hentschel, U., Pradel, G., 2013. 16S rRNA gene-based identification of Elizabethkingia meningoseptica (Flavobacteriales:Flavobacteriacease) as a dominant midgut bacterium of the Asain malaria vector Anopheles stephensi (Diptera: Culicidae) with antimicrobial activites. J. Med. Entomol. 50, 404–414.

Nicolas, E., Reichhart, J., Hoffmann, J., Lemaitre, B., 1998. In vivo regulation of the IkB homologue cactus during the immune response of Drosophila. J. Biol. Chem. 273, 10463–10469.

Nnakumusana, E., 1986. The effect of Coelomomyces indicus on the fecundity and longevity of Anopheles gambiae, Culex fatigans and Aedes aegypti exposed to infection at each larval instar. Int. J. Trop. Insect Sci. 7, 139–142.

Noden, B., Vaughan, J., Pumpuni, C., Beier, J., 2011. Mosquito ingestion of antibodies against mosquito midgut microbiota improves conversion of ookinetes to oocysts for Plasmodium falciparum, but not P. yoelii. Parasitol. Int. 60, 440–446.

Ocampo, C., Caicedo, P., Jaramillo, G., Ursic Bedoya, R., Baron, O., Serrato, I., Cooper, D., Lowenberger, C., 2013. Differential expression of apoptosis related genes in selected strains of Aedes aegypti with different susceptibilities to dengue virus. PLoS One 8, e61187.

Oliveira, J., Goncalves, R., Lara, F., Dias, F., Gandara, A., Menna-Barreto, R., Edwards, M., Laurindo, F., Silva-Neto, M., Sorgine, M., Oliveira, P., 2011. Blood meal-derived heme decreases ROS levels in the midgut of Aedes aegypti and allows proliferation of intestinal microbiota. PLoS Pathog. 7, e1001320.

Oliveira, G., Lieberman, J., Barillas-Mury, C., 2012. Epithelial nitration by a peroxidase/NOX5 system mediates mosquito antiplasmodial immunity. Science 335, 856–859.

Ondiaka, S., Masinde, E., Koenraadt, C., Takken, W., Mukabana, W., 2015. Effects of fungal infection on feeding and survival of Anopheles gambiae (Diptera: Culicidae) on plant sugars. Parasit. Vectors 8, 35.

Osei-Amo, S., Hussain, M., O'Neill, S., Asgari, S., 2012. Wolbachia-induced aae-miR-12 miRNA negatively regulates the expression of MCT1 and MCM6 genes in Wolbachia-infected mosquito cell line. PLoS One 7, e50049.

Osei-Poku, J., Mbogo, C., Palmer, W., Jiggins, F., 2012. Deep sequencing reveals extensive variation in the gut microbiota of wild mosquitoes from Kenya. Mol. Ecol. 21, 5138–5150.

Osta, M., Christophides, G., Kafatos, F., 2004. Effects of mosquito genes on Plasmodium development. Science 303, 2030–2032.

Pan, X., Zhou, G., Wu, J., Bian, G., Lu, P., Raikhel, A., Xi, Z., 2011. Wolbachia induces reactive oxygen species (ROS)-dependent activation of the Toll pathway to control dengue virus in the mosquito Aedes aegypti. Proc. Natl. Acad. Sci. U.S.A. 109, E23–E31.

Pantuwatana, S., Sattabongkot, J., 1990. Comparison of development of Bacillus thuringiensis subsp. israelensis and Bacillus sphaericus in mosquito larvae. J. Invertebr. Pathol. 55, 189–201.

Paradkar, P., Trinidad, L., Voysey, R., Duchemin, J., Walker, P., 2012. Secreted Vago restricts West Nile virus infection in Culex mosquito cells by activating the Jak-STAT pathway. Proc. Natl. Acad. Sci. U.S.A. 109, 18915–18920.

Paradkar, P., Duchemin, J., Voysey, R., Walker, P., 2014. Dicer-2-dependent activation of Culex Vago occurs via the TRAF-Rel2 signaling pathway. PLoS Negl. Trop. Dis. 8, e2823.

Paris, M., Tetreau, G., Laurent, F., Lelu, M., Despres, L., David, J., 2010. Persistence of Bacillus thuringiensis israelensis (Bti) in the environment induces resistance to multiple Bti toxins in mosquitoes. Pest Manag. Sci. 67, 122–128.

Paris, M., David, J., Despres, L., 2011. Fitness costs of resistance to Bti toxins in the dengue vector Aedes aegypti. Ecotoxicology 20, 1184–1194.

Pei, D., Jiang, J., Yu, W., Kukutla, P., Uentillie, A., Xu, J., 2015. The waaL gene mutation compromised the inhabitation of Enterobacter sp. Ag1 in the mosquito gut environment. Parasit. Vectors 8, 437.

Pereira, E., Sarquis, M., Ferreira-Keppler, R., Hamada, N., Alencar, Y., 2009. Filamentous fungi associated with mosquito larvae (Diptera: Culicidae) in municipalities of the Brazilian Amazon. Neotrop. Entomol. 38, 352–359.

Perera, R., Riley, C., Isaac, G., Hopf-Jannasch, A., Moore, R., Weitz, K., Pasa-Tolic, L., Metz, T., Adamec, J., Kuhn, R., 2012. Dengue virus infection perturbs lipid homeostasis in infected mosquito cells. PLoS Pathog. 8, e1002584.

Pidiyar, V., Jangid, K., Patole, M., Shouche, Y., 2004. Studies on cultured and uncultured microbiota of wild Culex quinquefasciatus mosquito midgut based on 16 s ribosomal RNA gene analysis. Am. J. Trop. Med. Hyg. 70, 597–603.

Possani, L., Zurita, M., Delepierre, M., Hernandez, F., Rodrguez, M., 1998. From Noxiustoxin to Shiva-3, a peptide toxic to the sporogonic development of Plasmodium berghei. Toxicon 36, 1683–1692.

Povelones, M., Waterhouse, R., Kafatos, F., Christophides, G., 2009. Leucine-rich repeat protein complex activates mosquito complement in defense against Plasmodium parasites. Science 324, 258–261.

Povelones, M., Bhagavatula, L., Yassine, H., Tan, L., Upton, L., Osta, M., Christophides, G., 2013. The CLIP-domain serine protease homolog SPCLIP1 regulates complement recruitment to microbial surfaces in the malaria mosquito Anopheles gambiae. PLoS Pathog. 9, e1003623.

Prabakaran, G., Paily, K., Padmanabhan, V., Hoti, S., Balaraman, K., 2002. Isolation of a Pseudomonas fluorescens metabolite/exotoxin active against both larvae and pupae of vector mosquitoes. Pest Manag. Sci. 59, 21–24.

Prakash, S., Singh, G., Soni, N., Sharma, S., 2010. Pathogenicity of Fusarium oxysporum against the larvae of Culex quinquefasciatus (Say) and Anopheles stephensi (Liston) in laboratory. Parasitol. Res. 107, 651–655.

Priyanka, Srivastava, J.N., Prakash, S., 2001. Chrysosporium tropicum efficacy against Anopheles stephensi larvae in the laboratory. J. Am. Mosq. Control Assoc. 17, 127–130.

Pumpuni, C., Demaio, J., Kent, M., Davis, J., Beier, J., 1996. Bacterial population dynamics in three anopheline species: the impact on Plasmodium sporogonic development. Am. J. Trop. Med. Hyg. 54, 214–218.

Qualls, W., Muller, G., Revay, E., Allan, S., Arheart, K., Beier, J., Smith, M., Scott, J., Kravchenko, V., Hausmann, A., Yefremova, Z., Xue, R., 2014. Evaluation of attractive toxic sugar bait (ATSB)-Barrier for control of vector and nuisance mosquitoes and its effect on non-target organisms in sub-tropical environments in Florida. Acta Trop. 131, 104–110.

Ramirez, J., Souza-Neto, J., Torres Cosme, R., Rovira, J., Ortiz, A., Pascale, J., Dimopoulos, G., 2012. Reciprocal tripartite interactions between the Aedes aegypti midgut microbiota, innate immune system and dengue virus influences vector competence. PLoS Negl. Trop. Dis. 6, e1561.

Ramirez, J., Garver, L., Brayner, F., Alves, L., Rodrigues, J., Molina-Cruz, A., Barillas-Mury, C., 2014a. The role of hemocytes in Anopheles gambiae antiplasmodial immunity. J. Innate Immun. 6, 119–128.

Ramirez, J., Short, S., Bahia, A., Saraiva, R., Dong, Y., Kang, S., Tripathi, A., Mlambo, G., Dimopoulos, G., 2014b. Chromobacterium Csp_P reduces malaria and dengue infection in vector mosquitoes and has entomopathogenic and in vitro anti-pathogen activities. PLoS Pathog. 10, e1004398.

Ramirez, J., de Almeida Oliveira, G., Calvo, E., Dalli, J., Colas, R., Serhan, C., Ribeiro, J., Barillas-Mury, C., 2015. A mosquito lipoxin/lipocalin complex mediates innate immune priming in Anopheles gambiae. Nat. Commun. 6, 7403.

Ramphul, U., Garver, L., Molina-Cruz, A., Canepa, G., Barillas-Mury, C., 2014. Plasmodium falciparum evades mosquito immunity by disrupting JNK-mediated apoptosis of invaded midgut cells. Proc. Natl. Acad. Sci. U.S.A. 112, 1273–1280.

Rances, E., Ye, Y., Woolfit, M., McGraw, E., O'Neill, S., 2012. The relative importance of innate immune priming in Wolbachia-mediated dengue interference. PLoS Pathog. 8, e1002548.

Rani, A., Sharma, A., Rajagopal, R., Adak, T., Bhatnagar, R., 2009. Bacterial diversity analysis of larvae and adult midgut micro-flora using culture-dependent and culture-independent methods in lab-reared and field-collected Anopheles stephensi-an Asian malarial vector. BMC Microbiol. 9, 96.

Reisen, W., Fang, Y., Martinez, V., 2006. Effects of temperature on the transmission of West Nile virus by Culex tarsalis (Diptera: Culicidae). J. Med. Entomol. 43, 309–317.

Ricci, I., Mosca, M., Valzano, M., Damiani, C., Scuppa, P., Rossi, P., Crotti, E., Cappelli, A., Ulissi, U., Capone, A., Esposito, F., Alma, A., Mandrioli, M., Sacchi, L., Bandi, C., Daffonchio, D., Favia, G., 2010. Different mosquito species host Wickerhamomyces anomalus (Pichia anomala): perspectives on vector-borne diseases symbiotic control. Antonie Van Leeuwenhoek 99, 43–50.

Ricci, I., Damiani, C., Scuppa, P., Mosca, M., Crotti, E., Rossi, P., Rizzi, A., Capone, A., Gonella, E., Ballarini, P., Chouaia, B., Sagnon, N., Esposito, F., Alma, A., Mandrioli, M., Sacchi, L., Bandi, C., Daffonchio, D., Favia, G., 2011. The yeast Wickerhamomyces anomalus (Pichia anomala) inhabits the midgut and reproductive system of the Asian malaria vector Anopheles stephensi. Environ. Microbiol. 13, 911–921.

Richards, S., Mores, C., Lord, C., Tabachnick, W., 2007. Impact of extrinsic incubation temperature and virus exposure on vector competence of Culex pipiens quinquefasciatus Say (Diptera: Culicidae) for West Nile virus. Vector Borne Zoonotic Dis. 7, 629–636.

Riehle, M., Moreira, C., Lampe, D., Lauzon, C., Jacobs-Lorena, M., 2007. Using bacteria to express and display anti-Plasmodium molecules in the mosquito midgut. Int. J. Parasitol. 37, 595–603.
Riehle, M., Xu, J., Lazzaro, B., Rottschaefer, S., Coulibaly, B., Sacko, M., Niare, O., Morlais, I., Traore, S., Vernick, K., 2008. Anopheles gambiae APL1 is a family of variable LRR proteins required for Rel1-mediated protection from the malaria parasite, Plasmodium berghei. PLoS One 3, e3672.
Rodrigues, J., Brayner, F., Alves, L., Dixit, R., Barillas-Mury, C., 2010. Hemocyte differentiation mediates innate immune memory in Anopheles gambiae mosquitoes. Science 329, 1353–1355.
Romi, R., Ravoniharimelina, B., Ramiakajato, M., Majori, G., 1993. Field trials of Bacillus thuringiensis H-14 and Bacillus sphaericus (strain 2362) formulations against Anopheles arabiensis in the central highlands of Madagascar. J. Am. Mosq. Control Assoc. 9, 325–329.
Rossi, P., Ricci, I., Cappelli, A., Damiani, C., Ulissi, U., Mancini, M., Valzano, M., Capone, A., Epis, S., Crotti, E., Chouaia, B., Scuppa, P., Joshi, D., Xi, Z., Mandrioli, M., Sacchi, L., O'Neill, S., Favia, G., 2015. Mutual exclusion of Asaia and Wolbachia in the reproductive organs of mosquito vectors. Parasit. Vectors 8, 278.
Sanchez-Vargas, I., Scott, J., Poole-Smith, B., Franz, A., Barbosa-Solomieu, V., Wilusz, J., Olson, K., Blair, C., 2009. Dengue virus type 2 infections of Aedes aegypti are modulated by the mosquito's RNA interference pathway. PLoS Pathog. 5, e1000299.
Sanders, H., Foy, B., Evans, A., Ross, L., Beaty, B., Olson, K., Gill, S., 2005. Sindbis virus induces transport processes and alters expression of innate immunity pathway genes in the midgut of the disease vector, Aedes aegypti. Insect Biochem. Mol. Biol. 35, 1293–1307.
Sardelis, M., Turell, M., Dohm, D., O'Guinn, M., 2001. Vector competence of selected North American Culex and Coquillettidia mosquitoes for West Nile virus. Emerg. Infect. Dis. 7, 1018–1022.
Schnitger, A., Yassine, H., Kafatos, F., Osta, M., 2009. Two C-type lectins cooperate to defend Anopheles gambiae against Gram-negative bacteria. J. Biol. Chem. 284, 17616–17624.
Scholte, E., Knols, B., Takken, W., 2006. Infection of the malaria mosquito Anopheles gambiae with the entomopathogenic fungus Metarhizium anisopliae reduces blood feeding and fecundity. J. Invertebr. Pathol. 91, 43–49.
Seye, F., Faye, O., Ndiaye, M., Njie, E., Marie Afoutou, J., 2009. Pathogenicity of the fungus, Aspergillus clavatus, isolated from the Locust, Oedaleus senegalensis, against larvae of the mosquitoes Aedes aegypti, Anopheles gambiae and Culex quinquefasciatus. J. Insect Sci. 9, 1–7.
Sharma, A., Dhayal, D., Singh, O., Adak, T., Bhatnagar, R., 2013. Gut microbes influence fitness and malaria transmission potential of Asian malaria vector Anopheles stephensi. Acta Trop. 128, 41–47.
Sharma, P., Sharma, S., Maurya, R., De, T., Thomas, T., Lata, S., Singh, N., Pandey, K., Valecha, N., Dixit, R., 2014. Salivary glands harbor more diverse microbial communities than gut in Anopheles culicifacies. Parasit. Vectors 7, 235.
Shin, S., Kokoza, V., Bian, G., Cheon, H., Kim, Y., Raikhel, A., 2005. REL1, a homologue of Drosophila dorsal, regulates toll antifungal immune pathway in the female mosquito Aedes aegypti. J. Biol. Chem. 280, 16499–16507.
Silverman, N., Zhou, R., Erlich, R., Hunter, M., Bernstein, E., Schneider, D., Maniatis, T., 2003. Immune activation of NF-Â B and JNK requires Drosophila TAK1. J. Biol. Chem. 278, 48928–48934.
Sim, S., Jupatanakul, N., Ramirez, J., Kang, S., Romero-Vivas, C., Mohammed, H., Dimopoulos, G., 2013. Transcriptomic profiling of diverse Aedes aegypti strains reveals

increased basal-level immune activation in dengue virus-refractory populations and identifies novel virus-vector molecular interactions. PLoS Negl. Trop. Dis. 7, e2295.
Singer, S., 1973. Insecticidal activity of recent bacterial isolates and their toxins against mosquito larvae. Nature 244, 110–111.
Singh, G., Prakash, S., 2011. Evaluation of culture filtrates of Culicinomyces clavisporus: Mycoadulticide for Culex quinquefasciatus, Aedes aegypti and Anopheles stephensi. Parasitol. Res. 110, 267–272.
Sinka, M., Bangs, M., Manguin, S., Rubio-Palis, Y., Chareonviriyaphap, T., Coetzee, M., Mbogo, C., Hemingway, J., Patil, A., Temperley, W., Gething, P., Kabaria, C., Burkot, T., Harbach, R., Hay, S., 2012. A global map of dominant malaria vectors. Parasit. Vectors 5, 69.
Skelton, E., Rances, E., Frentiu, F., Kusmintarsih, E., Iturbe-Ormaetxe, I., Caragata, E., Woolfit, M., O'Neill, S., 2015. A native Wolbachia endosymbiont does not limit dengue virus infection in the mosquito Aedes notoscriptus (Diptera: Culicidae). J. Med. Entomol. 53, 401–408.
Skovmand, O., Sanogo, E., 1999. Experimental formulations of Bacillus sphaericus and B. thuringiensis israelensis against Culex quinquefasciatus and Anopheles gambiae (Diptera: Culicidae) in Burkina Faso. J. Med. Entomol. 36, 62–67.
Sluss, H., Han, Z., Barrett, T., Davis, R., Ip, Y., 1996. A JNK signal transduction pathway that mediates morphogenesis and an immune response in Drosophila. Genes Dev. 10, 2745–2758.
Smith, R., Eappen, A., Radtke, A., Jacobs-Lorena, M., 2012. Regulation of anti-plasmodium immunity by a LITAF-like transcription factor in the malaria vector Anopheles gambiae. PLoS Pathog. 8, e1002965.
Smith, R., Vega-Rodriguez, J., Jacobs-Lorena, M., 2014. The Plasmodium bottleneck: malaria parasite losses in the mosquito vector. Mem. Inst. Oswaldo Cruz 109, 644–661.
Souza-Neto, J., Sim, S., Dimopoulos, G., 2009. An evolutionary conserved function of the JAK-STAT pathway in anti-dengue defense. Proc. Natl. Acad. Sci. U.S.A. 106, 17841–17846.
Stathopoulos, S., Neafsey, D., Lawniczak, M., Muskavitch, M., Christophides, G., 2014. Genetic dissection of Anopheles gambiae gut epithelial responses to Serratia marcescens. PLoS Pathog. 10, e1003897.
Stewart, Z., Oxborough, R., Tungu, P., Kirby, M., Rowland, M., Irish, S., 2013. Indoor application of attractive toxic sugar bait (ATSB) in combination with mosquito nets for control of pyrethroid-resistant mosquitoes. PLoS One 8, e84168.
Steyn, A., Roets, F., Botha, A., 2015. Yeasts associated with Culex pipiens and Culex theileri mosquito larvae and the effect of selected yeast strains on the ontogeny of Culex pipiens. Microb. Ecol. 71, 747–760.
Straif, S., Mbogo, C., Toure, A., Walker, E., Kaufman, M., Toure, Y., Beier, J., 1998. Midgut bacteria in Anopheles gambiae and An. funestus (Diptera: Culicidae) from Kenya and Mali. J. Med. Entomol. 35, 222–226.
Tajedin, L., Hashemi, J., Abaei, M., Hosseinpour, L., Rafei, F., Basseri, H., 2009. Study on fungal flora in the midgut of the larva and adult of the different populations of the malaria vector Anopheles stephensi. Iran. J. Arthropod Borne Dis. 3, 36–40.
Terenius, O., de Oliveira, C., Pinheiro, W., Tadei, W., James, A., Marinotti, O., 2008. 16S rRNA gene sequences from bacteria associated with adult Anopheles darlingi (Diptera: Culicidae) mosquitoes. J. Med. Entomol. 45, 172–175.
Terenius, O., Lindh, J., Eriksson-Gonzales, K., Bussiere, L., Laugen, A., Bergquist, H., Titanji, K., Faye, I., 2012. Midgut bacterial dynamics in Aedes aegypti. FEMS Microbiol. Ecol. 80, 556–565.
Thiery, I., Nicolas, L., Rippka, R., Tandeau de Marsac, N., 1991. Selection of cyanobacteria isolated from mosquito breeding sites as a potential food source for mosquito larvae. Appl. Environ. Microbiol. 57, 1354–1359.

Turell, M., 1989. Effect of environmental temperature on the vector competence of Aedes fowleri for rift valley fever virus. Res. Virol. 140, 147–154.

Vaidyanathan, R., Scott, T., 2006. Apoptosis in mosquito midgut epithelia associated with West Nile virus infection. Apoptosis 11, 1643–1651.

Valero-Jimenez, C., Debets, A., van Kan, J., Schoustra, S., Takken, W., Zwaan, B., Koenraadt, C., 2014. Natural variation in virulence of the entomopathogenic fungus Beauveria bassiana against malaria mosquitoes. Malar. J. 13, 479.

Valiente Moro, C., Tran, F., Nantenaina Raharimalala, F., Ravelonandro, P., Mavingui, P., 2013. Diversity of culturable bacteria including Pantoea in wild mosquito Aedes albopictus. BMC Microbiol. 13, 70.

Valzano, M., Cecarini, V., Cappelli, A., Capone, A., Bozic, J., Cuccioloni, M., Epis, S., Petrelli, D., Angeletti, M., Eleuteri, A., Favia, G., Ricci, I., 2016. A yeast strain associated to Anopheles mosquitoes produces a toxin able to kill malaria parasites. Malar. J. 15, 21.

van den Hurk, A., Hall-Mendelin, S., Pyke, A., Frentiu, F., McElroy, K., Day, A., Higgs, S., O'Neill, S., 2012. Impact of Wolbachia on infection with Chikungunya and Yellow Fever viruses in the mosquito vector Aedes aegypti. PLoS Negl. Trop. Dis. 6, e1892.

Vlachou, D., Schlegelmilch, T., Christophides, G., Kafatos, F., 2005. Functional genomic analysis of midgut epithelial responses in Anopheles during Plasmodium invasion. Curr. Biol. 15, 1185–1195.

Walker, T., Johnson, P., Moreira, L., Iturbe-Ormaetxe, I., Frentiu, F., McMeniman, C., Leong, Y., Dong, Y., Axford, J., Kriesner, P., Lloyd, A., Ritchie, S., O'Neill, S., Hoffmann, A., 2011. The wMel Wolbachia strain blocks dengue and invades caged Aedes aegypti populations. Nature 476, 450–453.

Wang, S., Jacobs-Lorena, M., 2013. Genetic approaches to interfere with malaria transmission by vector mosquitoes. Trends Biotechnol. 31, 185–193.

Wang, Y., Gilbreath, T., Kukutla, P., Yan, G., Xu, J., 2011. Dynamic gut microbiome across life history of the malaria mosquito Anopheles gambiae in Kenya. PLoS One 6, e24767.

Wang, H., Gort, T., Boyle, D., Clem, R., 2012a. Effects of manipulating apoptosis on Sindbis Virus infection of Aedes aegypti mosquitoes. J. Virol. 86, 6546–6554.

Wang, S., Ghosh, A., Bongio, N., Stebbings, K., Lampe, D., Jacobs-Lorena, M., 2012b. Fighting malaria with engineered symbiotic bacteria from vector mosquitoes. Proc. Natl. Acad. Sci. U.S.A. 109, 12734–12739.

Wang, Y., Wang, Y., Zhang, J., Xu, W., Zhang, J., Huang, F., 2013. Ability of TEP1 in intestinal flora to modulate natural resistance of Anopheles dirus. Exp. Parasitol. 134, 460–465.

Wang, Y., Hu, Y., Xing, L., Jiang, H., Hu, S., Raikhel, A., Zou, Z., 2015. A critical role for CLSP2 in the modulation of antifungal immune response in mosquitoes. PLoS Pathog. 11, e1004931.

Warr, E., Das, S., Dong, Y., Dimopoulos, G., 2008. The Gram-negative bacteria-binding protein gene family: its role in the innate immune system of Anopheles gambiae and in anti-Plasmodium defence. Insect Mol. Biol. 17, 39–51.

Weiser, J., 1984. A mosquito-virulent Bacillus sphaericus in adult Simulium damnosum from northern Nigeria. Zentralbl. Mikrobiol. 139, 57–60.

Werren, J., 1997. Biology of Wolbachia. Annu. Rev. Entomol. 42, 587–609.

Williamson, D., Tully, J., Rosen, L., Rose, D., Whitcomb, R., Abalain-Colloc, M., Carle, P., Bove, J., Smyth, J., 1996. Spiroplasma diminutum sp. nov., from Culex annulus mosquitoes collected in Taiwan. Int. J. Syst. Bacteriol. 46, 229–233.

Wotton, R., Chaloner, D., Yardley, C., Merritt, R., 1997. Growth of Anopheles mosquito larvae on dietary microbiota in aquatic surface microlayers. Med. Vet. Entomol. 11, 65–70.

Xi, Z., Ramirez, J., Dimopoulos, G., 2008. The Aedes aegypti Toll pathway controls dengue virus infection. PLoS Pathog. 4, e1000098.

Yassine, H., Kamareddine, L., Osta, M., 2012. The mosquito melanization response is implicated in defense against the entomopathogenic fungus Beauveria bassiana. PLoS Pathog. 8, e1003029.

Yassine, H., Kamareddine, L., Chamat, S., Christophides, G., Osta, M., 2014. A serine protease homolog negatively regulates TEP1 consumption in systemic infections of the malaria vector Anopheles gambiae. J. Innate Immun. 6, 806–818.

Young, M., Undeen, A., Dame, D., Wing, S., 1990. The effect of Bacillus spaericus upon the susceptibility of Anopheles quadrimaculatus to Plasmodium berghei. J. Am. Mosq. Control Assoc. 6, 139–140.

Zambon, R., Nandakumar, M., Vakharia, V., Wu, L., 2005. The Toll pathway is important for an antiviral response in Drosophila. Proc. Natl. Acad. Sci. U.S.A. 102, 7257–7262.

Zhang, G., Hussain, M., O'Neill, S., Asgari, S., 2013. Wolbachia uses a host microRNA to regulate transcripts of a methyltransferase, contributing to dengue virus inhibition in Aedes aegypti. Proc. Natl. Acad. Sci. U.S.A. 110, 10276–10281.

Zhou, D., Zhang, D., Ding, G., Shi, L., Hou, Q., Ye, Y., Xu, Y., Zhou, H., Xiong, C., Li, S., Yu, J., Hong, S., Yu, X., Zou, P., Chen, C., Chang, X., Wang, W., Lv, Y., Sun, Y., Ma, L., Shen, B., Zhu, C., 2014. Genome sequence of Anopheles sinensis provides insight into genetics basis of mosquito competence for malaria parasites. BMC Genomics 15, 42.

Zou, Z., Shin, S., Alvarez, K., Bian, G., Kokoza, V., Raikhel, A., 2008. Mosquito RUNX4 in the immune regulation of PPO gene expression and its effect on avian malaria parasite infection. Proc. Natl. Acad. Sci. U.S.A. 105, 18454–18459.

Zouache, K., Raharimalala, F., Raquin, V., Tran-Van, V., Raveloson, L., Ravelonandro, P., Mavingui, P., 2010. Bacterial diversity of field-caught mosquitoes, Aedes albopictus and Aedes aegypti, from different geographic regions of Madagascar. FEMS Microbiol. Ecol. 75, 377–389.

CHAPTER TEN

Mosquito Sensory Systems

C. Montell*, L.J. Zwiebel†,‡

*Neuroscience Research Institute, University of California, Santa Barbara, CA, United States
†Vanderbilt University, Nashville, TN, United States
‡Vanderbilt Brain Institute, Program in Developmental Biology, and Institutes of Chemical Biology and Global Health, Vanderbilt University Medical Center, Nashville, TN, United States

Contents

Abstract

In this chapter, we will highlight recent progress in describing the current state of our understanding of the diverse array of sensory pathways across the family *Culicidae*. These will include both contact (gustatory) and volatile (olfactory)-based chemosensory systems as well as auditory, mechanosensory, thermosensory and visual modalities. We will discuss both the cellular and the molecular basis underlying these senses as well as how they control the lifecycle, behaviour and vectorial capacity of mosquitoes. We will

Advances in Insect Physiology, Volume 51
ISSN 0065-2806
http://dx.doi.org/10.1016/bs.aiip.2016.04.007

discuss the impact of recent data derived from next-generation sequencing to our understanding of the evolution and function of these diverse sensory systems.

1. INTRODUCTION

In light of their role as vectors for the transmission of a range of important human and animal diseases, mosquitoes represent one of the greatest worldwide threats to global health. In fact, mosquitoes might well be considered the most dangerous animals on the planet. *Anopheles gambiae* and other Anopheline mosquitoes spread malaria, which is the most prevalent insect-borne disease worldwide. Each year, *Plasmodium falciparum* and other malaria pathogens infect hundreds of millions and cause the death of hundreds of thousands of people (Enayati and Hemingway, 2010; Guinovart et al., 2006). Arboviruses (arthropod-borne viruses) represent another significant class of mosquito-borne pathogens. Of these, dengue virus, which is responsible for both dengue fever and dengue haemorrhagic fever (DF/DHF), is rapidly emerging as a significant worldwide problem since it appeared on the public health landscape after the World War II (Bhatt et al., 2013). While DF/DHF was first described in the 18th century, today upwards of ~500,000,000 people are infected with dengue virus annually by the highly invasive *Aedes aegypti* mosquito, which is the main vector for this and other arboviruses (Bhatt et al., 2013). Together with other Aedine mosquitoes *Ae. aegypti* also transmits a wide range of arboviruses, including the causative agents for Chikungunya, Yellow and Zika fevers. In addition to their immediate symptomatic effects on infected individuals, which range from aches, fever to broad haemorrhagic episodes and death, public health agencies are increasingly becoming aware of a diverse range of very serious postinfection pathologies that include birth defects such as microcephaly, neurological conditions such as Guillain–Barre syndrome and many others (Cao-Lormeau et al., 2016; Chang et al., 2016).

Diseases spread by mosquitoes are not restricted to the developing world; indeed several are on the rise in the United States and other parts of the developed world. An example is West Nile Virus, which was first identified in New York in 1999. It is now documented in all 48 contiguous states (Petersen et al., 2013). With the increasing reality of climate change and its associated global warming that is predicted to occur over the coming century and beyond, it is highly likely that the catastrophic effects of insect-borne diseases are going to get worse across both the developed and developing world.

Disease transmission by mosquitoes is based on the reproductive requirements of females that take blood meals from humans and other animals in order to complete their gonadotropic cycles. In order to carry out this critical (from a public health perspective) component as well as nearly all the other aspects of both male and female life stages mosquitoes must utilize a diverse and highly sophisticated set of sensory systems. These systems receive and pass along essential information across a range of sensory modalities that is processed to drive the navigation and preference decisions that in many ways underlie the vectorial capacity of mosquitoes. Here we provide an overview of the current status of our mechanistic understanding of the senses in mosquito vectors, with a particular emphasis on the peripheral sensory neurons and receptors.

1.1 The Role of Sensory Systems in Mosquito Biology

Mosquito life cycles are conducted in a variety of environments and conditions; accordingly, they have evolved an array of sensory systems with which to sense and respond to a complex set of environmental stimuli and biological cues. These signals are received by a range of sensory cells, most notably neurons of the peripheral nervous system that transduce external signals into neuronal activity or inhibition through a variety of mechanisms. This information is subsequently collected, integrated and otherwise processed by downstream regions of the central nervous system (CNS) in order for mosquitoes to make the salient behavioural decisions that are required to successfully complete a set of discrete lifecycle tasks. Here we describe the sensory processes that impact the major elements of the mosquito lifecycle.

2. PREADULT LIFE STAGES

While the vast majority of study has been confined to adult stages where disease transmission and reproduction occur, mosquitoes engage in a robust range of preadult activities that often necessitate sensory acuity. Insofar as basic development is concerned, all mosquitoes exhibit complete metamorphosis and require aquatic habitats, although some species can lay their eggs in moist soil. After hatching they pass through four larval instars and a pupal stage during which the transformation into an adult takes place.

2.1 Prelarval Stages

Mosquito embryogenesis starts almost immediately after the eggs have been laid. It takes about 2–10 days or more until the embryos are fully developed, and the length of time required is dependent almost entirely on temperature (Clements, 1992b). For example, at 30°C, *Culex pipiens pipiens* larvae hatch 1 day after the eggs have been laid, at 20°C and 10°C it takes 3 and 10 days, respectively. However, at 4°C, embryonic development cannot be completed. Apart from this presumably purely metabolic response to temperature, and in light of the evidence that mosquito embryos neither feed nor display any autonomous mobility, it is likely that mosquito embryos do not have any significant sensory capability.

2.2 Delayed Hatching (Diapause)

Mosquitoes depend on multiple sensory systems to stimulate diapause and delay the process of hatching, which spans the last embryonic and first larval stage. The initiation and timing of larval hatching coincides with the ability to accurately sense the presence of ideal developmental conditions. There is a wide range of natural ecology and hatching behaviours exhibited by diverse mosquito taxa that inhabit tropical, temperate and other regions that illustrate the extent to which the hatching behaviour of each species is adapted to several aspects of the biological and abiotic conditions in their respective breeding sites. Therefore, it is not surprising that the entry to and exit from diapause is an extremely variable characteristic that is likely to be influenced by the ability to accurately sense several factors that include the length and temperature of (day/night) light/dark cycles, water level and temperature. For example, premature hatching during cold weather would greatly delay the development of the larvae, since the process is very temperature dependent. In addition, there is evidence that the level of dissolved oxygen is another critical parameter affecting hatching. In many species hatching occurs only when conditions are stagnant and the oxygen content starts to decrease very rapidly (due to bacterial degradation processes). The decreasing level of oxygen in the water signals the unhatched larvae that the water will remain stagnant at the time of their hatching, thus ensuring that the risk of being washed away has diminished. Moreover, the bacterial action causing decomposition of organic material ensures that the larvae will have an adequate food supply. Once initiated, the whole process of hatching takes only a few minutes (Clements, 1992b).

2.3 Larval Stages

The larval body of mosquitoes is divided into three distinct parts: (a) the head with mouthparts, eyes and antennae; (b) the broader thorax and (c) the abdomen which is composed of seven almost identical segments and three modified posterior segments (Fig. 1). While it is likely that sensory neurons are dispersed across all three parts, the principal sensory appendages are clustered on the head where the eyes carry out visual transduction, the mouthparts respond to mechanosensory and gustatory cues while the sensory cone and peg organ of the antennae mediate olfaction and gustation, respectively (Nicastro et al., 1998).

A major component of larval life stages is feeding, and depending upon the species and ecology, the location and selection (preference) of nutrients requires the transduction and processing of numerous sensory cues comprising chemical, mechanical, temperature and visual stimuli (Merritt et al., 1992). Broadly speaking, larval food consists of microorganisms, algae, protozoa, invertebrates and detritus. On the basis of their feeding behaviour mosquito larvae may be classified into filter feeders, browsers or predators. The filter feeders typically hang on the water surface filtering the water column beneath the surface by beating their head brushes (lateral palatal brushes) towards the oral cavity.

Mosquito larvae usually do not discriminate what they ingest, although the size of the particles is usually less than 50 mm. Larvae can also move slowly in the water column while filter feeding. The browsers (eg, most *Aedes* and *Ochlerotatus* species) collect food by resuspension, scraping or shredding particles, microorganisms, algae and protozoa from the surface of submerged substrates or the microbial film at the air–water interface (the *Anopheles* larvae). Predacious larvae of the genera *Toxorhynchites*, *Aedes*, *Psorophora* and *Culex* can also feed upon insects and often other mosquito larvae.

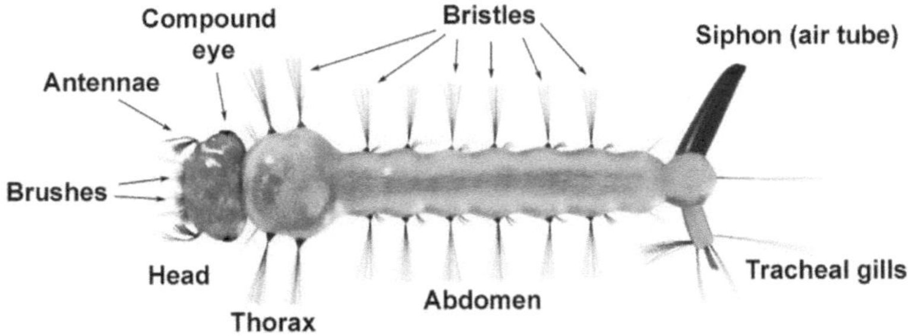

Fig. 1 Graphical depiction of the sensory organs of an *Aedes aegypti* fourth instar larvae. The main sensory structures distributed on the head, thorax and abdomen are shown.

The development of larvae is temperature dependent and as mentioned earlier, there are great differences in the optimum temperature for the development of different mosquito species. Once again, this is presumed to involve metabolic rather than sensory processes. Larvae moult four times at intervals, before reaching the pupal stage. At each moult, the head capsule is increased to the full size characteristic of the next instar, whereas the body grows continuously.

2.4 Pupae

Mosquito pupae are also aquatic and in most species this stage usually last about 2 days. However, as was the case for other developmental stages, this period may be reduced or extended, at higher or lower temperatures, respectively, which again is presumably a metabolic, rather than sensory response. Unlike larvae, pupae do not feed and instead are primarily occupied by the complex and poorly understood (in the mosquito) developmental programme of metamorphosis. That said, mosquito pupae are very mobile; they sense and are disturbed by mechanical and/or visual cues which results in a robust dive escape response after which positive buoyancy brings them back to the surface. In the absence of feeding and other externally oriented activity, the sensory demands on pupae are most likely restricted to the visual and mechanosensory stimuli for escape purposes.

3. ADULTS

3.1 Emergence

Eclosion (moulting) is the final stage of adult metamorphosis, and within a few minutes as the soft cuticle becomes fully sclerotized (hardened) the newly emerged adult is typically able to fly. Following emergence, both male and female adults are ready to begin their life cycle again through mating, feeding and oviposition, although male mosquitoes are not sexually mature at emergence as they typically require ~1 day before they are ready to mate.

3.2 Nectar Feeding

Plant sugars such as oral nectar, damaged fruits and honeydew are the main energy source during the adult life of all males and for most females (apart from blood-fed or gravid stages); in a limited number of mosquito species, it is the only food for both sexes (Foster, 1995). Plants emit an enormous diversity of volatile compounds, comprising many different chemical groups, including terpenoids, alcohols and aldehydes (Lucas-Barbosa et al., 2011; Nyasembe and Torto, 2014). While field studies with naturally occurring flowers and fruits have shown that *Anopheles* males and females can sense and prefer certain plant species over others (eg, Gouagna et al., 2010; Muller et al., 2010), a limited number of plant species that are attractive to *Anopheles* mosquitoes has been chemically analysed in terms of their volatile composition (Nikbakhtzadeh et al., 2014; Nyasembe et al., 2012). One chemosensory study identified six components that were consistently detected by the antennae of *An. gambiae* females (Nyasembe et al., 2012). It is likely the location and selection of plants for feeding involves mainly visual and chemical (olfactory/gustatory) cues that differ from species to species as well as both geographically and temporally along with their respective nectar sources (Foster and Hancock, 1994).

3.3 Mating

Mating of most mosquitoes takes place soon after female emergence as they seek out and enter swarms of flying males, which can consist of a few to up to several thousand male mosquitoes. Swarming (eurygamy) is not necessary for all species, and some species may indeed mate without it (stenogamy). There is a wide range of sensory modalities that dictate the species-specific elements of when and where males form a mating swarm. Furthermore, a complex array of auditory, chemosensory and visual signalling is likely to contribute to the ability of females to locate the swarms.

Significant attention has yielded considerable data in precisely defining the acoustic interactions that take place between male and female during mating, a process that is sometimes referred to as "duetting" (Cator et al., 2009; Gibson et al., 2010). Indeed, the highly articulated antennae of males are especially receptive to sound generated by the female. The flagellum of the antenna starts to vibrate and stimulate the Johnston's organ that is located in the swollen second segment (pedicel) of the antenna (Clements, 1999; McIver, 1982). From a chemosensory point of view, there is evidence that contact pheromones may also be entailed in the mating behaviour and aggregation (lekking); pheromones involved in formation of male swarms have recently been described for *Aedes* (Cabrera and Jaffe, 2007; Fawaz et al., 2014). There is almost certainly a significant visual component and indeed, there is considerable literature linking a range of visual markers (Allan et al., 1987; Takken et al., 2006) with the location of mating swarms.

However, much of the evidence is based on field-based observations that lack definitive characteristics and controls.

Copulation requires a complex merging of the male and female reproductive structures. Usually it takes less than half a minute for the male to deposit the spermatozoa in the female, whereupon the sperm move into the spermathecae (Clements, 1992b). In contrast to males, which may mate many times, females store sufficient sperm to fertilize several egg batches without further copulation. Interestingly, a recent study demonstrated that mosquito spermatozoa themselves express odorant receptors (ORs) and display robust responses to diverse chemosensory that are hypothesized to activate dormant sperm as well potentially act in a chemotactic capacity (Pitts et al., 2014). As is true for many animal species, male mosquitoes devote most of their adult life to a cycle of feeding and mating. After insemination, the search for a host to obtain a blood meal is the next important phase in the reproductive life of the female.

3.4 Ranging and Host-Seeking Behaviour

An elemental aspect of mosquito vectorial capacity is the blood meal, during which pathogen transmission occurs and which is required by all anautogenous mosquitoes in order to obtain a rich source of protein and other essential components to complete oogenesis. To accomplish this, females (males do not blood feed) have developed a suite of complex host-seeking behaviours to locate and select a potential blood meal host. Primarily, the location of the host is based on olfactory, visual and thermal stimuli (Fig. 2). Females possess numerous classes of antennal and other types of chemosensory receptors that respond to host odours. The host-seeking process may differ within species depending on the season and the availability of certain hosts. However, it can usually be divided into several distinct phases:

(a) Nonoriented ranging flight behaviour that enhances the likelihood of the female coming into contact with stimuli derived from a potential host. These stimuli are typically volatile host-derived odorants that are known as kairomones (as they benefit only the blood meal seeking mosquito receiving these signals), which are detected by the female mosquito's olfactory system.

(b) Oriented host location behaviour resulting from long-distance, olfactory-based contact with host stimuli. Typically, the concentration of these stimuli increases as the mosquito and host come closer together.

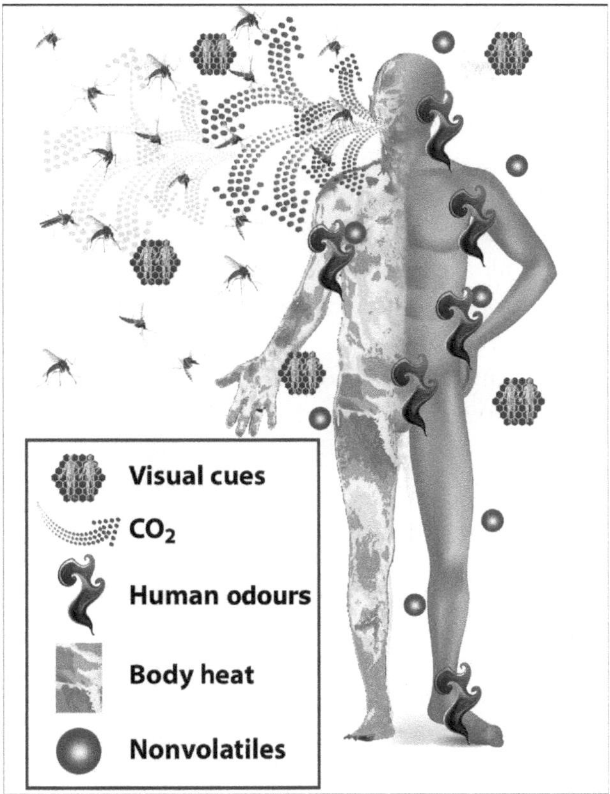

Fig. 2 Graphic illustrating the sensory cues used by mosquito vectors to target human hosts depicting visual cues, CO_2, odours, body heat and nonvolatiles. (See the colour plate.)

Interaction or synergism of the components of the host odour in attracting a given species is a very complex process that developed in the course of evolution between the insect and the target organisms. The female mosquito flies upwind in a zigzag pattern that holds the mosquito within the plume and brings it closer to the odour source (Dekker and Cardé, 2011).

(c) Selection and directed attraction to a suitable candidate host, once the female has identified it in her immediate vicinity. While still largely olfactory driven, as distance to host diminishes, the role of visual, thermosensory and other stimuli increases. The compound eyes serve to discriminate between form, movement, light intensity, contrast and colour. Mosquitoes respond particularly to blue, black and red colours, whereas least attraction is caused by white and yellow; they can easily detect temperature differences of 0.2°C, and water vapour in short-range orientation–attraction may also play a role (Lehane, 1991).

(d) Alighting upon the potential host provides still higher concentrations of volatile kairomone, as access to low volatile odorants that can only be received at close range, direct contact to gustatory cues as well as heightened thermal and visual signalling.

(e) Probing involves direct contact by the labellum, stylets and other mouthparts of female mosquitoes to activate mechanosensory as well as chemosensory pathways that play critical roles in the successful completion of the blood meal.

Mosquitoes differ in many aspects of their feeding and resting (the period during which blood meals are digested and oogenesis is completed) behaviour due to species-specific sensory adaptations. These characteristics include those that mainly feed/rest indoors (endophagic/endophilic) or outdoors (exophagic/exophilic). Ornithophily is expressed when females prefer to feed on birds, zoophily is used when they feed on other animals (zoophilic species) and the term anthropophily is used when they prefer to feed on humans (anthropophilic species). It is altogether reasonable to speculate that, along with visual cues, a broad range of chemo-, thermo- and mechanosensory inputs provide salient information to shape many, if not, all aspects of these critically important characteristics of the mosquito lifecycle.

3.5 Oviposition

The successful completion and subsequent digestion of blood meals afford anautogenous female mosquitoes the opportunity to complete oogenesis. At this point they are considered gravid and their behaviour shifts from resting to the active location and selection of suitable aqueous oviposition sites (Clements, 1992a). Apart from attraction to the humidity and texture derived from bodies of water (Huang et al., 2005), the selection of oviposition sites is mediated by visual cues (Allan et al., 1987; Huang et al., 2007). The large range of general and species-specific semiochemicals that elicit either attractive or aversive responses are collectively known as oviposition pheromones or cues. The origin of oviposition promoting semiochemicals can be maternally derived (pheromones) in the form of apical droplets left at the tip of eggs by ovipositing Culex females (Laurence and Pickett, 1982) or as breakdown products of bacterial origin as well as those directly emitted from plant-based or other biological material (cues) that are associated with the chemical ecology of particular sites (reviewed in Takken and Knols, 1999). In addition, recent studies suggest that overcrowding of larval habitats, resulting in suboptimal rearing conditions, give rise to a range of chemical signals that deter gravid females from further oviposition in those sites (Suh et al., 2016).

4. MOSQUITO SENSORY SYSTEMS AND DISEASE TRANSMISSION

It is clear that the complete spectrum of sensory modalities plays a role at one point or another during the mosquito lifecycle. More often than not, multiple sensory systems act together to provide an array of salient information that is processed and is required to mediate critical behavioural and developmental processes. In addition to the immediate effect of those processes upon the life of the individual insect, the success of mosquito sensory biology has a significant collateral impact on other animals.

In this light, it is not difficult to appreciate that the diverse spectrum of mosquito sensory systems plays a decisive role in establishing and maintaining the vectorial capacity of mosquitoes that transmit pathogens that cause both animal and human disease (Macdonald, 1957). Indeed, as discussed earlier, the range of sensory inputs that impact nearly every phase of mosquito lifecycles helps to establish the developmental, nutritional and reproductive success that is required to maintain sufficient vector population levels to drive overall vectorial capacity. Beyond establishing sufficient reproductive and developmental success to maintain vector population levels, the most salient contribution to a mosquito's vectorial capacity that is directly affected by sensory inputs is almost certainly the determination of blood meal host preference by adult females. For example, the highly anthropophilic blood meal-preference displayed by *An. gambiae* and a limited number of other Anophelines significantly contributes to those species ability to transmit human malaria as compared to bovine-feeding zoophilic Anophelines (as reviewed in Neafsey et al., 2015; Zwiebel and Takken, 2004). Moreover, the dramatic impact of distinctive and rapidly evolving host preference behaviours is not restricted to Anopheline malaria vectors; a similar anthropophilic specialization has been observed in the arbovirus vector mosquito *Ae. aegypti* (Eldridge, 1990; Mattingly, 1957; Tabachnick, 1991).

Taken together, it is difficult to overstate the role of sensory systems in mosquito biology and their effect on vectorial capacity. The ability to sense and respond to an immense spectrum of unitary or, as is more often the case, a combination of auditory, chemical, mechanical, thermal, tactile and visual stimuli is largely responsible for the global proliferation and evolutionary success of mosquito taxa. Regrettably, this has also contributed to the emergence of the ancient and ever increasing role of mosquitoes as unparalleled vectors of human and veterinary disease.

4.1 Gustation

Despite the considerable efforts extended to examine the role and molecular basis of responses to volatile-based olfactory semiochemicals, insofar as the behaviour of vector insects in general and mosquitoes in particular (reviewed in Suh et al., 2014), there has been a relative paucity of attention directed towards the role of soluble, contact-based chemosensory stimuli. Behavioural observations of feeding and the feeding-based acceptance/rejection rates of sugars, acids, ATP and other electrolytes were carried out for *Ae. aegypti* over a half century ago (Galun et al., 1963; Salama, 1966) and have since been augmented by a significant body of similar organismal preference studies (as reviewed in Foster, 1995). While it is likely these behaviours may indeed be a result of sensory responses to soluble tastants, little has been formally done to directly and specifically examine gustation (taste) in mosquitoes.

At the molecular and cellular level, vertebrates and invertebrates differ in many aspects of the morphology and signal transduction mechanisms of their gustatory system. In contrast to mammalian taste receptor cells (TRCs) that are localized across a range of different papillae of the tongue and palate epithelium (Chandrashekar et al., 2006), insect TRCs and their support cells are segregated into distinct gustatory hairs (bristles) on both the tarsi (legs) and proboscis/rostrum (Fig. 3), which are the principal external contact chemosensory appendages (Stocker, 1994). In addition, insect gustatory receptors (GRs) are also found in male and female reproductive organs as well as many other internal locations (eg, enteroendocrine cells) where they act as sensors of nutritional and metabolic state, as reviewed in Liman et al., 2014. Moreover, at the molecular level, mammalian taste receptors for the attractive taste modalities (sweet and umami) are members of the small T1R family of G-protein coupled receptors (GPCRs) that is distantly related to several types of metabotropic receptors. Bitter taste is mediated by the more divergent T2R family of GPCRs, and salt/sour activates TRCs directly via specialized ion channels (Liman et al., 2014). While originally thought

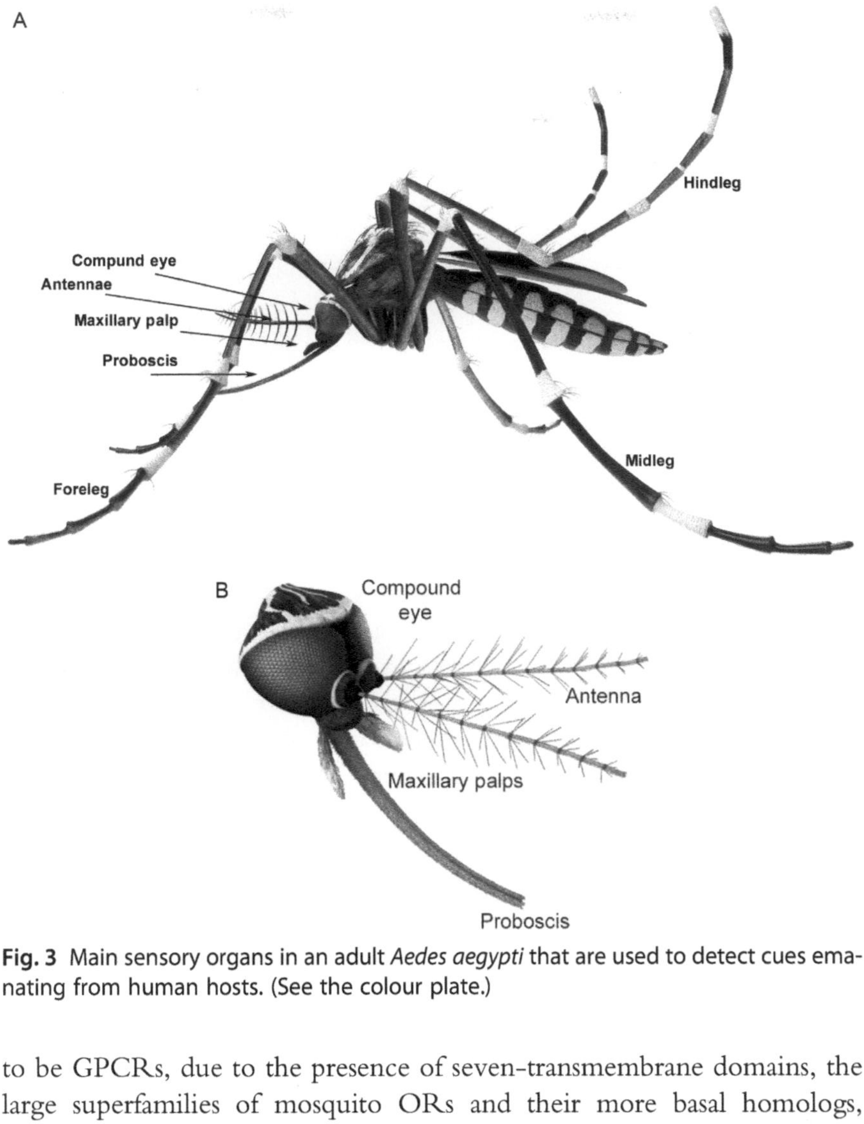

Fig. 3 Main sensory organs in an adult *Aedes aegypti* that are used to detect cues emanating from human hosts. (See the colour plate.)

to be GPCRs, due to the presence of seven-transmembrane domains, the large superfamilies of mosquito ORs and their more basal homologs, the GRs (Hill et al., 2002) represent an ancient "superfamily of animal chemoreceptors" (reviewed in Missbach et al., 2014; Robertson, 2015) that do not appear to act as GPCRs, but rather as ligand-gated ion channels (Jones et al., 2011; Sato et al., 2008, 2011; Shim et al., 2015; Wicher et al., 2008). In addition to OR and GR chemoreceptors, insect taste receptors encompass pickpocket epithelial sodium channels as well as ionotropic (IR) and transient receptor potential (TRP) channels (reviewed in Freeman and Dahanukar, 2015; Liman et al., 2014).

While more progress (insofar as number of receptors/species) has been achieved in the functional characterization of insect ORs (reviewed in Ray et al., 2014), a nevertheless significant number and subtypes of insect GRs (most notably from *Drosophila*) have been successfully characterized in several heterologous systems such as *Drosophila* olfactory neurons, *Xenopus* oocytes as well as insect and mammalian cell culture. These include triad of GRs that act as volatile CO_2 receptors in capitate peg A (cpA) neurons on maxillary palp olfactory receptor neurons (ORNs) in *An. gambiae* (Lu et al., 2007) and *Ae. aegypti*.

Beyond those triad CO_2 receptors only a single fructose receptor from *An. gambiae AgGr25*, which is a *DmGr43a* homolog has been functionally characterized via ectopic expression in *Drosophila* ab1C CO_2 sensing neurons (Sato et al., 2011).

4.2 Olfaction

The discrimination and interpretation of what is an essentially unlimited spectrum of volatile semiochemicals impacts numerous behavioural decisions that contribute to mosquito success and in many cases, survival. As detailed earlier, olfaction plays an essential role throughout the life cycle of mosquitoes, which respond to a diverse array of biological and environmental chemical signals/cues to locate and select food, mates, oviposition sites and avoid predators. There is a large body of organismal studies across a range of field and laboratory paradigms that have comprehensively characterized numerous olfactory-based aspects of vector behaviour that have been detailed over the years by several major reviews (Bowen, 1991; Bowen et al., 1988; Foster, 1995; Takken, 1991; Takken and Knols, 1999). In that light we shall not revisit those important studies but rather focus on more recent efforts that have employed genomics, transcriptomics and other state-of-the-art approaches.

At the cellular and molecular level, in order to interpret what has become known as "odour space" mosquitoes utilize a range of signal transduction components that centre around several large families of chemosensory receptors that are housed in a diverse array of hair-like structures called sensilla that are nonrandomly distributed across peripheral appendages such as the antennae, maxillary palps or labials (reviewed in Guidobaldi et al., 2014) (Fig. 3). The number and type of sensilla present on mosquito, and other insect olfactory appendages vary according to species, their developmental stage (eg, larvae vs adults) and gender. While it remains unclear how sensilla-specific ultrastructure contributes to the functionality of each sensilla type, the high degree of diversity may reflect significant evolutionary consequences for high sensitivity in odour trapping and phylogenetic and/or developmental constraints evolved with the physical environment (Hansson and Stensmyr, 2011). Odorants are thought to diffuse through numerous pores located on the sensilla surface after which they enter an aqueous lymph that must be traversed in order to reach the spectrum of molecular receptors present on the dendrites of ORNs (Steinbrecht, 1997). Activation of a diverse set of chemosensory receptors by odorants provides a range of both excitatory and inhibitory inputs for the generation of ORN action potentials that are further processed across several layers of downstream neural pathways. At the same time, the timely termination and desensitization of peripheral signalling are required to promote sensitivity and selectivity of ORN-based signalling, although the precise elucidation of the underlying mechanisms responsible for this essential component is still undefined.

4.2.1 Odorant-Binding Proteins and Odorant Degrading Proteins

Because odorants are typically hydrophobic chemicals, they require solubilization to traverse the sensillum lymph before they can reach chemosensory receptors on ORN dendritic membranes. This role is presumed to be carried out by odorant-binding proteins (OBPs; Vogt and Riddiford, 1981). OBPs are globular, water-soluble proteins (14–20 kDa) that contain six α-helical domains. In mosquitoes, classic OBPs are endowed with six cysteine residues. However, there are multiple cysteine variants, including PlusC (12 cysteines), MinusC (4 cysteines) and atypical (elongated C-terminus) (Vieira and Rozas, 2011). The total number of OBPs for *An. gambiae*, *Ae. aegypti* and *Cu. quinquefasciatus* is 69, 111 and 109, respectively (Manoharan et al., 2013). OBPs are typically the most highly expressed genes in peripheral olfactory tissues and accordingly are well studied in a wide range of insects (as reviewed in Pelosi et al., 2006).

Despite this attention there are multiple models regarding the mode of action of OBPs and for pheromone-binding proteins (PBPs), which are a specialized OBP subclass restricted to very narrow interactions around known semiochemicals that have been described in some insects but not, as yet mosquitoes. In addition, odorant-degrading enzymes (ODEs) play important roles in the termination of odour-based signal transduction.

In particular, several esterases and cytochrome P450s act as ODEs and pheromone-degrading enzymes and are abundant in insect sensilla lymph where they rapidly degrade odorants and pheromones (reviewed in Leal, 2013).

4.2.2 Odorant Receptors

After intensive effort, a large family of over 60 *ORs* was first identified and characterized in *Drosophila melanogaster* by the laboratory of John Carlson using novel bioinformatics-based methods (Clyne et al., 1999). ORs are localized on the dendritic membrane of *Drosophila* ORNs, and unexpectedly have an inverted topology where the N-terminus is intracellular relative to mammalian ORs (Benton et al., 2006). Subsequent work established that mosquito ORs are ion channels and are composed of heteromeric complexes with two subunits of unknown stoichiometry. One subunit is highly conserved and found in all OR complexes, while the other is highly divergent and variable (Jones et al., 2005; Pitts et al., 2004). In light of its highly conserved sequence and constitutive presence within OR complexes, the first of these subunits is now known as the OR coreceptor (ORCO; Vosshall and Hansson, 2011). The other OR subunit(s) are highly divergent both in terms of number and primary amino acid sequence across insect taxa and are often termed tuning ORs, reflecting the hypothesis they are responsible for recognizing odorant ligands. Consistent with mammalian paradigms, typically only one tuning OR/ORCO complex is expressed in each ORN, although coexpression of two tuning *An. gambiae Or* genes has been occasionally observed together with *Orco* (Karner et al., 2015).

Two models have been put forth for the molecular mechanisms responsible for insect OR-based signal transduction. In the first model, OR/ORCO complexes form odorant-activated IR cation channels (Sato et al., 2008). In the second, the OR/ORCO complex is both an IR cation channel that is activated by fast, odour-dependent pathways as well as being affected by a slow, metabotropic mechanism involving G-proteins and adenylyl cyclases (Wicher et al., 2008). In addition to mechanistic considerations, the comprehensive functional study of heterologously expressed OR/ORCO complexes has revealed the response profiles of individual tuning ORs to biologically relevant insect odorants.

Comprehensive "deorphanization" studies have been carried out in both *D. melanogaster* (Hallem and Carlson, 2006; Hallem et al., 2004; Kreher et al., 2005) and *An. gambiae* (Carey et al., 2010; Kwon et al., 2006; Lu et al., 2007; Wang et al., 2010; Xia et al., 2008). These analyses were conducted against panels of biologically relevant odours and various OR/ORCO complexes using the "empty neuron" paradigm established in *Drosophila* or by employing *Xenopus* oocytes to conduct two electrode voltage clamp experiments. Taken together, these studies indicate the distinctive odour space sampled by *D. melanogaster*, which focuses on esters, and *An. gambiae*, which focuses more on aromatics. This represents an example of species-specific evolution of OR repertoires that are specialized for detecting and discriminating odours that convey ecologically relevant information utilized for mediating critical behaviours including oviposition, nutrient localization and, in the case of *An. gambiae*, vertebrate host seeking for blood meals (Carey et al., 2010). In the years since the initial studies, several other mosquito OR complexes have been similarly deorphanized (Bohbot et al., 2011, 2013; Hughes et al., 2010, 2014; McBride et al., 2014; Pelletier et al., 2010).

These studies reveal a general principle of insect odour coding that is consistent with mammalian models whereby combinatorial signal transduction by multiple tuning ORs respond to wide range of chemicals. Specifically, individual odorants can activate specific groups of receptors, while individual receptors respond to overlapping groups of odorants. Some receptors broadly respond to large number of odorants to act as "generalists" while some "specialist" ORs, respond to unitary or small sets of odorants (Carey and Carlson, 2011). ORs specially tuned to pheromones are "ultra-specialists" and interact with other membrane components including the sensory neuron membrane protein SNMP and PBP accessory proteins.

Although in vivo ORCO is believed to act exclusively as a coreceptor that is not directly involved in odorant binding/recognition, recent high-throughput screens undertaken in *An. gambiae* have identified a novel class of small molecules that act as ORCO agonists and antagonists (Jones et al., 2011, 2012; Taylor et al., 2012). In light of ORCO's wide conservation and ubiquitous expression across insect taxa (Jones et al., 2005; Krieger et al., 2003; Larsson et al., 2004; Pitts et al., 2004) the discovery of powerful and insect-selective ORCO agonists has important implications in the design of novel, broadly effective insect control methods that utilize an excito-repellent paradigm (Jones et al., 2011; Taylor et al., 2012).

The availability of a suite of ORCO-specific reagents has facilitated studies that reveal several fundamental aspects of mosquito OR structure–function that impact our understanding of the role of ORCO and tuning ORs in mediating ORN signalling in *An. gambiae* and other mosquitoes (Pask et al., 2013a,b). These studies reveal an additional level of odour coding beyond odorant recognition wherein each tuning OR contributes to the formation of a spectrum of diverse OR ion channels each with unique IR characteristics. This suggests that odour coding exists in peripheral ORNs that respond differentially depending upon their OR/ORCO complex (Pask et al., 2011). Lastly, the functional characterization of ORCO and a subset of ORs that act in the chemical-based activation of insect sperm preceding fertilization represents a new aspect in our appreciation of the importance of ORs in insect life cycles. Moreover, these latter findings raise the possibility of additional novel target sites for insect control, by disrupting important signalling pathways in the reproductive cycle of disease vectors or pests (Pitts et al., 2014).

The origin of the *Or* gene family in mosquitoes and other insects is presumably the result of adaptations to terrestrial life (Robertson, 2015) as well as the emergence of winged insects adapting to the rapid spread and evolution of vegetation (Missbach et al., 2014). In general, the molecular evolution of tuning *Or* gene families reflects species-specific expansions and contractions resulting in highly divergent *Or* gene families with no apparent orthologs across insect orders (reviewed in Hansson and Stensmyr, 2011). Indeed, insect *Or* gene family repertoires range from 10 *Or* genes in lice (Phthiraptera) (Kirkness et al., 2010) to over 400 in ants (Hymenoptera) (Zhou et al., 2012). The extensive gene gain and loss are thought to be a consequence of adaptation of a species to a specific environmental condition reflecting different biological needs in the life history of each insect species (Zhou et al., 2015).

Mosquito GRs, which are discussed earlier and later, predate the divergence of crustacean and insects and are generally denoted as taste or contact receptors belonging to the same general superfamily as ORs (Hill et al., 2002). While there is a relative paucity of information on GR specific soluble ligands in the vector mosquitoes, a trio of highly conserved GRs coexpressed in the cpA non-ORN neuron specific to the maxillary palp of *An. gambiae* (*AgGrs 22, 23, 24*) and *Ae. aegypti* (*AeGrs 1, 2, 3*) collectively act as a CO_2 chemoreceptor (Erdelyan et al., 2012; Lu et al., 2007; McMeniman et al., 2014), which plays an important role in host-seeking behaviours (Gillies, 1980). In addition to CO_2, *An. gambiae* and *Ae. aegypti* cpA neurons also respond to a number of other odorants many of which are associated with human skin emanations (Lu et al., 2007; Tauxe et al., 2013) suggesting that these GR triads are sensitive to more than CO_2. This supposition is strengthened by genetic studies using the *Drosophila* CO_2 responsive ab1C neuron to express the *AgGr22/AeGr1* ortholog (and *AgGr23/AeGr2* paralog) *DmGr21a* as well as the *AgGr22/AeGr1* ortholog *DmGr63a* (Kwon et al., 2007). Interestingly, the *Drosophila* ab1C neuron also responds to a subset of odorants associated with human skin volatiles in a *DmGr63a*-dependent manner suggesting that in *An. gambiae* and *Ae. aegypti* mosquitoes those responses are similarly dependent upon *AgGr22* and *AeGr1*, respectively (Tauxe et al., 2013).

4.2.3 IR Receptors

IRs are the most ancient family of chemoreceptors and were first described in *D. melanogaster* coeloconic sensilla where OR/ORCO complexes are usually not present. IRs have overall predicted topologies reminiscent of mammalian ionotropic glutamate receptors (iGluRs), but share virtually no amino acid homology with iGluRs (Benton et al., 2009). Consistent with the structural relationship, *Drosophila* IRs act as cation channels, some of which are ligand gated (Abuin et al., 2011; Zhang et al., 2013). In *Drosophila* IRs specifically respond to amines or acid-based odorants (Abuin et al., 2011; Ai et al., 2010) and support a model in which IR-based signalling is required for odorant classes that are not covered by ORs (Silbering et al., 2011). While dramatically understudied relative to ORs, highly conserved families of IRs have been identified in the genomes of many insect taxa (Croset et al., 2010) as well as all mosquitoes examined to date (Neafsey et al., 2015). From a functional perspective, only a single IR has been characterized in a mosquito; *Ir76b* has been shown to mediate butylamine responses in *An. gambiae* (Liu et al., 2010).

4.3 Vision

The visual system impacts on an array of behaviours exhibited by mosquito vectors, ranging from host seeking to mating, oviposition and others. The visual system comes into play once the mosquitoes are within 5–15 m of humans (Bidlingmayer, 1994). Interestingly, visual-guided movement is heightened by the detection of CO_2, but only in females (van Breugel et al., 2015). This odour-driven modulation takes place even if the CO_2 precedes exposure to the visual cues by several seconds. It appears that the CO_2-induced increase in attractive to visual cues provides a mechanism to enhance the likelihood of approaching warm objects, such as human skin. Thus, visual, olfactory and thermal stimuli derived from human hosts all appear to operate synergistically (McMeniman et al., 2014; van Breugel et al., 2015).

The visual organs in adult mosquitoes are compound eyes, each of which consists of ~200–300 repeat units, referred to as ommatidia.

Each ommatidium contains eight photoreceptor cells. Light detection takes place in the microvillar portion of the photoreceptor cells, called the rhabdomeres. In *Ae. aegypti* and *Ae. gambiae* the cell bodies of the R1–6 cells are situated at the periphery of the ommatidia, and the rhabdomeres project to the centre, where they are fused (Hu et al., 2009). In contrast, the cell body of the R8 occupies the centre of the ommatidia, and the rhabdomere faces outwards so that it is fused with the ring of R1–6 rhabdomeres. Finally, the cell body of the R7 cell is located near the distal end of each ommatidium, between R5 and R6, and extends its rhabdomeres over the top of a portion of the ring of rhabdomeres. The fusion of all of the rhabdomeres within one ommatidium enables this large structure to share the same visual axis and increases image brightness. This is distinct from the open rhabdomere system in *Drosophila*, in which the rhabdomeres within an ommatidium are separate and act as independent light guides.

The phototransduction cascades in mosquitoes have not been defined. However, they are presumed to be similar to *Drosophila* visual transduction, which has been described in detail (Montell, 2012). In the fruit fly, the light activated rhodopsins couple to a trimeric G-protein (Gq), which in turn activates a phospholipase Cβ encoded by the *norpA* locus (Bloomquist et al., 1988). Following hydrolysis of phosphatidylinositol 4,5-bisphosphate (PIP_2), the phototransduction pathway culminates with opening of the TRP and TRPL channels (Montell, 1989; Phillips et al., 1992). This signalling cascade is remarkably similar to phototransduction in mammalian intrinsically sensitive retinal ganglion cells (Berson et al., 2002; Hattar et al., 2002; Xue et al., 2011).

The number of opsin genes encoded in the mosquito genomes is unusually large. The *Ae. aegypti* genome encodes 10 opsins, while 11 opsins predicted to be encoded by *An. gambiae* (Hill et al., 2002). Based on relationships to six well-characterized *Drosophila* rhodopsins, Rh1–Rh6, seven mosquito rhodopsins are most likely long wavelength opsins (>500 nm). In addition, each of these mosquitoes encode a short wavelength opsin (400–500 nm), an ultraviolet opsin (<400 nm), an Rh7 class opsin, and one or two pteropsins (Velarde et al., 2005), which are more closely related to the nonvisual pineal vertebrate ancient and teleost multiple tissue opsins, than to other insect opsins. These latter pteropsins are missing from *Drosophila*. The largest group of photoreceptor cells (R1–6) express Op1, while most of the remaining rhodopsins are expressed in different R7 and R8 populations that demarcate the retina into dorsal, central, ventral stripe and ventral regions (Hu et al., 2009, 2011).

4.4 Audition

Hearing is critical for mosquito mating, as males are attracted to the sounds generated by conspecific females that are in flight. The mosquito ears are long antenna, which are decorated with many flagella hairs extending out laterally from the main antennal shaft (Fig. 3). Because the flagella become successively shorter nearest the antennal tip, the more extensive feathery arrangement in males resembles a Christmas tree. These light structures are deflected by acoustic stimulation, and therefore may be considered the mosquito equivalent of the eardrum. The flagella are tuned to respond most effectively with resonant frequencies that closely match the sounds produced by the wing beats of the opposite sex (380 and 230 Hz in males and females, respectively; Gopfert et al., 1999). Both male and female mosquitoes alter their flight tones once they approach to within short distances on the order of a few centimetres. Surprisingly, the two conspecific sexes then match by establishing harmonic convergence and do so at frequencies approximately threefold higher than the beating of the female wings (eg, 1200 Hz) (Cator et al., 2009; Gibson and Russell, 2006; Jackson et al., 2009; Pennetier et al., 2010; Warren et al., 2009). This matching supports the model that the auditory responses in male and female mosquitoes contribute to mating success.

The vibrations of the flagella activate mechanosensory neurons in a chordotonal organ in the proximal region of the antennal called the Johnston's organ (Johnston, 1855). These neurons serve as the mosquito equivalent of hair cells in the mammalian auditory system. In male mosquitoes, the Johnston's organ is comprised of ~15,000 sensory neurons (Boo and Richards, 1975b)—a number far greater than the ~500 in *Drosophila.* The male Johnston's organ responds to movements of just 7 nm at the antennal tip, a sensitivity so exquisite that it equates to a 0.0001 degree change in angular movement of the flagella (Gopfert and Robert, 2000). The female Johnston's organ includes roughly half as many mechanosensory neurons as males (Boo and Richards, 1975b) and responds to flagellar rotations as small as 0.0005 degree (Gopfert and Robert, 2000). The sensitivities of the mosquito Johnston's organ exceed that of all other arthropods that have been analysed.

The basic unit in chordotonal organs such the Johnston's organ is scolopidia (Kavlie and Albert, 2013). Each scolopidium contains multiple types of support cells, as well as one or more sensory neurons. In mosquitoes, such as male *Ae. aegypti*, the Johnston's organ includes four types of scolopidia (A–D), although >97% are type A, which include two bipolar sensory neurons (Boo and Richards, 1975a). In the fruit fly about half the ∼500 scolopidia function in audition, while the remaining are important in the sensation of sustained mechanical stimuli due to wind and gravity.

In mosquitoes, the cation channels that are involved in hearing, wind and geotaxis are unknown. In fruit flies the best candidates for auditory transduction channels are NOMPC (TRPN subfamily), and Nan and Iav (TRPV subfamily). Several observations support the model that NOMPC is the primary auditory transduction channel. First, it is localized in to the tips of nearly all mechanosensory neurons in the Johnston's organ (Lee et al., 2010; Liang et al., 2011). Second, mutations in *nompC* disrupt mechanical amplification in vivo (Eberl et al., 2000; Effertz et al., 2011; Göpfert et al., 2006). Third, NOMPC is a mechanically activated channel (Gong et al., 2013; Yan et al., 2013). Fourth, the N-terminal array of 29 ankyrin repeats in NOMPC (Walker et al., 2000) define the elusive "gating spring" that induces opening of the channel gate in response to force imposed on the sensory cells (Howard and Bechstedt, 2004; Zhang et al., 2015). Nan and Iav are also mechanosensitive channels expressed in sensory neurons in the Johnston's organ (Gong et al., 2004; Kim et al., 2003). Most importantly, they may be important for auditory transduction as loss of either of these channels profoundly impairs sound evoked potentials (Gong et al., 2004; Kim et al., 2003).

The precise mechanism through which NOMPC, Nan and Iav contribute to hearing is not fully resolved. According to one model, NOMPC is the primary mechanosensor and also contributes to mechanical amplification, while Nan/Iav propagates the electrical signal (Effertz et al., 2011, 2012). These latter channels may also control feedback gain (Göpfert et al., 2006). However, there is also evidence arguing that Nan/Iav are the sound activated channels and that NOMPC functions in amplification (Lehnert et al., 2013).

4.5 Thermal Sensation

Female mosquitoes that spread infectious disease are attracted to the temperature of human skin during periods when they need blood to produce eggs. This behaviour appears to depend on sensilla on the antenna, which simultaneously assess changes in temperature through antagonistic warm and cool activated neurons (Gingl et al., 2005). These neurons may allow mosquitoes to identify human skin, despite large differences in ambient temperatures, by sensing relative differences in temperature (Corfas and Vosshall, 2015).

The TRPA1 channel is one candidate for allowing mosquitoes to sense skin temperatures. TRPA1 is expressed in the antennae of *An. gambiae* and is activated in the thermal range associated with human skin (Wang et al., 2009). Furthermore, knockout of TRPA1 in *Ae. aegypti* impairs the discrimination between skin temperatures and higher temperatures. However, TRPA1 does not fully explain thermal attraction to skin temperature since the mutant animals are still attracted to warmth.

In addition to the requirements for sensing host temperature, survival depends on the ability of larvae and adult mosquitoes to employ thermosensory cues to identify the most hospitable thermal landscape in their aquatic and terrestrial environments. In addition, ambient temperature has a profound effect on the rate of development. Adult *An. gambiae* prefer temperatures in the temperate range, while late (fourth instar)-stage larvae prefer two distinct temperature separated by 6°C (eg, 27°C and 33°C) (Liu and Zwiebel, 2013). However, the two preferred temperatures shift depending on the thermal rearing conditions during embryonic development (Liu and Zwiebel, 2013). Consistent with a role for antennae in adult thermotaxis, the larval antennae also functions in temperature sensation. Moreover, TRPA1 is one critical sensor for detecting excessively warm temperatures in both late larvae and adults (Liu and Zwiebel, 2013; Wang et al., 2009).

Currently, the sensors that allow mosquitoes to detect temperatures other than mildly hot remain to be identified. Based on work on fruit flies, sensation of cold, cool and very high temperatures are likely to be sensed through direct activation of TRP channels (Fowler and Montell, 2013). However, subtle changes in temperature in the comfortable range may be mediated through activation of thermal amplification cascades that are initiated by rhodopsin and culminates with activation of TRP channels (Kwon et al., 2008; Shen et al., 2011).

4.6 The Impact of Genomics and RNA Sequencing on the Study of Mosquito Sensory Systems

In light of their impact as disease vectors it was not surprising that mosquitoes were early targets of whole genome sequence (WGS) efforts; the 2002 publication of the genome for *An. gambiae* represented a milestone in vector genomics. This paved the way for comprehensive studies of gene function and evolution that spans the entire genome (Holt et al., 2002). Following the release of the *An. gambiae* genome, the genome sequences of *Ae. aegypti* and the West Nile mosquito, *C. quinquefasciatus*, were published

(Arensburger et al., 2010; Nene et al., 2007). The sequencing of these initial vector genomes was expensive efforts as well as time and labour intensive. Following the advent of and steady improvements in next-generation sequencing technologies, WGS has become an increasingly common undertaking, such that the number of available vector genomes has grown significantly over the last decade. Indeed, the annotated genomes of an additional 16 *Anopheles* species were recently published (Neafsey et al., 2015). The availability of these genomic resources has been a critical component in identifying mosquito sensory genes that span the entire spectrum of sensory modalities discussed here. The capacity to utilize sophisticated bioinformatics to interrogate ever increasing genome data sets has been instrumental in the identification of sensory genes that belong to both conserved and divergent gene families. These state-of-the-art resources have facilitated the accumulation of considerable insight into the sensory repertoires of vector as well as nonvector mosquitoes.

To illustrate the extraordinary value of genomic resources, we would point to the extensive series of interspecific analyses of the chemoreceptor superfamilies from 18 species of Anopheline mosquitoes that were recently completed (Neafsey et al., 2015) along with those from *Ae. aegypti* and the nonblood-feeding mosquito *Toxorhynchites amboinensis* (Zhou et al., 2014a). As a result, we already have a comprehensive understanding of the phylogenetic relationships among mosquito *Or*, *Ir* and *Gr* gene families. For example, these data indicate that there are no less than 26 one-to-one *Gr* orthologs between *Ae. aegypti* and *An. gambiae* sl. most of which also extend to *Culex quinquefasciatus* (Zhou et al., 2014a). These highly conserved *Grs* represent the most likely candidate screening targets for modulators that will be active across all mosquitoes.

Furthermore, the advent of massively parallel sequencing of RNA molecules (RNA-seq) has also significantly augmented genome-wide analyses of mosquito sensory systems by providing highly quantitative transcript abundance data, as well as a wealth of sequence, isoform and expression information for the vast majority of encoded genes in a vector species. Importantly, because RNA-seq largely captures only fully spliced transcripts, an informative de novo transcriptome assembly of RNA sequences can be generated affordably and analysed efficiently, even in the absence of an assembled genome. For mosquitoes, de novo transcriptome assemblies were successfully applied to *An. funestus* well in advance of the availability of a genome sequence (Crawford et al., 2010), and the transcriptome assembly of the nonblood-feeding genus, *Toxorhynchites,* has revealed extensive evidence for its phylogenetic relationship relative to the two fully sequenced major *Culicinae* vectors, *Ae. aegypti* and *C. pipiens* (Zhou et al., 2014b).

Inasmuch as components of the chemosensory signal transduction machinery are generally highly localized within specific appendages such as the antennae, maxillary palps and labellum, tissue-specific transcriptome profiles have become essential for measuring transcript abundances that would otherwise be diluted and potentially undetectable in whole body preparations of RNA. In contrast to the relative paucity of studies on gustatory appendages, chemoreceptor abundance quantitation by RNA-seq has been comprehensively profiled in many of the major olfactory appendages of *An. gambiae* (Pitts et al., 2011), *C. quinquefasciatus* (Leal et al., 2013), *Ae. aegypti* (McBride et al., 2014) and *An. quadriannulatus* (Rinker et al., 2013b).

An initial RNA-seq study of *An. gambiae* antennae showed that levels of chemoreceptors were lower overall in males than in females. While this observation is largely accounted for by the antennal sexual dimorphism that exists between blood-feeding females and the nonblood-feeding males, the RNA expression data also revealed the surprising finding that the variety and relative abundances of the chemoreceptors were otherwise remarkably conserved between the sexes (Pitts et al., 2011). Also, in *An. gambiae*, an RNA-seq time course study revealed the relationship between chemoreceptor transcriptional differences and reproductive biology. Here, individual OR transcript profiles in female antennae changed only slightly on the taking of a blood meal; however, the cumulative effects of these small changes when viewed collectively found that there was an overall shift in olfactory responsiveness in advance of ovipositing (Rinker et al., 2013a).

Similarly, subtle distinctions in OR transcriptome profiles were reported in two studies that compared mosquito species that were phenotypically divergent in terms of their host preferences. The first comparison between *An. gambiae* and its less-anthropophilic sibling species *An. quadriannulatus* pointed to an overall enrichment of transcripts for multiple ORs in *An. gambiae* that are predicted to act in concert to enhance responsiveness to human-associated odours (Rinker et al., 2013b). In the second study, differences in the antennal transcriptome profiles of two subspecies of *Ae. aegypti* that differ in their preferences for feeding on humans were examined. The results pointed to notable differences in both the abundance and functional response of a single OR, suggesting that it might be linked to the anthropophily exhibited by one subspecies (McBride et al., 2014; Zhou et al., 2014b). In addition, a recently published neurotranscriptome of *Ae. aegypti* provides detailed RNA-seq data on a broad range of sensory-related transcripts across nearly all tissues from the central and peripheral nervous system with the notable exception being non-CNS visual tissues such as the mosquito compound eyes (Matthews et al., 2016; Sparks et al., 2013). The increasing depth of these publically available data sets creates a unique opportunity for conducting comprehensive assessments of the molecular genetics of mosquito sensory systems that will play a central role in efforts to expand our understanding of these complex and interconnected systems.

4.7 Future Perspectives

As work progresses across the sensory modalities discussed earlier, it is clear that our understanding of mosquito sensory systems will continue to expand. Deeper knowledge of the sensory systems of multiple species will provide valuable opportunities to make comparisons and offer to illuminate our understanding of mosquito behaviour. An important challenge is to leverage those insights into the identification of the receptors and ion channels that control the sensory systems towards the development of creative approaches to vector control, thereby leading to the reduction of vector-borne disease. Given the recent demonstration of the efficacy of the CRISPR/Cas9 (Basu et al., 2015; Dong et al., 2015; Kistler et al., 2015) and gene drive systems in mosquitoes (Adelman and Tu, 2016; Gantz and Bier, 2015; Hammond et al., 2016; Pennisi, 2015), there are now many exciting new possibilities for controlling insect vectors through gene manipulation. However, judicious application of gene drive approaches will require building in safeguards to minimize unintended consequences resulting from the release of these modified mosquitoes.

REFERENCES

Abuin, L., Bargeton, B., Ulbrich, M.H., Isacoff, E.Y., Kellenberger, S., Benton, R., 2011. Functional architecture of olfactory ionotropic glutamate receptors. Neuron 69, 44–60.

Adelman, Z.N., Tu, Z., 2016. Control of mosquito-borne infectious diseases: sex and gene drive. Trends Parasitol. 32, 219–229.

Ai, M., Min, S., Grosjean, Y., Leblanc, C., Bell, R., Benton, R., Suh, G.S.B., 2010. Acid sensing by the *Drosophila* olfactory system. Nature 468, 691–695. U112.

Allan, S.A., Day, J.F., Edman, J.D., 1987. Visual ecology of biting flies. Annu. Rev. Entomol. 32, 297–316.

Arensburger, P., Megy, K., Waterhouse, R.M., Abrudan, J., Amedeo, P., Antelo, B., Bartholomay, L., Bidwell, S., Caler, E., Camara, F., Campbell, C.L., Campbell, K.S., Casola, C., Castro, M.T., Chandramouliswaran, I., Chapman, S.B., Christley, S., Costas, J., Eisenstadt, E., Feschotte, C., Fraser-Liggett, C., Guigo, R., Haas, B., Hammond, M., Hansson, B.S., Hemingway, J., Hill, S.R., Howarth, C., Ignell, R., Kennedy, R.C., Kodira, C.D., Lobo, N.F., Mao, C., Mayhew, G., Michel, K., Mori, A., Liu, N., Naveira, H., Nene, V., Nguyen, N., Pearson, M.D.,

Pritham, E.J., Puiu, D., Qi, Y., Ranson, H., Ribeiro, J.M., Roberston, H.M., Severson, D.W., Shumway, M., Stanke, M., Strausberg, R.L., Sun, C., Sutton, G., Tu, Z.J., Tubio, J.M., Unger, M.F., Vanlandingham, D.L., Vilella, A.J., White, O., White, J.R., Wondji, C.S., Wortman, J., Zdobnov, E.M., Birren, B., Christensen, B.M., Collins, F.H., Cornel, A., Dimopoulos, G., Hannick, L.I., Higgs, S., Lanzaro, G.C., Lawson, D., Lee, N.H., Muskavitch, M.A., Raikhel, A.S., Atkinson, P.W., 2010. Sequencing of *Culex quinquefasciatus* establishes a platform for mosquito comparative genomics. Science 330, 86–88.

Basu, S., Aryan, A., Overcash, J.M., Samuel, G.H., Anderson, M.A., Dahlem, T.J., Myles, K.M., Adelman, Z.N., 2015. Silencing of end-joining repair for efficient site-specific gene insertion after TALEN/CRISPR mutagenesis in *Aedes aegypti*. Proc. Natl. Acad. Sci. U.S.A. 112, 4038–4043.

Benton, R., Sachse, S., Michnick, S.W., Vosshall, L.B., 2006. Atypical membrane topology and heteromeric function of *Drosophila* odorant receptors in vivo. PLoS Biol. 4, e20.

Benton, R., Vannice, K., Gomezdiaz, C., Vosshall, L., 2009. Variant ionotropic glutamate receptors as chemosensory receptors in Drosophila. Cell 136, 149–162.

Berson, D.M., Dunn, F.A., Takao, M., 2002. Phototransduction by retinal ganglion cells that set the circadian clock. Science 295, 1070–1073.

Bhatt, S., Gething, P.W., Brady, O.J., Messina, J.P., Farlow, A.W., Moyes, C.L., Drake, J.M., Brownstein, J.S., Hoen, A.G., Sankoh, O., Myers, M.F., George, D.B., Jaenisch, T., Wint, G.R.W., Simmons, C.P., Scott, T.W., Farrar, J.J., Hay, S.I., 2013. The global distribution and burden of dengue. Nature, 1–5.

Bidlingmayer, W.L., 1994. How mosquitoes see traps: role of visual responses. J. Am. Mosq. Control Assoc. 10, 272–279.

Bloomquist, B.T., Shortridge, R.D., Schneuwly, S., Perdew, M., Montell, C., Steller, H., Rubin, G., Pak, W.L., 1988. Isolation of a putative phospholipase C gene of *Drosophila*, *norpA*, and its role in phototransduction. Cell 54, 723–733.

Bohbot, J.D., Jones, P.L., Wang, G., Pitts, R.J., Pask, G.M., Zwiebel, L.J., 2011. Conservation of indole responsive odorant receptors in mosquitoes reveals an ancient olfactory trait. Chem. Senses 36, 149–160.

Bohbot, J.D., Durand, N.F., Vinyard, B.T., Dickens, J.C., 2013. Functional development of the octenol response in *Aedes aegypti*. Front. Physiol. 4 (39), 1–8.

Boo, K.S., Richards, A.G., 1975a. Fine structure of scolopidia in Johnston's organ of female *Aedes aegypti* compared with that of the male. J. Insect Physiol. 21, 1129–1139.

Boo, K.S., Richards, A.G., 1975b. Fine structure of scolopidia in the Johnston's organ of male *Aedes aegypti* (L.) (Diptera: Culicidae). Int. J. Insect Morphol. Embryol. 4, 549–566.

Bowen, M.F., 1991. The sensory physiology of host-seeking behavior in mosquitoes. Annu. Rev. Entomol. 36, 139–158.

Bowen, M.F., Davis, E.E., Haggart, D.A., 1988. A behavioral and sensory analysis of host-seeking behavior in the diapausing mosquito *Culex pipiens*. J. Insect Physiol. 34, 805–813.

Cabrera, M., Jaffe, K., 2007. An aggregation pheromone modulates lekking behavior in the vector mosquito *Aedes aegypti* (Diptera: Culicidae). J. Am. Mosq. Control Assoc. 23, 1–10.

Cao-Lormeau, V.M., Blake, A., Mons, S., Lastere, S., Roche, C., Vanhomwegen, J., Dub, T., Baudouin, L., Teissier, A., Larre, P., Vial, A.L., Decam, C., Choumet, V., Halstead, S.K., Willison, H.J., Musset, L., Manuguerra, J.C., Despres, P., Fournier, E., Mallet, H.P., Musso, D., Fontanet, A., Neil, J., Ghawche, F., 2016. Guillain-Barre syndrome outbreak associated with Zika virus infection in French Polynesia: a case-control study. Lancet 387 (10027), 1531–1539.

Carey, A.F., Carlson, J.R., 2011. Insect olfaction from model systems to disease control. Proc. Natl. Acad. Sci. U.S.A. 108, 12987–12995.

Carey, A.F., Wang, G., Su, C.Y., Zwiebel, L.J., Carlson, J.R., 2010. Odorant reception in the malaria mosquito *Anopheles gambiae*. Nature 464, 66–71.

Cator, L.J., Arthur, B.J., Harrington, L.C., Hoy, R.R., 2009. Harmonic convergence in the love songs of the dengue vector mosquito. Science 323, 1077–1079.

Chandrashekar, J., Hoon, M.A., Ryba, N.J.P., Zuker, C.S., 2006. The receptors and cells for mammalian taste. Nature 444, 288–294.

Chang, C., Ortiz, K., Ansari, A., Gershwin, M.E., 2016. The Zika outbreak of the 21st century. J. Autoimmun. 68, 1–13.

Clements, A., 1992a. The Biology of Mosquitoes: Development, Nutrition and Reproduction. CABI Publishing, New York, NY.

Clements, A.N., 1992b. The Biology of Mosquitoes. Chapman & Hall, London, Glasgow, New York, Tokyo, Melbourne, Madras.

Clements, A., 1999. The Biology of Mosquitoes: Sensory, Reception, and Behaviour. CABI Publishing, New York, NY.

Clyne, P.J., Warr, C.G., Freeman, M.R., Lessing, D., Kim, J., Carlson, J.R., 1999. A novel family of divergent seven-transmembrane proteins: candidate odorant receptors in *Drosophila*. Neuron 22, 327–338.

Corfas, R.A., Vosshall, L.B., 2015. The cation channel TRPA1 tunes mosquito thermotaxis to host temperatures. eLife 4, pii: e11750. http://dx.doi.org/10.7554/eLife.11750.

Crawford, J.E., Guelbeogo, W.M., Sanou, A., Traoré, A., Vernick, K.D., Sagnon, N.A.F., Lazzaro, B.P., 2010. De novo transcriptome sequencing in *Anopheles funestus* using Illumina RNA-seq technology. PLoS One 5, e14202.

Croset, V., Rytz, R., Cummins, S., Budd, A., Brawand, D., 2010. Ancient protostome origin of chemosensory ionotropic glutamate receptors and the evolution of insect taste and olfaction. PLoS Genet. 6, e1001064.

Dekker, T., Cardé, R.T., 2011. Moment-to-moment flight manoeuvres of the female yellow fever mosquito (*Aedes aegypti* L.) in response to plumes of carbon dioxide and human skin odour. J. Exp. Biol. 214, 3480–3494.

Dong, S., Lin, J., Held, N.L., Clem, R.J., Passarelli, A.L., Franz, A.W., 2015. Heritable CRISPR/Cas9-mediated genome editing in the yellow fever mosquito, *Aedes aegypti*. PLoS One 10, e0122353.

Eberl, D.F., Hardy, R.W., Kernan, M.J., 2000. Genetically similar transduction mechanisms for touch and hearing in *Drosophila*. J. Neurosci. 20, 5981–5988.

Effertz, T., Wiek, R., Göpfert, M.C., 2011. NompC TRP channel is essential for *Drosophila* sound receptor function. Curr. Biol. 21, 592–597.

Effertz, T., Nadrowski, B., Piepenbrock, D., Albert, J.T., Göpfert, M.C., 2012. Direct gating and mechanical integrity of *Drosophila* auditory transducers require TRPN1. Nat. Neurosci. 15, 1198–1200.

Eldridge, B.F., 1990. Evolutionary relationships among California serogroup viruses (Bunyaviridae) and Aedes mosquitoes (Diptera: Culicidae). J. Med. Entomol. 27, 738–749.

Enayati, A., Hemingway, J., 2010. Malaria management: past, present, and future. Annu. Rev. Entomol. 55, 569–591.

Erdelyan, C.N.G., Mahood, T.H., Bader, T.S.Y., Whyard, S., 2012. Functional validation of the carbon dioxide receptor genes in *Aedes aegypti* mosquitoes using RNA interference. Insect Mol. Biol. 21, 119–127.

Fawaz, E.Y., Allan, S.A., Bernier, U.R., Obenauer, P.J., Diclaro 2nd, J.W., 2014. Swarming mechanisms in the yellow fever mosquito: aggregation pheromones are involved in the mating behavior of *Aedes aegypti*. J. Vector Ecol. 39, 347–354.

Foster, W.F., 1995. Mosquito sugar feeding and reproductive energetics. Annu. Rev. Entomol. 40, 443–447.

Foster, W.A., Hancock, R.G., 1994. Nectar-related olfactory and visual attractants for mosquitoes. J. Am. Mosq. Control Assoc. 10, 288–296.

Fowler, M.A., Montell, C., 2013. *Drosophila* TRP channels and animal behavior. Life Sci. 92, 394–403.
Freeman, E.G., Dahanukar, A., 2015. Molecular neurobiology of *Drosophila* taste. Curr. Opin. Neurobiol. 34, 140–148.
Galun, R., Avi-Dor, Y., Bar-Zeev, M., 1963. Feeding response in *Aedes aegypti*: stimulation by adenosine triphosphate. Science 142, 1674–1675.
Gantz, V.M., Bier, E., 2015. The mutagenic chain reaction: a method for converting heterozygous to homozygous mutations. Science (New York, NY) 348, 442–444.
Gibson, G., Russell, I., 2006. Flying in tune: sexual recognition in mosquitoes. Curr. Biol. 16, 1311–1316.
Gibson, G., Warren, B., Russell, I.J., 2010. Humming in tune: sex and species recognition by mosquitoes on the wing. J. Assoc. Res. Otolaryngol. 11, 527–540.
Gillies, M.T., 1980. The role of carbon dioxide in host-finding by mosquitoes (Diptera: Culicidae): a review. Bull. Entomol. Res. 70, 525–532.
Gingl, E., Hinterwirth, A., Tichy, H., 2005. Sensory representation of temperature in mosquito warm and cold cells. J. Neurophysiol. 94, 176–185.
Gong, Z., Son, W., Chung, Y.D., Kim, J., Shin, D.W., Mcclung, C.A., Lee, Y., Lee, H.W., Chang, D.-J., Kaang, B.-K., Cho, H., Oh, U., Hirsh, J., Kernan, M.J., Kim, C., 2004. Two interdependent TRPV channel subunits, inactive and Nanchung, mediate hearing in *Drosophila*. J. Neurosci. 24, 9059–9066.
Gong, J., Wang, Q., Wang, Z., 2013. NOMPC is likely a key component of *Drosophila* mechanotransduction channels. Eur. J. Neurosci. 38, 2057–2064.
Gopfert, M.C., Robert, D., 2000. Nanometre-range acoustic sensitivity in male and female mosquitoes. Proc. R. Soc. B Biol. Sci. 267, 453–457.
Gopfert, M.C., Briegel, H., Robert, D., 1999. Mosquito hearing: sound-induced antennal vibrations in male and female *Aedes aegypti*. J. Exp. Biol. 202, 2727–2738.
Göpfert, M.C., Albert, J.T., Nadrowski, B., Kamikouchi, A., 2006. Specification of auditory sensitivity by *Drosophila* TRP channels. Nat. Neurosci. 9, 999–1000.
Gouagna, L.-C., Poueme, R.S., Dabiré, K.R., Ouédraogo, J.-B., Fontenille, D., Simard, F., 2010. Patterns of sugar feeding and host plant preferences in adult males of *An. gambiae* (Diptera: Culicidae). J. Vector Ecol. 35, 267–276.
Guidobaldi, F., May-Concha, I.J., Guerenstein, P.G., 2014. Morphology and physiology of the olfactory system of blood-feeding insects. J. Physiol., Paris, 108 (2–3), 96–111. http://dx.doi.org/10.1016/j.jphysparis.2014.04.006. Epub 2014, May 15.
Guinovart, C., Navia, M.M., Tanner, M., Alonso, P.L., 2006. Malaria: burden of disease. Curr. Mol. Med. 6, 137–140.
Hallem, E.A., Carlson, J.R., 2006. Coding of odors by a receptor repertoire. Cell 125, 143–160.
Hallem, E., Ho, M.G., Carlson, J.R., 2004. The molecular basis of odor coding in the *Drosophila* antenna. Cell 117, 965–979.
Hammond, A., Galizi, R., Kyrou, K., Simoni, A., Siniscalchi, C., Katsanos, D., Gribble, M., Baker, D., Marois, E., Russell, S., Burt, A., Windbichler, N., Crisanti, A., Nolan, T., 2016. A CRISPR-Cas9 gene drive system targeting female reproduction in the malaria mosquito vector *Anopheles gambiae*. Nat. Biotechnol. 34, 78–83.
Hansson, B.S., Stensmyr, M.C., 2011. Evolution of insect olfaction. Neuron 72, 698–711.
Hattar, S., Liao, H.W., Takao, M., Berson, D.M., Yau, K.W., 2002. Melanopsin-containing retinal ganglion cells: architecture, projections, and intrinsic photosensitivity. Science 295, 1065–1070.
Hill, C.A., Fox, A.N., Pitts, R.J., Kent, L.B., Tan, P.L., Chrystal, M.A., Cravchik, A., Collins, F.H., Robertson, H.M., Zwiebel, L.J., 2002. G protein-coupled receptors in *Anopheles gambiae*. Science 298, 176–178.
Holt, R.A., Subramanian, G.M., Halpern, A., Sutton, G.G., Charlab, R., Nusskern, D.R., Wincker, P., Clark, A.G., Ribeiro, J.M., Wides, R., Salzberg, S.L., Loftus, B.,

Yandell, M., Majoros, W.H., Rusch, D.B., Lai, Z., Kraft, C.L., Abril, J.F., Anthouard, V., Arensburger, P., Atkinson, P.W., Baden, H., De Berardinis, V., Baldwin, D., Benes, V., Biedler, J., Blass, C., Bolanos, R., Boscus, D., Barnstead, M., Cai, S., Center, A., Chatuverdi, K., Christophides, G.K., Chrystal, M.A., Clamp, M., Cravchik, A., Curwen, V., Dana, A., Delcher, A., Dew, I., Evans, C.A., Flanigan, M., Grundschober-Freimoser, A., Friedli, L., Gu, Z., Guan, P., Guigo, R., Hillenmeyer, M.E., Hladun, S.L., Hogan, J.R., Hong, Y.S., Hoover, J., Jaillon, O., Ke, Z., Kodira, C., Kokoza, E., Koutsos, A., Letunic, I., Levitsky, A., Liang, Y., Lin, J.J., Lobo, N.F., Lopez, J.R., Malek, J.A., Mcintosh, T.C., Meister, S., Miller, J., Mobarry, C., Mongin, E., Murphy, S.D., O'brochta, D.A., Pfannkoch, C., Qi, R., Regier, M.A., Remington, K., Shao, H., Sharakhova, M.V., Sitter, C.D., Shetty, J., Smith, T.J., Strong, R., Sun, J., Thomasova, D., Ton, L.Q., Topalis, P., Tu, Z., Unger, M.F., Walenz, B., Wang, A., Wang, J., Wang, M., Wang, X., Woodford, K.J., Wortman, J.R., Wu, M., Yao, A., Zdobnov, E.M., Zhang, H., Zhao, Q., et al., 2002. The genome sequence of the malaria mosquito *Anopheles gambiae*. Science 298, 129–149.

Howard, J., Bechstedt, S., 2004. Hypothesis: a helix of ankyrin repeats of the NOMPC-TRP ion channel is the gating spring of mechanoreceptors. Curr. Biol. 14, R224–R226.

Hu, X., England, J.H., Lani, A.C., Tung, J.J., Ward, N.J., Adams, S.M., Barber, K.A., Whaley, M.A., O'tousa, J.E., 2009. Patterned rhodopsin expression in R7 photoreceptors of mosquito retina: implications for species-specific behavior. J. Comp. Neurol. 516, 334–342.

Hu, X., Whaley, M.A., Stein, M.M., Mitchell, B.E., O'tousa, J.E., 2011. Coexpression of spectrally distinct rhodopsins in *Aedes aegypti* R7 photoreceptors. PLoS One 6, e23121.

Huang, J., Walker, E.D., Giroux, P.Y., Vulule, J., Miller, J.R., 2005. Ovipositional site selection by *Anopheles gambiae*: influences of substrate moisture and texture. Med. Vet. Entomol. 19, 442–450.

Huang, J., Walker, E.D., Vulule, J., Miller, J.R., 2007. The influence of darkness and visual contrast on oviposition by *Anopheles gambiae* in moist and dry substrates. Physiol. Entomol. 32, 34–40.

Hughes, D.T., Pelletier, J., Luetje, C.W., Leal, W.S., 2010. Odorant receptor from the southern house mosquito narrowly tuned to the oviposition attractant skatole. J. Chem. Ecol. 36, 797–800.

Hughes, D.T., Wang, G., Zwiebel, L.J., Luetje, C.W., 2014. A determinant of odorant specificity is located at the extracellular loop 2-transmembrane domain 4 interface of an *Anopheles gambiae* odorant receptor subunit. Chem. Senses 39, 761–769.

Jackson, J.C., Windmill, J.F., Pook, V.G., Robert, D., 2009. Synchrony through twice-frequency forcing for sensitive and selective auditory processing. Proc. Natl. Acad. Sci. U.S.A. 106, 10177–10182.

Johnston, C., 1855. Original communications: auditory apparatus of the Culex mosquito. Q. J. Microsc. Sci. 1, 97–102.

Jones, W.D., Nguyen, T.-A.T., Kloss, B., Lee, K.J., Vosshall, L.B., 2005. Functional conservation of an insect odorant receptor gene across 250 million years of evolution. Curr. Biol. 15, R119–R121.

Jones, P.L., Pask, G.M., Rinker, D.C., Zwiebel, L.J., 2011. Functional agonism of insect odorant receptor ion channels. Proc. Natl. Acad. Sci. U.S.A. 108, 8821–8825.

Jones, P.L., Pask, G.M., Romaine, I.M., Taylor, R.W., Reid, P.R., Waterson, A.G., Sulikowski, G.A., Zwiebel, L.J., 2012. Allosteric antagonism of insect odorant receptor ion channels. PLoS One 7, e30304.

Karner, T., Kellner, I., Schultze, A., Breer, H., Krieger, J.R., 2015. Co-expression of six tightly clustered odorant receptor genes in the antenna of the malaria mosquito *Anopheles gambiae*. Front. Ecol. Evol. 3, Article number: 26. http://dx.doi.org/10.3389/fevo.2015.00026.

Kavlie, R.G., Albert, J.T., 2013. Chordotonal organs. Curr. Biol. 23, R334–R335.
Kim, J., Chung, Y.D., Park, D.-Y., Choi, S., Shin, D.W., Soh, H., Lee, H.W., Son, W., Yim, J., Park, C.-S., Kernan, M.J., Kim, C., 2003. A TRPV family ion channel required for hearing in *Drosophila*. Nature 424, 81–84.
Kirkness, E.F., Haas, B.J., Sun, W.L., Braig, H.R., Perotti, M.A., Clark, J.M., Lee, S.H., Robertson, H.M., Kennedy, R.C., Elhaik, E., Gerlach, D., Kriventseva, E.V., Elsik, C.G., Graur, D., Hill, C.A., Veenstra, J.A., Walenz, B., Tubio, J.M.C., Ribeiro, J.M.C., Rozas, J., Johnston, J.S., Reese, J.T., Popadic, A., Tojo, M., Raoult, D., Reed, D.L., Tomoyasu, Y., Krause, E., Mittapalli, O., Margam, V.M., Li, H.M., Meyer, J.M., Johnson, R.M., Romero-Severson, J., Vanzee, J.P., Alvarez-Ponce, D., Vieira, F.G., Aguade, M., Guirao-Rico, S., Anzola, J.M., Yoon, K.S., Strycharz, J.P., Unger, M.F., Christley, S., Lobo, N.F., Seufferheld, M.J., Wang, N.K., Dasch, G.A., Struchiner, C.J., Madey, G., Hannick, L.I., Bidwell, S., Joardar, V., Caler, E., Shao, R.F., Barker, S.C., Cameron, S., Bruggner, R.V., Regier, A., Johnson, J., Viswanathan, L., Utterback, T.R., Sutton, G.G., Lawson, D., Waterhouse, R.M., Venter, J.C., Strausberg, R.L., Berenbaum, M.R., Collins, F.H., Zdobnov, E.M., Pittendrigh, B.R., 2010. Genome sequences of the human body louse and its primary endosymbiont provide insights into the permanent parasitic lifestyle. Proc. Natl. Acad. Sci. U.S.A. 107, 12168–12173.
Kistler, K.E., Vosshall, L.B., Matthews, B.J., 2015. Genome engineering with CRISPR-Cas9 in the mosquito Aedes aegypti. Cell Rep. 11, 51–60.
Kreher, S.A., Kwon, J.Y., Carlson, J.R., 2005. The molecular basis of odor coding in the *Drosophila* larva. Neuron 46, 445–456.
Krieger, J., Klink, O., Mohl, C., Raming, K., Breer, H., 2003. A candidate olfactory receptor subtype highly conserved across different insect orders. J. Comp. Physiol. A Neuroethol. Sens. Neural Behav. Physiol. 189, 519–526.
Kwon, H.-W., Lu, T., Rützler, M., Zwiebel, L.J., 2006. Olfactory responses in a gustatory organ of the malaria vector mosquito *Anopheles gambiae*. Proc. Natl. Acad. Sci. U.S.A. 103, 13526–13531.
Kwon, J.Y., Dahanukar, A., Weiss, L.A., Carlson, J.R., 2007. The molecular basis of CO_2 reception in *Drosophila*. Proc. Natl. Acad. Sci. U.S.A. 104, 3574–3578.
Kwon, Y., Shim, H.-S., Wang, X., Montell, C., 2008. Control of thermotactic behavior via coupling of a TRP channel to a phospholipase C signaling cascade. Nat. Neurosci. 11, 871–873.
Larsson, M.C., Domingos, A.I., Jones, W.D., Chiappe, M.E., Amrein, H., Vosshall, L.B., 2004. Or83b encodes a broadly expressed odorant receptor essential for *Drosophila* olfaction. Neuron 43, 703–714.
Laurence, B.R., Pickett, J.A., 1982. An oviposition pheromone for *Culex quinquefasiatus*. Bull. Entomol. Res. 73, 283–290.
Leal, W.S., 2013. Odorant Reception in insects: roles of receptors, binding proteins, and degrading enzymes. Annu. Rev. Entomol. 58, 373–391.
Leal, W.S., Choo, Y.-M., Xu, P., Da Silva, C.S., Ueira-Vieira, C., 2013. Differential expression of olfactory genes in the southern house mosquito and insights into unique odorant receptor gene isoforms. Proc. Natl. Acad. Sci. U.S.A. 110, 18704–18709.
Lee, J., Moon, S., Cha, Y., Chung, Y.D., 2010. Drosophila TRPN(=NOMPC) channel localizes to the distal end of mechanosensory cilia. PLoS One 5, e11012.
Lehane, M., 1991. Biology of Blood-Sucking Insects. Chapman & Hall, London.
Lehnert, B.P., Baker, A.E., Gaudry, Q., Chiang, A.S., Wilson, R.I., 2013. Distinct roles of TRP channels in auditory transduction and amplification in *Drosophila*. Neuron 77, 115–128.
Liang, X., Madrid, J., Saleh, H.S., Howard, J., 2011. NOMPC, a member of the TRP channel family, localizes to the tubular body and distal cilium of *Drosophila* campaniform and chordotonal receptor cells. Cytoskeleton (Hoboken) 68, 1–7.

Liman, E.R., Zhang, Y.V., Montell, C., 2014. Peripheral coding of taste. Neuron 81, 984–1000.
Liu, C., Zwiebel, L.J., 2013. Molecular characterization of larval peripheral thermosensory responses of the malaria vector mosquito *Anopheles gambiae*. PLoS One 8, e72595.
Liu, C., Pitts, R., Bohbot, J., Jones, P., Wang, G., 2010. Distinct olfactory signaling mechanisms in the malaria vector mosquito *Anopheles gambiae*. PLoS Biol. 8, e1000467.
Lu, T., Qiu, Y.T., Wang, G., Kwon, J.Y., Rutzler, M., Kwon, H.W., Pitts, R.J., Van Loon, J.J., Takken, W., Carlson, J.R., Zwiebel, L.J., 2007. Odor coding in the maxillary palp of the malaria vector mosquito *Anopheles gambiae*. Curr. Biol. 17, 1533–1544.
Lucas-Barbosa, D., Van Loon, J.J.A., Dicke, M., 2011. The effects of herbivore-induced plant volatiles on interactions between plants and flower-visiting insects. Phytochemistry 72, 1647–1654.
Macdonald, G., 1957. The Epidemiology and Control of Malaria. Oxford University Press, London.
Manoharan, M., Chong, M.N.F., Vaitinadapoule, A., Frumence, E., Sowdhamini, R., Offmann, B., 2013. Comparative genomics of odorant binding proteins in *Anopheles gambiae*, *Aedes aegypti*, and *Culex quinquefasciatus*. Genome Biol. Evol. 5, 163–180.
Matthews, B.J., Mcbride, C.S., Degennaro, M., Despo, O., Vosshall, L.B., 2016. The neurotranscriptome of the *Aedes aegypti* mosquito. BMC Genomics 17, 32.
Mattingly, P.F., 1957. Genetical aspects of the *Aedes aegypti* problem. I. Taxonom: and bionomics. Ann. Trop. Med. Parasitol. 51, 392–408.
Mcbride, C.S., Baier, F., Omondi, A.B., Spitzer, S.A., Lutomiah, J., Sang, R., Ignell, R., Vosshall, L.B., 2014. Evolution of mosquito preference for humans linked to an odorant receptor. Nature 515, 222–227.
Mciver, S.B., 1982. Sensilla mosquitoes (Diptera: Culicidae). J. Med. Entomol. 19, 489–535.
Mcmeniman, C.J., Corfas, R.A., Matthews, B.J., Ritchie, S.A., Vosshall, L.B., 2014. Multimodal integration of carbon dioxide and other sensory cues drives mosquito attraction to humans. Cell 156, 1060–1071.
Merritt, R., Dadd, R., Walker, E., 1992. Feeding behavior, natural food, and nutritional relationships of larval mosquitoes. Annu. Rev. Entomol. 37, 349–374.
Missbach, C., Dweck, H.K., Vogel, H., Vilcinskas, A., Stensmyr, M.C., Hansson, B.S., Grosse-Wilde, E., 2014. Evolution of insect olfactory receptors. eLife 3, e02115.
Montell, C., 1989. Molecular genetics of *Drosophila* vision. Bioessays 11, 43–48.
Montell, C., 2012. *Drosophila* visual transduction. Trends Neurosci. 35, 356–363.
Muller, G.C., Beier, J.C., Traore, S.F., Toure, M.B., Traore, M.B., Bah, S., Doumbia, S., Schlein, Y., 2010. Successful field trial of attractive toxic sugar bait (ATSB) plant-spraying methods against malaria vectors in the Anopheles gambiae complex in Mali, West Africa. Malar. J. 9, 210.
Neafsey, D.E., Waterhouse, R.M., Abai, M.R., Aganezov, S.S., Alekseyev, M.A., Allen, J.E., Amon, J., Arca, B., Arensburger, P., Artemov, G., Assour, L.A., Basseri, H., Berlin, A., Birren, B.W., Blandin, S.A., Brockman, A.I., Burkot, T.R., Burt, A., Chan, C.S., Chauve, C., Chiu, J.C., Christensen, M., Costantini, C., Davidson, V.L., Deligianni, E., Dottorini, T., Dritsou, V., Gabriel, S.B., Guelbeogo, W.M., Hall, A.B., Han, M.V., Hlaing, T., Hughes, D.S., Jenkins, A.M., Jiang, X., Jungreis, I., Kakani, E.G., Kamali, M., Kemppainen, P., Kennedy, R.C., Kirmitzoglou, I.K., Koekemoer, L.L., Laban, N., Langridge, N., Lawniczak, M.K., Lirakis, M., Lobo, N.F., Lowy, E., Maccallum, R.M., Mao, C., Maslen, G., Mbogo, C., Mccarthy, J., Michel, K., Mitchell, S.N., Moore, W., Murphy, K.A., Naumenko, A.N., Nolan, T., Novoa, E.M., O'loughlin, S., Oringanje, C., Oshaghi, M.A., Pakpour, N., Papathanos, P.A., Peery, A.N., Povelones, M., Prakash, A., Price, D.P., Rajaraman, A., Reimer, L.J., Rinker, D.C., Rokas, A., Russell, T.L., Sagnon, N., Sharakhova, M.V., Shea, T., Simao, F.A., Simard, F.,

Slotman, M.A., Somboon, P., Stegniy, V., Struchiner, C.J., Thomas, G.W., Tojo, M., Topalis, P., Tubio, J.M., Unger, M.F., Vontas, J., Walton, C., Wilding, C.S., Willis, J.H., Wu, Y.C., Yan, G., Zdobnov, E.M., Zhou, X., Catteruccia, F., Christophides, G.K., Collins, F.H., Cornman, R.S., et al., 2015. Mosquito genomics. Highly evolvable malaria vectors: the genomes of 16 Anopheles mosquitoes. Science 347 (6217), 1258522.

Nene, V., Wortman, J.R., Lawson, D., Haas, B., Kodira, C., Tu, Z.J., Loftus, B., Xi, Z., Megy, K., Grabherr, M., Ren, Q., Zdobnov, E.M., Lobo, N.F., Campbell, K.S., Brown, S.E., Bonaldo, M.F., Zhu, J., Sinkins, S.P., Hogenkamp, D.G., Amedo, P., Arsenburger, P., Atkinson, P.W., Bidwell, S., Biedler, J., Birney, E., Bruggner, R.V., Costas, J., Coy, M.R., Crabtree, J., Crawford, M., Debruyn, B., Decaprio, D., Eiglmeier, K., Eisenstadt, E., El-Dorry, H., Gelbart, W.M., Gomes, S.L., Hammond, M., Hannick, L.I., Hogan, J.R., Holmes, M.H., Jaffe, D., Johnston, S.J., Kennedy, R.C., Koo, H., Kravitz, S., Kriventseva, E.V., Kulp, D., Labutti, K., Lee, E., Li, S., Lovin, D.D., Mao, C., Mauceli, E., Menck, C.F., Miller, J.R., Montgomery, P., Mori, A., Nascimento, A.L., Naveira, H.F., Nusbaum, C., O'leary, S., Orvis, J., Pertea, M., Quesneville, H., Reidenbach, K.R., Rogers, Y.H., Roth, C.W., Schneider, J.R., Schatz, M., Shumway, M., Stanke, M., Stinson, E.O., Tubio, J.M.C., Vanzee, J.P., Verjovski-Almeida, S., Werner, D., White, O., Wyder, S., Zeng, Q., Zhao, Q., Zhao, Q., Hill, C.A., Raikhel, A.S., Soares, M.B., Knudson, D.L., Lee, N.H., Galagan, J., Salzberg, S.L., Paulsen, L.T., Dimopoulos, G., Collins, F.H., Bruce, B., Fraser-Liggett, C.M., Severson, D.W., 2007. Genome sequence of *Aedes aegypti*, a major arbovirus vector. Science 316 (5832), 1718–1723.

Nicastro, D., Melzer, R., Hruschka, H., Smola, U., 1998. Evolution of small sense organs: sensilla on the larval antennae traced back to the origin of the Diptera. Naturwissenschaften 85, 501–505.

Nikbakhtzadeh, M.R., Terbot 2nd, J.W., Otienoburu, P.E., Foster, W.A., 2014. Olfactory basis of floral preference of the malaria vector *Anopheles gambiae* (Diptera: Culicidae) among common African plants. J. Vector Ecol. 39, 372–383.

Nyasembe, V.O., Torto, B., 2014. Volatile phytochemicals as mosquito semiochemicals. Phytochem. Lett. 8, 196–201.

Nyasembe, V.O., Teal, P.E.A., Mukabana, W.R., Tumlinson, J.H., Torto, B., 2012. Behavioural response of the malaria vector *Anopheles gambiae* to host plant volatiles and synthetic blends. Parasit. Vectors 5, 234.

Pask, G.M., Jones, P.L., Rutzler, M., Rinker, D.C., Zwiebel, L.J., 2011. Heteromeric Anopheline odorant receptors exhibit distinct channel properties. PLoS One 6, e28774.

Pask, G.M., Bobkov, Y.V., Corey, E.A., Ache, B.W., Zwiebel, L.J., 2013a. Blockade of insect odorant receptor currents by amiloride derivatives. Chem. Senses 38, 221–229.

Pask, G.M., Romaine, I.M., Zwiebel, L.J., 2013b. The molecular receptive range of a lactone receptor in *Anopheles gambiae*. Chem. Senses 38, 19–25.

Pelletier, J., Hughes, D.T., Luetje, C.W., Leal, W.S., 2010. An odorant receptor from the southern house mosquito *Culex pipiens quinquefasciatus* sensitive to oviposition attractants. PLoS One 5, e10090.

Pelosi, P., Zhou, J.J., Ban, L.P., Calvello, M., 2006. Soluble proteins in insect chemical communication. Cell. Mol. Life Sci. 63, 1658–1676.

Pennetier, C., Warren, B., Dabire, K.R., Russell, I.J., Gibson, G., 2010. Singing on the wing as a mechanism for species recognition in the malarial mosquito *Anopheles gambiae*. Curr. Biol. 20, 131–136.

Pennisi, E., 2015. SCIENCE AND SOCIETY. Gene drive turns mosquitoes into malaria fighters. Science 350, 1014.

Petersen, L.R., Carson, P.J., Biggerstaff, B.J., Custer, B., Borchardt, S.M., Busch, M.P., 2013. Estimated cumulative incidence of West Nile virus infection in US adults, 1999–2010. Epidemiol. Infect. 141, 591–595.
Phillips, A.M., Bull, A., Kelly, L.E., 1992. Identification of a *Drosophila* gene encoding a calmodulin-binding protein with homology to the *trp* phototransduction gene. Neuron 8, 631–642.
Pitts, R.J., Rinker, D.C., Jones, P.L., Rokas, A., Zwiebel, L.J., 2011. Transcriptome profiling of chemosensory appendages in the malaria vector *Anopheles gambiae* reveals tissue- and sex-specific signatures of odor coding. BMC Genomics 12, 271.
Pitts, R.J., Fox, A.N., Zwiebel, L.J., 2004. A highly conserved candidate chemoreceptor expressed in both olfactory and gustatory tissues in the malaria vector *Anopheles gambiae*. Proc. Natl. Acad. Sci. U.S.A. 101, 5058–5063.
Pitts, R.J., Liu, C., Zhou, X., Malpartida, J.C., Zwiebel, L.J., 2014. Odorant receptor-mediated sperm activation in disease vector mosquitoes. Proc. Natl. Acad. Sci. U.S.A. 111, 2566–2571.
Ray, A., Van Der Goes Van Naters, W., Carlson, J.R., 2014. Molecular determinants of odorant receptor function in insects. J. Biosci. 39, 555–563.
Rinker, D.C., Pitts, R.J., Zhou, X., Suh, E., Rokas, A., Zwiebel, L.J., 2013a. Blood meal-induced changes to antennal transcriptome profiles reveal shifts in odor sensitivities in *Anopheles gambiae*. Proc. Natl. Acad. Sci. U.S.A. 110, 8260–8265.
Rinker, D.C., Zhou, X., Pitts, R.J., Consortium, A.G.C., Rokas, A., Zwiebel, L.J., 2013b. Antennal transcriptome profiles of anopheline mosquitoes reveal human host olfactory specialization in *Anopheles gambiae*. BMC Genomics 14, 749.
Robertson, H.M., 2015. The insect chemoreceptor superfamily is ancient in animals. Chem. Senses 40, 609–614.
Salama, H.S., 1966. The function of mosquito taste receptors. J. Insect Physiol. 12, 1051–1060.
Sato, K., Pellegrino, M., Nakagawa, T., Nakagawa, T., Touhara, L.B.V.K., 2008. Insect olfactory receptors are heteromeric ligand-gated ion channels. Nature 452, 1002–1006.
Sato, K., Tanaka, K., Touhara, K., 2011. Sugar-regulated cation channel formed by an insect gustatory receptor. Proc. Natl. Acad. Sci. U.S.A. 108, 11680–11685.
Shen, W.L., Kwon, Y., Adegbola, A.A., Luo, J., Chess, A., Montell, C., 2011. Function of rhodopsin in temperature discrimination in *Drosophila*. Science (New York, NY) 331, 1333–1336.
Shim, J., Lee, Y., Jeong, Y.T., Kim, Y., Gee, M.G., Montell, C., Moon, S.J., 2015. The full repertoire of *Drosophila* gustatory receptors for detecting an aversive compound. Nat. Commun. 6, 8867.
Silbering, A.F., Rytz, R., Grosjean, Y., Abuin, L., Ramdya, P., Jefferis, G., Benton, R., 2011. Complementary function and integrated wiring of the evolutionarily distinct drosophila olfactory subsystems. J. Neurosci. 31, 13357–13375.
Sparks, J.T., Vinyard, B.T., Dickens, J.C., 2013. Gustatory receptor expression in the labella and tarsi of *Aedes aegypti*. Insect Biochem. Mol. Biol. 43, 1161–1171.
Steinbrecht, R., 1997. Pore structures in insect olfactory sensilla: a review of data and concepts. Int. J. Insect Morphol. Embryol. 26, 229–245.
Stocker, R.F., 1994. The organization of the chemosensory system in *Drosophila melanogaster*: a review. Cell Tissue Res. 275, 3–26.
Suh, E., Bohbot, J., Zwiebel, L.J., 2014. Peripheral olfactory signaling in insects. Curr. Opin. Insect Sci. 6, 86–92.
Suh, E., Choe, D.H., Saveer, A.M., Zwiebel, L.J., 2016. Suboptimal larval habitats modulate oviposition of the malaria vector mosquito *Anopheles coluzzii*. PLoS One 11, e0149800.
Tabachnick, W.J., 1991. The evolutionary relationships among arboviruses and the evolutionary relationships of their vectors provides a method for understanding vector-host interactions. J. Med. Entomol. 28, 297–298.

Takken, W., 1991. The role of olfaction in host-seeking of mosquitoes: a review. Insect Sci. Appl. 12, 287–295.
Takken, W., Knols, B.G.J., 1999. Odor-mediated behavior of Afrotropical malaria mosquitoes. Annu. Rev. Entomol. 44, 131–157.
Takken, W., Costantini, C., Dolo, G., Hassanali, A., Sagnon, N.F., Osir, E., 2006. Mosquito mating behaviour. In: Kitsos, L., Knols, B. (Eds.), Bridging Laboratory and Field Research for Genetic Control of Disease Vectors. Kluwer Academic Press, Dordrecht, pp. 183–188.
Tauxe, G.M., Macwilliam, D., Boyle, S.M., Guda, T., Ray, A., 2013. Targeting a dual detector of skin and CO_2 to modify mosquito host seeking. Cell 155, 1365–1379.
Taylor, R.W., Romaine, I.M., Liu, C., Murthi, P., Jones, P.L., Waterson, A.G., Sulikowski, G.A., Zwiebel, L.J., 2012. Structure-activity relationship of a broad-spectrum insect odorant receptor agonist. ACS Chem. Biol. 7, 1647–1652.
Van Breugel, F., Riffell, J., Fairhall, A., Dickinson, M.H., 2015. Mosquitoes use vision to associate odor plumes with thermal targets. Curr. Biol. 25, 2123–2129.
Velarde, R.A., Sauer, C.D., Walden, K.K., Fahrbach, S.E., Robertson, H.M., 2005. Pteropsin: a vertebrate-like non-visual opsin expressed in the honey bee brain. Insect Biochem. Mol. Biol. 35, 1367–1377.
Vieira, F.G., Rozas, J., 2011. Comparative genomics of the odorant-binding and chemosensory protein gene families across the Arthropoda: origin and evolutionary history of the chemosensory system. Genome Biol. Evol. 3, 476–490.
Vogt, R.G., Riddiford, L.M., 1981. Pheremone binding and inactivation by moth antennae. Nature 293, 161–163.
Vosshall, L.B., Hansson, B.S., 2011. A unified nomenclature system for the insect olfactory coreceptor. Chem. Senses 36, 497–498.
Walker, R.G., Willingham, A.T., Zuker, C.S., 2000. A *Drosophila* mechanosensory transduction channel. Science 287, 2229–2234.
Wang, G., Qiu, Y.T., Lu, T., Kwon, H.W., Pitts, R.J., Van Loon, J.J., Takken, W., Zwiebel, L.J., 2009. *Anopheles gambiae* TRPA1 is a heat-activated channel expressed in thermosensitive sensilla of female antennae. Eur. J. Neurosci. 30, 967–974.
Wang, G., Carey, A.F., Carlson, J.R., Zwiebel, L.J., 2010. Molecular basis of odor coding in the malaria vector mosquito *Anopheles gambiae*. Proc. Natl. Acad. Sci. U.S.A. 107, 4418–4423.
Warren, B., Gibson, G., Russell, I.J., 2009. Sex recognition through midflight mating duets in *Culex* mosquitoes is mediated by acoustic distortion. Curr. Biol. 19, 485–491.
Wicher, D., Fer, R.S., Bauernfeind, R., Stensmyr, M.C., Heller, R., Hansson, S.H.H.B.S., 2008. Drosophila odorant receptors are both ligand-gated and cyclic-nucleotide-activated cation channels. Nature 452, 1007–1011.
Xia, Y., Wang, G., Buscariollo, D., Pitts, R.J., Wenger, H., Zwiebel, L.J., 2008. The molecular and cellular basis of olfactory-driven behavior in *Anopheles gambiae* larvae. Proc. Natl. Acad. Sci. U.S.A. 105, 6433–6438.
Xue, T., Do, M.T., Riccio, A., Jiang, Z., Hsieh, J., Wang, H.C., Merbs, S.L., Welsbie, D.S., Yoshioka, T., Weissgerber, P., Stolz, S., Flockerzi, V., Freichel, M., Simon, M.I., Clapham, D.E., Yau, K.W., 2011. Melanopsin signalling in mammalian iris and retina. Nature 479, 67–73.
Yan, Z., Zhang, W., He, Y., Gorczyca, D., Xiang, Y., Cheng, L.E., Meltzer, S., Jan, L.Y., Jan, Y.N., 2013. *Drosophila* NOMPC is a mechanotransduction channel subunit for gentle-touch sensation. Nature 493, 221–225.
Zhang, Y.V., Ni, J., Montell, C., 2013. The molecular basis for attractive salt-taste coding in *Drosophila*. Science 340, 1334–1338.
Zhang, W., Cheng, L.E., Kittelmann, M., Li, J., Petkovic, M., Cheng, T., Jin, P., Guo, Z., Gopfert, M.C., Jan, L.Y., Jan, Y.N., 2015. Ankyrin repeats convey force to gate the NOMPC mechanotransduction channel. Cell 162, 1391–1403.

Zhou, X., Slone, J.D., Rokas, A., Berger, S.L., Liebig, J., Ray, A., Reinberg, D., Zwiebel, L.J., 2012. Phylogenetic and transcriptomic analysis of chemosensory receptors in a pair of divergent ant species reveals sex-specific signatures of odor coding. PLoS Genet. 8, e1002930.

Zhou, D., Zhang, D., Ding, G., Shi, L., Hou, Q., Ye, Y., Xu, Y., Zhou, H., Xiong, C., Li, S., Yu, J., Hong, S., Yu, X., Zou, P., Chen, C., Chang, X., Wang, W., Lv, Y., Sun, Y., Ma, L., Shen, B., Zhu, C., 2014a. Genome sequence of *Anopheles sinensis* provides insight into genetics basis of mosquito competence for malaria parasites. BMC Genomics 15, e42.

Zhou, X., Rinker, D.C., Pitts, R.J., Rokas, A., Zwiebel, L.J., 2014b. Divergent and conserved elements comprise the chemoreceptive repertoire of the nonblood-feeding mosquito *Toxorhynchites amboinensis*. Genome Biol. Evol. 6, 2883–2896.

Zhou, X., Rokas, A., Berger, S.L., Liebig, J., Ray, A., Zwiebel, L.J., 2015. Chemoreceptor evolution in hymenoptera and its implications for the evolution of eusociality. Genome Biol. Evol. 7, 2407–2416.

Zwiebel, L.J., Takken, W., 2004. Olfactory regulation of mosquito-host interactions. Insect Biochem. Mol. Biol. 34, 645–652.

CHAPTER ELEVEN

Molecular Physiology of Mosquito Diapause

D.L. Denlinger*, P.A. Armbruster†
*The Ohio State University, Columbus, OH, United States
†Georgetown University, Washington, DC, United States

Contents

Abstract

The dormant state of diapause is exploited by numerous mosquito species to survive seasonal periods of environmental stress. We discuss embryonic, larval and adult diapauses in mosquitoes and probe the molecular and physiological distinctions that comprise the diapause phenotype. Diapause evokes diverse and unique attributes including behavioural changes, arrested development, enhanced stress tolerance, fat accumulation and suppressed metabolic rates. Like most insects, mosquitoes in temperate latitudes precisely monitor daylength as a cue for diapause initiation. We examine the role of circadian clock genes in this response and trace downstream hormonal pathways involved in the diapause response. Insulin and the FOXO transcription factor signalling pathways appear to be keys for generating the diapause phenotype in adult females of *Culex pipiens* and perhaps other species. Elucidating the molecular regulation of diapause-associated physiology may provide a basis to identify novel targets for the control of mosquito vectors.

Advances in Insect Physiology, Volume 51
ISSN 0065-2806
http://dx.doi.org/10.1016/bs.aiip.2016.05.002

1. INTRODUCTION

A diapause strategy is integral to the success of mosquitoes in bridging seasonal constraints of winter in temperate latitudes and, most likely, dry seasons in the tropics. While mosquitoes that lack the capacity for diapause are geographically restricted to locations that provide a moderate environment throughout the year, insertion of diapause into the life cycle has enabled penetration of mosquitoes into regions with seasonally unfavourable environments. Diapause appears to be a trait that has evolved independently numerous times in mosquito lineages, as evidenced by its scattered appearance in closely related species. Diapause not only provides a critical mechanism for mosquitoes to survive an inimical environment, but it also enables disease agents vectored by mosquitoes (eg, West Nile and LaCrosse viruses) to bridge adverse seasons and initiate infections in their vertebrate hosts the following year.

Several reviews provide excellent background on the ecology of insect diapause (Danks, 1987; Tauber et al., 1986), and more physiologically oriented reviews discuss its hormonal regulation (Denlinger et al., 2012), diapause energetics (Hahn and Denlinger, 2011) and the dynamic progression of diapause (Kostal, 2006). A comprehensive review of *Culex pipiens* includes valuable insights on diapause in this important vector species (Vinogradova, 2000), and our recent review offers an overview of mosquito diapause (Denlinger and Armbruster, 2014). In this review, we focus more narrowly on the molecular physiology of mosquito diapause and discuss features that comprise the diapause phenotype, variation in the response and the signalling pathway leading from perception of daylength to generation of the diapause phenotype.

2. DIAPAUSE PHENOTYPE

Among insects as a whole, diapause can occur in any of the life stages. In a few cases, mainly in insects at high latitudes that require more than a single season to complete development, diapause can occur in more than one life stage. In mosquitoes, we are aware of no cases of pupal diapause, and all instances of embryonic diapause involve an arrest at the stage of the pharate first-instar larva; at that point in development embryogenesis has been completed and the mosquito remains as a 'hidden larva' within the chorion of the egg, waiting to hatch. Earlier stages of embryonic diapause are well known in other insect taxa, but we are not aware of any such cases among mosquitoes. A few species rely on larval diapause, but diapause in mosquitoes is much more common in pharate first-instar larvae and adults (Fig. 1, Denlinger and Armbruster, 2014). Although we discuss the phenotypes of embryonic, larval and adult mosquito diapause separately later, there are, of course, features that are common to all of these diapause stages.

In all cases, diapause represents a form of arrested development. This does not necessarily imply a complete halt to development: development may continue but proceed at a rate far below that observed in nondiapausing individuals. A cell cycle arrest is an expected common trait, as is some level of metabolic suppression and an enhanced stress response. Unlike quiescence, that is an arrest of development in direct response to an environmental challenge, diapause is an anticipated response programmed well in advance of its actual onset, thus allowing the insect to make preparative steps before the unfavourable season begins. For example, many temperate mosquitoes that enter overwintering diapause are responsive to short daylengths during a photosensitive period (a developmental window in which the insect is responsive to photoperiod) that precedes the developmental arrest.

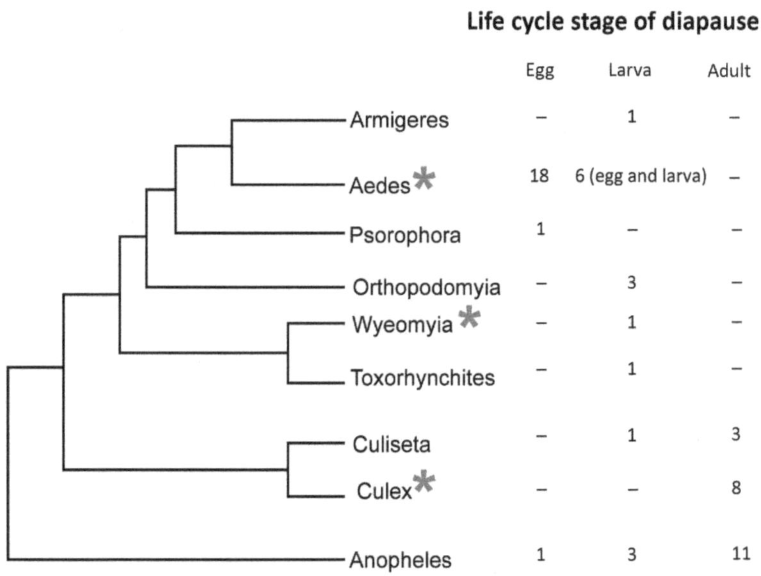

Fig. 1 Phylogenetic distribution of life cycle stages of diapause in nine Culicid genera based on Denlinger and Armbruster (2014). * indicates genera containing one species for which substantial information exists on the transcriptional basis of diapause.

Diapause thus impacts a broad spectrum of the insect's life and is best viewed as an alternative form of the life cycle (Denlinger, 2002).

2.1 Embryonic Diapause

Among mosquitoes, embryonic diapause is restricted almost exclusively to species from the genus *Aedes* (Fig. 1), with the only documented exceptions occurring in *Anopheles walkeri* (Hurlbut, 1938) and *Psorophora ferox* (Pinger and Eldridge, 1977). Photoperiod is the widely recognized token cue that initiates embryonic diapause, and it is the adult female that is photosensitive. As described earlier, in all cases of mosquito embryonic diapause, the developmental stage of arrest is the pharate larva inside the chorion of the egg. Thus, for many mosquito species that undergo embryonic diapause, the short daylengths perceived by the adult female lead to production of a diapause-inducing signal that is transmitted to the egg, persists throughout embryological development, and then initiates an arrest at the pharate larval stage. Though most embryonic mosquito diapauses are facultative (programmed by environmental signals) some, especially diapauses of univoltine species from high latitudes, are obligate (ie, genetically programmed without influence from the environment) (Denlinger and Armbruster, 2014). We do not know the specific signal that causes developmental arrest at the pharate larval stage in either obligate or facultative diapause.

The most thoroughly studied example of embryonic diapause in mosquitoes is that of the Asian tiger mosquito, *Aedes albopictus*. This highly invasive species has an unusually broad latitudinal distribution that includes both tropical and temperate regions (Hawley, 1988). The tropical populations do not undergo a photoperiodic diapause (Hawley, 1988; Urbanski et al., 2010), but it is at least possible that, like other tropical insects (Denlinger, 1986), they may enter aestivation ('summer' or 'dry season' dormancy) in response to a token cue other than photoperiod. In contrast, temperate populations of *Ae. albopictus* undergo a robust photoperiodic diapause. During embryonic development, when preparation for diapause is occurring, formation of the serosal cuticle, early segmentation of the embryo, pigmentation of the ocelli and formation of the egg burster all take longer in diapause-destined embryos relative to their nondiapause-destined counterparts (Lacour et al., 2014). Females of *Ae. albopictus* (Fonseca et al., 2015), *Aedes atropalpus* (Means, 1979) and *Aedes mariae* (Coluzzi et al., 1975) adjust their oviposition behaviour when laying diapause eggs. In the case of

Ae. mariae, females select more protected oviposition sites under diapause conditions and in the case of *Ae. atropalpus* females shift from laying nondiapause eggs on the water surface to ovipositing diapause eggs onto a solid rock substrate. Autumnal production of diapause eggs occurs far in advance of the onset of low winter temperatures. In the Washington, DC area (38° N), we have observed that diapause eggs of *Ae. albopictus* first appear as early as the beginning of September and continue to be produced until the end of October. This early season entry into diapause emphasizes the anticipatory nature of the diapause response and is noteworthy because the warmth of early autumn likely imposes a substantial energetic cost as reserves utilized during this period could otherwise be allocated to overwinter survival or to growth and reproduction the following spring.

Recent transcriptome sequencing studies have provided insight into transcriptional components of the diapause response in *Ae. albopictus* (Poelchau et al., 2011, 2013a,b,c). For example, similar to many other insects (Hahn and Denlinger, 2011), *Ae. albopictus* relies on lipid metabolism to meet the energetic demands of diapause. Diapause eggs are larger and contain approximately 30% more lipids than nondiapause eggs (Reynolds et al., 2012). This enhanced lipid content appears to be due to increased maternal provisioning, as adult females maintained under diapause-inducing conditions increase expression of *fatty acid synthase*, a gene that encodes an enzyme involved in lipid accumulation (Huang et al., 2015). Production of diapause eggs incurs a reproductive cost in the form of reduced size-specific fecundity of adult females. Although females reared under diapause-inducing conditions produce larger eggs with more total lipids, fewer eggs are produced relative to nondiapause conditions. In addition to maternal provisioning, the increased lipid content of diapause eggs also appears to be due to enhanced energy conservation during diapause preparation. Late stage developing embryos (6 days postoviposition) destined for diapause exhibit decreased expression of several genes involved in lipid catabolism, including *lipase 2*, *lipase 3*, *lipase 4*, *acyl-coA dehydrogenase 4* and *isovaleryl-coA dehydrogenase*. Increased expression of *lipid storage droplet protein 2* in early-stage developing embryos (3 days postoviposition) under diapause conditions implies that increased lipid storage is also important for diapause-associated lipid accumulation (Reynolds et al., 2012). Although less thoroughly studied than lipid metabolism, the citric acid cycle is transcriptionally downregulated late in embryological development (6 days postoviposition), reflecting an additional mechanism of energy conservation during diapause preparation.

In *Ae. albopictus*, altered lipid metabolism also contributes to stress tolerance during diapause by increasing desiccation resistance. This occurs through the increased synthesis of long-chain hydrocarbons that are important components of waxes that mediate water loss from the egg. *Fatty acid CoA elongase* (*face*) transcripts are more abundant in mature (stage V) oocytes of females producing diapause vs nondiapause eggs (Urbanski et al., 2010), and *face* expression is also higher in late-stage embryos under diapause conditions (Reynolds et al., 2012). Diapause eggs are coated with approximately 30% more surface hydrocarbons than nondiapause eggs, and water loss rates of diapause eggs are approximately half that of nondiapause eggs. In contrast to differences in hydrocarbon quantity, hydrocarbon composition, assessed by hydrocarbon chain length, does not differ between the two types of eggs. Under both conditions, chain length of egg surface hydrocarbons ranges from 19 to 51, and the abundance of specific chain lengths does not differ (Urbanski et al., 2010).

Diapause leads to increased cold hardiness in a wide range of insects, including those that undergo embryonic diapause (Denlinger, 1991), but mechanisms of cold hardiness during embryonic diapause of mosquitoes have not been studied thoroughly. Though diapausing embryos of the silk moth, *Bombyx mori*, rely on the polyol sorbitol for cold hardiness (Yamashita, 1996), diapausing eggs of the cricket, *Dianemobius nigrofasciatus*, do not use polyols as cryoprotectants (Goto et al., 2008). In *Ae. albopictus*, a combination of cold acclimation and diapause induction decreases the lower lethal temperature of eggs from an Indiana, Illinois strain to −12°C relative to a lower lethal temperature of −8°C under nondiapause, noncold acclimated conditions (Hanson and Craig, 1995). The only polyol detected in a preliminary metabolomics analysis of *Ae. albopictus* eggs 14 days postoviposition was threitol, and this polyol was actually present in lower concentrations during diapause. We did, however, detect elevated levels of amino acids that may serve as cryoprotectants. Lysine, the amino acid showing the greatest relative increase in our metabolomics analysis, also increases in response to low temperature in *Drosophila melanogaster* (Overgaard et al., 2007), and lysine is as effective as the sugar trehalose in conferring protection against low temperature injury (Mohammed et al., 2007). The elevated levels of amino acids we detected in preliminary metabolomics analysis of diapause eggs are consistent with RNA-seq results indicating differential expression of transcripts encoding proteins involved in amino acid metabolism in both adult (Huang et al., 2015) and egg stages (Poelchau et al., 2013c). Increased expression of *delta-(9)-desaturase* under diapause conditions in late-stage embryos may also contribute to cold hardiness by generating unsaturated fatty acids, a feature that enhances membrane fluidity at low temperatures.

Finally, at least two genes identified as differentially expressed in the embryonic diapause of *Ae. albopictus* have also been implicated in the molecular underpinnings of diapause in other species. For example, *pepck* (*phosphoenolpyruvate carboxykinase*), a rate-limiting enzyme in gluconeogenesis, has higher expression levels during diapause preparation in *Ae. albopictus* at both early and late stages of embryological development (Poelchau et al., 2013a). *Pepck* is also upregulated in association with the pupal diapauses of both the flesh fly, *Sarcophaga crassipalpis* (Ragland et al., 2010), and the apple maggot, *Rhagoletis pomonella* (Ragland et al., 2011), as well as dauer formation in the nematode *Caenorhabditis elegans* (McElwee et al., 2006). A somewhat paradoxical exception in which *pepck* is upregulated at diapause termination is discussed later; the more common pattern of *pepck* upregulation during diapause likely reflects a shift towards anaerobic metabolism during diapause (Hahn and Denlinger, 2011).

Similarly, the *pcna* (*proliferating cell nuclear antigen*) transcript, encoding a protein central to cell cycle progression, is differentially expressed in association with early diapause in embryos of *Ae. albopictus* (Poelchau et al., 2013a). This transcript is downregulated during larval diapause in the fly *Chymomyza costata* (Kostal et al., 2009) and in the pupal diapauses of both *S. crassipalpis* (Tammariello and Denlinger, 1998) and *R. pomonella* (Ragland et al., 2011). Involvement of *pepck* and *pcna* in regulation of diapause physiology in diverse species that undergo diapause at different life stages suggests that natural selection has acted on these genes repeatedly during the independent evolution of diapause in diverse insect lineages (Huang et al., 2015; Poelchau et al., 2013a).

2.2 Larval Diapause

Although larval diapause is less common than egg or adult diapause in mosquitoes, it is the most phylogenetically widespread stage of diapause having been described in at least one species from six genera (Fig. 1). In addition, there are six aedine species in which the developmental arrest of diapause can occur at both the embryo and larval stage (Denlinger and Armbruster, 2014). In mosquito species that undergo larval diapause, the diapause stage varies both within and between species. For example, *Anopheles barberi* enters a photoperiodically induced larval diapause in which most individuals undergo diapause at the second larval instar, but some do so as third instars

(Copeland and Craig, 1989). In the pitcher-plant mosquito, *Wyeomyia smithii*, southern (ancestral) populations enter diapause as fourth-instar larvae, while northern (derived) populations diapause as third-instar larvae or may enter a 'backup' fourth-instar diapause (Bradshaw and Lounibos, 1977; Lounibos and Bradshaw, 1975). Both southern populations of *W. smithii* (Bradshaw and Lounibos, 1977) and a Pennsylvania population of the predatory mosquito *Toxorhynchites rutilus* (Bradshaw and Holzapfel, 1977) diapause as fourth-instar larvae, and in both species the third instar of diapause-destined larvae is prolonged relative to nondiapause-destined larvae, allowing for an extended prediapause feeding period and increased resource accumulation. As is true for other stages of diapause, most mosquito larval diapauses are induced by photoperiod (Denlinger and Armbruster, 2014).

Most of what is known about the molecular physiology of larval diapause in mosquitoes comes from a microarray study of diapause termination in the pitcher-plant mosquito, *W. smithii* (Emerson et al., 2010). Many genes identified in this study are likely to be functionally involved in initiation of growth and development following termination of the third-instar diapause; such genes include the cell adhesion factor *Paxillan* (*Pax*), a *Myosin-binding subunit* (*Mbs*) transcript, seven transcripts encoding cuticle proteins (*Cpr49ab*, *Lcp65ac*, *Cpr65az*, *Cpr65av*, *Ccp84ad*, *Acp1*, *Acp65Aa*) and two ribosomal protein transcripts (*RpS23*, *RpS25*).

Differentially expressed genes involved in metabolism include two *sorbitol dehydrogenase* paralogs, *Sodh-1* and *Sodh-2*, which are expressed at high levels during diapause but are then downregulated at diapause termination. As noted earlier, sorbitol functions as a cryoprotectant for many overwintering insects (Lee and Denlinger, 1991). Diapause increases cold hardiness in *W. smithii* (Evans and Brust, 1972), thus changes in sorbitol metabolism at diapause termination are likely related to a reduced need for cold tolerance and altered energy allocation associated with vernal growth and development. Decreased expression of *Sodh* in association with diapause termination has also been demonstrated in the commercial silk moth, *Bombyx mori*, and in the hemimetabolous bug, *Pyrrhocoris apterus* (Kostal et al., 2008; Niimi et al., 1993). As discussed earlier, *pepck* encodes a protein that catalyses the rate-limiting step in gluconeogenesis and is upregulated during diapause in diverse insect groups. However, in *W. smithii*, *pepck* is upregulated at diapause termination. Diapausing larvae of *W. smithii* overwinter in water-filled pitcher-plant leaves under conditions that are likely hypoxic. Thus, upregulation of *pepck* at diapause termination may reflect a mechanism to utilize gluconeogenic precursors such as lactate and alanine, which are expected to accumulate as a result of anaerobic respiration during diapause (Emerson et al., 2010). The metabolic costs of larval diapause in *W. smithii* extend to the subsequent adult stage, manifested as reduced reproductive output of adults that have undergone larval diapause (Bradshaw et al., 1998).

Two identified genes may contribute to the 'photoperiodic switch' involved in measuring daylength and regulating the transition between direct development and larval diapause in *W. smithii*. The first, *Wsppdrg1* (*Wyeomyia smithii photoperiodic response gene 1*), is annotated as a low complexity cuticular protein, and, based on a microarray study, is upregulated at diapause termination (Emerson et al., 2010). Intriguingly, *Wsppdrg1* is also located on a quantitative traits locus (QTL) for diapause timing (critical photoperiod, CPP) (Mathias et al., 2007) and is associated with CPP in an intercross line between a northern and southern population (Emerson et al., 2010). Although annotation of this gene suggests it encodes a cuticular protein, and hence implies a structural role, Emerson et al. (2010) hypothesize a functional role involved in regulating photoperiodism through pleiotropic effects related to light sensing or signalling in the compound eye. The second gene proposed as a potential component of photoperiodic regulation of diapause in *W. smithii* is the circadian clock gene *timeless*. Expression of *timeless* varies among populations of *W. smithii* in association with differences in diapause timing (CPP) (Mathias et al., 2005), and the *timeless* locus interacts epistatically with a QTL affecting CPP (Mathias et al., 2007). Intriguingly, as discussed later, *timeless* has also been implicated in regulating diapause induction in both *Cx. pipiens* and *Ae. albopictus*.

2.3 Adult Diapause

Adult diapauses are known for at least three mosquito genera, *Culiseta*, *Culex* and *Anopheles* (Fig. 1), but the best-characterized adult mosquito diapause is that of the Northern house mosquito *Cx. pipiens* (Vinogradova, 2000). This diapause is specific to the female since males die in the autumn, shortly after mating. Thus, the mated female is the only stage capable of overwintering in temperate latitudes. She responds to the short daylengths of late summer and early autumn during a photosensitive stage that begins late in larval life and persists through the first few days of adulthood (Eldridge, 1968; Sanburg and Larsen, 1973; Spielman and Wong, 1973). Low rearing temperatures increase the diapause incidence. Prolongation of the final larval instar is the first obvious sign of the diapause fate of the adult (Zhang and Denlinger, 2011b). This developmental delay extends the period of larval feeding and hence accounts for the larger size of diapausing adults. After a fixed period of latency, the diapause can be terminated and development resumes in response to long daylength and elevated temperature. The relative importance of these two factors in breaking diapause in the spring has not been carefully examined, but the fact that adults from the same geographic area emerge earlier in the spring from warmer underground overwintering sites suggests that the response to temperature may be more critical than the seasonal increase in daylength (Vinogradova, 2000).

One of the most conspicuous features of diapause in *Cx. pipiens* is the halt in ovarian development. Oocytes of nondiapause-destined females grow quickly following adult eclosion and reach a resting stage, at an oocyte length of approximately 90 μm, in 3 days (Christopher's Stage IV or higher, Christopher, 1911). The female then takes a bloodmeal and the oocytes continue to grow until they reach maturity. By contrast, oocytes of diapause-destined females fail to show this initial growth spurt and halt development at 50–70 μm (Christopher's Stage I or II). The oocytes do slowly increase in size during diapause but it takes 20 weeks to attain the 90 μm length noted within 3 days in nondiapausing females (Readio and Meola, 1985).

Another obvious feature of diapause is failure of the female to seek a bloodmeal. Unlike her nondiapausing counterparts, females programmed for diapause do not take a bloodmeal but instead feed extensively on nectar and other sugar resources (Bowen, 1992; Mitchell, 1983). If a female is forced to take a bloodmeal the blood is not used for egg maturation but instead passes through the digestive tract unused (Mitchell and Briegel, 1989; Zhang and Denlinger, 2011b). Sugar consumed by the diapause-destined female is quickly converted to fat (Sim and Denlinger, 2009b; Zhou and Miesfeld, 2009), resulting in obese females with twice the fat reserves of their nondiapausing counterparts. Feeding is not the only behaviour altered by diapause: diapause also elicits a distinct behaviour designed to find caves, culverts, basements and other underground locations that provide protected sites for overwintering (Vinogradova, 2000).

Flight muscles quickly degrade in some insects that enter adult diapause, but not in mosquitoes. Flight muscles remain intact in diapausing adults of *Cx. pipiens*, enabling adults to remain mobile, a feature allowing them to constantly seek appropriate humidity levels within the hibernaculum during diapause. Yet, levels of *beta-tubulin* expression and microtubule abundance are considerably lower in flight muscles of diapausing females, suggesting that flight capacity during diapause is somewhat reduced (Kim and Denlinger, 2009).

Metabolic rates in diapausing mosquitoes are characteristically lower than in nondiapausing females of *Cx. pipiens* (Zhou and Miesfeld, 2009), but the rates are not as low as in some other insects that diapause as adults, and generally depression of the metabolic rate in diapausing adults is not as dramatic as it is in diapausing pupae (Danks, 1987).

Diapause evokes an obvious switch in the molecular machinery involved in digestion. Genes involved in sugar metabolism and fat accumulation are highly upregulated in diapausing females, while genes involved in bloodmeal digestion are turned off (Robich and Denlinger, 2005), eg, *fatty acid synthase*, a gene associated with lipid accumulation, is highly upregulated in early diapause, while *trypsin* and *chymotrypsin-like protein*, genes involved in digesting a bloodmeal, are strongly downregulated. Expression patterns of 31 fat-related genes throughout diapause (Sim and Denlinger, 2009b) reveal a dynamic progression in the expression profiles of genes associated with lipid metabolism: genes linked to fat accumulation, including *fatty acid synthase-1*, *fas-3* and *fatty acid-binding protein*, are significantly upregulated in early diapause, while transcripts associated with energy utilization (beta-oxidation) are strongly downregulated. But, towards the end of diapause as the female prepares to begin the active phase of her life, these expression profiles flip: genes involved in beta-oxidation are highly upregulated, while those involved in fat accumulation are downregulated. Transcript profiles nicely parallel the metabolic flux dynamics recorded in *Cx. pipiens* during diapause (Zhou and Miesfeld, 2009).

Cold and desiccation tolerance are enhanced in diapausing females of *Cx. pipiens* (Benoit and Denlinger, 2007; Rinehart et al., 2006), responses that likely reflect increases in levels of polyols or other cryoprotectants, although the basis for this enhanced tolerance has not been investigated. Heat shock proteins, well known to be involved in cold tolerance in several other diapausing species (Rinehart et al., 2007), do not appear to be involved in cold tolerance of diapausing adults of *Cx. pipiens* (Rinehart et al., 2006), but several cytoskeletal adjustments are linked to diapause and cold exposure. Two *actins* highly upregulated during diapause are linked to a redistribution of polymerized actin in the midgut, an effect that is most pronounced in diapausing females that have been exposed to low temperature (Kim et al., 2006). Rather than being evenly distributed along the muscle fibres of the midgut, the polymerized actin is conspicuously clustered at the intersections of the muscle fibres that envelop the midgut, a response that possibly helps to fortify the cytoskeleton during winter. A proteomic examination of diapausing adults of *Cx. pipiens* also uncovered a number of changes in protein abundance that may be associated with cytoskeletal reorganization. One protein in particular, identified as a pupal cuticle protein, shows a major increase in abundance early in diapause (Li and Denlinger, 2009). This protein possibly plays a role in desiccation or some other form of cuticular stress resistance; it also offers potential as a reliable biomarker for early diapause (Li and Denlinger, 2009). Genes encoding immune responses (Robich et al., 2007) and the antioxidant enzymes *catalase* and *superoxide dismutase-2* (Sim and Denlinger, 2011) are upregulated in *Cx. pipiens* diapause. Thus a suite of genes involved in protective responses against stress are components of the diapause syndrome, a finding that is true for diapause in most insects.

Several genes involved in regulatory processes are diapause-controlled in *Cx. pipiens*, based on suppressive subtractive hybridization (SSH) (Robich et al., 2007). Among these, expression of *ribosomal protein S3a* (*rpS3a*), is dramatically shut down in early diapause, and RNA interference (RNAi) directed against this small ribosomal subunit in nondiapausing mosquitoes arrests follicle development, thus mimicking the diapause state (Kim et al., 2010). This response can be reversed with application of juvenile hormone III (JHIII), an endocrine trigger that terminates diapause, thus suggesting that a shut down in *rpS3a* is a component of the diapause response. A similar response is evident for *ribosomal protein S2* (*rpS2*): this transcript is downregulated during early diapause and RNAi directed against this gene in nondiapausing females elicits a diapause-like response on ovarian development that can be rescued with JHIII (Kim and Denlinger, 2010). Thus, both *rpS3a* and *rsS2* may be implicated in regulating the ovarian shutdown associated with diapause, but the expression profiles of these two transcripts are somewhat different: expression of both transcripts is halted 5–10 days after adult eclosion in diapausing females and expression then rebounds, but the shutdown in expression of *rpS3a* is rather brief, lasting only a few days, while the shutdown in expression of *rsS2* is considerably longer, lasting at least 1 month. This temporal difference in expression likely reflects different roles for these two genes, but both are candidates for eliciting the halt in development that characterizes diapause in *Cx. pipiens*. *Couch potato*, a gene associated with hypoactivity, is another possible regulatory gene associated with diapause in *Cx. pipiens* (Zhang and Denlinger, 2011a). Transcripts encoding this gene are elevated throughout diapause and decline when diapause is terminated environmentally or by application of JHIII. This gene is linked to diapause in several species of insects, although the direction of the response varies among species. Another known regulatory gene that may contribute to diapause in *Cx. pipiens* is the gene encoding prothoracicotropic hormone (PTTH), the neuropeptide that stimulates ecdysone synthesis (Zhang and Denlinger, 2011b). A delay in larval expression of *ptth* in diapause-programmed mosquitoes contributes to a lengthening of larval life (and hence larger pupal and adult sizes) in diapausing *Cx. pipiens*. Additionally, expression of *ptth* remains elevated throughout adult diapause and declines as the end of diapause approaches, thus suggesting important roles for this neuropeptide in regulating both the timing of larval development as well as the duration of diapause.

An RNA-seq comparison of diapausing and nondiapausing females of *Cx. pipiens* identified 41 transcripts having different abundances in these two types of females (Kang et al., 2016). Differences detected using DAVID analysis were most pronounced for categories of biological process, cellular components and molecular function, especially transcripts related to glycolysis. Further refinement using manual investigation revealed ontologies related to the juvenile hormone (JH) pathway, anaerobic metabolism, innate immunity and cold tolerance; results that are consistent with predicted diapause-related responses and regulatory mechanisms.

In addition to the increasing number of genes with annotated function that are associated with adult diapause in *Cx. pipiens*, many diapause-regulated sequences found with SSH (Robich et al., 2007) and RNA-Seq (Kang et al., 2016) have identities that remain unknown. Since expression patterns of many such genes are dramatically altered by diapause and different phases of diapause, it is obvious that additional players will ultimately be implicated in diapause and its regulation. The whole diapause syndrome encompasses responses that begin well before the onset of adult diapause and include a wide range of behavioural, physiological and molecular adjustments that are distinct to diapause.

How well observations on the diapause of *Cx. pipiens* mirror events associated with adult diapause in other mosquitoes remains to be seen, but the overall attributes of adult diapause in different species of *Culex*, *Culiseta* and *Anopheles* mosquitoes from high latitudes appear to be quite similar. A block of ovarian maturation is a common thread. Although diapause is widely reported in males of nonmosquito species, we are unaware of diapause occurring in adult male mosquitoes from temperate latitudes. Like *Cx. pipiens*, males of other temperate zone mosquito species mate in the autumn and then die, leaving fertilized females to bridge the winter in diapause. Likewise, only females of *Culiseta inornata* survive a dry season summer diapause (aestivation) in California, but they may enter this arrest after having already completed one gonotrophic cycle (Barnard and Mulla, 1978), a feature not observed in *Cx. pipiens*. As we discuss below, we expect that some attributes of *Anopheles* mosquitoes that bridge tropical dry seasons in a dormant state may indeed differ from the diapause patterns noted in mosquitoes from higher latitudes.

2.4 Aestivation

Although diapause is most commonly associated with overwintering of temperate insects (hibernal diapause), many tropical insects survive seasonally recurring harsh conditions such as dry seasons by entering a 'dry season' or 'summer' diapause referred to as aestivation (Denlinger, 1986). Understanding the ecology and physiology of aestivation in mosquitoes has significant practical importance due to the large health burden of diseases transmitted by tropical species. The issue of aestivation in mosquitoes, particularly the malaria vector *Anopheles gambiae*, has been of longstanding interest but is poorly understood. However, recent groundbreaking work on the malaria vector *Anopheles coluzzii* (formerly the 'M molecular form' of *An. gambiae*) from the Sahel region has begun to fill this important knowledge gap. Two lines of evidence support the hypothesis that *An. coluzzii* undergoes an adult stage aestivation to survive the prolonged (3–8 months) dry season during which no surface water and larval breeding sites are available. The first line of evidence is the dramatic increase in abundance of *An. coluzzii* adults 3–7 days after the first rains at the end of the dry season (Dao et al., 2014). The second is the recapture of a marked female at the end of the dry season, 7 months after she was marked and released (Lehmann et al., 2010).

Mosquitoes captured during the dry season vs wet season point to both similarities and differences between the physiology of hibernal diapause and aestivation (Huestis and Lehmann, 2014). For example, cues that initiate aestivation in *An. coluzzii* have not yet been conclusively determined, although there is suggestive evidence that photoperiod has at least a small effect on some populations (Wagoner et al., 2014). Similar to the adult hibernal diapause of *Cx. pipiens* described earlier, adults of *An. coluzzii* are able to remain active during aestivation, at least for brief periods. Repeated annual surveillance consistently indicates two short periods of dry season abundance during which mosquitoes presumably leave aestivation shelters to seek nutrition by blood- or sugar-feeding (Dao et al., 2014). But, in contrast to the hibernal diapause of *Cx. pipiens*, blood-feeding rates in *An. coluzzii* do not change during the dry season relative to the wet season, and small mating swarms are occasionally observed during the dry season (Yaro et al., 2012). Furthermore, the reproductive physiology of *An. coluzzii* is not completely shut down during the dry season, but is dramatically depressed, as reflected by reduced oviposition rates, egg batch sizes and intermediate frequencies of gonotrophic dissociation (Yaro et al., 2012). The fact that mating swarms are sporadically observed during the dry season suggests that both sexes survive the inclement season, a contrast to the female only diapause strategy noted for overwintering mosquitoes at higher latitudes. Although flight activity is reduced during the dry season, metabolic activity does not show a consistent decrease (Huestis et al., 2012). Like diapausing *Cx. pipiens*, dry season *An. coluzzii* exhibit increased desiccation resistance, but in *An. coluzzii* this effect is elicited by changes in composition rather than overall quantity of cuticular hydrocarbons (Arcaz et al., 2016). Intriguingly, dry season *An. coluzzii* also exhibit altered thorax and spiracle size allometry, although the effect of these changes on desiccation resistance is less clear. Lack of consistent physiological parallels between hibernal diapause and aestivation, in combination with the more permissive (warmer) temperatures during a tropical dry season vs a temperate winter, have led Huestis and Lehmann (2014) to hypothesize that behavioural changes are likely to be a more important component of the 'aestivation syndrome' relative to the 'diapause syndrome'.

3. DIAPAUSE DYNAMICS

The perception that diapause is a static state is far from correct. Not only are preparative steps detectable before the actual onset of diapause, but diapause itself is a dynamic state that is especially evident by monitoring changes in gene expression throughout diapause. As shown in Fig. 2, profiles of genes expressed in diapausing and nondiapausing pharate larvae of *Ae. albopictus* are quite distinct in early diapause, but the profiles gradually converge as diapause proceeds (Poelchau et al., 2013c), suggesting a progression in gene expression that leads eventually to diapause termination. Early in the adult diapause of *Cx. pipiens*, there is a major conversion of glucose into lipids, a flux that declines progressively during diapause (Zhou and Miesfeld, 2009). This trend is also reflected in patterns of gene expression: early in diapause genes associated with fat accumulation are highly

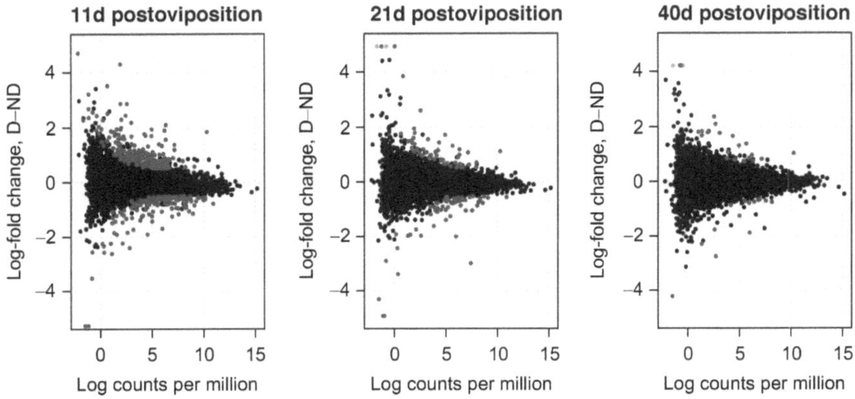

Fig. 2 Profiles of gene changes throughout diapause in embryos of *Ae. albopictus*, from Poelchau et al. (2013c). Differential gene expression at 11 (early diapause), 21 (middle diapause) and 40 days postoviposition (late diapause). Each point represents an individual gene plotted as log-fold change under diapause (D) relative to nondiapause (ND) conditions vs abundance after TMM normalization. Positive values represent genes with higher expression in diapause conditions and negative values indicate genes with higher expression in nondiapause conditions. *Red* indicates significant differential expression (FDR-corrected $p<0.05$, absolute log twofold change >0.5), *orange* represents genes unique to either diapause or nondiapause. (See the colour plate.)

upregulated and then decline in expression as diapause progresses (Sim and Denlinger, 2009b). Concurrently, genes associated with beta-oxidation are suppressed early in diapause but are then upregulated as the end of diapause approaches. Shifts in utilization patterns of fats and other energy reserves are commonly noted during diapause (Hahn and Denlinger, 2011), as are shifts in oxygen consumption patterns, hormone sensitivity and responsiveness to diapause-terminating stimuli (Denlinger et al., 1988). Such changes reflect the dynamic progression in diapause, referred to as diapause development (Andrewartha, 1952).

What ultimately capacitates the insect to terminate diapause and be competent to initiate development is still largely unknown. Insects usually complete diapause (become competent to develop) long before they actually begin to reinitiate development. Following a fixed period of diapause, most insects enter a period of postdiapause quiescence that is maintained by prevailing unfavourable environmental conditions. For example, submersion in water will not elicit egg hatching in *Ae. albopictus* during diapause, but when diapause is completed the pharate larvae will hatch immediately if submerged in water. As long as the eggs are dry, the pharate larvae will remain inside the chorion in a postdiapause quiescence. Likewise, adults of *Cx. pipiens* are fully competent to terminate diapause in northern latitudes by mid-winter, but they remain in a postdiapause quiescence during the cold months until the springtime rise in temperature and increase in daylength prompt initiation of blood feeding and reproduction (Vinogradova, 2000). Attaining the capacity to respond to environmental signals to initiate development marks the transition between diapause and postdiapause quiescence.

4. GENETIC BASIS FOR THE CAPACITY TO DIAPAUSE

Despite substantial recent progress in elucidating transcriptional components of the diapause response in mosquitoes and other insects, identifying the genetic (DNA sequence level) basis of diapause remains almost entirely unresolved. A promising approach for addressing this issue is to perform QTL mapping experiments by crossing a strain capable of undergoing diapause with a strain not capable of diapause. For example, Mori et al. (2007) generated two mapping populations by crossing individuals from a diapausing population of *Cx. pipiens* with the nondiapausing mosquito *Culex quinquefasciatus*. Crosses of 96 and 192 F_2 individuals were scored for the diapause phenotype (follicle size under short-day conditions) and genotyped at 9 and 13 restriction fragment length polymorphism loci, respectively. Results indicate four QTLs located across all three chromosomes. Individual QTLs explain 11–23% of the phenotypic variation in the mapping populations, indicating moderate effect sizes. However, because a relatively small number of markers were used in this analysis, the QTL intervals ranged from 6 to 30 centimorgans (cM) and each QTL likely contains multiple causative loci affecting the diapause capacity. Hybrid crosses between *Cx. pipiens*, which has the capacity for diapause, and *Cx. quinquefasciatus*, the southern species that lacks the capacity for diapause, indicate that diapause status of the mother is more important than that of the father in determining the diapause capacity of the progeny (Meuti et al., 2015b).

The results from *Cx. pipiens* imply a polygenic basis for the capacity to undergo photoperiodic diapause. This conclusion is consistent with results of genetic mapping experiments for diapause timing (CPP) in the pitcher-plant mosquito, *W. smithii*, results that also imply a polygenic and complex genetic architecture for diapause (Bradshaw et al., 2012; Mathias et al., 2007).

The advent of high-throughput, next-generation sequencing technologies has made it increasingly feasible to score thousands of genetic polymorphisms for nonmodel organisms using methods such as restriction site-associated DNA (RAD tag) markers (Baird et al., 2008). The availability of whole-genome sequences for many mosquitoes (Giraldo-Calderon et al., 2015), including that of the diapausing species *Ae. albopictus* (Chen et al., 2015), opens new opportunities for dissecting the genetics of diapause. These advances should make it possible to substantially increase the resolution of genetic mapping of diapause genes. One of us (Armbruster) is currently applying these methods to identify genes associated with insect diapause, using genetic mapping performed by crossing individuals from a nondiapausing (tropical) population of *Ae. albopictus* to individuals from a diapausing (temperate) population.

5. ENVIRONMENTAL CUES

Most insects inhabiting temperate latitudes, including mosquitoes, rely on photoperiod as the primary environmental signal for programming diapause. Photoperiod is referred to as a 'token' stimulus because it represents a signal predicting future inimical environmental conditions, rather than a direct environmental stressor. Accordingly, the photosensitive stage at which the insect responds to a diapause-inducing token stimulus occurs well in advance of the stage of developmental arrest, as in *Ae. albopictus*, for example, where the photosensitive adult female programmes the developmental arrest of her eggs. The male's photoperiodic experience plays no role in this decision (Pumpuni, 1989). Temperature often interacts with photoperiod to influence the diapause decision. Low temperatures during the photosensitive stage generally enhance diapause induction in response to short photoperiods, while high temperatures can avert entry into diapause (Anderson, 1968; Parker, 1985; Pumpuni et al., 1992). In *Cx. pipiens*, both short daylengths and low temperatures are required for the programming of diapause (Eldridge, 1968; Sim and Denlinger, 2009a). As discussed later, circadian clock genes appear to be involved in interpreting daylength, but the specific mechanisms by which this occurs and how the interpretation of short daylength is translated into a diapause-inducing signal remains poorly understood.

Like temperature, food availability can modify the diapause-inducing effects of short daylength. In both *Ae. albopictus* and *Ae. atropalpus*, females with reduced access to food as larvae produce a higher incidence of diapause eggs than females provisioned with optimal levels of larval food (Beach, 1978; Pumpuni et al., 1992). As discussed earlier, the environmental factors that induce aestivation in tropical mosquitoes remain unresolved, although based on studies of other tropical insects potentially important environmental factors include photoperiod (at latitudes beyond 5° North or South of the equator), temperatures at specific times of day, food composition and availability, population density, and moisture or humidity levels (Denlinger, 1986).

Unlike diapause induction, specific environmental cues are not commonly used as triggers for diapause termination. After a fixed period of latency has transpired, the insect becomes competent to break diapause and enters a period of postdiapause quiescence. It will then remain in that quiescent state until favourable environmental conditions return, eg, elevated temperatures of spring in *Cx. pipiens* or flooding of eggs in *Ae. albopictus*. There are, however, exceptions that rely directly on photoperiodic cues not only for diapause induction but also for diapause termination, eg, the pitcher-plant mosquito *W. smithii* relies on the lengthening of daylength in the spring to terminate its larval diapause (Bradshaw and Lounibos, 1972), and there is at least circumstantial evidence that the increase of daylength in the spring may facilitate termination of adult diapause in *Cx. pipiens* as well.

6. CENTRAL ROLE FOR THE CIRCADIAN CLOCK GENES

Mounting evidence supports a role for circadian (daily) clock genes in regulating the photoperiodic (seasonal) clock involved in diapause (Goto et al., 2010; Meuti and Denlinger, 2013). The major clock genes and their interactions in mosquitoes are depicted in Fig. 3. In nondiapausing females of *Cx. quinquefasciatus* (Gentile et al., 2009) and *Cx. pipiens* (Meuti et al., 2015a), the canonical clock genes *period* and *timeless* peak in expression near the onset of the scotophase, as noted in *D. melanogaster*. Circadian clock genes continue to cycle robustly throughout diapause (Meuti et al., 2015a). Thus, unlike the halt in clock gene cycling noted in hibernating European hamsters (Revel et al., 2007), the clock continues to tick during diapause in *Cx. pipiens*, a response that likely enables diapausing mosquitoes to continue monitoring daylength throughout the winter and to then use the lengthening of days in spring as a cue for initiating reproduction.

One of the challenges in assigning a role for the clock in a photoperiodic response such as diapause is that clock genes play roles in a number of diverse

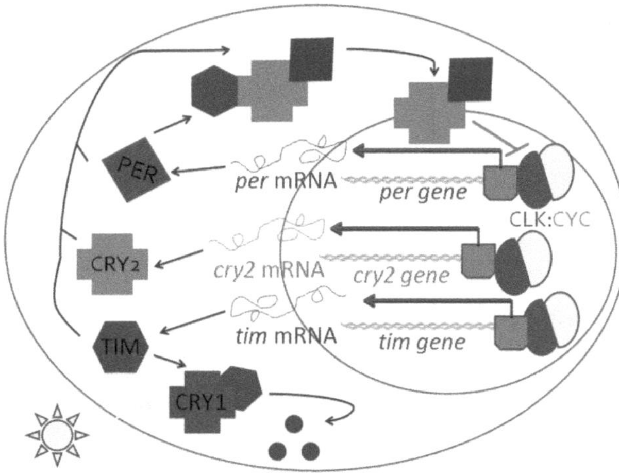

Fig. 3 Interactions of circadian clock genes in mosquitoes. Within the nucleus the proteins CLOCK (CLK) and CYCLE (CYC) form a heterodimer that acts as a transcriptional activator by binding to the E-box promoter region of the *period* (*per*), *timeless* (*tim*) and *cryptochrome2* (*cry2*) genes. The mRNAs encoded by these genes are translated in the cytoplasm where their protein products dimerize. The heterotrimer of PER, TIM and CRY2 then translocate into the nucleus where both PER and CRY2 likely inhibit the action of CLK:CYC, thereby suppressing the transcription of *per*, *tim* and *cry2*. The CRYPTOCHROME1 (CRY1) protein degrades TIM and itself in the presence of light. This results in increasing levels of *per*, *tim* and *cry2* mRNA throughout the day when CLK:CYC activity is uninhibited and decreasing levels of these transcripts during night. This scheme differs from the better-studied clock system in *Drosophila*, an insect that lacks involvement of CRY2. In mosquitoes, PER appears to assist CRY2 in nuclear translocation, whereas TIM helps to stabilize PER and CRY2. *Adapted from Meuti, M.E., Denlinger, D.L., 2013. Evolutionary links between circadian clocks and photoperiodic diapause in insects. Int. Comp. Biol. 53, 131–143.* (See the colour plate.)

physiological processes in addition to their role in time measurement. For example, the protein encoded by *cycle* functions not only in operation of the clock but also as a protein that forms a heterodimer with the JH receptor protein methoprene-tolerant, thus CYCLE has a role beyond the simple operation of the clock (Shin et al., 2012). Likewise, the clock genes *cycle* and *clock* play critical roles in midgut function that are distinct from their roles in regulation of the clock (Bajgar et al., 2013). Thus, it is not always evident whether disruption of the diapause response is directly related to clock function or to one of the other functions of the clock genes (Emerson et al., 2009).

But, in *Cx. pipiens*, we think it is most likely that the time-keeping function of the clock genes is essential for the programming of diapause.

We arrive at this conclusion because RNAi used to knockdown any of the negative circadian regulators *period, timeless* or *cryptochrome2*, causes diapause-programmed mosquitoes to avert diapause (Meuti et al, 2015a). If any of these clock genes are disabled, the clock appears to malfunction, resulting in diapause aversion. Interestingly, knockdown of *period* in diapause-programmed females affects both ovarian development (increase size of oocytes to nondiapause status) as well as lipid deposition (decrease lipid accumulation to nondiapause status), but knockdown of either *timeless* or *cryptochrome2* affects oocyte size without affecting lipid acquisition. The minimum interpretation is that different attributes of the diapause phenotype may be attained through the action of different pathways leading from specific components of the clock. There is cross-talk between the clock genes as evidenced by the fact that knocking down *timeless* or *cryptochrome2* also reduces expression levels of *period*. The reverse, however, is not true: knocking down *period* does not influence expression of the other clock genes.

Knocking down *pigment-dispersing factor*, a clock-associated gene, causes females of *Cx. pipiens* programmed for nondiapause to enter a diapause-like state, as demonstrated by reduced ovarian growth and higher accumulation of fat (Meuti et al., 2015a). The protein encoded by *pigment-dispersing factor* helps to maintain synchronous cycling among individual circadian clock neurons (Shafer and Yao, 2014), and in the case of *Cx. pipiens* this may imply that synchrony among the neurons has been lost. The diapause decision can be altered in either direction by manipulation of the clock and clock-associated genes in *Cx. pipiens*, thus successful models for interpreting the programing of diapause will require understanding how the clock and the clock-associated gene *pigment-dispersing factor* interact and connect to the downstream hormonal signalling pathways implicated in the diapause decision.

In adult females of *Ae. albopictus*, the clock genes *timeless* and *cryptochrome 1* are upregulated under a diapause-inducing environment (Huang et al., 2015). Along with *timeless*, the clock gene *period* has been cloned and sequenced in *Ae. albopictus* (Summa et al., 2012), thus enabling future work on the relationship of the clock and a mosquito that enters an embryonic diapause. *Timeless* has also been implicated in larval diapause of the pitcher-plant mosquito, *W. smithii* (Mathias et al., 2005). Expression levels of *timeless* are higher in southern populations, suggesting a possible clock-related function in determining the observed variation in diapause response across a geographic gradient. Recent publication of the *W. smithii* transcriptome includes an impressive collection of circadian clock and related genes (Tormey et al., 2015), thus setting the stage for probing details of clock function in this important mosquito model for larval diapause.

7. HORMONAL CONTROL PATHWAYS

Little work has been done on the hormonal control of embryonic diapause and no such work has been reported for larval diapause in mosquitoes. Application of the JH analogue pyripoxyfen terminates diapause in eggs of *Ae. albopictus*, but exposure of larvae and adults to this same JH analogue does not prevent production of diapause eggs (Suman et al., 2015). Embryonic diapause is of special interest because some sort of maternal control is likely involved, as shown in embryonic diapauses of other insects (Denlinger et al., 2012). In the silkworm, *Bombyx mori*, the neuropeptide diapause hormone (DH) acts on the mother's ovaries to influence carbohydrate metabolism, which in turn influences the diapause fate of her eggs, but DH appears not to be present in Diptera, so there is no reason to suspect it is involved in mosquito diapause. Ecdysteroids play a critical role in embryonic diapause (pharate larval stage, as in mosquitoes) of the gypsy moth, *Lymantria dispar*: ecdysteroids from the mother are abundant in newly deposited eggs and a decline in the maternally packaged ecdysteroids is essential for diapause termination. As long as the ecdysteroid titer remains high, pharate larvae of the gypsy moth remain locked in diapause (Lee and Denlinger, 1997). In larval diapause the developmental halt is usually a consequence of the prothoracic gland's failure to produce the moult-inducing steroid hormone ecdysone, and this is the most likely scenario regulating larval diapause in mosquitoes as well.

Most cases of adult diapause are characterized as a JH deficiency syndrome (Denlinger et al., 2012), and this is also true for adult diapause in *Cx. pipiens*. In response to short-day signals, the corpora allata, the endocrine gland that produces JH, remain small and fail to synthesize JH (Kang et al., 2014; Readio et al., 1999). The fact that diapause can be terminated readily with application of JH or one of the JH analogues further demonstrates the role of JH as a regulator of diapause in *Cx. pipiens* (Spielman, 1974). Two neuropeptides, allatotropin and allatostatin, regulate output of JH from the corpus allatum: allatotropin promotes JH synthesis while the allatostatins exert an inhibitory role. Levels of mRNA encoding allatostatin C, the allatostatin present in *Cx. pipiens*, do not differ in diapausing and nondiapausing females, but the mRNA encoding allatotropin is much less abundant in diapausing females (Kang et al., 2014), suggesting that one of the important regulators of the corpora allata's synthetic activity is the upstream neuropeptide allatotropin. Thus, the endocrine axis leading from production of allatotropin in the brain to synthesis of JH by the corpora allata appears to be shut down during diapause. When diapause is broken, allatotropin is synthesized and released, leading to synthesis of JH by the corpora allata and the subsequent maturation of the ovaries, as well as initiation of blood feeding and other attributes of nondiapausing adults.

Growing evidence also points to a critical role for insulin signalling as a regulator of adult diapause in *Cx. pipiens* (Sim and Denlinger, 2008, 2013b), a scenario that is also true for dauer formation in the nematode *C. elegans* (Lee et al., 2001) as well as diapause regulation in several other insect species (Sim and Denlinger, 2013b). Knocking down *insulin receptor* (*InR*) or *insulin-like peptide-1* (*ilp*-1) with RNAi mimics the diapause state in females environmentally programmed by long daylengths for nondiapause development (Sim and Denlinger, 2008, 2009a). Such females fail to develop their ovaries and accumulate fat reserves characteristic of diapause. Like other invertebrates, *Cx. pipiens* has multiple ILPs, but knockdown of *ilp-1* has thus far been the only one showing this effect. Knockdown of two other *ilps* that are abundant in adult females of *Cx. pipiens*, *ilp-2* and *ilp-5*, does not alter the diapause response.

The fact that application of JH or a JH analogue can rescue females that are in a diapause-like state as a consequence of *InR* or *ilp-1* knockdown suggests that insulin signalling exerts a direct effect on JH synthesis. An interpretation consistent with these observations is that the environmental programming of diapause shuts down insulin signalling, which in turn has a negative effect on JH synthesis, as depicted in Fig. 4. Together a shutdown in both the JH and insulin signalling pathways results in manifestation of diapause. Whether insulin signalling impacts the corpora allata directly or acts by blocking synthesis, release or action of the neuropeptide allatotropin is not yet clear.

7.1 Foxo as a Candidate for Generating the Diapause Phenotype

Though substantial evidence points to JH and insulin signalling pathways as being critical for eliciting the diapause response in adult mosquitoes, the next question is how do these pathways generate the diapause phenotype? As discussed earlier, diapause is a complex phenotype, consisting of multiple behavioural, physiological and morphological responses. We argue that

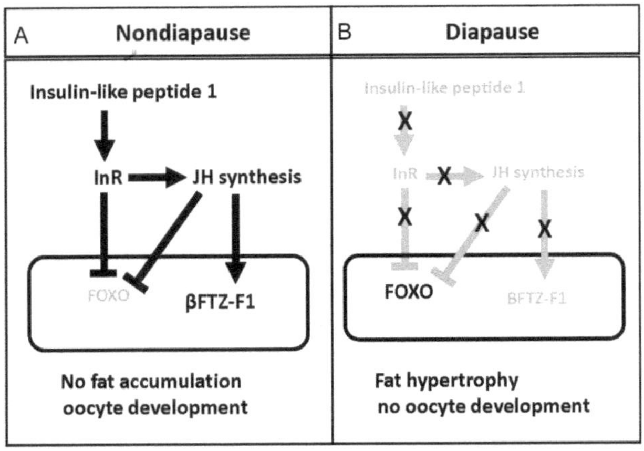

Fig. 4 The role of the insulin signalling pathway in regulating adult diapause in the mosquito *Culex pipiens.* (A) The nondiapause phenotype is prompted by long daylength, which promotes synthesis of insulin-like peptide 1, leading to stimulation of juvenile hormone (JH) synthesis and concurrent suppression of forkhead transcription factor (Foxo). (B) The diapause phenotype is promoted by short daylength and results in a shutdown of the insulin signalling pathway, failure to stimulate JH synthesis and concurrent release of the suppression of Foxo. Abbreviations: *InR*, insulin receptor; *bFTZ-F1*, beta nuclear receptor fushi tarazu factor 1. *Based on Sim, C., Denlinger, D.L., 2008. Insulin signaling and FOXO regulate the overwintering diapause of the mosquito* Culex pipiens. *Proc. Natl. Acad. Sci. U.S.A. 105, 6777–6781; Sim, C., Denlinger, D.L., 2009a. A shut-down in expression of an insulin-like peptide, ILP-1, halts ovarian maturation during the overwintering diapause of the mosquito* Culex pipiens. *Insect Mol. Biol. 18, 325–332; Sim, C., Denlinger, D.L., 2009b. Transcription profiling and regulation of fat metabolism genes in diapausing adults of the mosquito* Culex pipiens. *Physiol. Genomics. 39, 202–209; Sim, C., Denlinger, D.L., 2013a. Juvenile hormone III suppresses forkhead of transcription factor in the fat body and reduces fat accumulation in the diapausing mosquito* Culex pipiens. *Insect Mol. Biol. 22, 1–11; Sim, C., Denlinger, D.L., 2013b. Insulin signaling and the regulation of insect diapause. Front. Physiol. 4, 189; adapted from Denlinger, D.L., Armbruster, P.A., 2014. Mosquito diapause. Ann. Rev. Entomol. 59, 73–93.*

the forkhead transcription factor, Foxo, is a candidate regulator of these diverse responses (Sim et al., 2015).

The transcription factor Foxo is downstream of insulin, and when the transcript encoding Foxo is knocked down using RNAi, mosquitoes programmed environmentally by short daylength fail to accumulate fat and show the shortened lifespan characteristic of nondiapause (Sim and Denlinger, 2008). Genes that encode antioxidant enzymes catalase and superoxide dismutase-2 are expressed more highly in diapausing mosquitoes; knockdown of *foxo* in *Cx. pipiens* also knocks down expression of the genes encoding these two protective enzymes (Sim and Denlinger,

2011), suggesting that expression of both genes is regulated by Foxo. Foxo has many targets, and the use of ChIP-Seq makes it possible to identify other genes possibly targeted by Foxo. Over 70 potential targets have been identified (Sim et al., 2015), and among these, 10 have thus far been examined in more depth to verify specificity of Foxo binding and to demonstrate that these genes are transcriptionally regulated during diapause. Genes thus far identified as putative Foxo targets in *Cx. pipiens* include genes involved in stress tolerance, metabolic pathways, lifespan extension, cell cycle and growth regulation and circadian rhythms. These results are consistent with the idea that Foxo is linked to a network of genes that contribute to the diverse features of the diapause phenotype. Our model (Fig. 4) suggests that in the presence of insulin (ie, the nondiapause state) Foxo is suppressed, while in the absence of insulin (ie, the diapause state) Foxo is activated, leading to fat accumulation, extended lifespan, antioxidant enzyme production and shifts in metabolism and morphology that characterize diapause. Certainly additional pathways are also likely to be involved, but Foxo emerges as at least one major signalling pathway that can help explain the complexity of this diapause phenotype.

A connection between the JH signalling pathway and Foxo is also evident by the fact that JHIII, the native JH present in adult mosquitoes, suppresses expression of *foxo*, as depicted in Fig. 4, resulting in the nondiapause attribute of low fat accumulation (Sim and Denlinger, 2013a). Thus, it appears that Foxo can elicit its downstream effect in generating the diapause phenotype only when neither insulin nor JH are present.

8. THE MISSING PIECES

There remain a number of missing pieces in our understanding of mosquito diapause. Eventually, it will be exciting to trace the entire pathway from the clock genes that measure daylength to manifestation of the diapause phenotype. Pieces of the pathway have been defined. A functional clock is essential, and in *Cx. pipiens* this eventually leads to the interconnected JH and insulin signalling pathways, through Foxo, and possibly other transcription factors, to activate diverse features of the diapause phenotype. But, how does the clock feed into the JH and insulin signalling pathways? How is the clock information stored in the brain during the interval between late larval life and early adulthood to be acted upon by the young adult? Although the black box is shrinking, significant gaps remain to be filled.

Another intriguing but unanswered question is how maternal effects in mosquitoes regulate diapause in the next generation. This is especially prevalent in embryonic diapauses that are programmed by photoperiod the mother receives. What information is perceived and how is this information transferred from the mother to her embryos? The rising recognition that epigenetic processes are involved in many types of environmental responses, including diapause in some higher Diptera (eg, Reynolds et al., 2013), suggests that such levels of regulation are likely to be involved in embryonic and potentially other forms of mosquito diapause as well.

Equally unknown is how the mosquito determines how long to remain in diapause and what environmental or physiological signals are used to prompt diapause termination and the onset of development. Clearly, a fixed period of latency must be experienced before the diapausing mosquito, be it an embryo, larva or adult, is competent to again initiate development, but exactly what those processes are remains to be determined. Energetic reserves are limited and it is thus likely that the mosquito will 'decide' to terminate diapause before all reserves are consumed (Hahn and Denlinger, 2011), but how does the insect measure its energetic reserves and how does this set-point trigger diapause termination?

Furthermore, in spite of the huge investment of research on malaria-transmitting mosquitoes living in the tropics, it is still unclear how they bridge the dry seasons. The fact that a marked female of *An. coluzzii* (previously M-form of *An. gambiae*) survived a prolonged dry season in the Sahel certainly suggests an intact diapause response (Lehmann et al., 2010), and phenotypic differences in populations from dry and rainy seasons (Arcaz et al., 2016) lend further credence to some sort of dry season survival mechanism. Yet, until it is possible to elicit such responses by manipulating the mosquito in controlled environmental conditions, it will be difficult to conclude with certainty whether diapause is a normal component of the life cycle in this tropical disease vector.

To know how vector species are capable of carrying viruses such as West Nile and LaCrosse from one season to the next it is essential to understand the mechanisms of diapause. The diapausing mosquito appears capable of storing the virus, halting its replication and then reinitiating replication and transmission the following year. How this is achieved is vital for understanding the transmission cycles of major mosquito-borne diseases. Little is currently known about interactions between diapausing vectors and the viruses they harbour.

Finally, how can knowledge of the molecular physiology underpinning diapause be used to develop novel forms of vector control? Because diapause is a fundamental requirement for overwinter survival of temperate populations of mosquitoes, disrupting diapause would be a highly effective mechanism to reduce populations of vectors such as *Ae. albopictus* and *Cx. pipiens*. Additionally, because the diapause response involves modulation of a wide variety of fundamental physiological processes including stress resistance, energy storage and the hormonal control of development, studies of diapause-associated physiology are likely to identify a wide range of novel targets. Genetic approaches to vector control are becoming increasingly feasible (McGraw and O'Neill, 2013), and the increasingly widespread use of CRISPR/Cas9 technology for genome editing of nonmodel organisms suggests that genetic control of vector mosquitoes by targeting diapause-associated physiological processes may be feasible in the near future.

ACKNOWLEDGEMENTS

This work was supported in part by NIH Grant 1R15AI111328-01 to P.A.A. and by NIH Grant 2R56-AI058279 to D.L.D. We appreciate Megan Meuti's thoughtful comments on the manuscript.

REFERENCES

Anderson, J.F., 1968. Influence of photoperiod and temperature on the induction of diapause in *Aedes atropalpus* (Diptera, Culicidae). Entomol. Exp. Appl. 11, 321–330.

Andrewartha, H.G., 1952. Diapause in relation to the ecology of insects. Biol. Rev. 27, 50–107.

Arcaz, A.C., Huestis, D.L., Dao, A., Yaro, A.S., Diallo, M., Andersen, J., Blomquist, G.J., Lehmann, T., 2016. Desiccation tolerance in Anopheles coluzzii: the effects of spiracle size and cuticular hydrocarbons. J. Exp. Biol. http://dx.doi.org/10.1242/jeb.135665, Epub ahead of print.

Baird, N.A., Etter, P.D., Atwood, T.S., Currey, M.C., Shiver, A.L., Lewis, Z.A., Selker, E.U., Cresko, W.A., Johnson, E.A., 2008. Rapid SNP discovery and genetic mapping using sequenced RAD markers. PLoS One 3, e3376.

Bajgar, A., Jindra, M., Dolezel, D., 2013. Autonomous regulation of the insect gut by circadian genes acting downstream of juvenile hormone signaling. Proc. Natl. Acad. Sci. U.S.A. 110, 4416–4421.

Barnard, D.R., Mulla, M.S., 1978. The ecology of *Culiseta inornata* in the Colorado Desert of California: seasonal abundance, gonotrophic status, and oviparity of adult mosquitoes. Ann. Entomol. Soc. Am. 71, 397–400.

Beach, R.F., 1978. The required day number and timely induction of diapause in geographic strains of the mosquito, *Aedes atropalpus*. J. Insect Physiol. 24, 449–455.

Benoit, J.B., Denlinger, D.L., 2007. Suppression of water loss during adult diapause in the northern house mosquito, *Culex pipiens*. J. Exp. Biol. 210, 217–226.

Bowen, M.F., 1992. Patterns of sugar feeding in diapausing and nondiapausing *Culex pipiens* (Diptera, Culicidae). J. Med. Entomol. 29, 843–844.

Bradshaw, W.E., Holzapfel, C.M., 1977. Interaction between photoperiod, temperature and chilling in dormant larvae of the tree-hole mosquito, *Toxorhnchites rutilus* Coq. Biol. Bull. 152, 147–158.

Bradshaw, W.E., Lounibos, L.P., 1972. Photoperiodic control of development in the pitcher plant mosquito, *Wyeomyia smithii*. Can. J. Zool. 50, 713–719.

Bradshaw, W.E., Lounibos, L.P., 1977. Evolution of dormancy and its photoperiodic control in pitcher-plant mosquitoes. Evolution 31, 546–567.

Bradshaw, W.E., Armbruster, P., Holzapfel, C.M., 1998. Fitness consequences of hibernal diapause in the pitcher-plant mosquito, *Wyeomyia smithii*. Ecology 79, 1458–1462.

Bradshaw, W.E., Emerson, K.J., Catchen, J.M., Cresko, W.A., Holzapfel, C.M., 2012. Footprints in time: comparative quantitative trait loci mapping of the pitcher-plant mosquito, *Wyeomyia smithii*. Proc. R. Soc. B 279, 4551–4558.

Chen, X.G., Jiang, X.T., Gu, J.B., Xu, M., Wu, Y., et al., 2015. Genome sequence of the Asian tiger mosquito, *Aedes albopictus*, reveals insights into its biology, genetics, and evolution. Proc. Natl. Acad. Sci. U.S.A. 112, E5907–E5915.

Christopher, S., 1911. The development of the egg follicles in Anophelines. Paludism 1, 73–88.

Coluzzi, M., Di Deco, M., Gironi, A., 1975. The influence of photoperiod on the selection of oviposition sites in *Aedes mariae* (Diptera:Culicidae). Parassitologia 17, 121–130.

Copeland, R.S., Craig, G.B.J., 1989. Winter cold influences the spatial and age distributions of the North American treehole mosquito *Anopheles barberi*. Oecologia 79, 287–292.

Danks, H.V., 1987. Insect Dormancy: An Ecological Perspective. Biological Survey of Canada, Ottawa.

Dao, A., Yaro, A.S., Diallo, M., Timbine, S., Huestis, D.L., Kassogue, Y., Traore, A.I., Sanogo, Z.L., Samake, D., Lehmann, T., 2014. Signatures of aestivation and migration in Sahelian malaria mosquito populations. Nature 516, 387–390.

Denlinger, D.L., 1986. Dormancy in tropical insects. Annu. Rev. Entomol. 31, 239–264.

Denlinger, D.L., 1991. Relationship between cold hardiness and diapause. In: Lee, R.E., Denlinger, D.L. (Eds.), Insects at Low Temperature. Chapman & Hall, New York, NY, pp. 174–198.

Denlinger, D.L., 2002. Regulation of diapause. Annu. Rev. Entomol. 47, 93–122.

Denlinger, D.L., Armbruster, P.A., 2014. Mosquito diapause. Annu. Rev. Entomol. 59, 73–93.

Denlinger, D.L., Giebultowicz, J., Adedokun, T., 1988. Insect diapause: dynamics of hormone sensitivity and vulnerability to environmental stress. In: Sehnal, F.A., Zabza, A., Denlinger, D.L. (Eds.), Endocrinological Frontiers in Physiological Insect Ecology. Wroclaw Technical University Press, Wroclaw, Poland, pp. 309–324.

Denlinger, D.L., Yocum, G.D., Rinehart, J.P., 2012. Hormonal control of diapause. In: Gilbert, L.I. (Ed.), Insect Endocrinology. Academic Press, San Diego, pp. 430–463.

Eldridge, B.F., 1968. The effect of temperature and photoperiod on blood-feeding and ovarian development in mosquitoes of *Culex pipiens* complex. Am. J. Trop. Med. Hyg. 17, 133–140.

Emerson, K.J., Bradshaw, W.E., Holzapfel, C.M., 2009. Complications of complexity: integrating environmental, genetic and hormonal control of insect diapause. Trends Genet. 25, 217–225.

Emerson, K.J., Bradshaw, W.E., Holzapfel, C.M., 2010. Microarrays reveal early transcriptional events during the termination of larval diapause in natural populations of the mosquito, *Wyeomyia smithii*. PLoS One 5, e9574.

Evans, K.W., Brust, R.A., 1972. Induction and termination of diapause in *W. smithii* (Diptera:Culicidae), and larval survival studies at low and sub-zero temperatures. Can. Entomol. 104, 1937–1950.

Fonseca, D.M., Kaplan, L.R., Heiry, R.A., Strickman, D., 2015. Density-dependent oviposition by female *Aedes albopictus* (Diptera:Culicidae) spreads eggs among containers during the summer but accumulates them in the fall. J. Med. Entomol. 52, 705–712.

Gentile, C., Rivas, G.B., Meireles-Filho, A.C., Lima, J.B., Peixoto, A.A., 2009. Circadian expression of clock genes in two mosquito disease vectors: cry2 is different. J. Biol. Rhythms 24, 444–451.

Giraldo-Calderon, G.I., Emrich, S.J., MacCallum, R.M., Maslen, G., Dialynas, E., Topalis, P., Ho, N., Gesing, S., Madey, G., Collins, F.H., Lawson, D., VectorBase, C., 2015. VectorBase: an updated bioinformatics resource for invertebrate vectors and other organisms related with human diseases. Nucleic Acids Res. 43, D707–D713.

Goto, S.G., Doi, K., Nakayama, S., Numata, H., 2008. Maternal control of cold and desiccation tolerance in eggs of the band-legged ground cricket *Dianemobius nigrofasciatus* in relation to embryonic diapause. Entomol. Res. 38, 17–23.

Goto, S.S., Shiga, S., Numata, H., 2010. Photoperiodism in insects: perception of light and the role of clock genes. In: Nelson, R.J., Denlinger, D.L., Somers, D.E. (Eds.), Photoperiodism: The Biological Calendar. Oxford University Press, Oxford, pp. 258–286.

Hahn, D.A., Denlinger, D.L., 2011. Energetics of insect diapause. Annu. Rev. Entomol. 56, 103–121.

Hanson, S.M., Craig, G.B.J., 1995. Relationship between cold hardiness and supercooling point in *Aedes albopictus* eggs. J. Am. Mosq. Control Assoc. 11, 35–38.

Hawley, W.A., 1988. The biology of *Aedes albopictus*. J. Am. Mosq. Control Assoc. 4, 1–39.

Huang, X., Poelchau, M.F., Armbruster, P., 2015. Global transcriptional dynamics of diapause induction in non-blood-fed and blood-fed *Aedes albopictus*. PLoS Negl. Trop. Dis. 9, e0003724.

Huestis, D.L., Lehmann, T., 2014. Ecophysiology of *Anopheles gambiae* s.l.: persistence in the Sahel. Infect. Genet. Evol. 28, 648–661.

Huestis, D.L., Yaro, A.S., Traore, A.I., Dieter, K.L., Nwagbara, J.I., Bowie, A.C., Adamou, A., Kassogue, Y., Diallo, M., Timbine, S., Dao, A., Lehmann, T., 2012. Seasonal variation in metabolic rate, flight activity and body size of *Anopheles gambiae* in the Sahel. J. Exp. Biol. 215, 2013–2021.

Hurlbut, H.S., 1938. Further notes on the overwintering of the eggs of *Anopheles walkeri* with a description of the eggs. J. Parasitol. 24, 521–526.

Kang, D.S., Denlinger, D.L., Sim, C., 2014. Suppression of allatotropin simulates reproductive diapause in the mosquito *Culex pipiens*. J. Insect Physiol. 64, 48–53.

Kang, D.S., Cotten, M.A., Denlinger, D.L., Sim, C., 2016. Comparative transcriptomics reveals key gene expression differences between diapausing and nondiapausing adults of Culex pipiens. PLoS One 11, e0154892.

Kim, M., Denlinger, D.L., 2009. Decrease in *expression* of *beta-tubulin* and microtubule abundance in flight muscles during diapause in adults of *Culex pipiens*. Insect Mol. Biol. 18, 295–302.

Kim, M., Denlinger, D.L., 2010. A potential role for ribosomal protein S2 in the gene network regulating reproductive diapause in the mosquito *Culex pipiens*. J. Comp. Physiol. B. 180, 171–178.

Kim, M., Robich, R.M., Rinehart, J.P., Denlinger, D.L., 2006. Upregulation of two actin genes and redistribution of actin during diapause and cold stress in the northern house mosquito, *Culex pipiens*. J. Insect Physiol. 52, 1226–1233.

Kim, M., Sim, C., Denlinger, D.L., 2010. RNA interference directed against ribosomal protein S3a suggests a link between this gene and arrested ovarian development during adult diapause in *Culex pipiens*. Insect Mol. Biol. 19, 27–33.

Kostal, V., 2006. Eco-physiological phases of insect diapause. J. Insect Physiol. 52, 113–127.

Kostal, V., Tollarova, M., Dolezel, D., 2008. Dynamism in physiology and gene diapause in a heteropteran transcription during reproductive bug, *Pyrrhocoris apterus*. J. Insect Physiol. 54, 77–88.
Kostal, V., Simunkova, P., Kobelkova, A., Shimada, K., 2009. Cell cycle arrest as a hallmark of insect diapause: Changes in gene transcription during diapause induction in the drosophilid fly, *Chymomyza costata*. Insect Biochem. Mol. Biol. 39, 875–883.
Lacour, G., Vernichon, F., Cadilhac, N., Boyer, S., Lagneau, C., Hance, T., 2014. When mothers anticipate: Effects of the prediapause stage on embryo development time and of maternal photoperiod on eggs of a temperate and a tropical strains of *Aedes albopictus* (Diptera:Culicidae). J. Insect Physiol. 71, 87–96.
Lee, R.E., Denlinger, D.L., 1991. Insects at Low Temperature. Chapman and Hall, New York.
Lee, K.-Y., Denlinger, D.L., 1997. A role for ecdysteroids in the induction and maintenance of the pharate first instar diapause of the gypsy moth, *Lymantria dispar*. J. Insect Physiol. 43, 289–296.
Lee, R.Y., Hench, J., Ruykun, G., 2001. Regulation of *C. elegans* DAF-16 and its human ortholog FKHRL1 by the daf-2 insulin-like signaling pathways. Curr. Biol. 11, 1950–1957.
Lehmann, T., Dao, A., Yaro, A.S., Adamou, A., Kassogue, T., et al., 2010. Aestivation of the African malaria mosquito, *Anopheles gambiae*, in the Sahel. Am. J. Trop. Med. Hyg. 83, 601–606.
Li, A., Denlinger, D.L., 2009. Pupal cuticle protein is abundant during early adult diapause in the mosquito *Culex pipiens*. J. Med. Entomol. 46, 1382–1386.
Lounibos, L.P., Bradshaw, W.E., 1975. A second diapause in *Wyeomyia smithii*: seasonal incidence and maintenance by photoperiod. Can. J. Zool. 53, 215–221.
Mathias, D., Jacky, L., Bradshaw, W.E., Holzapfel, C.M., 2005. Geographic and developmental variation in expression of the circadian rhythm gene, *timeless*, in the pitcher-plant mosquito, *Wyeomyia smithii*. J. Insect Physiol. 51, 661–667.
Mathias, D., Jacky, L., Bradshaw, W.E., Holzapfel, C.M., 2007. Quantitative trait loci associated with photoperiodic response and stage of diapause in the pitcher-plant mosquito, *Wyeomyia smithii*. Genetics 176, 391–402.
McElwee, J.J., Schuster, E., Blanc, E., Thornton, J., Gems, D., 2006. Diapause-associated metabolic traits reiterated in long-lived *daf-2* mutants in the nematode *Caenorhabditis elegans*. Mech. Ageing Dev. 127, 922–936.
McGraw, E.A., O'Neill, S.L., 2013. Beyond insecticides: new thinking on an ancient problem. Nat. Rev. Microbiol. 11, 181–193.
Means, R.G., 1979. Mosquitoes of New York: Part 1. The genus Aedes (Meigen) with identification keys to genera of Culicidae, 430a Bulletin New York State Museum, Albany, NY.
Meuti, M.E., Denlinger, D.L., 2013. Evolutionary links between circadian clocks and photoperiodic diapause in insects. Integr. Comp. Biol. 53, 131–143.
Meuti, M.E., Stone, M., Ikeno, T., Denlinger, D.L., 2015a. Functional circadian clock genes are essential for the overwintering diapause of the Northern house mosquito, *Culex pipiens*. J. Exp. Biol. 218, 412–422.
Meuti, M.E., Short, C.A., Denlinger, D.L., 2015b. Mom matters: diapause characteristics of *Culex pipiens-Cx. quinquefasciatus* (Diptera:Culicidae) hybrid mosquitoes. J. Med. Entomol. 52, 131–137.
Mitchell, C.J., 1983. Differentiation of host-seeking behavior from blood-feeding behavior in overwintering *Culex pipiens* (Diptera, Culicidae) and observations on gonotrophic dissociation. J. Med. Entomol. 20, 157–163.
Mitchell, C.J., Briegel, H., 1989. Fate of bloodmeal in force-fed, diapausing *Culex pipiens* (Diptera, Culicidae). J. Med. Entomol. 26, 332–341.

Mohammed, A.R., Coombes, A.G.A., Perrie, Y., 2007. Amino acids as cryoprotectants for liposomal delivery systems. Eur. J. Pharm. Sci. 30, 406–413.

Mori, A., Romero-Severson, J., Severson, D.W., 2007. Genetic basis for reproductive diapause is correlated with life history traits within the *Culex pipiens* complex. Insect Mol. Biol. 16, 515–524.

Niimi, T., Yamashita, O., Yaginuma, T., 1993. A cold-inducible Bombyx gene encoding a protein similar to mammalian sorbitol dehydrogenase—yolk nuclei-dependent gene-expression in diapause eggs. Eur. J. Biochem. 213, 1125–1131.

Overgaard, J., Malmendal, A., Sorensen, J.G., Bundy, J.G., Loeschcke, V., Nielsen, N.C., Holmstrup, M., 2007. Metabolomic profiling of rapid cold hardening and cold shock in *Drosophila melanogaster*. J. Insect Physiol. 53, 1218–1232.

Parker, B.M., 1985. Effects of photoperiod on the induction of embryonic diapause in *Aedes taeniorhynchus* (Diptera:Culicidae). J. Med. Entomol. 22, 392–397.

Pinger, R.R., Eldridge, B.F., 1977. The effect of photoperiod on diapause induction in *Aedes canadensis* and *Psorophora ferox* (Diptera:Culicidae). Ann. Entomol. Soc. Am. 70, 437–441.

Poelchau, M.F., Reynolds, J.A., Denlinger, D.L., Elsik, C.G., Armbruster, P.A., 2011. A *de nova* transcriptome of the Asian tiger mosquito, *Aedes albopictus*, to identify candidate transcripts for diapause preparation. BMC Genomics 12, 619.

Poelchau, M.F., Reynolds, J.A., Elsik, C.G., Denlinger, D.L., Armbruster, P.A., 2013a. Deep sequencing reveals complex mechanisms of diapause preparation in the invasive mosquito, *Aedes albopictus*. Proc. R. Soc. B 280, 20130143.

Poelchau, M.F., Reynolds, J.A., Denlinger, D.L., Elsik, C.G., Armbruster, P.A., 2013b. Transcriptome sequencing as a platform to elucidate molecular components of the diapause response in *Aedes albopictus*. Physiol. Entomol. 38, 173–181.

Poelchau, M.F., Reynolds, J.A., Elsik, C.G., Denlinger, D.L., Armbruster, P.A., 2013c. RNA-Seq reveals early distinctions and late convergence of gene expression between diapause and quiescence in the Asian tiger mosquito, *Aedes albopictus*. J. Exp. Biol. 216, 4082–4090.

Pumpuni, C.B., 1989. Factors influencing photoperiodic control of egg diapause in Aedes albopictus (Skuse). Ph.D. thesis, University of Notre Dame.

Pumpuni, C.B., Knepler, J., Craig, G.B., 1992. Influence of temperature and larval nutrition on the diapause inducing photoperiod of *Aedes albopictus*. J. Am. Mosq. Control Assoc. 14, 223–227.

Ragland, G.J., Denlinger, D.L., Hahn, D.A., 2010. Mechanisms of suspended animation are revealed by transcript profiling of diapause in the flesh fly. Proc. Natl. Acad. Sci. U.S.A. 17, 14909–14914.

Ragland, G.J., Egan, S.P., Feder, J.L., Berlocher, S.H., Hahn, D.A., 2011. Developmental trajectories of gene expression reveal candidates for diapause termination: a key life-history transition in the apple maggot fly *Rhagoletis pomonella*. J. Exp. Biol. 214, 3948–3959.

Readio, J., Meola, R., 1985. Two stages of juvenile hormone-mediated growth of secondary follicles in *Culex pipiens*. J. Insect Physiol. 31, 559–562.

Readio, J., Chen, M.H., Meola, R., 1999. Juvenile hormone biosynthesis in diapausing and nondiapausing *Culex pipiens* (Diptera:Culicidae). J. Med. Entomol. 36, 355–360.

Revel, F.G., Herwig, A., Garidou, M.L., Dardente, H., Menet, J.S., Masson-Peet, M., Simonneaux, V., Saboureau, M., Pevet, P., 2007. The circadian clock stops ticking during deep hibernation in the European hamster. Proc. Natl. Acad. Sci. U.S.A. 104, 13816–13820.

Reynolds, J.A., Poelchau, M.F., Rahman, Z., Armbruster, P., Denlinger, D.L., 2012. Transcript profiling reveals mechanisms for lipid conservation during diapause in the mosquito, *Aedes albopictus*. J. Insect Physiol. 58, 966–973.

Reynolds, J.A., Clark, J., Diakoff, S.J., Denlinger, D.L., 2013. Transcriptional evidence for small RNA regulation of pupal diapause in the flesh fly, *Sarcophaga bullata*. Insect Biochem. Mol. Biol. 43, 982–989.

Rinehart, J.P., Robich, R.M., Denlinger, D.L., 2006. Enhanced cold and desiccation tolerance in diapausing adults of *Culex pipiens*, and a role for *Hsp70* in response to cold shock but not as a component of the diapause program. J. Med. Entomol. 43, 713–722.

Rinehart, J.P., Li, A., Yocum, G.D., Robich, R.M., Hayward, S.A.L., Denlinger, D.L., 2007. Up-regulation of heat shock proteins is essential for cold survival during insect diapause. Proc. Natl. Acad. Sci. U.S.A. 104, 11130–11137.

Robich, R.M., Denlinger, D.L., 2005. Diapause in the mosquito *Culex pipiens* evokes a metabolic switch from blood feeding to sugar gluttony. Proc. Natl. Acad. Sci. U.S.A. 102, 15912–15917.

Robich, R.M., Rinehart, J.P., Kitchen, L.J., Denlinger, D.L., 2007. Diapause-specific gene expression in the northern house mosquito, *Culex pipiens* L., identified by suppressive subtractive hybridization. J. Insect Physiol. 53, 235–245.

Sanburg, L.L., Larsen, J.R., 1973. Effect of photoperiod and temperature on ovarian development in *Culex pipiens pipiens*. J. Insect Physiol. 19, 1173–1190.

Shafer, O.T., Yao, Z., 2014. Pigment-dispersing factor signaling and circadian rhythms in insect locomotor activity. Curr. Opin. Insect Sci. 1, 73–80.

Shin, S.W., Zou, Z., Saha, T.T., Raikhel, A.S., 2012. bHLH-PAS heterodimer of methoprene-tolerant and Cycle mediates circadian expression of juvenile hormone-induced mosquito genes. Proc. Natl. Acad. Sci. U.S.A. 109, 16576–16581.

Sim, C., Denlinger, D.L., 2008. Insulin signaling and FOXO regulate the overwintering diapause of the mosquito *Culex pipiens*. Proc. Natl. Acad. Sci. U.S.A. 105, 6777–6781.

Sim, C., Denlinger, D.L., 2009a. A shut-down in expression of an insulin-like peptide, ILP-1, halts ovarian maturation during the overwintering diapause of the mosquito *Culex pipiens*. Insect Mol. Biol. 18, 325–332.

Sim, C., Denlinger, D.L., 2009b. Transcription profiling and regulation of fat metabolism genes in diapausing adults of the mosquito *Culex pipiens*. Physiol. Genomics 39, 202–209.

Sim, C., Denlinger, D.L., 2011. Catalase and superoxide dismutase-2 enhance survival and protect ovaries during overwintering diapause in the mosquito *Culex pipiens*. J. Insect Physiol. 57, 628–634.

Sim, C., Denlinger, D.L., 2013a. Juvenile hormone III suppresses forkhead of transcription factor in the fat body and reduces fat accumulation in the diapausing mosquito *Culex pipiens*. Insect Mol. Biol. 22, 1–11.

Sim, C., Denlinger, D.L., 2013b. Insulin signaling and the regulation of insect diapause. Front. Physiol. 4, 189.

Sim, C., Kang, D.S., Kim, S., Bai, X., Denlinger, D.L., 2015. Identification of FOXO targets that generate the diverse features of the diapause phenotype in the mosquito *Culex pipiens*. Proc. Natl. Acad. Sci. U.S.A. 112, 3811–3816.

Spielman, A., 1974. Effect of synthetic juvenile hormone on ovarian diapause ofCulex pipiens mosquitoes. J. Med. Entomol. 11, 223–225.

Spielman, A., Wong, J., 1973. Environmental control of ovarian diapause in *Culex pipiens*. Ann. Entomol. Soc. Am. 66, 905–907.

Suman, D.S., Wang, Y., Gaugler, R., 2015. The insect growth regulator pyriproxyfen terminates egg diapause in the Asian tiger mosquito, *Aedes albopictus*. PLoS One 10, 12.

Summa, K., Urbanski, J.M., Zhao, X., Poelchau, M.F., Armbruster, P., 2012. Cloning and sequence analysis of the circadian clock genes *period* and *timeless* in the Asian tiger mosquito, *Aedes albopictus* (Diptera:Culicidae). J. Med. Entomol. 49, 777–782.

Tammariello, S.P., Denlinger, D.L., 1998. G0/G1 cell cycle arrest in the brain of *Sarcophaga crassipalpis* during pupal diapause and the expression pattern of the cell cycle regulator, *proliferating cell nuclear antigen*. Insect Biochem. Mol. Biol. 28, 83–89.

Tauber, M.J., Tauber, C.A., Masaki, S., 1986. Seasonal Adaptations of Insects. Oxford University Press, New York.

Tormey, D., Colbourne, J.K., Mockaitis, K., Choi, J.-H., Lopez, J., Burkhart, J., Bradshaw, W., Holzapfel, C., 2015. Evolutionary divergence of core and post-translational circadian clock genes in the pitcher-plant mosquito, *Wyeomyia smithii*. BMC Genomics 16, 754.

Urbanski, J.M., Benoit, J.B., Michaud, M.R., Denlinger, D.L., Armbruster, P., 2010. The molecular physiology of increased egg desiccation resistance during diapause in the invasive mosquito, *Aedes albopictus*. Proc. R. Soc. B 277, 2683–2692.

Vinogradova, E.B., 2000. *Culex Pipiens Pipiens* Mosquitoes: Taxonomy, Distribution Ecology, Physiology, Genetics, Applied Importance and Control. Pensoft Publisher, Sofia.

Wagoner, K.M., Lehmann, T., Huestis, D.L., Ehrmann, B.M., Cech, N.B., Wasserberg, G., 2014. Identification of morphological and chemical markers of dry- and wet-season conditions in female *Anopheles gambiae* mosquitoes. Parasit. Vectors 7, 13.

Yamashita, O., 1996. Diapause hormone of the silkworm, *Bombyx mori*: structure, gene expression and function. J. Insect Physiol. 42, 669–679.

Yaro, A.S., Traore, A., Huestis, D.L., Adamou, A., Timine, S., Kassogue, Y., Diallo, M., Dao, A., Troare, S.F., Lehmann, T., 2012. Dry season reproductive depression of *Anopheles gambiae* in the Sahel. J. Insect Physiol. 58, 1050–1059.

Zhang, Q., Denlinger, D.L., 2011a. Elevated *couch potato* transcripts associated with adult diapause in the mosquito *Culex pipiens*. J. Insect Physiol. 57, 620–627.

Zhang, Q., Denlinger, D.L., 2011b. Molecular structure of the prothoracicotropic hormone gene in the northern house mosquito, *Culex pipiens*, and its expression analysis in association with diapause and blood feeding. Insect Mol. Biol. 20, 201–213.

Zhou, G., Miesfeld, R.L., 2009. Energy metabolism during diapause in *Culex pipiens* mosquitoes. J. Insect Physiol. 55, 40–46.

CHAPTER TWELVE

Nitrogen Metabolism in Mosquitoes

New Insights into the Nitrogen Metabolism in Blood-Fed Mosquitoes

N. Petchampai, P.Y. Scaraffia

Vector-Borne Infectious Disease Research Center, School of Public Health and Tropical Medicine, Tulane University, New Orleans, LA, United States

Contents

Abstract

Female mosquitoes metabolize nitrogen compounds by exploiting metabolic pathways that were unpredicted based on our current knowledge of mammalian nitrogen metabolism. Within the last decade, researchers have begun to study mosquito metabolism with isotopically labelled compounds, mass spectrometry and RNA interference. The application of classical and cutting-edge techniques to discover cross-talk signalling mechanisms between metabolic pathways has added layers of complexity, as well as fascinating insight, into our understanding of nitrogen metabolism. This chapter provides an overview of the most recent findings reported in mosquitoes with the emphasis on *Aedes aegypti* females. The implementation of integrated approaches to elucidate

Advances in Insect Physiology, Volume 51
ISSN 0065-2806
http://dx.doi.org/10.1016/bs.aiip.2016.04.002

the multiplicity of factors that govern nitrogen metabolism, which could be useful for uncovering potential metabolic targets for mosquito control, is also discussed in this chapter.

ABBREVIATIONS

ALAT alanine aminotransferase
ALLC allantoicase
ALLN allantoinase
DsRNA double-stranded RNA
GDH glutamate dehydrogenase
GltS glutamate synthase
GS glutamine synthetase
LCS L-cycloserine
MRM multiple reaction monitoring
MS mass spectrometry
P5CR pyrroline-5-carboxylate reductase
P5CS pyrroline-5-carboxylate synthase
Rh50-1 *Rhesus* 50 glycoprotein
RNAi RNA interference
UO urate oxidase
XDH xanthine dehydrogenase

1. INTRODUCTION

Mosquitoes are undoubtedly the most devastating vectors of human diseases causing almost a million deaths annually (World Health Organization, 2016). *Aedes aegypti* is one of the mosquito species that can transmit several viruses including dengue (Choumet and Desprès, 2015), yellow fever (Barrett and Higgs, 2007; Beasley et al., 2015), chikungunya (Horwood and Buchy, 2015) and Zika, the latest virus that has recently become a global health threat (Attar, 2016; Fauci and Morens, 2016). Despite worldwide attempts to control mosquitoes, the rise in insecticide-resistant mosquitoes and industrialization that results in climate change and increased international travel has made the current strategies insufficient for controlling these vectors. A better understanding of mosquito metabolism is crucial for the discovery of novel targets that can be used in the development of more powerful tools to control mosquito populations and eradicate the diseases they transmit.

A. aegypti females take a vertebrate blood meal that could be almost twice their own weight. This represents a tremendous metabolic and osmotic challenge due to the excessive amount of water, ions and nutrients present in a single blood meal. Blood-fed females undergo a massive diuresis, during which excess ions and water are excreted at high rates (Benoit et al., 2014; Beyenbach, 2003; Beyenbach and Piermarini, 2011; Drake et al., 2015; Larsen et al., 2014; Pacey and O'Donnell, 2014; Williams et al., 1983). Additionally, an extensive variety of transcripts are differentially expressed in *A. aegypti* tissues in response to blood feeding (Bonizzoni et al., 2011; Hou et al., 2015; Price et al., 2011; Roy et al., 2015; Sanders et al., 2003). Midgut epithelial cells form a peritrophic matrix, which not only provides a spatial organization of digestive processes, but also serves as a physical and biochemical barrier to abrasive food particles, pathogens and ingested toxins (Hegedus et al., 2009; Shao et al., 2001). Within the peritrophic matrix, blood digestion takes place and the proteins are catabolized into peptides and amino acids by the action of endo- and exopeptidases. Genes encoding digestive enzymes have been broadly studied in the midgut of *A. aegypti* females (Table 1).

Since 1991 a total of 12 serine proteases, 11 carboxypeptidases and 5 aminopeptidases have been identified, but only a few have been biochemically characterized (Barillas-Mury et al., 1991; Bian et al., 2008; Brackney et al.,

Table 1 Digestive Proteases Identified in the Midgut of *A. aegypti*

Proteases	Accession #	Predicted Enzymatic Activity	References
Serine proteases			
AaLT	M77814	Collagenase-like	Barillas-Mury et al. (1991), Brackney et al. (2010), Isoe et al. (2009a), and Rascón et al. (2011)
AaET	X64362	Trypsin-like	Brackney et al. (2010), Isoe et al. (2009a), Kalhok et al. (1993), Noriega et al. (1996), and Rascón et al. (2011)
Aa5G1	X64363	Trypsin-like	Kalhok et al. (1993)
AaCHYMO	U56423	Chymotrypsin-like	Jiang et al. (1997)
AaJA15	AY957559	Chymotrypsin-like	Bian et al. (2008)
AaSP I	GQ398043	Trypsin-like	Brackney et al. (2010) and Isoe et al. (2009a)
AaSP II	GQ398044	Trypsin-like	Brackney et al. (2010) and Isoe et al. (2009a)

Continued

Table 1 Digestive Proteases Identified in the Midgut of *A. aegypti*—cont'd

Proteases	Accession #	Predicted Enzymatic Activity	References
AaSP III	GQ398045	Trypsin-like	Brackney et al. (2010) and Isoe et al. (2009a)
AaSP IV	GQ398046	Trypsin-like	Brackney et al. (2010) and Isoe et al. (2009a)
AaSP V	GQ398047	Trypsin-like	Brackney et al. (2010) and Isoe et al. (2009a)
AaSP VI	GQ398048	Trypsin-like	Brackney et al. (2010), Isoe et al. (2009a), and Rascón et al. (2011)
AaSP VII	GQ398049	Trypsin-like	Brackney et al. (2010), Isoe et al. (2009a), and Rascón et al. (2011)
Carboxypeptidases			
CPA-I	AY590487	Carboxypeptidase A	Edwards et al. (2000) and Isoe et al. (2009b)
CPA-II	AY590488	Carboxypeptidase A	Isoe et al. (2009b)
CPA-III	AY590489	Carboxypeptidase A	Isoe et al. (2009b)
CPA-V	AY590491	Carboxypeptidase A	Isoe et al. (2009b)
CPA-VI	AY590492	Carboxypeptidase A	Isoe et al. (2009b)
CPA-VII	AY590493	Carboxypeptidase A	Isoe et al. (2009b)
CPB-I	AY590494	Carboxypeptidase B	Isoe et al. (2009b)
CPB-II	AY590495	Carboxypeptidase B	Isoe et al. (2009b)
CPB-III	AY590496	Carboxypeptidase B	Isoe et al. (2009b)
CPB-IV	EF423586	Carboxypeptidase B	Isoe et al. (2009b)
CPB-V	EF423587	Carboxypeptidase B	Isoe et al. (2009b)
Aminopeptidases			
APN	AF378117	Aminopeptidase N	
APN-II	AY785154	Aminopeptidase N	
APN-III	AY785155	Aminopeptidase N	
APN-IV	AY785156	Aminopeptidase N	
PN-V	AY785157	Aminopeptidase N	

2010; Edwards et al., 2000; Isoe et al., 2009a,b; Jiang et al., 1997; Kalhok et al., 1993; Noriega et al., 1996; Rascón et al., 2011). Structural and kinetic studies on these proteases could reveal unique catalytic properties that can be useful for the identification of mosquito-selective inhibitors capable of disrupting proteolysis. With the availability of the complete genome sequence in *A. aegypti* (Nene et al., 2007) and the successful application of RNA interference (RNAi) to knockdown genes, several factors including protein components in the coatomer protein I vesicle transport system (Isoe et al., 2011), mosquito insulin receptor (Gulia-Nuss et al., 2011) and microRNAs (Bryant et al., 2010; Liu et al., 2014; Lucas et al., 2015) have been identified as critical factors involved in controlling key functions during blood digestion and vitellogenesis. Additionally, the effects of ingested vertebrate-derived factors on mosquito immune response (Pakpour et al., 2014 and herein references) have added new insights into our limited knowledge on the signalling mechanisms that govern multiple vital events in blood-fed females and pathogen transmission.

Through the intensive digestive process, female mosquitoes absorb dietary nutrients, synthesize new compounds and efficiently eliminate waste (Briegel, 1985, 1986; Clements, 1992; Lehane, 2005; Zhou et al., 2005, 2007). Transport of nutrients and waste is a topic that has been extensively investigated in recent years but has not yet been fully uncovered (Attardo et al., 2006; Benoit et al., 2014; Boudko et al., 2015; Carpenter et al., 2012; Evans et al., 2009; Hansen et al., 2011; Harvey et al., 2009; Weihrauch et al., 2012). Degradation of haemoglobin, the most abundant protein in blood meal, results in the release of a high concentration of haem, which can be toxic to the cells (Graça-Souza et al., 2006). In mosquitoes the majority of haem released from haemoglobin catabolism is bound to the peritrophic matrix and later excreted to ameliorate haem toxicity (Pascoa et al., 2002). A recent study in the *A. aegypti* cell line Aag2 demonstrated the capability of haem to modulate genes related to redox metabolism indicating that haem acts as a harmful signalling molecule in mosquito cells (Bottino-Rojas et al., 2015).

Although it is well-known that anautogenous female mosquitoes need a blood meal to secure nutrients for egg production, it is remarkable that only 10% of the amino acids released from blood digestion are incorporated into eggs (4% as proteins and 6% as lipids); approximately 20% of them are utilized to synthesize maternal reserves: carbohydrates (2%), lipids (8%) and proteins (10%), whereas 70% of the amino acids produced during blood protein catabolism are completely oxidized for energy production and excreted

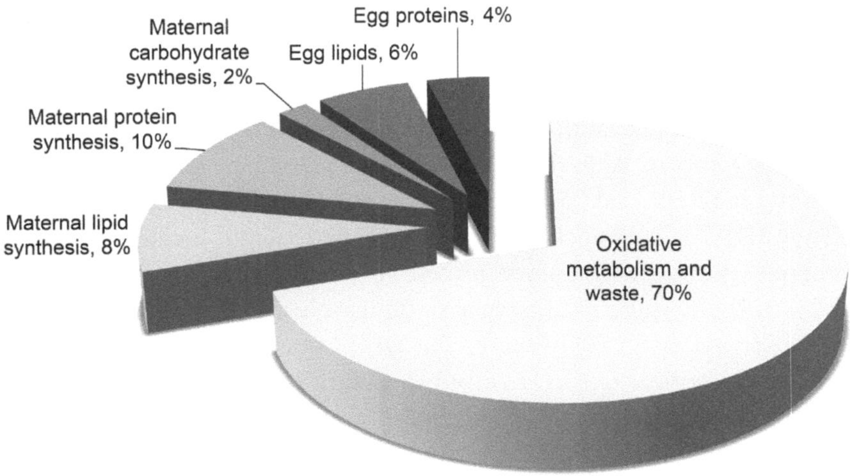

Fig. 1 Schematic representation of the fate of blood meal protein amino acids at the end of a gonotrophic cycle in *A. aegypti* mosquitoes. *Data obtained from Zhou, G., Flowers, M., Friedrich, K., Horton, J., Pennington, J., Wells, M.A., 2004. Metabolic fate of [^{14}C]-labeled meal protein amino acids in* Aedes aegypti *mosquitoes. J. Insect Physiol. 50, 337–349.* (See the colour plate.)

as CO_2 and other waste (Fig. 1, see also Zhou et al., 2004). This remarkable amino acid oxidation causes a massive deamination that results in a huge production of ammonia. The term ammonia is defined here as NH_3, NH_4^+ or a combination of the two. Ammonia cannot be stored inside cells, and therefore nitrogen balance must be regulated to prevent ammonia toxicity. In the absence of a urea cycle (Krebs and Henseleit, 1932) for ammonia disposal, mosquitoes can directly excrete or metabolize ammonia efficiently into other nitrogen compounds as discussed in the next sections of this chapter.

An overview of the most recent discoveries in nitrogen metabolism in mosquitoes with emphasis on *A. aegypti* females is presented in this chapter. The implementation of traditional and innovative approaches for studying ammonia metabolism and its mechanisms of regulation in mosquitoes is also discussed.

2. RECENT DISCOVERIES IN NITROGEN METABOLISM OF *A. aegypti* MOSQUITOES BY USING INTEGRATED APPROACHES

A. aegypti females have evolved efficient mechanisms for coping with the toxic ammonia concentrations that are released during a blood meal's digestion. This is a critical detoxification process. If this process is somehow disrupted, the ammonia levels could increase to lethal concentrations in the tissues and cause mortality in the females. A summary of the most relevant findings about how mosquitoes metabolize excess nitrogen, specifically ammonia, is discussed below.

2.1 Proline Serves as a Temporary Sink of Nitrogen and as a Source of Energy

It was proposed that *A. aegypti* females utilize Pro, the predominant amino acid in the haemolymph of adult mosquitoes (Henn et al., 1998; Su and Mulla, 1997; Uchida et al., 1990), as a temporary nitrogen sink to store ammonia (Fig. 2) arising from deamination of blood meal protein amino acids (Goldstrohm et al., 2003). This Pro role was later demonstrated by using different approaches that involve unlabelled and ^{15}N-labelled ammonium chloride solutions and specific inhibitors (Scaraffia et al., 2005, 2006, 2008, 2010). It was also found that Pro powers *A. aegypti* flight through a Pro-Ala cycle (Fig. 3), even though it was thought that mosquitoes only use carbohydrates as a source of energy during flight (Scaraffia and Wells, 2003). The Pro-Ala cycle was first discovered in tsetse flies (Bursell,

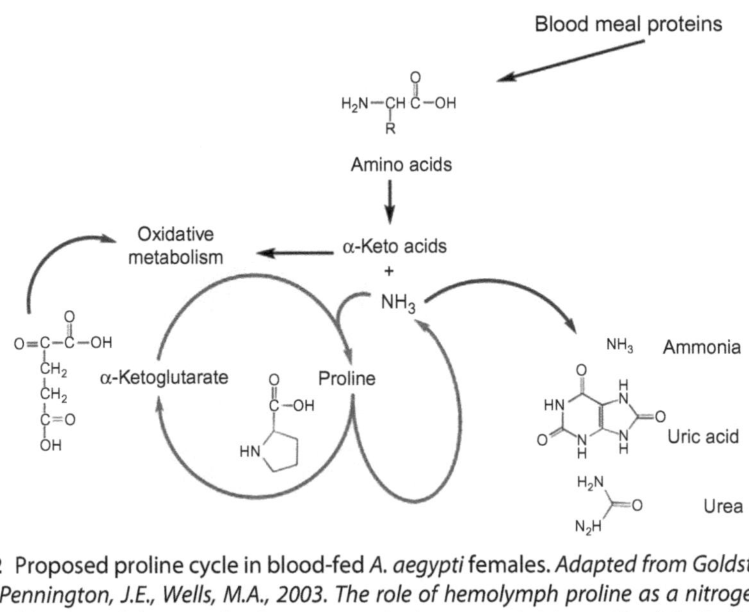

Fig. 2 Proposed proline cycle in blood-fed *A. aegypti* females. *Adapted from Goldstrohm, D.A., Pennington, J.E., Wells, M.A., 2003. The role of hemolymph proline as a nitrogen sink during blood meal digestion by the mosquito* Aedes aegypti. *J. Insect Physiol. 49, 115–121.*

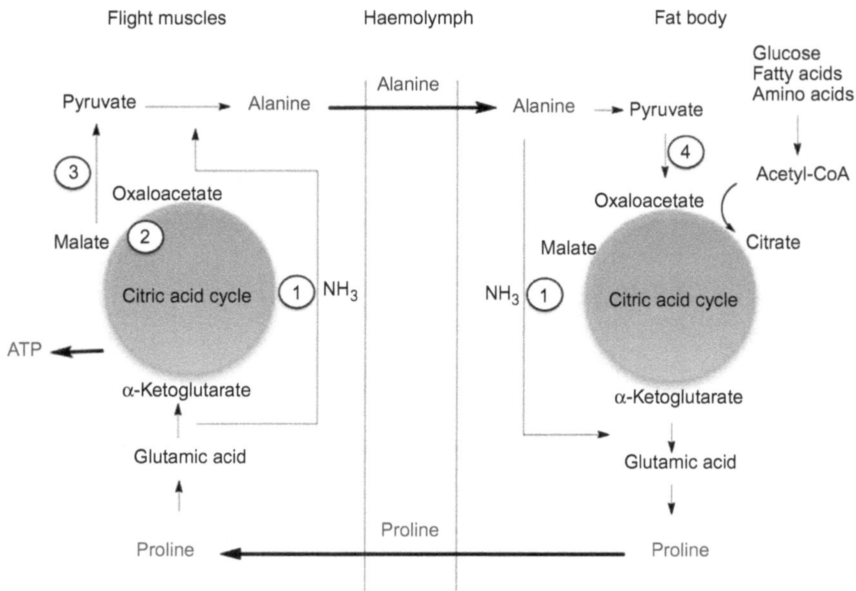

Fig. 3 Schematic representation of proline-alanine cycle in *A. aegypti* females. Key enzymes: (1) alanine aminotransferase, (2) malate dehydrogenase, (3) NAD-linked "malic" enzyme, (4) pyruvate carboxylase. *Reproduced with permission from Scaraffia, P.Y., Wells, M.A., 2003. Proline can be utilized as an energy substrate during flight of* Aedes aegypti *females. J. Insect Physiol. 49, 591–601.*

1966, 1975, 1977) and later in other insects (Gäde and Auerswald, 2002 and herein references). When Pro serves as a source of energy in mosquitoes, Ala and Gln are both involved in shuttling amino groups between flight muscles and the fat body to avoid toxic ammonia effects. Enzymes related to amino acid metabolism go through significant changes in activity levels in the cephalothorax and thorax of both males and females after the moult, indicating that the capacity to metabolize Pro is not fully developed in immature stages (Scaraffia and Wells, 2003). Additionally, significant changes of Pro concentrations in the haemolymph and thorax of female mosquitoes during flight, as well as high activities of the enzymes involved in Pro metabolism in thorax, fat body and midgut from both adult males and females are consistent with Pro serving as a shuttle to move acetyl units between tissues during flight to support Pro-Ala cycle (Scaraffia and Wells, 2003). These findings are also well-correlated with recent data about respiration mediated by Pro oxidation in the mitochondria of *A. aegypti* flight muscles (Soares et al., 2015).

2.2 Glutamine Synthetase and Glutamate Synthase (GS/GltS) Pathway Is Active

When *A. aegypti* females are given access to sucrose solutions containing ammonium chloride, Gln and Pro concentrations increase markedly in the haemolymph (Scaraffia et al., 2005), suggesting that mosquitoes are able to metabolize ammonia mainly through the synthesis of these two amino acids along with the excretion of ammonia, uric acid and urea. A mechanism for ammonia fixation and assimilation in mosquitoes involving GS and GltS was suggested based on inhibitor studies, and the fat body was implicated as the main tissue involved in the ammonia detoxification in mosquito females (Scaraffia et al., 2005). GltS is an amidotransferase that catalyses the interconversion of Gln and α-ketoglutarate into two Glu molecules. An NADH-dependent GltS was cloned from *A. aegypti* and sequenced. This sequence encodes an open reading frame of 2084 amino acids and constitutes the first sequence of GltS obtained from a metazoan (Scaraffia et al., 2006). Its expression was analysed in fat body and midgut from sugar- and blood-fed *A. aegypti* females (Scaraffia et al., 2005). The existence and importance of a GS/GltS pathway in mosquitoes was later established by using a modern approach: isotopically labelled compounds and mass spectrometry (MS) techniques. The fragmentation mechanism of derivatized Gln (dimethylformamidine glutamine isobutyl ester) was first investigated (Zhang et al., 2005). Then an MS method was developed and optimized in order to identify and quantify both ^{15}N-labelled and unlabelled compounds at a series of different neutral losses by performing selected reaction monitoring, also known as multiple reaction monitoring (MRM) scans in a triple–quadrupole mass spectrometer (Scaraffia et al., 2006, 2008, 2010).

2.3 *A. aegypti* Mosquitoes Detoxify Ammonia Through Three Phases

To better understand how the N from ammonia is incorporated into different nitrogen compounds, whole-body ammonia metabolism in *A. aegypti* was analysed over many time points post $^{15}NH_4Cl$ feeding, which varied between 0 and 96 h. Mosquitoes were immersed in liquid nitrogen and then processed for MS analysis (Fig. 4). For the detection of unlabelled [^{14}N]-Pro and labelled [^{15}N]-Pro, samples were derivatized as isobutyl esters, whereas for the identification of unlabelled [^{14}N]-Glu, [^{14}N]-Ala and [2-5-$^{14}N_2$]-Gln, and labelled [^{15}N]-Glu, [^{15}N]-Ala, [5-^{15}N]-Gln, [2-^{15}N]-Gln and [2-5-$^{15}N_2$]-Gln, mosquito samples were derivatized as

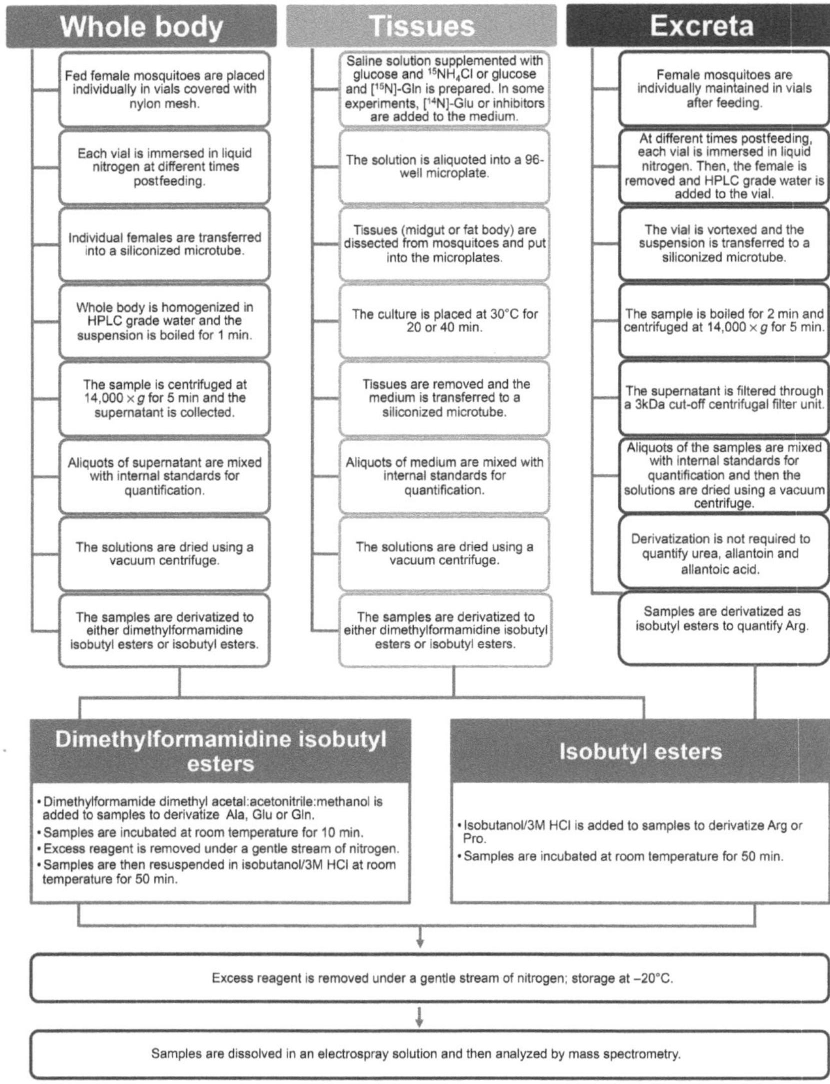

Fig. 4 A simplified protocol of mosquito whole body, tissue and excreta preparation for mass spectrometry analysis. *Detailed mosquito sample preparation is described in Bush, D.R., Wysocki, V.H., Scaraffia, P.Y., 2012. Study of the fragmentation of arginine isobutyl ester applied to arginine quantification in* Aedes aegypti *mosquito excreta. J. Mass Spectrom. 47, 1364–1371; Isoe, J., Scaraffia, P.Y., 2013. Urea synthesis and excretion in* Aedes aegypti *mosquitoes are regulated by a unique cross-talk mechanism. PLoS One 8, e65393; Scaraffia, P.Y., Zhang, Q., Thorson, K., Wysocki, V.H., Miesfeld, R.L., 2010. Differential ammonia metabolism in* Aedes aegypti *fat body and midgut tissues. J. Insect Physiol. 56, 1040–1049; Scaraffia, P.Y., Tan, G., Isoe, J., Wysocki, V.H., Wells, M.A., Miesfeld, R.L., 2008. Discovery of an alternate metabolic pathway for urea synthesis in adult* Aedes aegypti *mosquitoes. Proc. Natl. Acad. Sci. U.S.A. 105, 518–523; Scaraffia, P.Y., Zhang, Q., Wysocki, V.H., Isoe, J., Wells, M.A., 2006. Analysis of whole body ammonia metabolism in* Aedes aegypti *using [^{15}N]-labeled compounds and mass spectrometry. Insect Biochem. Mol. Biol. 36, 614–622.*

dimethylformamidine isobutyl esters. The quantification of amino acids was performed by MRM using a specific neutral loss that occurs with these amino acid derivatives. Thus, for the quantification of Pro, Glu and Ala, a neutral loss of 102 Da was used, whereas for the quantification of unlabelled or labelled Gln, a neutral loss of 73 or 74 Da was used depending on whether the ^{15}N label is involved in the neutral loss or not (Table 2). Through MRM, multiple analytes are analysed in the same experiment with high sensitivity and selectivity by detecting the precursor ion-product ion pairs that are specific to each analyte (Hager and Yves Le Blanc, 2003; Kitteringham et al., 2009; Mazzalupo and Scaraffia, 2014; Wang, 2009; Xia et al., 2003; Yost and Enke, 1979).

The results obtained from the kinetics of ^{15}N incorporation from $^{15}NH_4Cl$ into several amino acids in whole body (Scaraffia et al., 2006), tissues (Scaraffia et al., 2010) and excreta from mosquitoes (Scaraffia et al., 2008) show that ammonia detoxification occurs in three phases: fixation, assimilation and excretion (Fig. 5). During the fixation phase, the $^{15}NH_3$ is rapidly incorporated into Gln by GS. This reaction fixes labelled ammonia ($^{15}NH_3$) into unlabelled [^{14}N]-Glu to yield [5-^{15}N]-Gln, which accumulates in large amounts during the first hour postfeeding. The assimilation phase turns on after the fixation phase starts and remains active throughout the entire time course (0–96 h). During the assimilation phase, part of the ^{15}N from [5-^{15}N]-Gln is metabolized by GltS to produce [^{15}N]-Glu, which is mainly converted to [^{15}N]-Pro and to a lesser extent to [^{15}N]-Ala. The [^{15}N]-Glu produced by GltS can also be used by GS to fix another molecule of $^{15}NH_3$ to produce [2, 5-$^{15}N_2$]-Gln, which can also be converted into [^{15}N]-Glu through GltS (Scaraffia et al., 2006). The $^{15}NH_3$ is also fixed to α-ketoglutarate by glutamate dehydrogenase (GDH) to produce [^{15}N]-Glu, which is mainly converted to [^{15}N]-Pro by pyrroline-5-carboxylate synthase (P5CS) and pyrroline-5-carboxylate reductase (P5CR). The presence of [^{15}N]-Ala indicates that [^{15}N]-Glu can be metabolized by alanine aminotransferase (ALAT). The use of specific inhibitors of GS and GltS corroborated that GS/GltS enhances the production of Glu to Pro. To explore how mosquito tissues respond to an ammonia challenge, fat body and midgut from *A. aegypti* females were exposed to 1 mM $^{15}NH_4Cl$ *in vitro* and analysed over different time points (Fig. 4). It was found that the fat body mainly detoxifies ammonia through the GS/GltS pathway, whereas the midgut mainly uses GS, GDH and ALAT (Scaraffia et al., 2010). It stands to reason that the coordinated aminations, transamidation and transamination reactions maintain the nitrogen homeostasis to avoid a

Table 2 Identification and Quantification of Amino Acids and Nitrogen Waste in Mosquitoes by Multiple Reaction Monitoring Scans

Nitrogen Compounds	Precursor Ion	Neutral Loss	Fragment Ion
Amino acids			
[^{14}N]-Ala, [^{15}N]-Ala, **[^{2}H$_4$]-Ala**	201*, 202*, 205*	102	99, 100, 103
[^{14}N]-Glu, [^{15}N]-Glu, **[^{2}H$_5$]-Glu**	315*, 316*, *320*	102	213, 214, 218
[^{14}N]-Pro, [^{15}N]-Pro, **[^{2}H$_7$]-Pro**	172**, 173**, 179**	102	70, 71, 77
[^{14}N]-Ala, [^{15}N]-Ala, **[^{2}H$_4$]-Ala**	201*, 202*, 205*	102	99, 100, 103
[2,5,^{14}N$_2$]-Gln	258*	73	185
[5,^{15}N]-Gln	259*	74	185
[2,^{15}N]-Gln	259*	73	186
[2,5-^{15}N$_2$]-Gln	260*	74	186
[^{2}H$_5$]-Gln	263*	73	190
[^{14}N$_4$]-Arg	231**	161, 115	70, 116
[^{15}N$_2$]-Arg	233**	163, 117	70, 116
Nitrogen waste			
[^{14}N$_4$]-Uric acid	169	17, 28, 43	152, 141, 126
[1,3-^{15}N$_2$]-Uric acid	171	18, 28, 44	153, 143, 127
[^{14}N$_4$]-Allantoin	159	43	116
[^{14}N$_4$]-Allantoic acid	177	43	134
[^{15}N]-UL-50%-Allantoin	161	43, 44	118, 117
[^{14}N$_2$]-Urea	61	17	44
[^{15}N]-Urea	62	17,18	45,44
[^{15}N$_2$]-Urea	63	18	45
[^{13}C-^{15}N$_2$]-Urea	64	18	46

*Unlabelled and labelled dimethylformamidine isobutyl ester of amino acids.
**Isobutyl ester of amino acids.
Internal standards are indicated in bold.
Reproduced with permission from Mazzalupo, S., Scaraffia, P.Y., 2014. Application of isotopically labeled compounds and tandem mass spectrometry for studying metabolic pathways in mosquitoes. In: Chandrasekar, R., Tyagi, B.K., Gui, Z.Z., Reeck, G.R. (Eds.), Short View on Insect Biochemistry and Molecular Biology. K-State Union, Kansas State University/International Book Mission, Manhattan, pp. 99–126.

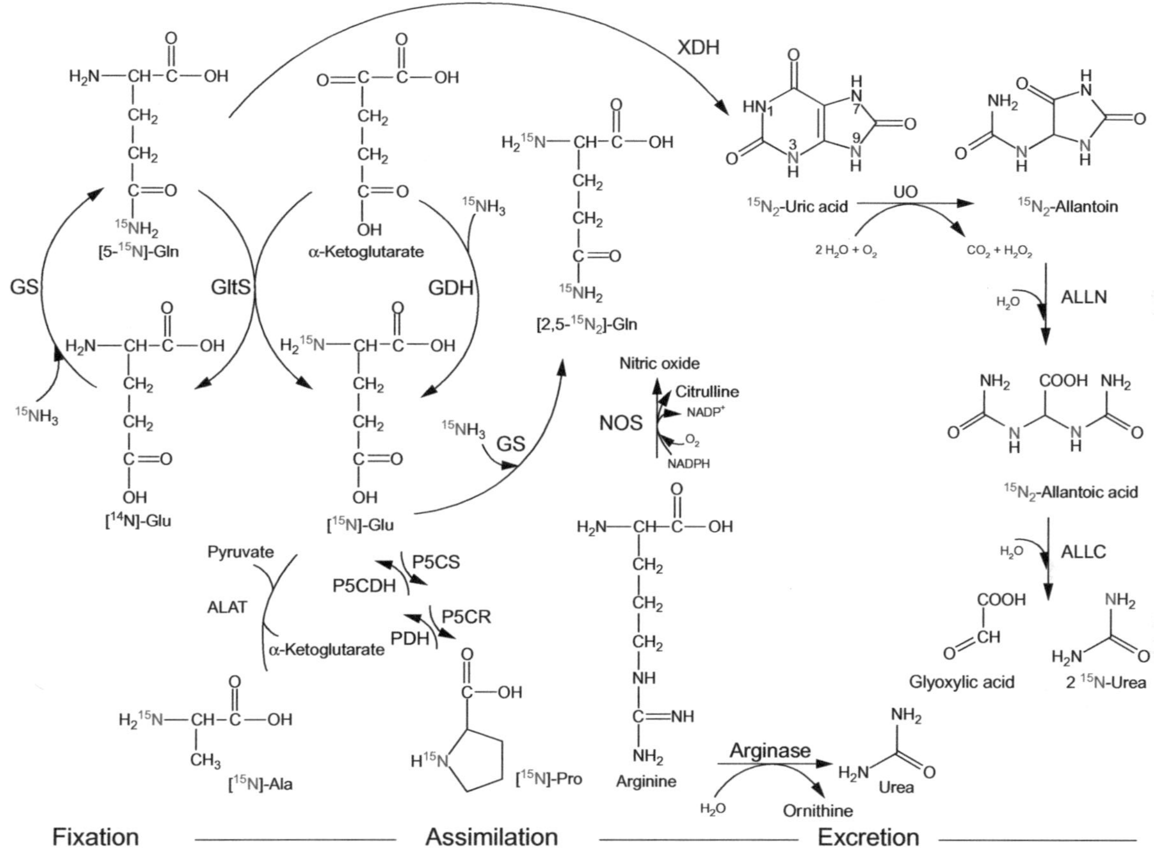

Fig. 5 See legend on next page.

lethal increase of ammonia in the mosquito tissues. The data reported also suggest that glucose plays an important role during ammonia detoxification (Scaraffia et al., 2010). To investigate this further, ^{13}C-isotopically labelled compounds and MS techniques have been used (Jiang et al., 2010; Ma et al., 2013). The [5-^{15}N]-Gln that is not metabolized by GltS can have other metabolic fates such as uric acid synthesis. Mosquito excreta sample preparation for MS analysis is illustrated in Fig. 4. Indeed, *A. aegypti* females can utilize the nitrogen of the amide group of two [5-^{15}N]-Gln molecules to synthesize one [3-9-$^{15}N_2$]-uric acid molecule (Scaraffia et al., 2008) following the metabolic pathways that terrestrial vertebrates and some insects use for synthesizing uric acid (Barrett and Friend, 1970; Levenberg et al., 1956; Sonne et al., 1956). Ammonia, haem and uric acid, along with urea and some amino acids (Briegel, 1986; von Dungern and Briegel, 2001a,b), are some of the nitrogen compounds that are excreted in mosquitoes, as well as in other insects (Cochran, 1985; O'Donnell, 2009; Singer, 2003; Wright, 1995).

Studies performed with MS and RNAi show that [$^{15}N_2$]-uric acid in *A. aegypti* can be excreted without modification, but also it can further be metabolized into [$^{15}N_2$]-allantoin, [$^{15}N_2$]-allantoic acid and two [^{15}N]-urea molecules (Scaraffia et al., 2008). Quantification of uric acid and the products of its degradation were performed by means of MRM. For the identification of uric acid, neutral losses of 17, 28 and 43 Da were utilized; for the quantification of allantoin and allantoic acid, a neutral loss of 43 Da was used, whereas for the quantification of urea, a neutral loss of 17 or 18 Da was employed (Table 2). Uric acid, allantoin, allantoic acid and urea have different solubilities in water, which suggests that the synthesis and excretion of these compounds in mosquitoes are dependent on water availability.

Fig. 5 Multiple metabolic pathways involved in *A. aegypti* ammonia metabolism. Pathways were identified by using isotopically labelled compounds, mass spectrometry, specific inhibitors and RNAi. Abbreviations used in this figure: GS, glutamine synthetase; GltS, glutamate synthase; GDH, glutamate dehydrogenase; ALAT, alanine aminotransferase; P5CS, pyrroline-5-carboxylate synthase; P5CR, pyrroline-5-carboxylate reductase; P5CDH, pyrroline-5-carboxylate dehydrogenase; PDH, proline dehydrogenase; XDH, xanthine dehydrogenase; NOS, nitric oxide synthase; UO, urate oxidase; ALLN, allantoinase; ALLC, allantoicase. *Reproduced with permission from Scaraffia, P.Y., 2016. Disruption of mosquito blood meal protein metabolism. In: Adelman, Z.N. (Ed.), Genetic Control of Malaria and Dengue. Elsevier Academic Press, San Diego, pp. 253–275.* (See the colour plate.)

2.4 An Amphibian-Like Metabolic Pathway for Urea Synthesis Is Functional

Based on studies reported in terrestrial vertebrates such as mammals, it was assumed that urea present in the mosquito excreta was produced by converting the excess dietary Arg into urea and ornithine through the reaction catalysed by arginase. However, by using a combination of ^{15}N-isotopically labelled compounds, inhibitors, bioinformatic analysis of *A. aegypti* genome sequence, and RNAi-mediated silencing of urate oxidase (UO) expression, it was demonstrated that a portion of this urea excreted is derived from an amphibian-like metabolic pathway (Hayashi et al., 2000), also known as uricolytic pathway. It was found that UO, allantoinase (ALLN), allantoicase (ALLC) are enzymes involved in the metabolic degradation of uric acid in mosquitoes (Scaraffia et al., 2008). In Malpighian tubules from *A. aegypti* female mosquitoes, UO and ALLN transcripts reach the maximum level at 24 h after blood feeding, whereas the maximum level of ALLC transcript is observed between 36 and 48 h after blood feeding. Additionally, induced expression of the UO transcript in response to an ammonia meal is observed in fat body and Malpighian tubules, at 1 and 6 h after feeding, respectively (Scaraffia et al., 2008). Interestingly, it was recently reported that in *Culex quinquefasciatus* whole body the level of UO transcript, also called uricase, is highly expressed at 20 h after blood feeding, which decreased by 60 h after feeding (Reid et al., 2015).

Silencing of UO by RNAi causes increased uric acid concentration in the whole body of *A. aegypti* mosquitoes after feeding a blood or [^{15}N]-ammonia meal. Moreover, a significant decrease of [^{15}N]-urea concentration in mosquito excreta is also observed (Scaraffia et al., 2008). These data demonstrate that the genes of the uricolytic pathway express functional enzymes in *A. aegypti* mosquitoes, and that urea is the end product of the pathway (Scaraffia et al., 2008). This finding has high significance since the production of urea in insects had only been attributed to arginase. Now, it is known that in fact, *A. aegypti* mosquitoes have two metabolic pathways for urea synthesis: argininolysis and uricolysis (Fig. 5, excretion phase). Both pathways contribute to the urea pool that is excreted. However, in the first case, unlike terrestrial vertebrates where Arg is synthesized in the urea cycle (Krebs and Henseleit, 1932), the action of arginase is limited to Arg from dietary sources or from endogenous protein turnover. This is because mosquitoes lack the gene encoding ornithine transcarbamylase, also called carbamoyltransferase (Zdobnov et al., 2002), which makes Arg an essential amino acid. In the second case, urea is derived from the catabolism of uric acid, which serves as a temporary sink for excess ammonia produced during the digestion of the blood.

Although it is possible that other mosquito species and perhaps other haematophagous insects use similar biochemical and molecular strategies to metabolize excess nitrogen as *A. aegypti*, the information available to date is still too scarce to verify that assumption. Recently, it was reported that arginase, as well as the uricolytic pathway present in *A. aegypti*, are putatively absent in the genome of haematophagous *Rhodnius prolixus*, the major vector of Chagas disease (Mesquita et al., 2015). These findings are well-correlated with the fact that no urea excretion has been reported in any triatomine insects to date. Orthologs for many of the genes discussed in this chapter are present in the primary mosquito vector responsible for malaria, *Anopheles gambiae*. Sequence identity between *A. aegypti*, *Drosophila melanogaster* and *An. gambiae*, as well as GenBank accession numbers were reported for several enzymes involved in ammonia metabolism (Holt et al., 2002; Isoe and Scaraffia, 2013; Marinotti et al., 2006; Mazzalupo et al., 2016; Nene et al., 2007; Ribeiro, 2003; Scaraffia et al., 2005, 2008; Severson et al., 2004; Zdobnov et al., 2002). Regarding genes encoding ammonia transporters, in *A. aegypti* two ammonia transporters, *Rhesus* 50 glycoproteins (AaRh50-1 and AaRh50-2), were cloned and sequenced (GenBank accession numbers: AY926463 and AY926464) but they have not yet been fully characterized (Mazzalupo et al., 2016). The transcripts of two ammonia transporters, AgAmt and AgRh50, are highly expressed in the antennae of *An. gambiae* (Pitts et al., 2014). Their transcripts were also found in whole bodies of both females and males and therefore it was postulated that both transporters play important roles in the peripheral chemosensory signal transduction pathways, as well as in ammonia clearance (Pitts et al., 2014). In the Asian tiger mosquito, *A. albopictus*, the ammonia transporter AalRh50 was cloned, and its expression also analysed in head, thorax, fat body, midgut, Malpighian tubules and ovary dissected from sugar- and blood-fed mosquitoes (Wu et al., 2010). Immunohistochemistry studies indicated that AalRh50 is mainly located in head, thorax, fat body and Malpighian tubules. An interesting temporal and spatial correlation between the expression of AalRh50 and several genes involved in *A. aegypti* ammonia metabolism was discussed, suggesting that both mosquito species could detoxify ammonia similarly (Wu et al., 2010). Comparative alignments of AalRh50 with other organisms including *A. aegypti*, *C. quinquefasciatus* and *An. gambiae* were also analysed by Wu et al. (2010). Based on transcriptome analysis, it was recently proposed that the pathways described in *A. aegypti* for ammonia detoxification could also be present in *A. albopicus* (Esquivel et al., 2014). However, further molecular, biochemical and physiological studies are needed to better understand how different mosquito species avoid ammonia toxicity.

3. NOVEL FINDINGS ON THE METABOLIC REGULATION OF AMMONIA METABOLISM IN MOSQUITOES

Although we now know that multiple metabolic pathways are involved in ammonia detoxification in *A. aegypti*, little is known about how these metabolic pathways are regulated.

3.1 Argininolysis and Uricolysis Are Regulated by a Unique Cross-Talk Signalling Mechanism

To investigate how urea synthesis and excretion in mosquitoes are regulated through argininolysis and uricolysis, the fragmentation patterns of Arg isobutyl ester were first studied by electrospray ionization tandem MS. Thus, Arg fragmentation pathways not previously described were characterized, and an MS method was optimized for a rapid, sensitive and accurate measurement of Arg in individual mosquito excreta by MRM scans (Bush et al., 2012; Isoe and Scaraffia, 2013). Arg concentrations were monitored in mosquito excreta from sugar- and blood-fed mosquitoes by using neutral losses of 161 and 115 Da (see Fig. 4 and Table 2). Arg is not present in sugar-fed female excreta. At 1 and 6 h after blood feeding, only a minimal Arg concentration is present in the blood-fed female excreta. However, by 12 h after feeding the Arg value is approximately 20 nmol/female. The concentration reaches its highest value between 36 and 48 h after feeding (about 60 nmol/female) and then remains constant through 120 h (Bush et al., 2012). These results demonstrate that Arg can be directly excreted during a gonotrophic cycle contributing to the disposal of excess nitrogen in blood-fed female mosquitoes.

By studying the metabolic regulation of urea synthesis by RNAi and MS techniques, it was found that arginase transcript levels in fat body and Malpighian tubules increase significantly when UO gene expression is silenced and *vice versa* (Fig. 6). However, the urea concentration in the excreta of blood-fed females injected with dsRNA against arginase or UO remains unchanged compared with dsRNA firefly luciferase-injected control mosquitoes. Knockdown of arginase increases allantoin and Arg concentration

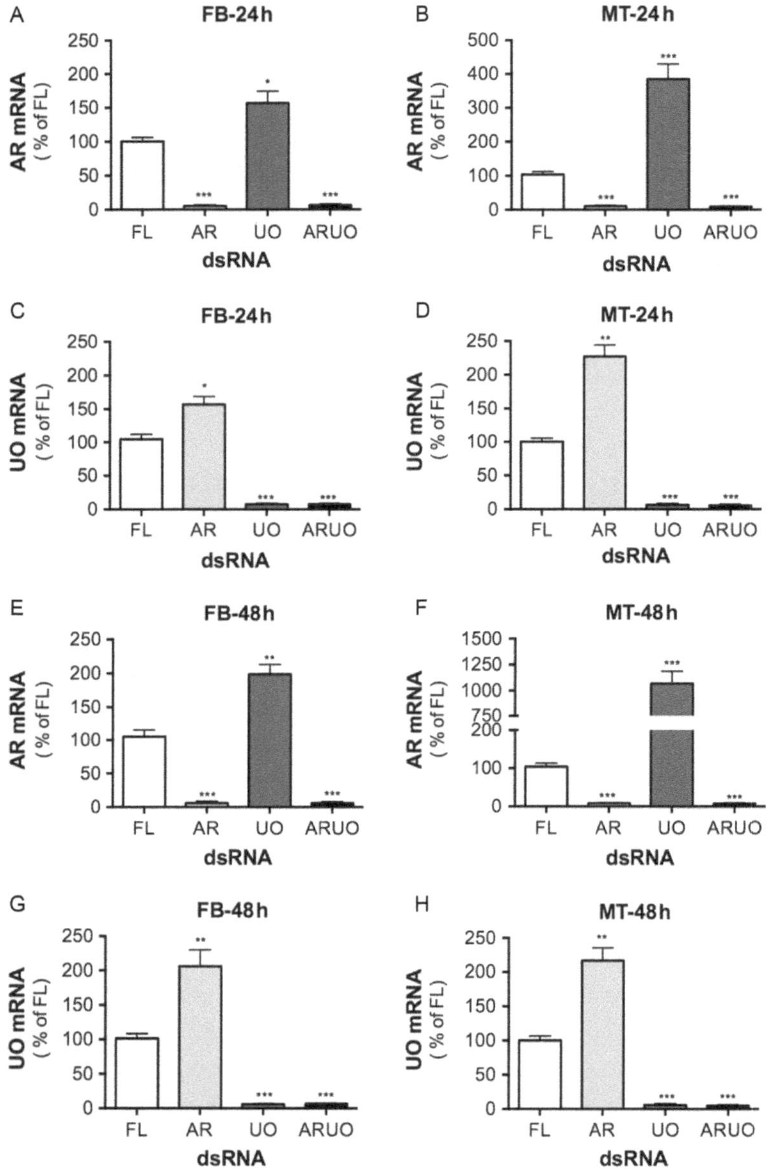

Fig. 6 Effect of arginase and/or urate oxidase knockdown on gene expression. *A. aegypti* females were injected with double-stranded RNA (dsRNA) firefly luciferase (FL), arginase (AR), urate oxidase (UO) or both (ARUO), and then fed a blood meal. (A–H) Relative abundance of AR and UO mRNA in mosquito fat body (FB) and Malpighian tubules (MT) of dsRNA-injected females at 24 h (A–D) and 48 h (E–H) after blood feeding. Data are expressed as the mean ± SEM of five independent samples. $^{*}p < 0.05$, $^{**}p < 0.001$, $^{***}p < 0.0001$ (when compared to dsRNA-FL by ANOVA). *Reproduced with permission from Isoe, J., Scaraffia, P.Y., 2013. Urea synthesis and excretion in* Aedes aegypti *mosquitoes are regulated by a unique cross-talk mechanism. PLoS One 8, e65393.*

in the excreta, whereas knockdown of UO decreases allantoin concentration. In correlation with these data, the silencing of both arginase and UO mainly leads to a decrease in urea and allantoin excretion, and an increase in Arg excretion, indicating that a cross-talk regulation between both argininolysis and uricolysis occurs in mosquito tissues (Isoe and Scaraffia, 2013). Furthermore, arginase and nitric oxide synthase inhibition increases UO transcript expression in tissues and increases urea, allantoin and Arg excretion. A transient delay in blood digestion and in the uptake of vitellogenin is also observed in mosquitoes with arginase and UO deficiency. In addition, it was reported that silencing of argininolysis and uricolysis decreases the expression of genes encoding GS1, GS2, GltS, GDH, P5CR1, P5CR2, P5CR3, XDH1 and XDH2 without affecting ALAT1, ALAT2 and P5CS transcript levels. Thus, the balanced operation of both argininolysis and uricolysis controls the synthesis and excretion of urea, as well as other nitrogen waste (Isoe and Scaraffia, 2013). These unexpected findings uncovered a new mode of metabolic regulation, not previously investigated in any organism that has two functional pathways for urea synthesis.

3.2 Alanine Aminotransferases (ALAT1 and ALAT2) Modulate Ammonia Transporter (Rh50-1) and Xanthine Dehydrogenase 1 (XDH1) Transcript Levels

A. aegypti females have two genes encoding ALAT (ALAT1 and ALAT2). Each transcript and protein has a distinct temporal and spatial expression in fat body, thorax, Malpighian tubules, ovary and midgut from sugar- and blood-fed mosquitoes (Mazzalupo et al., 2016). Exposure to L-cycloserine (LCS), a well-known inhibitor of ALAT (Beuster et al., 2011), causes impairment of motor activity and mortality in *A. aegypti* females in a time- and dose-dependent manner. An amplification of the LCS effects is observed when mosquito blood meals are supplemented with high doses of LCS and glucose or when females are starved prior to the treatment, suggesting impairments in carbohydrate and ammonia metabolism in a time-dependent manner. It was also proposed that surviving mosquitoes use alternative pathways to deal with ALAT disruption (Belloni and Scaraffia, 2014). Unexpectedly, impairment of ALAT 1 and/or ALAT 2 by RNAi has no impact on mosquito survival, indicating that a possible off-target LCS effect led to a high mortality observed in the previous study. Mosquitoes with an ALAT deficiency slightly reduce motor activity but significantly delay blood digestion. High levels of three major serine proteases (5GL, CxLT and LT) are observed at 36 h after blood feeding instead of 24 h, while at 48 h 5GL and LT are still abundant in the midgut of those mosquitoes with ALAT dsRNA-mediated knockdown (Mazzalupo et al., 2016). The prolonged expression of the digestive proteases is well-correlated with a delay observed in the synthesis and excretion of nitrogen waste products such as uric acid and haem. Furthermore, depletion of each ALAT or both together causes a massive but temporary accumulation of uric acid in the posterior region of the midgut (Fig. 7) at 6, 12, 18, 24, 36 and 48 h after blood feeding (Mazzalupo et al., 2016).

Several lines of evidence suggest that the transient storage of uric acid, as well as the significant increase of ammonia transporter Rh50-1 and XDH1 transcripts observed in fat body, thorax and Malpighian tubules from mosquitoes with ALAT deficiency are compensatory responses to prevent ammonia and free radical damage caused by a disrupted nitrogen metabolism. Additionally, oviposition is retarded and more importantly the number of eggs laid is significantly reduced in females injected with dsRNA against ALAT1 or ALAT2 when compared to mosquito control (Mazzalupo et al., 2016).

Recent studies also report impaired blood meal digestion in *A. aegypti* mosquitoes. These studies include silencing of the mosquito insulin receptor (Gulia-Nuss et al., 2011) and certain recently identified microRNAs (Bryant et al., 2010; Liu et al., 2014; Lucas et al., 2015). The regulatory factors in these pathways could be involved in the regulation of enzymes involved in ammonia metabolism, which are upregulated in response to blood feeding (Mazzalupo et al., 2016; Scaraffia et al., 2005, 2008). These pathways may also indirectly lead to alterations in ALAT function or merely phenocopy a decrease in ALAT activity. Future experiments might give evidence that disruption of insulin signalling and/or target of rapamycin pathways changes levels of many of the enzymes discussed in this chapter.

The unexpected and novel phenotypes described after silencing genes encoding enzymes involved in ammonia metabolism in *A. aegypti* females highlight the tremendous impact that an altered nitrogen metabolism causes to several aspects of mosquito physiology including digestion, excretion and reproduction. The results indicate that blood-fed mosquitoes have efficiently evolved compensatory mechanisms to balance changes associated with a deficient nitrogen metabolism. These findings also emphasize the need of a better understanding of nitrogen metabolism and its mechanisms of regulation in mosquitoes. New discoveries in this field could help scientists to identify novel targets to block or disrupt nitrogen metabolism in these haematophagous vectors.

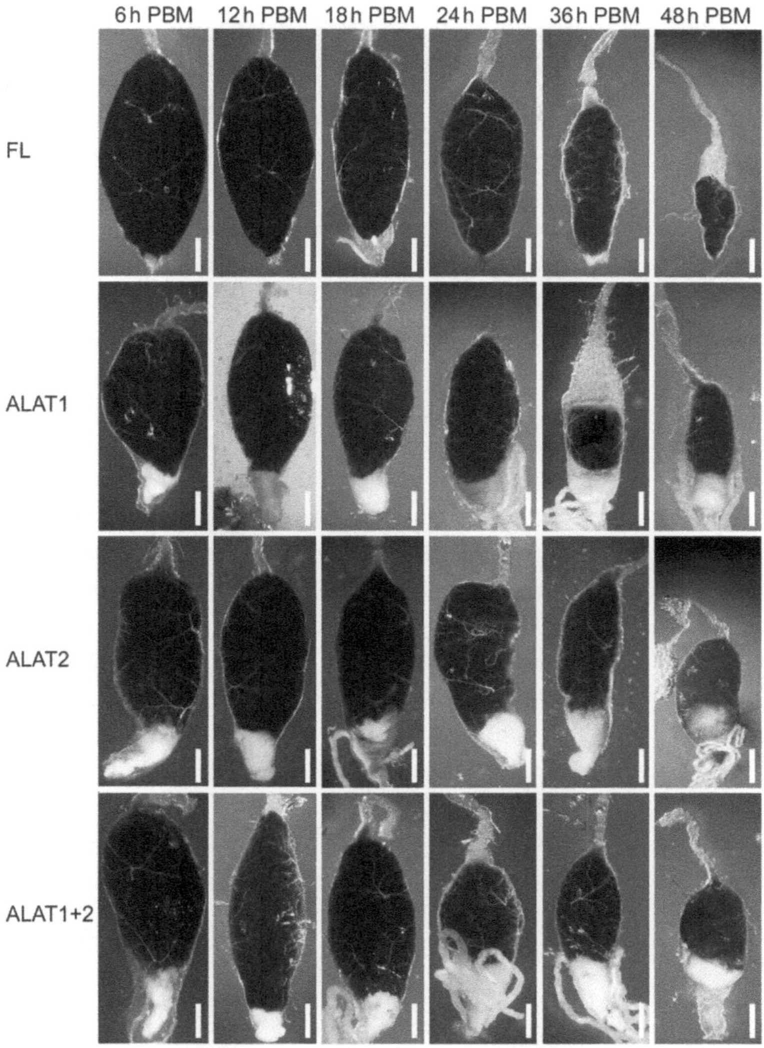

Fig. 7 Uric acid accumulation in the midgut of mosquitoes with alanine aminotransferase (ALAT) deficiency. Midguts were dissected from dsRNA firefly luciferase (FL) or ALAT injected mosquitoes at different time post blood meal (PBM). The posterior end of the midgut is down in this orientation. Scale bars, 500 μm. *Reproduced with permission from Mazzalupo, S., Isoe, J., Belloni, V., Scaraffia, P.Y., 2016. Effective disposal of nitrogen waste in blood-fed* Aedes aegypti *mosquitoes requires alanine aminotransferase. FASEB J. 30, 111–120.* (See the colour plate.)

4. CONCLUDING NOTES

Blood-fed mosquitoes have powerful mechanisms to metabolize nitrogen compounds released from blood meal digestion. Studies performed using traditional and state-of-the-art techniques have revealed that female mosquitoes detoxify ammonia by multiple metabolic pathways, which were unpredicted based on our current mammalian nitrogen knowledge. It was discovered that GltS, UO, ALLN and ALLC are functional in *A. aegypti* mosquitoes and that these enzymes play an important role during ammonia detoxification. The application of classical and cutting-edge techniques has also deciphered novel cross-talk signalling mechanisms involved in the metabolic regulation of nitrogen. This knowledge converts mosquitoes into an extraordinary biological model to explore ammonia metabolism, a physiological process that is critical to all organisms known including humans. The implementation of the integrated approaches discussed in this chapter can be an excellent tool to further investigate the metabolic interaction between mosquitoes and the pathogens they transmit. Recent outbreaks of infectious diseases transmitted by mosquitoes put the health of human populations at serious risk, and therefore it is imperative to increase our understanding of mosquito metabolism in order to swiftly implement new strategies to reduce mosquito populations. The knowledge gained from the past decade can certainly contribute towards paving the way for the successful identification of metabolic and/or regulatory targets that could be used for designing better biological, chemical and/or genetic strategies for mosquito control.

ACKNOWLEDGEMENTS

The authors thank Dr Stacy Mazzalupo for her critical readings and valuable comments. This work was financially supported by the Corine Adams Baines Professorship Award and NIH Grant R01AI088092 (to P.Y.S.).

REFERENCES

Attar, N., 2016. ZIKA virus circulates in new regions. Nat. Rev. Microbiol. 14, 62.

Attardo, G.M., Hansen, I.A., Shiao, S.H., Raikhel, A.S., 2006. Identification of two cationic amino acid transporters required for nutritional signaling during mosquito reproduction. J. Exp. Biol. 209, 3071–3078.

Barillas-Mury, C., Graf, R., Hagedorn, H.H., Wells, M.A., 1991. cDNA and deduced amino acid sequence of the blood meal-induced trypsin from the mosquito, Aedes aegypti. Insect Biochem. 21, 825–831.

Barrett, F.M., Friend, W.G., 1970. Uric acid synthesis in *Rhodnius prolixus*. J. Insect Physiol. 16, 121–129.

Barrett, A.D., Higgs, S., 2007. Yellow fever: a disease that has yet to be conquered. Annu. Rev. Entomol. 52, 209–229.

Beasley, D.W., McAuley, A.J., Bente, D.A., 2015. Yellow fever virus: genetic and phenotypic diversity and implications for detection, prevention and therapy. Antiviral Res. 115, 48–70.

Belloni, V., Scaraffia, P.Y., 2014. Exposure to L-cycloserine incurs survival costs and behavioral alterations in *Aedes aegypti* females. Parasit. Vectors 7, 373.

Benoit, J.B., Hansen, I.A., Szuter, E.M., Drake, L.L., Burnett, D.L., Attardo, G.M., 2014. Emerging roles of aquaporins in relation to the physiology of blood-feeding arthropods. J. Comp. Physiol. B. 184, 811–825.

Beuster, G., Zarse, K., Kaleta, C., Thierbach, R., Kiehntopf, M., Steinberg, P., Schuster, S., Ristow, M., 2011. Inhibition of alanine aminotransferase *in silico* and *in vivo* promotes mitochondrial metabolism to impair malignant growth. J. Biol. Chem. 286, 22323–22330.

Beyenbach, K.W., 2003. Transport mechanisms of diuresis in Malpighian tubules of insects. J. Exp. Biol. 206, 3845–3856.

Beyenbach, K.W., Piermarini, P.M., 2011. Transcellular and paracellular pathways of transepithelial fluid secretion in Malpighian (renal) tubules of the yellow fever mosquito *Aedes aegypti*. Acta Physiol. 202, 387–407.

Bian, G., Raikhel, A.S., Zhu, J., 2008. Characterization of a juvenile hormone-regulated chymotrypsin-like serine protease gene in *Aedes aegypti* mosquito. Insect Biochem. Mol. Biol. 38, 190–200.

Bonizzoni, M., Dunn, W.A., Campbell, C.L., Olson, K.E., Dimon, M.T., Marinotti, O., James, A.A., 2011. RNA-seq analyses of blood-induced changes in gene expression in the mosquito vector species, *Aedes aegypti*. BMC Genomics 12, 82.

Bottino-Rojas, V., Talyuli, O.A., Jupatanakul, N., Sim, S., Dimopoulos, G., Venancio, T.M., Bahia, A.C., Sorgine, M.H., Oliveira, P.L., Paiva-Silva, G.O., 2015. Heme signaling impacts global gene expression, immunity and dengue virus infectivity in *Aedes aegypti*. PLoS One 10, e0135985.

Boudko, D.Y., Tsujimoto, H., Rodriguez, S.D., Meleshkevitch, E.A., Price, D.P., Drake, L.L., Hansen, I.A., 2015. Substrate specificity and transport mechanism of amino-acid transceptor Slimfast from *Aedes aegypti*. Nat. Commun. 6, 8546.

Brackney, D.E., Isoe, J., Black IV, W.C., Zamora, J., Foy, B.D., Miesfeld, R.L., Olson, K.E., 2010. Expression profiling and comparative analyses of seven midgut serine proteases from the yellow fever mosquito, *Aedes aegypti*. J. Insect Physiol. 56, 736–744.

Briegel, H., 1985. Mosquito reproduction: incomplete utilization of the blood meal protein for oögenesis. J. Insect Physiol. 31, 15–21.

Briegel, H., 1986. Protein catabolism and nitrogen partitioning during oögenesis in the mosquito *Aedes aegypti*. J. Insect Physiol. 32, 455–462.

Bryant, B., Macdonald, W., Raikhel, A.S., 2010. microRNA *miR-275* is indispensable for blood digestion and egg development in the mosquito *Aedes aegypti*. Proc. Natl. Acad. Sci. U. S. A 107, 22391–22398.

Bursell, E., 1966. Aspects of the flight metabolism of tsetse flies (*Glossina*). Comp. Biochem. Physiol. 19, 809–818.

Bursell, E., 1975. Substrates of oxidative metabolism in dipteran flight muscle. Comp. Biochem. Physiol. 52B, 235–238.

Bursell, E., 1977. Synthesis of proline by fat body of the tsetse fly (*Glossina morsitans*): metabolic pathways. Insect Biochem. 7, 427–434.

Bush, D.R., Wysocki, V.H., Scaraffia, P.Y., 2012. Study of the fragmentation of arginine isobutyl ester applied to arginine quantification in *Aedes aegypti* mosquito excreta. J. Mass Spectrom. 47, 1364–1371.

Carpenter, V.K., Drake, L.L., Aguirre, S.E., Price, D.P., Rodriguez, S.D., Hansen, I.A., 2012. SLC7 amino acid transporters of the yellow fever mosquito *Aedes aegypti* and their role in fat body TOR signaling and reproduction. J. Insect Physiol. 58, 513–522.

Choumet, V., Desprès, P., 2015. Dengue and other flavivirus infections. Rev. Sci. Tech. Off. Int. Epiz. 34, 467–478.

Clements, A.N., 1992. The biology of mosquitoes. Development, Nutrition and Reproduction, vol. 1. Chapman & Hall, New York.

Cochran, D.G., 1985. Nitrogen excretion in cockroaches. Annu. Rev. Entomol. 30, 29–49.

Drake, L.L., Rodriguez, S.D., Hansen, I.A., 2015. Functional characterization of aquaporins and aquaglyceroporins of the yellow fever mosquito, *Aedes aegypti*. Sci. Rep. 5, 7795.

Edwards, M.J., Moskalyk, L.A., Donelly-Doman, M., Vlaskova, M., Noriega, F.G., Walker, V.K., Jacobs-Lorena, M., 2000. Characterization of a carboxypeptidase A gene from the mosquito, *Aedes aegypti*. Insect Mol. Biol. 9, 33–38.

Esquivel, C.J., Cassone, B.J., Piermarini, P.M., 2014. Transcriptomic evidence for a dramatic functional transition of the Malpighian tubules after a blood meal in the Asian tiger mosquito *Aedes albopictus*. PLoS Negl. Trop. Dis. 8, e2929.

Evans, A.M., Aimanova, K.G., Gill, S.S., 2009. Characterization of a blood-meal-responsive proton-dependent amino acid transporter in the disease vector, *Aedes aegypti*. J. Exp. Biol. 212, 3263–3271.

Fauci, A.S., Morens, D.M., 2016. Zika virus in the Americas—yet another arbovirus threat. N. Engl. J. Med. 374, 601–604.

Gäde, G., Auerswald, L., 2002. Beetles' choice-proline for energy output: control by AKHs. Comp. Biochem. Physiol. B. Biochem. Mol. Biol. 132, 117–129.

Goldstrohm, D.A., Pennington, J.E., Wells, M.A., 2003. The role of hemolymph proline as a nitrogen sink during blood meal digestion by the mosquito *Aedes aegypti*. J. Insect Physiol. 49, 115–121.

Graça-Souza, A.V., Maya-Monteiro, C., Paiva-Silva, G.O., Braz, G.R., Paes, M.C., Sorgine, M.H., Oliveira, M.F., Oliveira, P.L., 2006. Adaptations against heme toxicity in blood-feeding arthropods. Insect Biochem. Mol. Biol. 36, 322–335.

Gulia-Nuss, M., Robertson, A.E., Brown, M.R., Strand, M.R., 2011. Insulin-like peptides and the target of rapamycin pathway coordinately regulate blood digestion and egg maturation in the mosquito *Aedes aegypti*. PLoS One 6, e20401.

Hager, J.W., Yves Le Blanc, J.C., 2003. Product ion scanning using a Q-q-$Q_{linear\ ion\ trap}$ (Q TRAP^{TM}) mass spectrometer. Rapid Commun. Mass Spectrom. 17, 1056–1064.

Hansen, I.A., Boudko, D.Y., Shiao, S.H., Voronov, D.A., Meleshkevitch, E.A., Drake, L.L., Aguirre, S.E., Fox, J.M., Attardo, G.M., Raikhel, A.S., 2011. AaCAT1 of the yellow fever mosquito, *Aedes aegypti*: a novel histidine-specific amino acid transporter from the SLC7 family. J. Biol. Chem. 286, 10803–10813.

Harvey, W.R., Boudko, D.Y., Rheault, M.R., Okech, B.A., 2009. NHE_{VNAT}: an H^+ V-ATPase electrically coupled to a Na^+:nutrient amino acid transporter (NAT) forms an Na^+/H^+ exchanger (NHE). J. Exp. Biol. 212, 347–357.

Hayashi, S., Fujiwara, S., Noguchi, T., 2000. Evolution of urate-degrading enzymes in animal peroxisomes. Cell Biochem. Biophys. 32, 123–129.

Hegedus, D., Erlandson, M., Gillott, C., Toprak, U., 2009. New insights into peritrophic matrix synthesis, architecture, and function. Annu. Rev. Entomol. 54, 285–302.

Henn, M.W., Schopf, R., Maier, W.A., Seitz, H.M., 1998. The amino acid composition of *Anopheles stephensi* (Diptera: Culicidae) infected with *Nosema algerae* (Microsporida: Nosematidae). J. Invertebr. Pathol. 71, 42–47.

Holt, R.A., Subramanian, G.M., Halpern, A., Sutton, G.G., Charlab, R., Nusskern, D.R., Wincker, P., Clark, A.G., Ribeiro, J.M., Wides, R., Salzberg, S.L., Loftus, B., Yandell, M., Majoros, W.H., Rusch, D.B., Lai, Z., Kraft, C.L., Abril, J.F., Anthouard, V., Arensburger, P., Atkinson, P.W., Baden, H., de Berardinis, V., Baldwin, D., Benes, V., Biedler, J., Blass, C., Bolanos, R., Boscus, D., Barnstead, M., Cai, S., Center, A., Chaturverdi, K., Christophides, G.K., Chrystal, M.A., Clamp, M., Cravchik, A., Curwen, V., Dana, A., Delcher, A., Dew, I., Evans, C.A.,

Flanigan, M., Grundschober-Freimoser, A., Friedli, L., Gu, Z., Guan, P., Guigo, R., Hillenmeyer, M.E., Hladun, S.L., Hogan, J.R., Hong, Y.S., Hoover, J., Jaillon, O., Ke, Z., Kodira, C., Kokoza, E., Koutsos, A., Letunic, I., Levitsky, A., Liang, Y., Lin, J.J., Lobo, N.F., Lopez, J.R., Malek, J.A., McIntosh, T.C., Meister, S., Miller, J., Mobarry, C., Mongin, E., Murphy, S.D., O'Brochta, D.A., Pfannkoch, C., Qi, R., Regier, M.A., Remington, K., Shao, H., Sharakhova, M.V., Sitter, C.D., Shetty, J., Smith, T.J., Strong, R., Sun, J., Thomasova, D., Ton, L.Q., Topalis, P., Tu, Z., Unger, M.F., Walenz, B., Wang, A., Wang, J., Wang, M., Wang, X., Woodford, K.J., Wortman, J.R., Wu, M., Yao, A., Zdobnov, E.M., Zhang, H., Zhao, Q., et al., 2002. The genome sequence of the malaria mosquito *Anopheles gambiae.* Science 298, 129–149.

Horwood, P.F., Buchy, P., 2015. Chikungunya. Rev. Sci. Tech. Off. Int. Epiz. 34, 479–489.

Hou, Y., Wang, X.L., Saha, T.T., Roy, S., Zhao, B., Raikhel, A.S., Zou, Z., 2015. Temporal coordination of carbohydrate metabolism during mosquito reproduction. PLoS Genet. 11, e1005309.

Isoe, J., Scaraffia, P.Y., 2013. Urea synthesis and excretion in *Aedes aegypti* mosquitoes are regulated by a unique cross-talk mechanism. PLoS One 8, e65393.

Isoe, J., Rascón Jr., A.A., Kunz, S., Miesfeld, R.L., 2009a. Molecular genetic analysis of midgut serine proteases in *Aedes aegypti* mosquitoes. Insect Biochem. Mol. Biol. 39, 903–912.

Isoe, J., Zamora, J., Miesfeld, R.L., 2009b. Molecular analysis of the *Aedes aegypti* carboxypeptidase gene family. Insect Biochem. Mol. Biol. 39, 68–73.

Isoe, J., Collins, J., Badgandi, H., Day, W.A., Miesfeld, R.L., 2011. Defects in coatomer protein I (COPI) transport cause blood feeding-induced mortality in yellow fever mosquitoes. Proc. Natl. Acad. Sci. U. S. A. 108, E211–E217.

Jiang, Q., Hall, M., Noriega, F.G., Wells, M.A., 1997. cDNA cloning and pattern of expression of an adult, female-specific chymotrypsin from *Aedes aegypti* midgut. Insect Biochem. Mol. Biol. 27, 283–289.

Jiang, W., Wysocki, V.H., Dodds, E.D., Miesfeld, R.L., Scaraffia, P.Y., 2010. Differentiation and quantification of C1 and C2 ^{13}C-labeled glucose by tandem mass spectrometry. Anal. Biochem. 404, 40–44.

Kalhok, S.E., Tabak, L.M., Prosser, D.E., Brook, W., Downe, A.E., White, B.N., 1993. Isolation, sequencing and characterization of two cDNA clones coding for trypsin-like enzymes from the midgut of *Aedes aegypti.* Insect Mol. Biol. 2, 71–79.

Kitteringham, N.R., Jenkins, R.E., Lane, C.S., Elliott, V.L., Park, B.K., 2009. Multiple reaction monitoring for quantitative biomarker analysis in proteomics and metabolomics. J. Chromatogr. B Analyt. Technol. Biomed. Life Sci. 877, 1229–1239.

Krebs, H.A., Henseleit, K., 1932. Analysis concerning urea formation in animal bodies. H-S Z Physiol. Chem. 210, 33–66.

Larsen, E.H., Deaton, L.E., Onken, H., O'Donnell, M., Grosell, M., Dantzler, W.H., Weihrauch, D., 2014. Osmoregulation and excretion. Compr. Physiol. 4, 405–573.

Lehane, M.J., 2005. The Biology of Blood-Sucking in Insects, second ed. Cambridge University Press, New York.

Levenberg, B., Hartman, S.C., Buchanan, J.M., 1956. Biosynthesis of the purines. X. Further studies *in vitro* on the metabolic origin of nitrogen atoms 1 and 3 of the purine ring. J. Biol. Chem. 220, 379–390.

Liu, S., Lucas, K.J., Roy, S., Ha, J., Raikhel, A.S., 2014. Mosquito-specific microRNA-1174 targets serine hydroxymethyltransferase to control key functions in the gut. Proc. Natl. Acad. Sci. U. S. A. 111, 14460–14465.

Lucas, K.J., Zhao, B., Roy, S., Gervaise, A.L., Raikhel, A.S., 2015. Mosquito-specific microRNA-1890 targets the juvenile hormone-regulated serine protease JHA15 in the female mosquito gut. RNA Biol. 12, 1383–1390.

Ma, X., Dagan, S., Somogyi, A., Wysocki, V.H., Scaraffia, P.Y., 2013. Low mass MS/MS fragments of protonated amino acids used for distinction of their ^{13}C-isotopomers in metabolic studies. J. Am. Soc. Mass Spectrom. 24, 622–631.

Marinotti, O., Calvo, E., Nguyen, Q.K., Dissanayake, S., Ribeiro, J.M., James, A.A., 2006. Genome-wide analysis of gene expression in adult *Anopheles gambiae*. Insect Mol. Biol. 15, 1–12.

Mazzalupo, S., Scaraffia, P.Y., 2014. Application of isotopically labeled compounds and tandem mass spectrometry for studying metabolic pathways in mosquitoes. In: Chandrasekar, R., Tyagi, B.K., Gui, Z.Z., Reeck, G.R. (Eds.), Short View on Insect Biochemistry and Molecular Biology. K-State Union, Kansas State University/ International Book Mission, Manhattan, pp. 99–126.

Mazzalupo, S., Isoe, J., Belloni, V., Scaraffia, P.Y., 2016. Effective disposal of nitrogen waste in blood-fed *Aedes aegypti* mosquitoes requires alanine aminotransferase. FASEB J. 30, 111–120.

Mesquita, R.D., Vionette-Amaral, R.J., Lowenberger, C., Rivera-Pomar, R., Monteiro, F.A., Minx, P., Spieth, J., Carvalho, A.B., Panzera, F., Lawson, D., Torres, A.Q., Ribeiro, J.M., Sorgine, M.H., Waterhouse, R.M., Montague, M.J., Abad-Franch, F., Alves-Bezerra, M., Amaral, L.R., Araujo, H.M., Araujo, R.N., Aravind, L., Atella, G.C., Azambuja, P., Berni, M., Bittencourt-Cunha, P.R., Braz, G.R., Calderón-Fernández, G., Carareto, C.M., Christensen, M.B., Costa, I.R., Costa, S.G., Dansa, M., Daumas-Filho, C.R., De-Paula, I.F., Dias, F.A., Dimopoulos, G., Emrich, S.J., Esponda-Behrens, N., Fampa, P., Fernandez-Medina, R.D., da Fonseca, R.N., Fontenele, M., Fronick, C., Fulton, L.A., Gandara, A.C., Garcia, E.S., Genta, F.A., Giraldo-Calderón, G.I., Gomes, B., Gondim, K.C., Granzotto, A., Guarneri, A.A., Guigó, R., Harry, M., Hughes, D.S., Jablonka, W., Jacquin-Joly, E., Juárez, M.P., Koerich, L.B., Latorre-Estivalis, J.M., Lavore, A., Lawrence, G.G., Lazoski, C., Lazzari, C.R., Lopes, R.R., Lorenzo, M.G., Lugon, M.D., Majerowicz, D., Marcet, P.L., Mariotti, M., Masuda, H., Megy, K., Melo, A.C., Missirlis, F., Mota, T., Noriega, F.G., Nouzova, M., Nunes, R.D., Oliveira, R.L., Oliveira-Silveira, G., Ons, S., Pagola, L., Paiva-Silva, G.O., Pascual, A., Pavan, M.G., Pedrini, N., Peixoto, A.A., Pereira, M.H., Pike, A., Polycarpo, C., Prosdocimi, F., Ribeiro-Rodrigues, R., Robertson, H.M., Salerno, A.P., Salmon, D., Santesmasses, D., Schama, R., Seabra-Junior, E.S., Silva-Cardoso, L., Silva-Neto, M.A., et al., 2015. Genome of *Rhodnius prolixus*, an insect vector of Chagas disease, reveals unique adaptations to hematophagy and parasite infection. Proc. Natl. Acad. Sci. U. S. A. 112, 14936–14941.

Nene, V., Wortman, J.R., Lawson, D., Haas, B., Kodira, C., Tu, Z.J., Loftus, B., Xi, Z., Megy, K., Grabherr, M., Ren, Q., Zdobnov, E.M., Lobo, N.F., Campbell, K.S., Brown, S.E., Bonaldo, M.F., Zhu, J., Sinkins, S.P., Hogenkamp, D.G., Amedeo, P., Arensburger, P., Atkinson, P.W., Bidwell, S., Biedler, J., Birney, E., Bruggner, R.V., Costas, J., Coy, M.R., Crabtree, J., Crawford, M., Debruyn, B., Decaprio, D., Eiglmeier, K., Eisenstadt, E., El-Dorry, H., Gelbart, W.M., Gomes, S.L., Hammond, M., Hannick, L.I., Hogan, J.R., Holmes, M.H., Jaffe, D., Johnston, J.S., Kennedy, R.C., Koo, H., Kravitz, S., Kriventseva, E.V., Kulp, D., Labutti, K., Lee, E., Li, S., Lovin, D.D., Mao, C., Mauceli, E., Menck, C.F., Miller, J.R., Montgomery, P., Mori, A., Nascimento, A.L., Naveira, H.F., Nusbaum, C., O'Leary, S., Orvis, J., Pertea, M., Quesneville, H., Reidenbach, K.R., Rogers, Y.H., Roth, C.W., Schneider, J.R., Schatz, M., Shumway, M., Stanke, M., Stinson, E.O., Tubio, J.M., Vanzee, J.P., Verjovski-Almeida, S., Werner, D., White, O., Wyder, S., Zeng, Q., Zhao, Q., Zhao, Y., Hill, C.A., Raikhel, A.S., Soares, M.B., Knudson, D.L., Lee, N.H., Galagan, J., Salzberg, S.L., Paulsen, I.T., Dimopoulos, G., Collins, F.H., Birren, B., Fraser-Liggett, C.M., Severson, D.W.,

2007. Genome sequence of *Aedes aegypti*, a major arbovirus vector. Science 316, 1718–1723.
Noriega, F.G., Wang, X.Y., Pennington, J.E., Barillas-Mury, C.V., Wells, M.A., 1996. Early trypsin, a female-specific midgut protease in *Aedes aegypti*: isolation, amino-terminal sequence determination, and cloning and sequencing of the gene. Insect Biochem. Mol. Biol. 26, 119–126.
O'Donnell, M.J., 2009. Too much of a good thing: how insects cope with excess ions or toxins in the diet. J. Exp. Biol. 212, 363–372.
Pacey, E.K., O'Donnell, M.J., 2014. Transport of H^+, Na^+ and K^+ across the posterior midgut of blood-fed mosquitoes (*Aedes aegypti*). J. Insect Physiol. 61, 42–50.
Pakpour, N., Riehle, M.A., Luckhart, S., 2014. Effects of ingested vertebrate-derived factors on insect immune responses. Curr. Opin. Insect Sci. 3, 1–5.
Pascoa, V., Oliveira, P.L., Dansa-Petretski, M., Silva, J.R., Alvarenga, P.H., Jacobs-Lorena, M., Lemos, F.J., 2002. *Aedes aegypti* peritrophic matrix and its interaction with heme during blood digestion. Insect Biochem. Mol. Biol. 32, 517–523.
Pitts, R.J., Derryberry Jr., S.L., Pulous, F.E., Zwiebel, L.J., 2014. Antennal-expressed ammonium transporters in the malaria vector mosquito *Anopheles gambiae*. PLoS One 9, e111858.
Price, D.P., Nagarajan, V., Churbanov, A., Houde, P., Milligan, B., Drake, L.L., Gustafson, J.E., Hansen, I.A., 2011. The fat body transcriptomes of the yellow fever mosquito *Aedes aegypti*, pre- and post-blood meal. PLoS One 6, e22573.
Rascón Jr., A.A., Gearin, J., Isoe, J., Miesfeld, R.L., 2011. *In vitro* activation and enzyme kinetic analysis of recombinant midgut serine proteases from the Dengue vector mosquito *Aedes aegypti*. BMC Biochem. 12, 43.
Reid, W.R., Zhang, L., Liu, N., 2015. Temporal gene expression profiles of pre blood-fed adult females immediately following eclosion in the southern house mosquito *Culex Quinquefasciatus*. Int. J. Biol. Sci. 11, 1306–1313.
Ribeiro, J.M., 2003. A catalogue of *Anopheles gambiae* transcripts significantly more or less expressed following a blood meal. Insect Biochem. Mol. Biol. 33, 865–882.
Roy, S., Saha, T.T., Johnson, L., Zhao, B., Ha, J., White, K.P., Girke, T., Zou, Z., Raikhel, A.S., 2015. Regulation of gene expression patterns in mosquito reproduction. PLoS Genet. 11, e1005450.
Sanders, H.R., Evans, A.M., Ross, L.S., Gill, S.S., 2003. Blood meal induces global changes in midgut gene expression in the disease vector, *Aedes aegypti*. Insect Biochem. Mol. Biol. 33, 1105–1122.
Scaraffia, P.Y., Wells, M.A., 2003. Proline can be utilized as an energy substrate during flight of *Aedes aegypti* females. J. Insect Physiol. 49, 591–601.
Scaraffia, P.Y., Isoe, J., Murillo, A., Wells, M.A., 2005. Ammonia metabolism in *Aedes aegypti*. Insect Biochem. Mol. Biol. 35, 491–503.
Scaraffia, P.Y., Zhang, Q., Wysocki, V.H., Isoe, J., Wells, M.A., 2006. Analysis of whole body ammonia metabolism in *Aedes aegypti* using [^{15}N]-labeled compounds and mass spectrometry. Insect Biochem. Mol. Biol. 36, 614–622.
Scaraffia, P.Y., Tan, G., Isoe, J., Wysocki, V.H., Wells, M.A., Miesfeld, R.L., 2008. Discovery of an alternate metabolic pathway for urea synthesis in adult *Aedes aegypti* mosquitoes. Proc. Natl. Acad. Sci. U. S. A. 105, 518–523.
Scaraffia, P.Y., Zhang, Q., Thorson, K., Wysocki, V.H., Miesfeld, R.L., 2010. Differential ammonia metabolism in *Aedes aegypti* fat body and midgut tissues. J. Insect Physiol. 56, 1040–1049.
Severson, D.W., DeBruyn, B., Lovin, D.D., Brown, S.E., Knudson, D.L., Morlais, I., 2004. Comparative genome analysis of the yellow fever mosquito *Aedes aegypti* with *Drosophila melanogaster* and the malaria vector mosquito *Anopheles gambiae*. J. Hered. 95, 103–113.

Shao, L., Devenport, M., Jacobs-Lorena, M., 2001. The peritrophic matrix of hematophagous insects. Arch. Insect Biochem. Physiol. 47, 119–125.

Singer, M.A., 2003. Dietary protein-induced changes in excretory function: a general animal design feature. Comp. Biochem. Physiol. B Biochem. Mol. Biol. 136, 785–801.

Soares, J.B., Gaviraghi, A., Oliveira, M.F., 2015. Mitochondrial physiology in the major arbovirus vector *Aedes aegypti*: substrate preferences and sexual differences define respiratory capacity and superoxide production. PLoS One 10, e0120600.

Sonne, J.C., Lin, I., Buchanan, J.M., 1956. Biosynthesis of the purines. IX. Precursors of the nitrogen atoms of the purine ring. J. Biol. Chem. 220, 369–378.

Su, T., Mulla, M.S., 1997. Quantitative determination of free amino acids in the hemolymph of autogenous and anautogenous strains of *Culex tarsalis* (Diptera: Culicidae). J. Med. Entomol. 34, 729–734.

Uchida, K., Ohmori, D., Yamakura, F., Suzuki, K., 1990. Changes in free amino acid concentration in the hemolymph of the female *Culex pipiens pallens* (Diptera: Culicidae), after a blood meal. J. Med. Entomol. 27, 302–308.

von Dungern, P., Briegel, H., 2001a. Enzymatic analysis of uricotelic protein catabolism in the mosquito *Aedes aegypti*. J. Insect Physiol. 47, 73–82.

von Dungern, P., Briegel, H., 2001b. Protein catabolism in mosquitoes: ureotely and uricotely in larval and imaginal *Aedes aegypti*. J. Insect Physiol. 47, 131–141.

Wang, J., 2009. Analysis of macrolide antibiotics, using liquid chromatography-mass spectrometry, in food, biological and environmental matrices. Mass Spectrom. Rev. 28, 50–92.

Weihrauch, D., Donini, A., O'Donnell, M.J., 2012. Ammonia transport by terrestrial and aquatic insects. J. Insect Physiol. 58, 473–487.

Williams, J.C., Hagedorn, H.H., Beyenbach, K.W., 1983. Dynamic changes in flow rate and composition of urine during the post-bloodmeal diuresis in *Aedes aegypti* (L.). J. Comp. Physiol. 153, 257–265.

World Health Organization, 2016. Vector-Borne Diseases [Online]. Geneva. Available: http://www.who.int/mediacentre/factsheets/fs387/en/.

Wright, P.A., 1995. Nitrogen excretion: three end products, many physiological roles. J. Exp. Biol. 198, 273–281.

Wu, Y., Zheng, X., Zhang, M., He, A., Li, Z., Zhan, X., 2010. Cloning and functional expression of Rh50-like glycoprotein, a putative ammonia channel, in *Aedes albopictus* mosquitoes. J. Insect Physiol. 56, 1599–1610.

Xia, Y.Q., Miller, J.D., Bakhtiar, R., Franklin, R.B., Liu, D.Q., 2003. Use of a quadrupole linear ion trap mass spectrometer in metabolite identification and bioanalysis. Rapid Commun. Mass Spectrom. 17, 1137–1145.

Yost, R.A., Enke, C.G., 1979. Triple quadrupole mass spectrometry for direct mixture analysis and structure elucidation. Anal. Chem. 51, 1251–1264.

Zdobnov, E.M., von Mering, C., Letunic, I., Torrents, D., Suyama, M., Copley, R.R., Christophides, G.K., Thomasova, D., Holt, R.A., Subramanian, G.M., Mueller, H.M., Dimopoulos, G., Law, J.H., Wells, M.A., Birney, E., Charlab, R., Halpern, A.L., Kokoza, E., Kraft, C.L., Lai, Z., Lewis, S., Louis, C., Barillas-Mury, C., Nusskern, D., Rubin, G.M., Salzberg, S.L., Sutton, G.G., Topalis, P., Wides, R., Wincker, P., Yandell, M., Collins, F.H., Ribeiro, J., Gelbart, W.M., Kafatos, F.C., Bork, P., 2002. Comparative genome and proteome analysis of *Anopheles gambiae* and *Drosophila melanogaster*. Science 298, 149–159.

Zhang, Q., Wysocki, V.H., Scaraffia, P.Y., Wells, M.A., 2005. Fragmentation pathway for glutamine identification: loss of 73 Da from dimethylformamidine glutamine isobutyl ester. J. Am. Soc. Mass Spectrom. 16, 1192–1203.

Zhou, G., Flowers, M., Friedrich, K., Horton, J., Pennington, J., Wells, M.A., 2004. Metabolic fate of [^{14}C]-labeled meal protein amino acids in *Aedes aegypti* mosquitoes. J. Insect Physiol. 50, 337–349.
Zhou, G., Scaraffia, P.Y., Wells, M.A., 2005. Vector nutrition and energy metabolism. In: Marquardt, W.C. (Ed.), Biology of Disease Vectors, second ed. Elsevier Academic Press, Burlington, pp. 211–215.
Zhou, G., Kohlhepp, P., Geiser, D., Frasquillo, M.C., Vazquez-Moreno, L., Winzerling, J.J., 2007. Fate of blood meal iron in mosquitoes. J. Insect Physiol. 53, 1169–1178.

CHAPTER THIRTEEN

Renal Excretory Processes in Mosquitoes

P.M. Piermarini
Ohio Agricultural Research and Development Center, The Ohio State University, Wooster, OH, United States

Contents

Abstract

During their complex life cycle, which includes aquatic larvae, terrestrial adults, and hematophagous females, mosquitoes face diverse osmoregulatory challenges. To counter these challenges mosquitoes utilize a dynamic renal excretory system, consisting of the Malpighian tubules and hindgut. The goal of this chapter is to review the contributions that the renal excretory system makes to the osmoregulatory physiology of mosquitoes, with a focus on the cellular and molecular mechanisms of ion and water transport in the Malpighian tubules. Recent advances made with molecular, reverse genetic, and pharmacological approaches are highlighted, and the potential of targeting the renal excretory system as a novel approach for the control of mosquito vectors is discussed.

1. INTRODUCTION

The renal excretory system or 'kidney' of mosquitoes is composed of two major epithelial tissues: the renal (Malpighian) tubules and hindgut (ileum and rectum). In general terms, the Malpighian tubules and hindgut are analogous in function to the proximal and distal components of the mammalian kidney, respectively; ie, the tubules produce urine, and

Advances in Insect Physiology, Volume 51
ISSN 0065-2806
http://dx.doi.org/10.1016/bs.aiip.2016.04.003

the hindgut attenuates the composition of the urine before it is excreted. The goal of this chapter is to review the molecular physiology of the mosquito 'kidney' in the context of osmoregulation, with a focus on the mechanisms of ion and water transport in the Malpighian tubules.

1.1 Roles of the Renal Excretory System in Mosquito Osmoregulation

The goal of this section is to provide an introduction to mosquito osmoregulation with an emphasis on the role of the renal excretory system. Several reviews were used as general resources (Benoit and Denlinger, 2010; Beyenbach, 2003; Beyenbach and Hagedorn, 2005; Bradley, 1987; Clements, 2000; Coast, 2009).

1.1.1 Larval Osmoregulation

Mosquito larvae are aquatic and can tolerate a narrow or wide range of environmental salinities from fresh water to sea water, depending on the species (Albers and Bradley, 2011; Bradley, 1987). However, the vast majority of mosquito larvae develop in fresh water, which is the focus here. In freshwater habitats, the mosquito hemolymph faces an osmotic influx of water through (1) the cuticle from the environment and (2) the midgut epithelium from imbibed water (Fig. 1A). Moreover, the hemolymph faces a diffusive

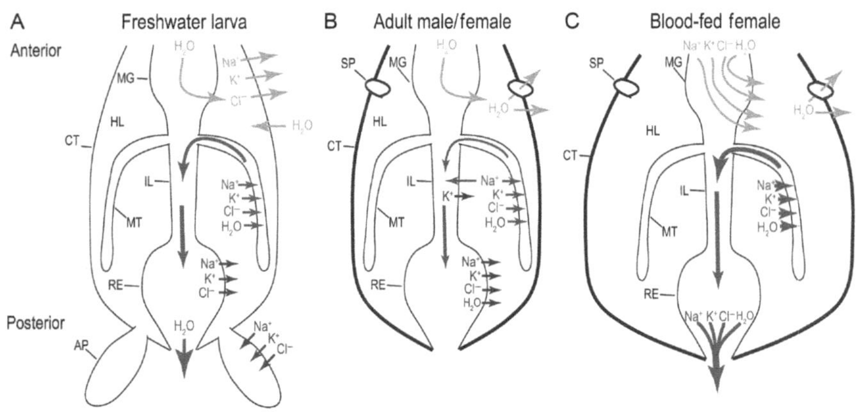

Fig. 1 Simplified illustrations of mosquito osmoregulation: (A) freshwater larva, (B) adult male or female, (C) blood fed female. The *orange* (*light grey* in the print version) *arrows* indicate diffusive/osmotic pathways for uptake from or loss to the environment. The *red* (*grey* in the print version) *arrows* indicate secretory/excretory processes, whereas the *blue* (*dark grey* in the print version) *arrows* indicate absorptive processes. *AP*, anal papillae; *CT*, cuticle; *HL*, hemolymph; *IL*, ileum; *MG*, midgut; *MT*, Malpighian tubules; *RE*, rectum; *SP*, spiracle. The thickened CT of adults reflects cuticular waterproofing.

loss of ions to the environment through the cuticle (Fig. 1A). To counter the osmotic influx of water, the Malpighian tubules secrete a 'primary' urine consisting of NaCl, KCl, and water that is isosmotic to the hemolymph (Fig. 1A). Secreted fluid that enters the hindgut flows through the ileum without much modification. When fluid reaches the rectum, ions are selectively reabsorbed into the hemolymph without water, resulting in a dilute urine that is ready for expulsion (Fig. 1A). Thus, the combined actions of the Malpighian tubules and hindgut allow for the excretion of water from the hemolymph with minimal losses of ions. To counter the ionic losses through the cuticle and urine, the anal papillae actively absorb ions from the environment (Fig. 1A).

1.1.2 Adult Osmoregulation

Adult mosquitoes are terrestrial insects that experience unique osmoregulatory challenges compared to larvae. In most terrestrial environments, the mosquito hemolymph faces desiccation through the cuticle and open respiratory spiracles via evaporative water loss (Fig. 1B), which can secondarily lead to concentration of hemolymph ions and other solutes. To counter and limit evaporative water losses, mosquitoes ingest free water and enhance the water-proofing properties of their cuticle, respectively (Fig. 1B). In the excretory system, the Malpighian tubules secrete a primary urine consisting of NaCl, KCl, and water that is isosmotic to the hemolymph (Fig. 1B). The ileum of adults has molecular and physiological potential to modify the cation content of the secreted fluid (Paluzzi et al., 2014; Patrick et al., 2006) (Fig. 1B). When the fluid reaches the rectum, ions and water are reabsorbed into the hemolymph, resulting in a minimal volume of urine that is available for expulsion (Fig. 1B). Thus, the combined actions of the Malpighian tubules and hindgut allow for the recycling of water and ions from the hemolymph to minimize excretory water losses.

1.1.2.1 Diuresis

While minimizing excretory water loss is the *modus operandi* of unfed adult male and female mosquitoes, the situation radically changes when females engorge on blood to complete their reproductive cycle. During a single feeding, adult female mosquitoes may ingest a volume of blood that matches or exceeds their own body mass within a few minutes. Vertebrate blood is hypoosmotic to the hemolymph of mosquitoes, which results in the osmotic absorption of H_2O through the midgut epithelium and thereby an acute expansion of extracellular fluid volume (Fig. 1C). In addition, the ingested plasma and blood cells contain significant loads of Na^+, K^+, and Cl^- that will be absorbed into the hemolymph through the midgut epithelium (Fig. 1C). To counter these challenges, the neuroendocrine system releases diuretic neuropeptides into the hemolymph that stimulate the Malpighian tubules to secrete primary urine at rates that far exceed those of nonblood fed mosquitoes; the urine primarily consists of NaCl, KCl, and water and is isosmotic to the hemolymph (Fig. 1C). The urine is quickly passed through the ileum and rectum for expulsion (Fig. 1C), facilitated by frequent peristaltic contractions of the hindgut. The rapid movement of urine through the hindgut provides the epithelial cells with little time to reabsorb ions or water, thereby allowing for the maximal excretion of water and ions (Fig. 1C). This rapid, dramatic excretion of urine is also known as the 'postprandial' diuresis and begins while the mosquito is still feeding upon the vertebrate host (Fig. 2A).

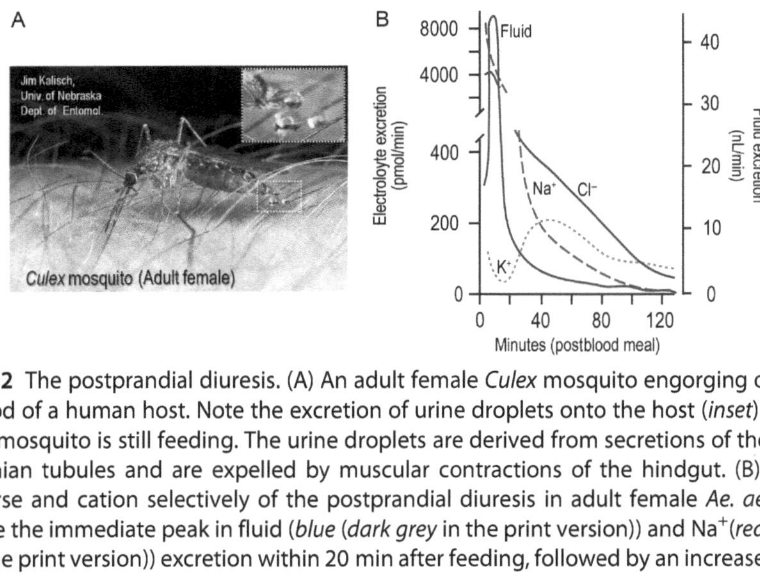

Fig. 2 The postprandial diuresis. (A) An adult female *Culex* mosquito engorging on the blood of a human host. Note the excretion of urine droplets onto the host (*inset*) while the mosquito is still feeding. The urine droplets are derived from secretions of the Malpighian tubules and are expelled by muscular contractions of the hindgut. (B) Time course and cation selectively of the postprandial diuresis in adult female *Ae. aegypti*. Note the immediate peak in fluid (*blue* (*dark grey* in the print version)) and Na^+(*red* (*grey* in the print version)) excretion within 20 min after feeding, followed by an increase of K^+ excretion from ~30 to 120 min. Cl^- (*green* (*grey* in the print version)) is the counter anion excreted with the cations. *Panel (A) Photo is courtesy of Dr Jim Kalisch (Department of Entomology, University of Nebraska). Panel (B) is modified from Beyenbach, K.W., Piermarini, P. M., 2011. Transcellular and paracellular pathways of transepithelial fluid secretion in Malpighian (renal) tubules of the yellow fever mosquito* Aedes aegypti. *Acta Physiol. 202, 387–407, with permission. Original data were from Williams, J.C., Hagedorn, H.H., Beyenbach, K.W., 1983. Dynamic changes in flow rate and composition of urine during the post blood meal diuresis in* Aedes aegypti. *J. Comp. Physiol. B 153, 257–266.*

In *Aedes aegypti*, the postprandial diuresis lasts until 1–2 h postblood meal (Williams et al., 1983). Notably, during the initial stages of the diuresis (0–30 min postblood meal), Na^+ is the dominant urinary cation excreted (Fig. 2B), which presumably reflects the initial rapid absorption of Na^+ from the ingested blood plasma into the hemolymph (Williams et al., 1983). However, between ~40–120 min postblood meal, K^+ is the dominant urinary cation excreted, which presumably reflects the absorption of intracellular K^+ from the ingested blood cells as they are digested (Williams et al., 1983). In *Anopheles gambiae*, the mosquito elicits a prediuresis that occurs during the first 5 min postblood meal in which ~33% of the ingested blood plasma is excreted directly by the hindgut without absorption into the hemolymph (Coast, 2009). The contributions of the Malpighian tubules and postprandial diuresis become apparent between ~5–30 min postblood meal, when K^+ is the dominant cation in the excreted fluid (Coast, 2009). Thus, the prediuresis of *An. gambiae* reduces the acute need for Na^+ secretion in the Malpighian tubules by limiting the amount of Na^+ that is initially absorbed into the hemolymph by the midgut.

1.1.2.2 Neuroendocrine Control of Diuresis

Neuroendocrine factors play key roles in modulating the activity of the mosquito renal excretory system to promote the postprandial diuresis in adult females. In particular, diuretic hormone 31 (DH_{31}), also known as the mosquito natriuretic factor and calcitonin-like peptide, is considered a critical contributor to the initial phases of the postprandial diuresis when Na^+ is the dominant cation excreted. In isolated Malpighian tubules of *Ae. aegypti* and *An. gambiae*, treatment with DH_{31} selectively increases the rates of transepithelial NaCl and fluid secretion (ie, a natriuresis) using cAMP as a secondary messenger (Beyenbach and Petzel, 1987; Coast et al., 2005; Petzel et al., 1985, 1987). Moreover, in isolated hindguts of *Ae. aegypti*, DH_{31} treatment increases the frequency of peristaltic contractions used for the expulsion of urine (Kwon and Pietrantonio, 2013). Furthermore, injection of the hemolymph of adult female *An. gambiae* mosquitoes with an antibody against DH_{31} impairs their capacity to excrete loads of Na^+ injected into their hemolymph (Coast, 2007).

Kinins are also considered important neuropeptides that contribute to diuresis. Mosquitoes express three kinin peptides (I, II, III) that are encoded by a single gene (Radford et al., 2004; Veenstra, 1994; Veenstra et al., 1997). In isolated Malpighian tubules of adult female *Ae. aegypti*, adding 'aedes'-kinins or heterologous kinins from roaches (ie, 'leuco'-kinins) to the peritubular bath nonselectively increases the rates of transepithelial NaCl, KCl, and fluid secretion (ie, a generic diuresis) using Ca^{2+} as the secondary messenger (Cady and Hagedorn, 1999b; Hayes et al., 1989; Pannabecker et al., 1993; Schepel et al., 2010; Veenstra et al., 1997; Wang et al., 1996; Yu and Beyenbach, 2002). Moreover, treatment of isolated hindguts of *Ae. aegypti* with aedeskinins increases the frequency of peristaltic contractions used for urine expulsion (Veenstra et al., 1997). Furthermore, aedeskinins stimulate urine excretion in vivo when injected into the hemolymph of adult female *Ae. aegypti* mosquitoes (Cady and Hagedorn, 1999a).

Lastly, diuretic hormone 44 (DH_{44}), also known as the corticotropic-releasing factor-like peptide, may also contribute to the postprandial diuresis. In isolated Malpighian tubules of adult female *Ae. aegypti* and *An. gambiae*, adding DH_{44} to the peritubular bath elicits a generic diuresis or a small natriuresis (compared to DH_{31}) using cAMP as a secondary messenger (Cady and Hagedorn, 1999b; Clark et al., 1998; Coast et al., 2005). The effects of DH_{44} treatment on contractions of the mosquito hindgut are unknown, but when injected into the hemolymph of adult female *Ae. aegypti* mosquitoes, it stimulates urine excretion in vivo (Cady and Hagedorn, 1999a).

With the above primer, we now focus the remaining sections of this review on the cellular and molecular mechanisms of ion and water transport in the Malpighian tubules of mosquitoes. For a recent comprehensive review of the cellular and molecular mechanisms of ion transport in the mosquito hindgut, we direct the reader to a chapter in a recent volume of *Advances in Insect Physiology* (Linser and Dinglasan, 2014). In general, our understanding of the cellular and molecular mechanisms of ion and water transport in the Malpighian tubules of mosquitoes is more advanced than that in the hindgut, because of (1) the simple gross and cellular anatomy of Malpighian tubules and (2) the ease of studying the physiology of Malpighian tubules in vitro, relative to the hindgut.

2. MALPIGHIAN TUBULES

2.1 Structure/Function

Mosquitoes possess five Malpighian tubules, regardless of species, life stage, or sex (Fig. 3A). Moreover, individual Malpighian tubules isolated from the same mosquito are very similar to one another in both form and function (Beyenbach et al., 1993). Thus, there does not appear to be anatomical or

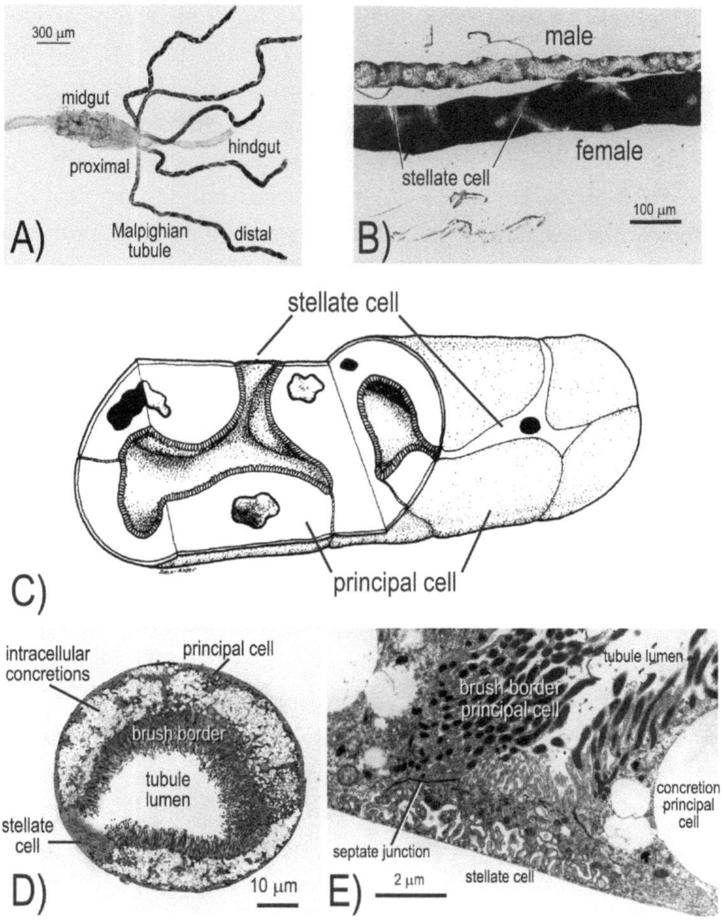

Fig. 3 Malpighian tubule anatomy in mosquitoes. (A) Isolated Malpighian tubules and alimentary canal of an adult female mosquito (*Ae. aegypti*). There are five Malpighian tubules, which each connect to the alimentary canal near the junction of the midgut and hindgut at their proximal ends and are blind-ended at their distal ends. (B) Sexual dimorphism in adult Malpighian tubules. The male tubule (*top*) is dwarfed in size by that of the female tubule (*bottom*). The principal cells accumulate mineralized concretions in their cytoplasm that block transmitted light and give them an opaque appearance. Stellate cells are transparent and are easily distinguished in the female tubule. (C) Illustration of a distal end of an isolated Malpighian tubule showing the two major cell types of the epithelium. Principal cells are fusiform; stellate cells contain a nucleated cell body and 3–4 arm-like projections that intercalate between principal cells. (D) Transverse section of an isolated Malpighian tubule observed on a transmission electron microscope (TEM) showing the general ultrastructure of a principal cell and an arm-like projection of a stellate cell. (E) Transverse section of an arm-like projection of a stellate cell and neighboring principal cells observed on a TEM. The principal cells are distinguished from the stellate cell by the presence of a mitochondrion-rich brush

(Continued)

functional specialization among individual tubules as occurs in *Drosophila melanogaster*, which possesses anatomically distinct anterior and posterior Malpighian tubules that are characterized by novel molecular and functional repertoires (Chintapalli et al., 2012; Wessing and Zierold, 1993). The general anatomy of Malpighian tubules does not differ greatly between larval and adult mosquitoes, but there is an obvious sexual dimorphism in adults. That is, the tubule length and diameter are, respectively, ~1.4 times and ~2 times greater in females compared to males (Fig. 3B), which correlates with a greater capacity for diuresis in the hematophagous females (Plawner et al., 1991). Independent of life stage and sex, each Malpighian tubule can be divided into two regions: (1) the distal segment (relative to the alimentary canal), which is blind-ended and comprises ~2/3 of the tubule length, and (2) the proximal segment, which is open-ended and comprises the remaining ~1/3 of the tubule length (Fig. 3A). The proximal segment attaches to the alimentary canal near the pylorus (junction of the midgut and hindgut) (Fig. 3A).

Both the distal and proximal segments of the Malpighian tubules consist entirely of epithelial cells, without nerve or muscle. The distal segment is composed of two epithelial cell types: principal cells and stellate cells (Fig. 3C). The principal cells are relatively large, fusiform cells that constitute the majority cell type of the epithelium and comprise over 90% of the tubule mass (Beyenbach and Piermarini, 2011; Cabrero et al., 2004; Piermarini and Gillen, 2015) (Fig. 3C). The stellate cells are relatively small, star-shaped cells that intercalate between principal cells, constitute the minority cell type of the epithelium, and comprise <10% of the tubule mass (Beyenbach and Piermarini, 2011; Cabrero et al., 2004; Piermarini and Gillen, 2015) (Fig. 3C). In addition to size and shape, principal cells are distinguished from stellate cells by (1) an extensive, luminal brush border membrane that houses mitochondria (Fig. 3D and E), (2) mineralized concretions throughout the

Fig. 3—Cont'd border and intracellular concretions. A septate junction is shown, which forms the paracellular pathway between the epithelial cells. *Panels (A, D, E) from Beyenbach, K.W., Piermarini, P. M., 2011. Transcellular and paracellular pathways of transepithelial fluid secretion in Malpighian (renal) tubules of the yellow fever mosquito* Aedes aegypti. *Acta Physiol. 202, 387–407, with permission. Panel (B) from Plawner, L., Pannabecker, T.L., Laufer, S., Baustian, M.D., Beyenbach, K.W., 1991. Control of diuresis in the yellow fever mosquito* Aedes aegypti*: evidence for similar mechanisms in the male and female. J. Insect Physiol. 37, 119–128, with permission. Panel (C) from Bradley, T.J., Stuart, A.M., Satir, P., 1982. The ultrastructure of the larval Malpighian tubules of a saline-water mosquito. Tissue Cell 14, 759–73, with permission.*

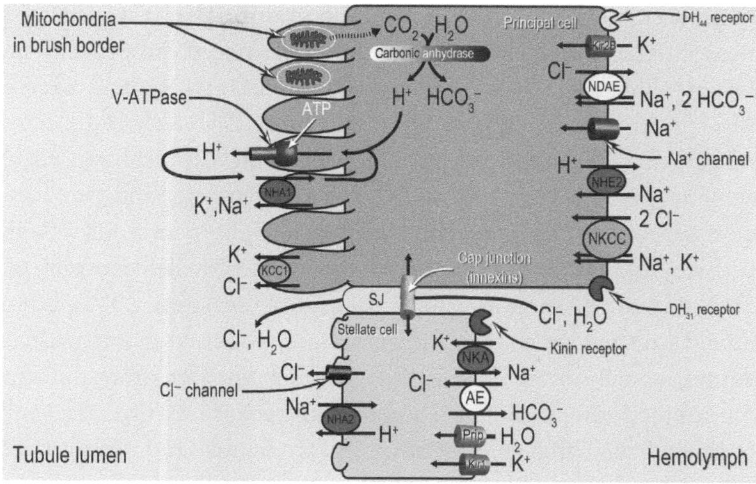

Fig. 4 Cellular and molecular mechanisms of transepithelial fluid secretion in mosquito Malpighian tubules (see Section 2.2 for details). The transport activities of NHA1 and NHA2 are hypothesized (see Section 2.2.3 for details). *AE*, anion exchanger; *ATP*, adenosine triphosphate; *DH*, diuretic hormone; *KCC*, K,Cl cotransporter; *Kir*, inward rectifier K^+ channel; *NDAE*, Na-driven anion exchanger; *NHA*, Na/H antiporter; *NHE*, Na/H exchanger; *NKA*, Na,K-ATPase; *NKCC*, Na,K,Cl cotransporter; *Prip*, *Pyroceoelia rufa* integral protein (an aquaporin); *SJ*, septate junction. *Modified from Esquivel, C.J., Cassone, B.J., Piermarini, P.M., 2016. A de novo transcriptome of the Malpighian tubules in non-blood-fed and blood-fed Asian tiger mosquitoes Aedes albopictus: insights into diuresis, detoxification, and blood meal processing. PeerJ 4, e1784, with permission.*

cytoplasm that give principal cells an opaque appearance compared to the translucent stellate cells (Fig. 3B, D, E), and (3) the expression of different suites of transport proteins (Fig. 4; discussed in Section 2.2) (Beyenbach and Piermarini, 2011; Piermarini and Gillen, 2015). The proximal segment is composed only of principal cells, which are smaller in size compared to those of the distal segment. It has been hypothesized that principal cells of the proximal segment contribute to solute reabsorption (Patrick et al., 2006), but direct physiological evidence is lacking.

2.2 Cellular and Molecular Mechanisms of Urine Production

The distal segments of mosquito Malpighian tubules are considered the primary sites of urine production via transepithelial fluid secretion. In brief, the epithelium actively transports Na^+ and K^+ from the hemolymph into the tubule lumen, which generates a favorable electrochemical gradient for Cl^- to follow passively. As NaCl and KCl are transported into the lumen, water follows its osmotic gradient resulting in urine that is isosmotic to the hemolymph. Classically, principal cells are considered the primary sites of active transcellular Na^+ and K^+ transport, whereas stellate cells are considered the primary sites of passive transcellular Cl^- and H_2O transport (Beyenbach and Piermarini, 2011). However, as will be discussed in the following sections, recent molecular, pharmacological, and physiological evidence appears to challenge this dichotomy. The paracellular (between cell) pathway is another potential route for transepithelial transport of Cl^- and H_2O via septate junctions (Beyenbach and Piermarini, 2011), but given the limited knowledge on the molecular components of this pathway in mosquitoes, it will not be discussed in this chapter. For more information on this subject, the reader is directed to recent reviews (Beyenbach, 2012; Beyenbach and Piermarini, 2011; Izumi and Furuse, 2014; Jonusaite et al., 2016).

Fig. 4 presents a model of the known or hypothesized cellular and molecular mechanisms that contribute to the production of urine in mosquito Malpighian tubules. The model is an amalgam based on various studies in Malpighian tubules of larval and adult female mosquitoes (*Ae. aegypti, Aedes albopictus, An. gambiae*) but is biased towards the distal segments of adult female *Ae. aegypti* mosquitoes, which have received the majority of experimental attention. However, the model should be widely applicable to various mosquito species and life stages, because (1) the qualitative expression of mRNAs encoding ion and water transport mechanisms is very similar between the Malpighian tubules of adult female *An. gambiae* and *Ae. albopictus* (Esquivel et al., 2016; Overend et al., 2015), and (2) in at least *An. gambiae*, there appears to be no major differences in the types of mRNAs encoding ion and water transporters that are expressed in the Malpighian tubules of larval and adult mosquitoes (Overend et al., 2015). Rather there are quantitative differences wherein mRNA expression levels of the ion and water transport mechanisms tend to be higher in the tubules of adult vs larval mosquitoes (Overend et al., 2015). Thus, the central mechanisms of ion and water transport in mosquito Malpighian tubules appear to be widely conserved among species and life stages, and it is through differential expression of these mechanisms that the physiology of the tubule can be appropriately 'tailored' to accommodate the particular challenge faced by the mosquito (Overend et al., 2015).

2.2.1 ATPases

The production of urine by Malpighian tubules is an active process that depends on the availability of intracellular ATP (Wu and Beyenbach, 2003).

In particular, ATP is used as a fuel to operate ion-motive ATPases, which are membrane-bound proteins that utilize the energy of ATP hydrolysis to translocate ions across plasma membranes, against their electrochemical gradient. Here, we focus on two particular ATPases known to play important roles in urine production by mosquito Malpighian tubules.

2.2.1.1 V-ATPase

The vacuolar or V-type H^+-ATPase (V-ATPase) is the dominant ATPase in the Malpighian tubule epithelium, contributing to over 50% of the total biochemical ATPase activity (Tiburcy et al., 2013; Weng et al., 2003). On the molecular level, the V-ATPase is a large protein complex consisting of two multisubunit sectors that together form the holoenzyme (Fig. 5A). The V_0 or membrane-bound sector translocates H^+ from the cytosol into a fluid compartment, and the V_1 or cytosolic sector hydrolyzes ATP, providing the energy for active H^+ transport (Beyenbach and Wieczorek, 2006). Transcriptomic studies in Malpighian tubules of *An. gambiae* and *Ae. albopictus* have detected the expression of a full complement of mRNAs encoding subunits that form the V_0 and V_1 sectors (Esquivel et al., 2014, 2016; Overend et al., 2015). In distal segments of Malpighian tubules in *Ae. aegypti* and *An. gambiae*, the immunoreactivities for subunits B and C of the V_1 sector localize to principal cells, but not stellate cells (Linser et al., 2012; Patrick et al., 2006; Piermarini et al., 2009; Tiburcy et al., 2013; Weng et al., 2003; Xiang et al., 2012) (Fig. 5B). The immunolabeling is usually associated with the apical brush border membrane of principal cells, but cytoplasmic labeling is also observed, which likely reflects the presence of V_1 sectors that are disassociated from the V_0 sectors (Tiburcy et al., 2013) (Fig. 5B).

The V-ATPase plays a pivotal physiological role in mosquito Malpighian tubules by establishing the necessary electrochemical gradients for transepithelial fluid secretion. That is, the V-ATPase selectively pumps H^+ from the cytosol to the tubule lumen, generating negative membrane voltages not only across the apical membrane, but also the basolateral membrane, of principal cells (Beyenbach et al., 2000). The membrane voltages are thought to contribute to (1) the secretion of Na^+ and/or K^+ into the tubule lumen via an apical cation–proton antiporter in principal cells, (2) the uptake of peritubular Na^+ and K^+ via basolateral ion channels in principal cells, and (3) the secretion of Cl^- into the tubule lumen via the paracellular pathway and/or apical Cl^- channels in stellate cells (Fig. 4). The critical role of the V-ATPase in transepithelial fluid secretion has been demonstrated by pharmacological studies with bafilomycin, a potent inhibitor of the V-ATPase. That is, in

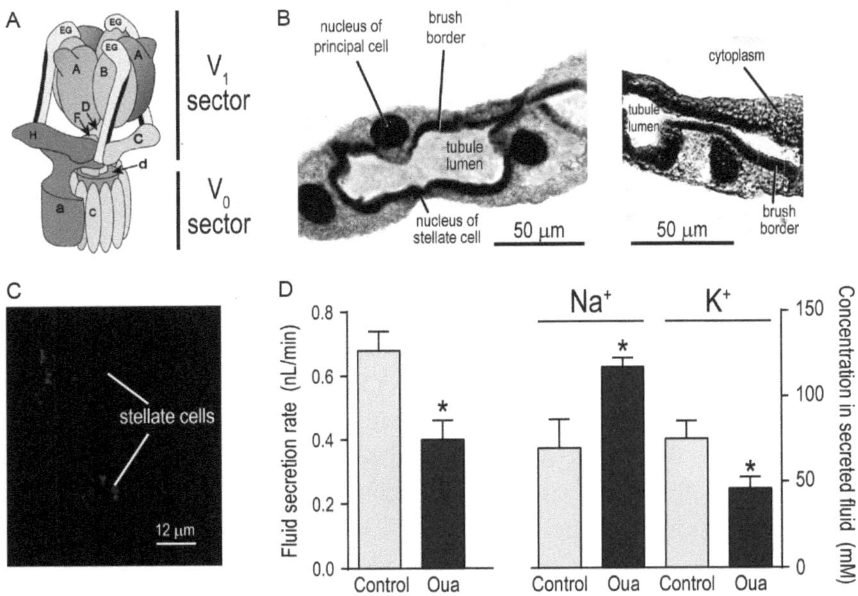

Fig. 5 ATPases in mosquito Malpighian tubules. (A) Schematic of the structure of the V-ATPase holoenzyme consisting of the membrane-bound V_0 sector and cytosolic V_1 sector. *Lowercase letters* indicate subunits of the V_0 sector; *uppercase letters* indicate subunits of the V_1 sector. The activity of the holoenzyme can be regulated via disassociation of the V_1 sector from V_0 sector. (B) Immunolocalization of subunit B of the V_1 sector in sections of Malpighian tubules isolated from adult female *Ae. aegypti*. Immunolabeling (*red stain*) primarily occurs in the brush border of principal cells but is also found in the cytoplasm of principal cells (see Section 2.2.1 for details). (C) Immunolocalization of the α-subunit of the NKA in Malpighian tubules isolated from larval *An. gambiae*. Immunolabeling (*blue*) occurs in stellate cells. (D) Effects of peritubular ouabain (Oua), an inhibitor of the NKA, on fluid secretion rates (*left*) and the concentrations of Na^+ and K^+ in the secreted fluid (*right*) in isolated Malpighian tubules of adult female *Ae. aegypti*. Values are means ± SEM ($N = 8$). *Asterisks* indicate significant difference ($P < 0.05$) from control tubules. *Panel (A) modified from Song, C.F., Papachristos, K., Rawson, S., Huss, M., Wieczorek, H., Paci, E., Trinick, J., Harrison, M.A., Muench, S.P., 2013. Flexibility within the rotor and stators of the vacuolar H^+-ATPase. PLoS One 8, e82207, with permission. Panel (B) modified from Piermarini, P.M., Weihrauch, D., Meyer, H., Huss, M., Beyenbach, K.W., 2009. NHE8 is an intracellular cation/H^+ exchanger in renal tubules of the yellow fever mosquito* Aedes aegypti. *Am. J. Physiol. Renal Physiol. 296, F730–F750, with permission. Panel (C) modified from Xiang, M.A., Linser, P.J., Price, D.A., Harvey, W.R., 2012. Localization of two Na^+- or K^+-H^+ antiporters, AgNHA1 and AgNHA2, in* Anopheles gambiae *larval Malpighian tubules and the functional expression of AgNHA2 in yeast. J. Insect Physiol. 58, 570–579, with permission. Panel (D) data from Hine, R.M., Rouhier, M.F., Park, S.T., Qi, Z., Piermarini, P.M., Beyenbach, K.W., 2014. The excretion of NaCl and KCl loads in mosquitoes: 1. Control data. Am. J. Physiol. Regul. Integr. Comp. Physiol. 307, R837–R849.* (See the colour plate.)

isolated Malpighian tubules of adult female *Ae. aegypti*, the addition of bafilomycin to the peritubular bath reduces the rates of transepithelial fluid secretion by ~80% and dramatically collapses the voltage across the epithelium (Beyenbach et al., 2000).

Given that the pumping of protons can only occur when the holoenzyme is assembled (ie, V_0 is bound to V_1), reversible disassembly of the V_0 and V_1 sectors is a potential mechanism for regulating the activity of the V-ATPase and thereby fluid secretion in Malpighian tubules. Recent studies in Malpighian tubules of *Ae. aegypti* have indeed shown that this is the case. In brief, under control unstimulated conditions, when the tubules secrete fluid at low rates, the immunolocalization of subunit C of the V_1 sector is highly variable within principal cells ranging from diffuse cytosolic labeling to prominent brush border labeling (Tiburcy et al., 2013). However, when the tubules are stimulated to secrete at high rates with DH_{31} or a membrane-permeant analog of cAMP (6-MP-cAMP) that duplicates the effects of DH_{31}, the immunolabeling of subunit C occurs exclusively in the brush border, suggesting high levels of holoenzyme formation during diuresis (Tiburcy et al., 2013). Consistent with the localization results, the biochemical and electrophysiological correlates of V-ATPase activity in the tubule epithelium both significantly increase after treatment with 6-MB-cAMP (Tiburcy et al., 2013). Thus, the activity of the V-ATPase and rates of transepithelial fluid secretion are dynamically and rapidly regulated in the tubule epithelium via reversible disassembly of the V_0 and V_1 sectors.

2.2.1.2 Na,K-ATPase

Compared to the V-ATPase, the Na,K-ATPase (NKA) is a relatively minor contributor (≤28%) to the total ATPase activity of Malpighian tubules (Tiburcy et al., 2013; Weng et al., 2003). On the molecular level, the NKA consists of α- and β-subunits. The α-subunit is catalytic and responsible for pumping Na^+ out of cells and K^+ into cells, whereas the β-subunit serves as a molecular chaperone and regulator of the α-subunit (Morth et al., 2011). In Malpighian tubules of *Ae. aegypti*, *Ae. albopictus*, and *An. gambiae* expression of mRNAs encoding both subunits has been detected (Esquivel et al., 2014, 2016; Overend et al., 2015; Patrick et al., 2006). In the distal segments of Malpighian tubules in *Ae. aegypti* and *An. gambiae*, immunoreactivity of the α-subunit localizes to the basolateral membranes of stellate cells, but not principal cells (Patrick et al., 2006; Xiang et al., 2012) (Fig. 5C). However, in proximal segments, immunolabeling of the α-subunit localizes to the basolateral membrane of principal cells (Patrick et al., 2006; Tsujimoto et al., 2013).

Despite its relatively low biochemical activity and restricted expression in the distal segment to stellate cells, pharmacological studies suggest that the NKA plays an important role in transepithelial fluid secretion. That is, in isolated Malpighian tubules of adult female *Ae. aegypti*, the addition of ouabain (an NKA inhibitor) to the peritubular bath reduces the rates of transepithelial fluid secretion by ~40–45% (Hegarty et al., 1991; Hine et al., 2014) (Fig. 5D). Moreover, ouabain decreases the K^+ concentration and increases the Na^+ concentration of the secreted fluid (Hine et al., 2014) (Fig. 5D). Thus, the NKA in stellate cells appears to contribute to the (1) basolateral uptake of peritubular K^+ for its eventual secretion across the apical membrane and (2) the basolateral extrusion of Na^+ from the epithelium into the peritubular fluid (Fig. 4).

2.2.2 Basolateral Mechanisms of Ion Uptake

As mentioned in Section 2.2, the Malpighian tubule epithelium transports ions and water from the hemolymph into the tubule lumen to produce urine. Important to this process is the uptake of Na^+, K^+, and Cl^- across the basolateral membrane from the hemolymph into the tubule epithelium. While the NKA of stellate cells is an example of one particular mechanism for K^+ uptake (see Section 2.2.1), several other mechanisms are also known to contribute to basolateral ion uptake. Here we discuss these mechanisms.

2.2.2.1 K^+ Channels

Barium-sensitive K^+ channels are a major pathway for the uptake of K^+ into the Malpighian tubule epithelium. Namely, in isolated Malpighian tubules of *Ae. aegypti*, barium (Ba^{2+}) blocks over 60% of the basolateral membrane conductance and inhibits the transepithelial secretion of fluid and K^+ by over 55% and 80%, respectively (Beyenbach and Masia, 2002; Scott et al., 2004). Notably, the effects of Ba^{2+} on tubule electrophysiology are potent, with obvious channel block observed when using as little as 50 μM Ba^{2+} in the peritubular bath (Masia et al., 2000), suggesting the involvement of inwards rectifier K^+ (Kir) channels.

Mosquitoes possess four distinct clades of genes that encode Kir channel subunits: Kir1, Kir2A, Kir2B, and Kir3. Some of these clades are represented by more than one gene depending on the species (eg, Kir2B is duplicated in *Ae. aegypti*; Kir2A and Kir3 are duplicated in *An. gambiae*) (Piermarini et al., 2013; Raphemot et al., 2014a). The Malpighian tubules of adult female *Ae. aegypti*, *Ae. albopictus*, and *An. gambiae* express representative mRNAs from each clade (*Kir1*, *Kir2A*, *Kir2B*, and *Kir3*); the expression levels of the *Kir2B* and *Kir3* mRNAs are more abundant than those of *Kir1* and *Kir2A* (Esquivel et al., 2016; Piermarini et al., 2013; Raphemot et al., 2014a) (Fig. 6A). The relative abundances of the *Kir1*, *Kir2B*, and *Kir3* mRNAs

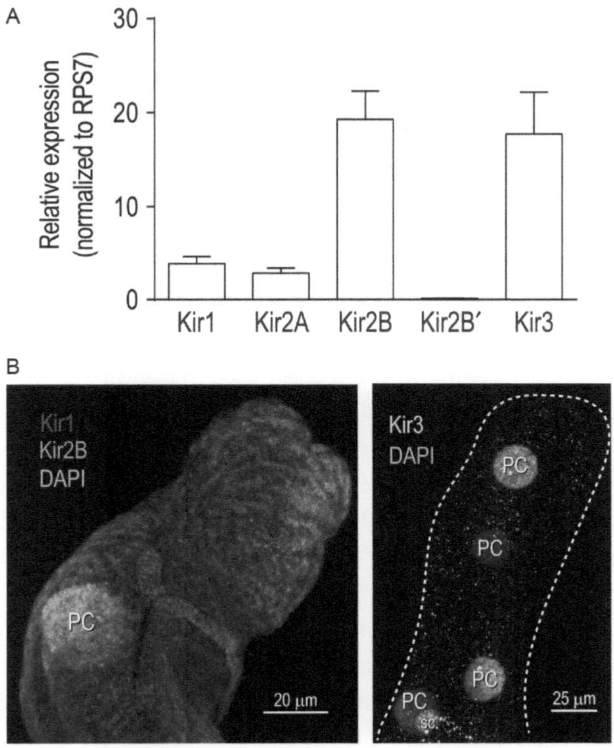

Fig. 6 Molecular and immunochemical expression of Kir subunits in mosquito Malpighian tubules. (A) Relative expression levels of Kir subunit mRNAs as determined by quantitative RT-PCR (qPCR) in isolated Malpighian tubules of adult female *Ae. aegypti*. Ribosomal protein S7 (RPS7) was used as a reference gene. The expression levels of *Kir2B* and *Kir3* are ~five times greater than those of *Kir1* and *Kir2A*. Values are means ± SEM ($N = 3$). Data are previously unpublished. (B) Immunolocalization of Kir1 (*magenta*), Kir2B (*green*), and Kir3 (*yellow*) subunits in whole mounts of isolated Malpighian tubules from adult female *Ae. aegypti*. Nuclei are labeled with DAPI (*cyan*). Note the segregated immunolabeling of Kir1 and Kir2B to the basolateral surface of principal and stellate cells, respectively (*left*), and the punctate intracellular localization of Kir3 throughout the epithelium (*right*). 'PC' and 'sc' indicate nuclei of principal and stellate cells, respectively. The *dashed line* (*right*) outlines the boundary of the tubule epithelium. *Modified from Piermarini, P.M., Dunemann, S.M., Rouhier, M.F., Calkins, T.L., Raphemot, R., Denton, J.S., Hine, R.M., Beyenbach, K.W., 2015. Localization and role of inward rectifier K channels in Malpighian tubules of the yellow fever mosquito* Aedes aegypti. *Insect Biochem. Mol. Biol. 67, 59–73, with permission.* (See the colour plate.)

are consistent with the localization of their respective immunoreactivities in the epithelium. That is, using subunit-specific antibodies, we have shown that Kir1 and Kir2B immunoreactivities, respectively, localize to the basolateral membranes of the minority stellate cells and majority principal cells in Malpighian tubules of adult female *Ae. aegypti* (Piermarini et al., 2015) (Fig. 6B). Kir3 immunolabeling occurs throughout the epithelium in intracellular compartments of both stellate and principal cells (Piermarini et al., 2015) (Fig. 6B). Although antibodies for Kir2A are not yet available, it is reasonable to hypothesize that it localizes to stellate cells given its similar mRNA expression levels as *Kir1* (Fig. 6A).

To functionally characterize the channels encoded by the Kir subunit mRNAs, we have cloned the cDNAs of *Kir1*, *Kir2B*, and *Kir3* in Malpighian tubules of *Ae. aegypti* and expressed them heterologously in *Xenopus* oocytes (Piermarini et al., 2013). The Kir3 subunit does not form functional homomeric K^+ channels in the plasma membrane of *Xenopus* oocytes, which is consistent with its localization to intracellular compartments of Malpighian tubules (Fig. 6B). In contrast, the Kir1 and Kir2B subunits each form homomeric channels in *Xenopus* oocytes that (1) mediate inward-rectifying currents for K^+, (2) are blocked by Ba^{2+}, and (3) preferentially conduct K^+ and Tl^+ over other monovalent cations (Piermarini et al., 2013). However, Kir1 distinguishes itself from Kir2B in that it (1) mediates robust inward K^+ currents in low extracellular [K^+], (2) is activated by extracellular high [Na^+], and (3) does not conduct Cs^+ (Piermarini et al., 2013). In collaboration with the Denton laboratory, we have also shown that the heterologously expressed Kir1 and Kir2B channels exhibit different sensitivities to three structurally unique small molecule modulators of mosquito Kir channels: VU573, VU590, and VU625. That is, (1) VU573 inhibits Kir1 but activates Kir2B, (2) VU590 inhibits Kir1 but does not affect Kir2B, and (3) VU625 preferentially inhibits Kir1 vs Kir2B (Raphemot et al., 2013, 2014b; Rouhier et al., 2014b). Taken together, these data suggest that the Kir2B channels in principal cells and Kir1 channels in stellate cells possess novel functional, regulatory, and pharmacological properties.

We have taken advantage of the novel pharmacologies of Kir1 and Kir2B to dissect their relative contributions to transepithelial K^+ transport in isolated Malpighian tubules of *Ae. aegypti*. In brief, inhibiting Kir1 with peritubular VU590 blocks 46% of the transepithelial K^+ secretion, while inhibiting both Kir1 and Kir2B with peritubular VU625 blocks 66% of the transepithelial K^+ secretion (Piermarini et al., 2015). Thus, Kir2B in principal cells contributes to only 20% of the transepithelial K^+ transport, whereas Kir1 in stellate cells contributes to 46% of the transepithelial K^+ transport. The data suggest that stellate cells, through the combined activities of Kir1 and the NKA (see Section 2.2.1), play major roles in the uptake of K^+ across the basolateral membrane of mosquito Malpighian tubules (Fig. 4).

2.2.2.2 Na^+ Channels

The molecular mechanisms of conductive Na^+ uptake across the basolateral membrane of mosquito Malpighian tubules are not as well characterized as those for K^+. Electrophysiological studies in the Malpighian tubules of adult female *Ae. aegypti* have demonstrated that (1) ~15% of the basolateral membrane conductance of principal cells is attributed to Na^+ (Beyenbach and Masia, 2002), and (2) treatment of the tubules with a membrane-permeant analog of cAMP (db-cAMP) or DH_{31} opens a Na^+ conductance in the basolateral membrane that is blocked by amiloride, a generic inhibitor of Na^+ channels and transporters (Beyenbach and Masia, 2002; Coast et al., 2005; Petzel et al., 1985; Sawyer and Beyenbach, 1985). Presumably, the conductance is mediated by Na^+ channels of the epithelial Na^+ channel/degenerin family, which are encoded by '*pickpocket*' genes in insects (Zelle et al., 2013), but molecular expression studies of Na^+ channels in mosquito Malpighian tubules are limited. We have detected the expression of at least one mRNA encoding a putative pickpocket (*ppk*) channel in the Malpighian tubule transcriptome of adult female *Ae. albopictus* (an ortholog of *ppk13* in *D. melanogaster* and AAEL006613 in *Ae. aegypti*) (Esquivel et al., 2014, 2016). In the Malpighian tubules of *An. gambiae*, the expression levels of the orthologous *ppk* gene (AGAP007945) are enriched compared to those in the whole insect (http://mozatlas.gen.cam.ac.uk/mozatlas/; http://Moztubules.org). Thus, it is reasonable to hypothesize that the *ppk13* ortholog of mosquitoes encodes a channel that contributes to the amiloride-sensitive Na^+ conductance in Malpighian tubules.

An alternate hypothesis is that the amiloride-sensitive Na^+ conductance is mediated by an electrogenic Na^+ transporter rather than a Na^+ channel. We have recently shown that the most abundantly expressed mRNA encoding a putative ion transport mechanism in the Malpighian tubules of adult female *Ae. albopictus* is a putative K^+-dependent Na^+/Ca^{2+} exchanger (NCKX) (Esquivel et al., 2016). In vertebrates, Na^+/Ca^{2+} exchangers mediate the electrogenic uptake of three Na^+ in exchange for one Ca^{2+} and are known to be amiloride sensitive (Terracciano and Hancox, 2013). Thus, the activity of an amiloride-sensitive electrogenic NCKX in the basolateral membrane of principal cells would be difficult to distinguish from that of an amiloride-sensitive Na^+ channel with standard electrophysiological approaches. Future studies that compare the membrane localization and functional/pharmacological properties of ppk13 and NCKX will be required to ascertain whether one or both of these mechanisms play a major role in the electroconductive uptake of Na^+ by mosquito Malpighian tubules.

2.2.2.3 Na/H Exchangers

Electroneutral Na/H exchangers (NHEs) of the solute carrier 9 (SLC9) superfamily of cation–proton antiporters are considered an important pathway for the basolateral uptake of Na^+ in the tubule epithelium. In unstimulated Malpighian tubules of adult female *Ae. aegypti*, peritubular amiloride has minor effects on the electrophysiology of the epithelium but substantially reduces the rates of transepithelial fluid and Na^+ secretion by ~60% and ~70%, respectively (Hegarty et al., 1992). Furthermore, peritubular EIPA, an amiloride analog with higher selectivity for NHEs vs Na^+ channels, potently inhibits (1) transepithelial fluid secretion and (2) the Na^+-dependent extrusion of H^+ from principal cells (Petzel, 2000), suggesting the presence of a basolateral amiloride/EIPA-sensitive NHE.

Mosquitoes possess three genes that encode putative NHEs: *NHE1*, *NHE2*, and *NHE3* (following the nomenclature of *D. melanogaster*) (Giannakou and Dow, 2001; Rheault et al., 2007). In Malpighian tubules of *Ae. albopictus* and *An. gambiae*, *NHE2* is the most abundant NHE mRNA expressed, followed by *NHE1* and *NHE3* (Esquivel et al., 2016; Overend et al., 2015). The Gill and Petzel laboratories cloned cDNAs for *NHE2* (an ortholog of mammalian *NHE3*) from the Malpighian tubules of *Ae. aegypti* (Hart et al., 2002; Pullikuth et al., 2006). As expected, NHE2 mediates Na/H exchange when expressed heterologously in fibroblast cells, and its immunoreactivity localizes to the basolateral membrane of principal cells (Pullikuth et al., 2006). However, NHE2 is weakly inhibited by amiloride and EIPA (Pullikuth et al., 2006), which raises questions as to whether NHE2 is responsible for the potent inhibitory effects of EIPA in the aforementioned physiological and pharmacological studies in *Ae. aegypti* Malpighian tubules.

NHE1, cloned from the Malpighian tubules of adult female *Ae. aegypti*, mediates EIPA-sensitive Na/H exchange when expressed heterologously in yeast, fibroblasts, and *Xenopus* oocytes, but it does not localize to the basolateral membrane of principal cells (Kang'ethe et al., 2007; Piermarini et al., 2009). The pharmacology of mosquito NHE3 is unknown, but it is an ortholog of mammalian intracellular NHEs, which makes it unlikely to localize to the plasma membrane. Thus, NHE2 remains the best candidate for the basolateral NHE in mosquito Malpighian tubules. However, without a potent pharmacological blocker, reverse genetics studies will likely be required to confirm the contributions of NHE2 to Na^+ uptake and pH_i regulation in principal cells.

2.2.2.4 Na,K,2Cl Cotransporters

Electroneutral Na,K,2Cl cotransporters (NKCCs) of the SLC12 superfamily of cation-coupled chloride cotransporters are considered important mechanisms of Na^+ and K^+ uptake across the basolateral membrane of Malpighian tubules in adult female mosquitoes. That is, in unstimulated Malpighian tubules of adult female *Ae. aegypti*, peritubular bumetanide (a preferential inhibitor of NKCCs) has insignificant effects on the electrophysiology of the epithelium and rates of transepithelial fluid secretion (Hegarty et al., 1991). However, bumetanide decreases the $[K^+]$ and increases the $[Na^+]$ in the secreted fluid, and when used in combination with Ba^{2+}, it inhibits nearly all transepithelial secretion of fluid, K^+, and Na^+, suggesting a prominent role of an NKCC together with Kir channels in K^+ uptake under resting conditions (Hegarty et al., 1991). Moreover, (1) when the tubules are treated with db-cAMP, bumetanide reverses the stimulation of fluid and Na^+ secretion, and (2) if tubules are preincubated with bumetanide, the natriuretic effects of db-cAMP are blocked (Hegarty et al., 1991). Thus, a NKCC appears to play an important role during natriuresis in Malpighian tubules of adult females. However, when Malpighian tubules of larval *Ae. aegypti* are stimulated to secrete fluid at diuretic rates with db-cAMP or 5-hydroxytryptamine (serotonin), bumetanide does not inhibit rates of transepithelial fluid, Na^+, or K^+ secretion (Clark and Bradley, 1996). Thus, if a NKCC contributes to fluid secretion in tubules of larval mosquitoes, then it is not activated by db-cAMP, and/or it is pharmacologically distinct from the bumetanide-sensitive NKCC in the tubules of adult females.

Mosquitoes possess three genes that encode putative SLC12-like NKCCs. One of these genes (AAEL006180 in *Ae. aegypti*; AGAP001557 in *An. gambiae*) shares a common phylogenetic ancestry with vertebrate NKCCs and is an ortholog of *ncc69* (CG4357) in *D. melanogaster* (Piermarini and Gillen, 2015). Ncc69 is a bona fide bumetanide-sensitive NKCC that contributes to transepithelial fluid and K^+ secretion in Malpighian tubules of *D. melanogaster* (Leiserson et al., 2011; Rodan et al., 2012). The other two genes (AAEL009888 and AAEL009886 in

Ae. aegypti; AGAP003275 and AGAP003274 in *An. gambiae*) are more distantly related to vertebrate NKCCs and their functional/pharmacological properties are unknown (Piermarini and Gillen, 2015). In collaboration with the Gillen laboratory, we have measured the expression of mRNAs for the *ncc69* ortholog and AAEL00988 in the Malpighian tubules of adult female *Ae. aegypti*; the expression levels for both are ~10-fold higher in the tubules than in the midgut (Piermarini and Gillen, 2015). In Malpighian tubules of larval and adult female *An. gambiae*, the expression levels of all three putative NKCC mRNAs are detectable, but not enriched compared to levels in the whole insect (Overend et al., 2015) (http://mozatlas.gen.cam.ac.uk/mozatlas/; http://Moztubules.org). Thus, it will be necessary to characterize the functional and pharmacological properties of the putative mosquito NKCCs and localize their membrane expression in the Malpighian tubule epithelium before it can be better hypothesized as to which transporters contribute to the aforementioned effects of bumetanide on tubule physiology.

2.2.2.5 Bicarbonate Transporters

In vertebrate epithelial cells expressing a V-ATPase in the apical membrane, a Cl/HCO_3 exchanger is often found in the opposing basolateral membrane to promote bicarbonate (HCO_3^-) reabsorption (eg, α-intercalated cells, osteoclasts) (Brown and Breton, 1996). The Cl/HCO_3 exchanger belongs to the SLC4 superfamily of bicarbonate transporters, which includes the Na^+-coupled bicarbonate transporters (NCBTs) and Na^+-independent Cl/HCO_3 anion exchangers (AEs) (Romero et al., 2013). Most SLC4 transporters are inhibited by stilbene derivatives, such as SITS and DIDS.

Initial studies by the Beyenbach lab found that SITS modestly, but significantly, inhibited rates of transepithelial fluid secretion in unstimulated Malpighian tubules of adult female *Ae. aegypti* (Hegarty et al., 1991), suggesting the pharmacological presence of a SLC4-like transporter in the epithelium. Nearly 20 years later, the Beyenbach lab cloned a cDNA encoding a SLC4-like AE from the Malpighian tubules of adult female *Ae. aegypti* and demonstrated that (1) the AE mediates DIDS-sensitive, electroneutral Cl/HCO_3 exchange when expressed heterologously in *Xenopus* oocytes (Fig. 7A) and (2) AE immunoreactivity localizes to the basolateral membrane of stellate cells, not principal cells (Fig. 7B) (Piermarini et al., 2010). Subsequent immunochemical studies by the Linser laboratory confirmed that stellate cells express the AE in Malpighian tubules of larval and adult female *An. gambiae* and *Ae. aegypti* (Linser et al., 2012).

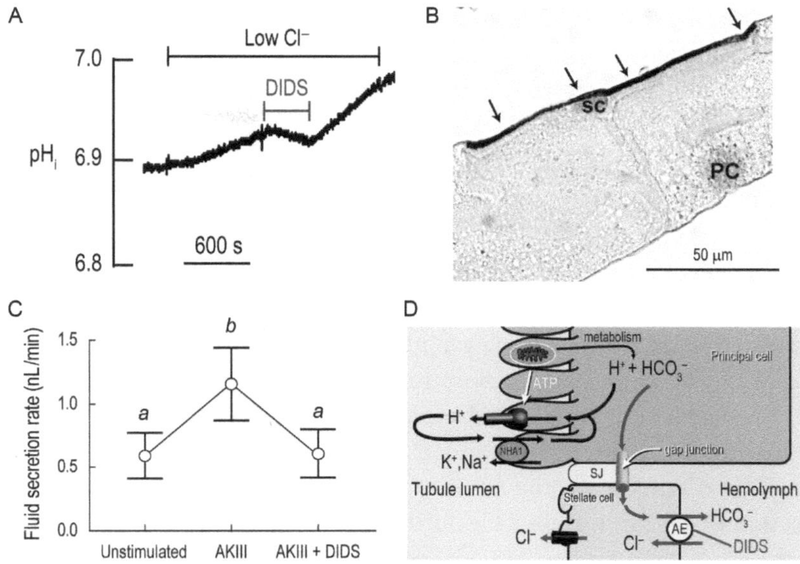

Fig. 7 The DIDS-sensitive AE. (A) Representative tracing of intracellular pH (pH_i) in a *Xenopus* oocyte heterologously expressing the AE cloned from the Malpighian tubules of adult female *Ae. aegypti*. The pH_i increases upon lowering extracellular [Cl^-], which is consistent with the activity of a Cl/HCO_3 exchanger. The increase of pH_i is reversibly blocked by DIDS. (B) immunolocalization of the AE in a section of a Malpighian tubule isolated from an adult female *Ae. aegypti*. Immunolabeling (*red* (*black* in the print version) *stain*) occurs exclusively along the basolateral membrane of a stellate cell (*arrows*). 'PC' and 'sc' indicate nuclei of principal and stellate cells, respectively. (C) Effects of peritubular DIDS on fluid secretion rates in isolated Malpighian tubules of adult female *Ae. aegypti* after stimulation with aedeskinin III (AKIII). DIDS reverses the effects of AKIII on fluid secretion. Values are means ± SEM ($N = 8$). *Lowercase letters* indicate statistical categorization of the means. (D) Simplified model of ion transport in mosquito Malpighian tubules to explain how DIDS may selectively inhibit the diuretic rates of fluid secretion. *Red* (*grey* in the print version) *arrows* indicate transport steps that are potentially affected by DIDS. In brief, when stimulated by diuretic peptides, the metabolism of the epithelium is expected to increase resulting in elevated production of H^+ and HCO_3^- in principal cells. The AE in stellate cells may act as a sink for principal cell HCO_3^- through gap junctions, which connect the cytoplasm of the two cells. Thus, inhibition of the AE with DIDS may lead to the accumulation of HCO_3^- in principal cells, which may buffer away intracellular H^+ used as a substrate by the V-ATPase and/or inhibit mitochondrial ATP synthesis (Beyenbach and Piermarini, 2011; Piermarini and Calkins, 2014; Piermarini et al., 2010). In addition, inhibition of the AE with DIDS may reduce the availability of intracellular Cl^- for secretion via apical Cl^- channels. *Panels (A–C) modified from Piermarini, P.M., Grogan, L.F., Lau, K., Wang, L., Beyenbach, K.W., 2010. A SLC4-like anion exchanger from renal tubules of the mosquito (*Aedes aegypti*): evidence for a novel role of stellate cells in diuretic fluid secretion. Am. J. Physiol. Regul. Integr. Comp. Physiol. 298, R642–R660, with permission.*

Interestingly, in isolated Malpighian tubules of adult female *Ae. aegypti* that are unstimulated, adding peritubular DIDS does not significantly affect the basal rates of transepithelial fluid secretion, suggesting a nominal role of the AE in fluid secretion under resting conditions (Piermarini et al., 2010). However, in the presence of diuretic peptides (ie, DH_{31} or aedeskinin III), DIDS reverses the stimulated rates of fluid secretion to levels that resemble unstimulated tubules (Fig. 7C), suggesting that the AE plays a selective role in mediating diuretic rates of fluid secretion when the metabolism of the epithelium is elevated (Piermarini et al., 2010). Assuming that stellate cells are connected to principal cells via gap junctions, it is possible that the AE provides a pathway for the reabsorption of HCO_3^- generated in principal cells during periods of elevated metabolism (ie, diuresis) (Beyenbach and Piermarini, 2011; Piermarini and Calkins, 2014; Piermarini et al., 2010) (Fig. 7D). Alternatively, the AE may contribute directly to the transcellular secretion of Cl^- in stellate cells during diuresis by providing a mechanism for the basolateral uptake of Cl^- into the epithelium (Fig. 7D). The two possibilities are not mutually exclusive.

In addition to the AE, Malpighian tubules of *Ae. aegypti* and *An. gambiae* express a NCBT, the Na^+-driven AE (NDAE) (Linser et al., 2012; Yamahiro et al., 2008). When expressed in *Xenopus* oocytes, the NDAE of *An. gambiae* mediates the cotransport of Na^+ and $2HCO_3^-$ in exchange for Cl^- (Linser et al., 2012). Although not confirmed experimentally, the mosquito NDAE is expected to be weakly sensitive to DIDS, because it lacks the conserved 'DIDS-binding' motifs found in other SLC4 transporters (Piermarini et al., 2010; Yamahiro et al., 2008). Immunolabeling studies in Malpighian tubules of larval *An. gambiae* indicate that NDAE immunoreactivity localizes to the basolateral membrane of principal cells, where it may contribute to the uptake of Na^+ and HCO_3^- from the hemolymph (Linser et al., 2012). In adult female *Ae. aegypti*, immunochemical experiments indicate that the AE is the dominant SLC4-like transporter expressed in the Malpighian tubules, whereas the NDAE is the dominant SLC4-like transporter expressed in the midgut (Piermarini and Gillen, 2015). Thus, a functional role of the NDAE in Malpighian tubules remains to be determined.

2.2.3 Apical Mechanisms of Ion Secretion

Section 2.2.2 summarized the various mechanisms by which ions enter the mosquito Malpighian tubule epithelium across the basolateral membrane. Here the complementary mechanisms of ion secretion across the apical membrane, from the cytosol into the tubule lumen, are discussed. Compared to numerous mechanisms of basolateral ion uptake (Section 2.2.2), only a few mechanisms of apical ion secretion have been identified to date.

2.2.3.1 Na/H Antiporters

The apical membrane of principal cells is hypothesized to be the site of a putative cation/H^+ exchange mechanism that would couple the electrochemical gradient established by the V-ATPase to the secretion of Na^+ and/or K^+ into the tubule lumen. The working hypothesis proposes that the exchanger mediates electrogenic Na^+/nH^+ and/or K^+/nH^+ exchange, which would allow the negative apical membrane voltage to drive the uptake of luminal H^+ in exchange for Na^+ and/or K^+ (Beyenbach and Piermarini, 2011; Wieczorek et al., 2009). However, this so-called Wieczorek exchanger—named after the pioneering work of Helmut Wieczorek and colleagues in the lepidopteran midgut (Wieczorek, 1992)—has proven elusive to identify.

To date, Na/H antiporters (NHAs) are the best candidates for the putative Wieczorek exchanger in mosquito Malpighian tubules. Although the NHAs belong to the SLC9 superfamily of cation–proton antiporters, they are distantly related to the aforementioned NHEs (ie, *NHE1*, *NHE2*, *NHE3*) in Section 2.2.2. Mosquitoes possess two genes encoding NHAs: *NHA1* and *NHA2* (Rheault et al., 2007). The mRNAs for both NHAs are expressed in Malpighian tubules of *An. gambiae*, but (1) the abundance of *NHA2* mRNAs in the tubules of both larvae and adults is at least 10 times greater than that of *NHA1* (Overend et al., 2015) (http://mozatlas.gen.cam.ac.uk/mozatlas/; http://Moztubules.org), (2) mRNA expression levels of *NHA2* are enriched in the tubules compared to the whole insect, whereas those of *NHA1* are not (Overend et al., 2015) (http://mozatlas.gen.cam.ac.uk/mozatlas/; http://Moztubules.org), (3) mRNA expression levels of *NHA2* are the highest in Malpighian tubules of adults compared to other tissues (http://mozatlas.gen.cam.ac.uk/mozatlas/), and (4) mRNA expression levels of *NHA1* are among the lowest in the Malpighian tubules compared to other tissues (http://mozatlas.gen.cam.ac.uk/mozatlas/) (Rheault et al., 2007). Similarly, in Malpighian tubules of adult female *Ae. albopictus*, mRNA expression levels of *NHA2* are ~10-fold greater than those of *NHA1* (Esquivel et al., 2016).

Consistent with the relative mRNA expression levels, NHA1 immunoreactivity in Malpighian tubules of larval *An. gambiae* appears to be expressed primarily in principal cells of the proximal segments vs those of the distal segments (Okech et al., 2008; Rheault et al., 2007). However, in these principal cells, NHA1 colocalizes with the V-ATPase in the apical brush border, consistent with the anticipated location of the Wieczorek exchanger (Xiang et al., 2012) (Fig. 4). On the other hand, NHA2 localizes to the apical membrane of stellate cells in the distal segments of Malpighian tubules in larval *An. gambiae* (Xiang et al., 2012) (Fig. 4). Thus, the mRNA expression and immunochemical data suggest that NHA2 is the dominant NHA in the distal secretory segment.

Functional characterizations of mosquito NHAs are limited. Heterologous expression of *An. gambiae* NHA2 in yeast suggests that it is capable of transporting Na^+, K^+, and Li^+ (Xiang et al., 2012). Parallel studies on mosquito NHA1 were not performed. Initial functional characterization studies of the orthologous NHAs of *D. melanogaster* expressed heterologously in yeast suggested that NHA1 and NHA2 preferentially transport K^+ and Na^+, respectively (Day et al., 2008). More recent heterologous expression studies in *Xenopus* oocytes have demonstrated that *D. melanogaster* NHA2 at least mediates Na/H exchange in an electroneutral manner (Chintapalli et al., 2015). Surprisingly, oocytes-expressing NHA1 appear to have an enhanced activity of H^+,Cl^- cotransport or Cl/OH exchange, but not Na/H exchange (Chintapalli et al., 2015). However, the potential for *D. melanogaster* NHA1 and NHA2 to mediate K/H exchange was not assessed, leaving the possibilities open that (1) NHA1 is a selective K/H exchanger, and (2) NHA2 can mediate K/H exchange in addition to Na/H exchange.

Assuming that NHA2 of mosquitoes functions similarly to that of *D. melanogaster*, then the apical membrane of stellate cells appears to at least possess an electroneutral mechanism of Na/H exchange, but as mentioned earlier, potential K/H exchange by NHA2 cannot be ruled out. Thus, NHA2 could potentially participate in the reabsorption of luminal Na^+ in exchange for H^+ secretion (Fig. 4) or the secretion of intracellular K^+ in exchange for luminal H^+. The role of NHA1 in principal cells is unclear, but it would appear to be well positioned for the secretion of intracellular Na^+ or K^+ in exchange for luminal H^+ (Fig. 4). However, given its abundance in the proximal segment, it cannot be ruled out that NHA1 plays a role in reabsorbing Na^+ or K^+ from the lumen in exchange for secreting H^+, which may help limit urinary cation loss in mosquito larvae living in fresh water. Additional functional and pharmacological characterizations of mosquito NHAs, as well as further immunochemical localizations in Malpighian tubules (eg, in adult females), are clearly needed to better assess the potential physiological roles they perform in mosquito Malpighian tubules.

Thus, despite much progress, the identity of the putative Wieczorek exchanger in mosquito Malpighian tubules remains an enigma.

2.2.3.2 K,Cl Cotransporters

After discovering a SLC4-like AE in stellate cells of Malpighian tubules of *Ae. aegypti* (see Section 2.2.2), the Beyenbach laboratory tested the hypothesis that the AE was functionally coupled to a K,Cl cotransporter (KCC) of the SLC12 superfamily for the recycling of Cl^- across the basolateral membrane, as occurs in α-intercalated cells of the mammalian renal collecting tubule (Jentsch, 2005). Consistent with this hypothesis, they cloned a cDNA encoding a putative KCC (*KCC1*) from the Malpighian tubules of adult female *Ae. aegypti* and demonstrated that when expressed in *Xenopus* oocytes KCC1 mediates Na^+-independent, Cl^--dependent uptake of $^{86}Rb^+$ (a radiotracer for K^+) that is activated by cell swelling and inhibited by DIOA (Fig. 8A), hallmarks of canonical KCC activity (Piermarini et al., 2011).

However, immunolabeling of KCC1 occurred in the brush border of principal cells and not in the basolateral membrane of stellate cells (Piermarini et al., 2011) (Fig. 8B). Given that KCCs typically mediate the export of K^+ and Cl^- from cells, it was reasoned that an apical KCC in Malpighian tubules may contribute to the electroneutral secretion of K^+ and Cl^- into the tubule lumen (Fig. 4). Indeed, addition of DIOA to the peritubular bath of isolated Malpighian tubules significantly inhibited unstimulated and diuretic rates of transepithelial fluid secretion (Piermarini et al., 2011). Moreover, the inhibition by DIOA was without effect on tubule electrophysiology, suggesting that KCC1 is a key component of an electroneutral pathway in principal cells of the distal segment for transcellular KCl secretion (Piermarini et al., 2011). It is tempting to speculate that the activity of KCC1 is functionally coupled to a bumetanide-sensitive NKCC in the basolateral membrane of principal cells, thereby allowing the basolateral entry of KCl via the NKCC to be matched by its apical exit via KCC1 (Fig. 4).

2.2.3.3 Cl^- Channels

Along with septate junctions of the paracellular pathway, stellate cells are considered important sites of transepithelial Cl^- secretion in mosquito Malpighian tubules (Fig. 4). Using patch clamping on the apical membrane of stellate cells from unstimulated Malpighian tubules of *Ae. aegypti*, the Beyenbach lab identified Cl^- conductances, consistent with a hypothesized role of these cells in Cl^- secretion (O'Connor and Beyenbach, 2001).

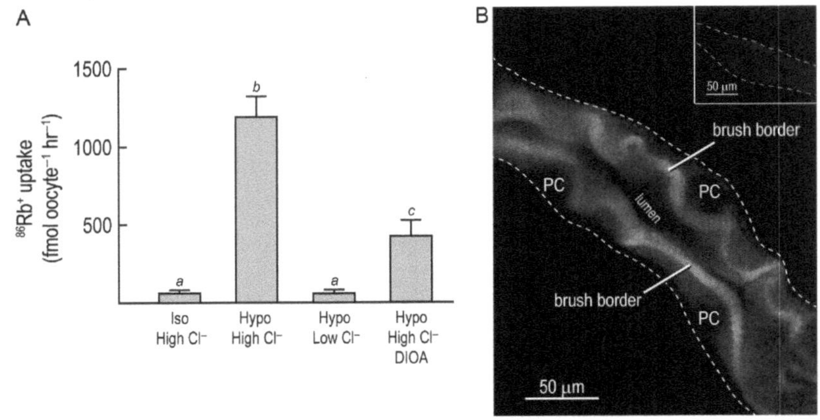

Fig. 8 The DIOA-sensitive KCC. (A) $^{86}Rb^+$ uptake rates in *Xenopus* oocytes heterologously expressing KCC1 cloned from the Malpighian tubules of adult female *Ae. aegypti*. The uptake of $^{86}Rb^+$, which is a tracer for K^+ transport, is stimulated by cell swelling, Cl^- dependent, and inhibited by DIOA. 'Iso' and 'Hypo' indicate incubation in isotonic and hypotonic buffers, respectively. The presence of Cl^- and DIOA in the buffers is also indicated. All buffers contain nominal Na^+. Values are means ± SEM (N = 15–22 oocytes). *Lowercase letters* indicate statistical categorization of the means. (B) Immunolocalization of KCC1 in a Malpighian tubule isolated from an adult female *Ae. aegypti*. Immunolabeling (*green* (*grey* in the print version)) occurs exclusively along the brush border of principal cells. 'PC' indicates representative nuclei of principal cells. The *inset* shows a representative negative control: ie, a tubule incubated with secondary antibody alone. The *dashed lines* outline the boundaries of the tubule epithelium. Previously unpublished images. *Panel (A) modified from Piermarini, P.M., Hine, R.M., Schepel, M., Miyauchi, J.T., Beyenbach, K.W., 2011. Role of an apical K,Cl cotransporter in urine formation by renal tubules of the yellow fever mosquito (*Aedes aegypti*). Am. J. Physiol. Regul. Integr. Comp. Physiol. 301, R1318-R1337, with permission.*

The primary Cl^- conductance measured (type I) was (1) blocked by niflumic acid and 2,2′-iminodibenzoic acid from the cytoplasmic side, (2) not affected by changes in [Ca^{2+}] on the cytoplasmic side, (3) permeable to organic anions, and (4) weakly selective for Cl^- (O'Connor and Beyenbach, 2001). These data suggest that the type I conductance may be mediated by volume-regulated anion channels, which at least in *D. melanogaster* are suspected to be bestrophin Cl^- channels (Chien and Hartzell, 2007; Stotz and Clapham, 2012).

Molecular expression studies on putative Cl^- channels in mosquito Malpighian tubules are limited. In Malpighian tubules of *An. gambiae* and *Ae. albopictus*, mRNAs encoding three putative Cl^--conducting channels (ClCs), a bestrophin, and a Cl^- intracellular channel (Clic) are expressed. However, the cellular/membrane localizations of these channels and relative contributions that they make to Cl^- transport in mosquito Malpighian tubules are unknown. A recent study in Malpighian tubules of *D. melanogaster* has implicated a prominent role of a ClC expressed on both the apical and basolateral membranes of stellate cells (ie, ClC-a) as a major contributor to the transcellular transport of Cl^- during stimulation by kinin neuropeptides (Cabrero et al., 2014). Thus, ClC channels would be excellent starting points for studying the molecular physiology of transcellular Cl^- transport in mosquito Malpighian tubules.

2.2.4 Mechanisms of Transcellular Water Transport

Sections 2.2.1, 2.2.2, and 2.2.3 described the cellular and molecular mechanisms of ion secretion from the hemolymph into the tubule lumen. As mentioned in Section 2.2, the transport of ions into the lumen produces an osmotic gradient allowing water to follow the movement of ions, which may occur through paracellular and transcellular pathways (Piermarini and Gillen, 2015). Here we focus on the mechanisms of transcellular water transport.

2.2.4.1 Aquaporins

The aquaporin water channels (AQPs) are presumed to be the principal mechanisms for transcellular water transport in mosquito Malpighian tubules. Before reviewing the evidence, it is important to mention that the nomenclature of genes encoding mosquito AQPs has been inconsistent within *Ae. aegypti* and between *Ae. aegypti* and *An. gambiae*. Thus, in Table 1, we suggest an alternative nomenclature based on recent comprehensive work of Finn and colleagues on the evolution of arthropod AQPs (Finn et al., 2015).

The Malpighian tubules of *Ae. aegypti*, *Ae. albopictus*, and *An. gambiae* are characterized by the abundant expression of four mRNAs encoding putative AQPs: *Drip*, *Prip*, *Eglp1*, and *Eglp2* (Drake et al., 2010, 2015; Esquivel et al., 2016; Marusalin et al., 2012; Overend et al., 2015). When expressed heterologously in *Xenopus* oocytes, Drip and Prip each form H_2O-selective channels; ie, they enhance the permeability of oocytes to H_2O, but not to glycerol and other small solutes (Drake et al., 2015; Duchesne et al., 2003; Liu et al., 2011). On the other hand, when Eglp1 is expressed heterologously in *Xenopus* oocytes, it dramatically enhances the permeability of the oocytes to glycerol, urea, erythritol, adonitol, mannitol, and trehalose, but not to H_2O, suggesting that it is an aquaglyceroporin that forms a generic channel for small organic solutes (Drake et al., 2015). Intriguingly, heterologous expression of Eglp2 in *Xenopus* oocytes enhances the permeability of

Table 1 Proposed Nomenclature Revision of Mosquito AQPs

Proposed Nomenclature (Finn et al., 2015)	*Ae. aegypti* Vectorbase Accession #	*Ae. aegypti* Previous Nomenclature (Drake et al., 2010, 2015)	*Ae. aegypti* Previous Nomenclature (Marusalin et al., 2012)	*An. gambiae* Vectorbase Accession #	*An. gambiae* Previous Nomenclature (Liu et al., 2011; Tsujimoto et al., 2013)
Drip	AAEL003512	AQP1	AQP1a	AGAP008842	
Prip	AAEL003550	AQP2	AQP1b	AGAP008443	AQP1
Bib	AAEL004741	AQP3	AQP2	AGAP008766, AGAP008767	
Eglp1	AAEL005001	AQP4	AQP3a	AGAP010326	
Eglp2	AAEL005008	AQP5	AQP3b	AGAP010325	
Aqp12L	AAEL014108	AQP6	AQP4	AGAP010878	

Drip, *Drosophila* integral protein; *Prip*, *Pyroceoelia rufa* integral protein; *Eglp*, entomoglyceroporin; *Aqp12L*, AQP12-like or unorthodox AQP.

the oocytes to both H_2O and certain solutes, such as erythritol, adonitol, and trehalose, suggesting that it is a multifunctional aquaglyceroporin that forms a channel for both H_2O and small solutes (Drake et al., 2015). Thus, mosquito Malpighian tubules appear to have a robust molecular capacity for AQP-mediated H_2O and organic solute transport.

In the Malpighian tubules of adult female *Ae. aegypti*, *Drip* mRNA expression and Drip-like immunoreactivity do not localize to the epithelial cells, but rather to the surrounding tracheal tubes that are tethered to the tubules (Duchesne et al., 2003; Pietrantonio et al., 2000). Thus, a direct contribution of Drip to transepithelial H_2O transport in Malpighian tubules is uncertain. However, in the Malpighian tubules of *An. gambiae*, Prip immunoreactivity localizes to the basolateral membranes of stellate cells in the distal segment and principal cells in the proximal segment (Liu et al., 2011; Tsujimoto et al., 2013), suggesting that it may contribute to the section of H_2O in the distal segment and reabsorption of H_2O in the proximal segment. The cellular and membrane localizations of Eglp1 and Eglp2 in mosquito Malpighian tubules are unknown.

Experiments utilizing RNA interference have demonstrated that AQPs play important roles in mosquito diuresis and desiccation tolerance. That is, in adult female *Ae. aegypti*, the knockdown of *Drip*, *Prip*, *Eglp1*, and *Eglp2* mRNAs, individually or all together, reduces the diuretic capacity of the mosquitoes (Drake et al., 2010). Moreover, in adult female *Ae. aegypti*, the knockdown of *Drip*, *Eglp1*, or *Eglp2* mRNAs enhances the tolerance of mosquitoes to a desiccation stress (Drake et al., 2015); a similar outcome is observed in adult female *An. gambiae* when *Prip* mRNA levels are knocked down (Liu et al., 2011). One potential interpretation of the above results is that lowering AQP expression reduces the capacity of the Malpighian tubules to produce urine and thereby limits the excretory loss of water through the renal system, which would reduce the diuretic capacity of mosquitoes but promote survival during a desiccation stress. Although the results are consistent with roles of AQPs in fluid secretion by Malpighian tubules, additional experiments are necessary to confirm that AQP mRNA knockdown indeed inhibits the capacity of isolated Malpighian tubules to secrete fluid.

2.2.5 Neuropeptide Receptors

Recent transcriptomic studies in the Malpighian tubules of *An. gambiae* and *Ae. albopictus* have identified the expression of mRNAs encoding a variety of putative G protein-coupled receptors for neuropeptides that may influence Malpighian tubule function (Esquivel et al., 2016; Overend et al., 2015).

Below, we focus on receptors for the three neuropeptides discussed in Section 1.1 that have diuretic effects in isolated mosquito Malpighian tubules: DH_{31}, kinin, and DH_{44}.

2.2.5.1 DH_{31} Receptors

The expression of mRNAs encoding a putative DH_{31} receptor has been detected in the Malpighian tubules of adult female *Ae. aegypti*, adult female *Ae. albopictus*, and larval and adult *An. gambiae* (Esquivel et al., 2016; Kwon et al., 2012; Overend et al., 2015). Binding of fluorescently labeled DH_{31} peptides and localization of DH_{31} receptor immunoreactivity in Malpighian tubules of *An. gambiae* and *Ae. aegypti*, respectively, indicate that the DH_{31} receptor is expressed exclusively in the basolateral membrane of principal cells in the distal segment (Kwon et al., 2012; Overend et al., 2015) (Fig. 9A). Notably, in *An. gambiae*, the DH_{31} receptor mRNA levels and binding of fluorescently labeled DH_{31} peptides are stronger in tubules of adults compared to larvae, suggesting a more important role of DH_{31} in regulating the physiology of adult Malpighian tubules (Overend et al., 2015). Intriguingly, in Malpighian tubules of *Ae. aegypti*, immunolabeling of the DH_{31} receptor does not occur in every principal cell of the distal segment (Fig. 9A), which may indicate the presence of functional subtypes of principal cells along the axial length of the distal segment (Kwon et al., 2012). Alternatively, this variegated pattern of expression may further support the notion of the tubule acting as a syncytium. In other words, given that up to three consecutive principal cells are electrically connected via gap junctions (Weng et al., 2008), the activation of a DH_{31} receptor in one principal cell could rapidly send the signal to cells that are up- and downstream of it, allowing for a prompt, coordinated response of the epithelium to the peptide without expression of the receptor in every principal cell (Piermarini and Calkins, 2014).

Knockdown of DH_{31} receptor mRNA expression in adult female *Ae. aegypti* mosquitoes via RNA interference dampens the immunolabeling of DH_{31} immunoreactivity in principal cells as expected (Kwon et al., 2012). Consequently, the capacity of isolated Malpighian tubules to mediate DH_{31}-stimulated transepithelial fluid secretion in vitro is reduced by ~40% (Kwon et al., 2012). Moreover, the capacity of mosquitoes to excrete urine in vivo within the first 60 min following a blood meal is reduced by over 20% (Kwon et al., 2012). Taken together, the above data indicate that the DH_{31} receptor in principal cells plays an important role in mediating the postprandial diuresis of adult female mosquitoes.

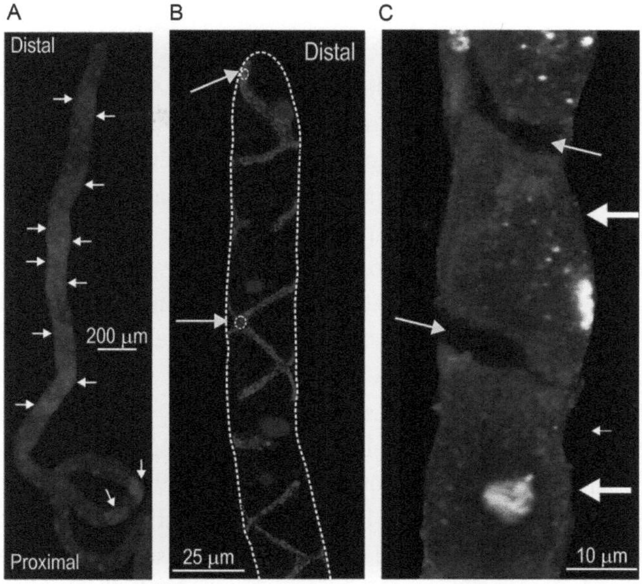

Fig. 9 Localization of neuropeptide receptors in mosquito Malpighian tubules. (A) DH_{31}-immunoreactivity (*red*) along the axial length of a Malpighian tubule isolated from an adult female *Ae. aegypti*. *Arrows* indicate representative principal cells expressing the receptor. Note that not all principal cells of the distal segment are labeled. Nuclei are labeled with DAPI (*blue*). (B) Binding of fluorescently labeled kinin peptides (*red*) to the Malpighian tubule of an adult female *An. gambiae*. *Yellow arrows* indicate representative stellate cells with bound peptide, indicating the presence of the kinin receptor. Principal cells do not bind detectable amounts of the peptide. Nuclei are labeled with DAPI (*blue*). The *dashed line* outlines the boundary of the tubule epithelium. (C) Binding of fluorescently labeled DH_{44} peptide (*red*) to the Malpighian tubule of an adult female *An. gambiae*. *White arrows* indicate representative principal cells with bound peptide, indicating the presence of the DH_{44} receptor. *Yellow arrows* indicate stellate cells that do not bind detectable amounts of peptide. Nuclei are labeled with DAPI (*white*). The author is grateful to Dr Kenneth Halberg (University of Glasgow) for providing the high resolution images used in panels (B) and (C). *Panel (A) modified from Kwon, H., Lu, H.L., Longnecker, M.T., Pietrantonio, P.V., 2012. Role in diuresis of a calcitonin receptor (GPRCAL1) expressed in a distal-proximal gradient in renal organs of the mosquito* Aedes aegypti *(L.). PLoS One 7, e50374, with permission. Panel (B) modified from Halberg, K.A., Terhzaz, S., Cabrero, P., Davies, S.A., Dow, J.A., 2015. Tracing the evolutionary origins of insect renal function. Nat. Commun. 6, 6800, with permission. Panel (C) modified from Overend, G., Cabrero, P., Halberg, K.A., Ranford-Cartwright, L.C., Woods, D.J., Davies, S.A., Dow, J.A.T., 2015. A comprehensive transcriptomic view of renal function in the malaria vector,* Anopheles gambiae. *Insect Biochem. Mol. Biol. 67, 47–58, with permission.* (See the colour plate.)

2.2.5.2 Kinin Receptors

The expression of mRNAs encoding a putative kinin receptor has been detected in the Malpighian tubules of adult female *Ae. aegypti*, adult female *Ae. albopictus*, and larval and adult *An. gambiae* (Esquivel et al., 2016; Overend et al., 2015; Pietrantonio et al., 2005). Electrophysiological studies on segments of isolated Malpighian tubules of *Ae. aegypti* with no apparent stellate cells were shown to be responsive to kinins, suggesting that principal cells were a site of kinin receptor expression (Yu and Beyenbach, 2004). However, binding of fluorescently labeled kinin peptides in Malpighian tubules of *An. gambiae* and localization of kinin receptor immunoreactivity in the Malpighian tubules of *Ae. aegypti* and *An. stephensi* indicate that molecular expression of the kinin receptor occurs exclusively in the basolateral membrane of stellate cells in the distal segment (Halberg et al., 2015; Lu et al., 2011; Radford et al., 2004) (Fig. 9B). Despite expression of the receptor in stellate cells, the kinins induce rapid posttranslational effects on proteins expressed in principal cells (eg, V-ATPase subunits) (Beyenbach et al., 2009). Thus, the effects of kinins on principal cell physiology are likely indirect and mediated by stellate cells, perhaps through direct intercellular signaling via gap junctions (Piermarini and Calkins, 2014).

Heterologous expression of the *Ae. aegypti* and *An. stephensi* kinin receptors in mammalian or insect cell lines indicates that a single kinin receptor can bind to all three kinin peptides found in mosquitoes (ie, I, II, and III) (Pietrantonio et al., 2005; Radford et al., 2004). Moreover, these studies confirm that the binding of each kinin to the receptor leads to an increase of intracellular Ca^{2+} (Pietrantonio et al., 2005; Radford et al., 2004), which is consistent with the signaling pathway observed in Malpighian tubules (Cady and Hagedorn, 1999b; Yu and Beyenbach, 2002). Knockdown of kinin receptor mRNA expression in adult female *Ae. aegypti* mosquitoes via RNA interference dampens the immunolabeling of kinin immunoreactivity in stellate cells as expected. Consequently, the diuretic capacity of the mosquitoes in vivo is reduced by over 50% within the first 60 min following a blood meal (Kersch and Pietrantonio, 2011). Thus, together with the DH_{31} receptor, the kinin receptor in stellate cells also plays a critical role in regulating the postprandial diuresis of adult female mosquitoes.

2.2.5.3 DH_{44} Receptors

The expression of mRNAs encoding putative DH_{44} receptors has been detected in the Malpighian tubules of adult female *Ae. aegypti* and larval and adult *An. gambiae* (Jagge and Pietrantonio, 2008; Overend et al., 2015). In at least *An. gambiae*, the binding of fluorescently labeled DH_{44}

peptides in Malpighian tubules indicates that the DH_{44} receptor is expressed exclusively in the basolateral membrane of principal cells in the distal segment, similar to the DH_{31} receptor (Overend et al., 2015) (Fig. 9C). In contrast to the DH_{31} receptor, mRNA levels of the DH_{44} receptor and binding of fluorescently labeled DH_{44} peptides are stronger in tubules of larval *An. gambiae* compared to adults, suggesting a potentially more relevant role of DH_{44} in regulating the physiology of larval Malpighian tubules (Overend et al., 2015). Consistent with this notion, using RNA-Seq, we were unable to detect the expression of a mRNA encoding a putative DH_{44} receptor in the Malpighian tubules of nonblood fed and blood fed adult female *Ae. albopictus* (Esquivel et al., 2016). However, in Malpighian tubules of adult female *Ae. aegypti*, the mRNA expression levels of the DH_{44} receptor exhibit their highest levels of expression at 24 h postblood meal (Jagge and Pietrantonio, 2008). Thus, the DH_{44} receptor may play species-specific roles in regulating the physiology of mosquito Malpighian tubules.

3. THE RENAL EXCRETORY SYSTEM AS A TARGET FOR MOSQUITO CONTROL

Given the vital role that the renal excretory system plays in osmotic and ionic homeostasis of mosquitoes, it has been proposed as a novel physiological target for the control of mosquitoes and other insect pests (Beyenbach et al., 2015; Cohen, 2013; Raphemot et al., 2013; Ruiz-Sanchez and O'Donnell, 2015). Since the Malpighian tubules are vital for alleviating the acute stresses to salt and water balance after a blood meal via the postprandial diuresis (eg, Fig. 1D), they are an especially attractive target for developing novel control agents (eg, insecticides) that may be used to selectively kill hematophagous adult female mosquitoes. Mosquitoes may even be slower to evolve resistance to a 'renal'-targeting mosquitocide compared to neurotoxic compounds (eg, DDT, pyrethroids), which exert a strong selective pressure on all life stages and lead to resistance within a few years of use (Read et al., 2009).

In collaboration with the Denton and Beyenbach laboratories, we have been attempting to develop insecticides that inhibit the renal functions of adult female mosquitoes. In particular, we are targeting the Kir1 channel with small molecule inhibitors; Kir1 is expressed in the stellate cells of Malpighian tubules and plays an important role in transepithelial K^+ and fluid secretion (Figs 4 and 6). In brief, we hypothesized that inhibition of Kir1 with small molecules would elicit renal failure by disrupting the regulation of hemolymph K^+ and volume homeostasis. Using a high-throughput screening assay developed by the Denton laboratory, we discovered three small molecule inhibitors of the *Ae. aegypti* Kir1 channel (VU573, VU590, and VU625); each molecule inhibits transepithelial fluid and K^+ secretion in isolated Malpighian tubules of adult female *Ae. aegypti* (Beyenbach et al., 2015; Raphemot et al., 2013, 2014b; Rouhier et al., 2014a).

When injected into the hemolymph of adult female *Ae. aegypti*, each of the molecules is toxic within 24 h (Raphemot et al., 2013, 2014b; Rouhier et al., 2014b). Moreover, the physiological performance of mosquitoes treated with the small molecules is consistent with impaired renal functions. That is, treatment of adult female *Ae. aegypti* with VU573, VU590, or VU625 inhibits their capacity to excrete a volume load to their hemolymph (Raphemot et al., 2013, 2014b; Rouhier et al., 2014a,b) (Fig. 10A). In addition, treatment of mosquitoes with VU573 or VU590 leads to dramatic abdominal bloating in adult females, consistent with an impaired capacity to regulate hemolymph volume (Raphemot et al., 2013;

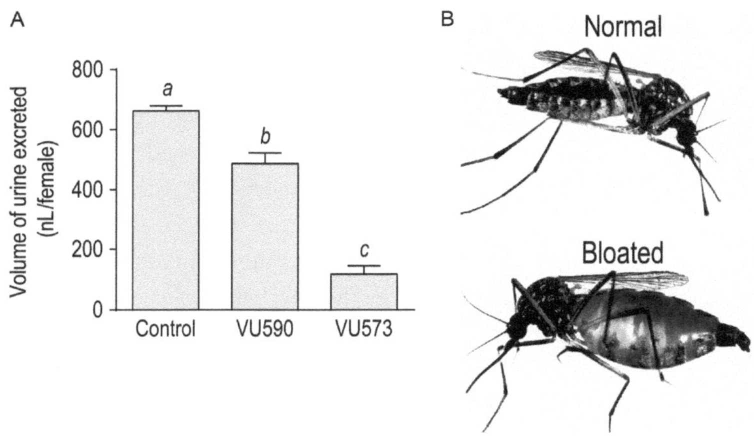

Fig. 10 Renal failure in adult female mosquitoes. (A) Amount of urine excreted by adult female *Ae. aegypti* mosquitoes 1 h after injection with 900 nL of a phosphate buffered saline (PBS) containing small molecule VU590 or VU573. Controls were injected with PBS. Values are means ± SEM (N = 11–23 independent trials). *Lowercase letters* indicate statistical categorization of the means. (B) Representative images of adult female *Ae. aegypti* with normal and bloated abdomens, 24 h after hemolymph injection with PBS or VU573, respectively. Similar abdominal bloating is observed after injection of VU590 (Rouhier et al., 2014b). *Panel (A) data are from Rouhier, M.F., Raphemot, R., Denton, J.S., Piermarini, P.M., 2014b. Pharmacological validation of an inward-rectifier potassium (Kir) channel as an insecticide target in the yellow fever mosquito* Aedes aegypti. *PLoS One 9, e100700. Panel (B) image is from Raphemot, R., Rouhier, M.F., Hopkins, C.R., Gogliotti, R.D., Lovell, K.M., Hine, R.M., Ghosalkar, D., Longo, A., Beyenbach, K.W., Denton, J.S., Piermarini, P.M., 2013. Eliciting renal failure in mosquitoes with a small-molecule inhibitor of inward-rectifying potassium channels. PLoS One 8, e64905, with permission.*

Rouhier et al., 2014b) (Fig. 10B). In some cases, the retention of fluid exceeds the carrying capacity of the abdominal wall, thereby compromising the integrity of the cuticle (ie, an exploded abdomen; Piermarini, unpublished observations). Moreover, mosquitoes treated with VU573 are more susceptible to hemolymph loads of K^+ vs Na^+, consistent with the block of renal Kir1 channels that mediate K^+ secretion (Raphemot et al., 2013; Rouhier et al., 2014a).

Thus, the aforementioned studies demonstrate proof of concept that disrupting the activities of the renal excretory system with small molecule inhibitors of Kir channels offers a potentially exciting new mechanism of action for mosquitocide development, compared to conventional molecules that target voltage-gated Na^+ channels and acetylcholinesterases in the nervous system. However, disrupting Kir channels and osmoregulatory functions of mosquito Malpighian tubules are just the tip of the iceberg. As shown in Fig. 4, numerous cellular and molecular mechanisms contribute to transepithelial fluid secretion and diuresis, which all offer potential targets to exploit for disrupting renal functions in mosquitoes. In particular, given the recent development of a high-throughput assay for discovering small molecule modulators of mosquito G protein-coupled receptors for dopamine (Meyer et al., 2012; Nuss et al., 2015), it may be feasible to develop analogous assays for discovering novel modulators of the DH_{31}, kinin, and DH_{44} receptors that could disrupt renal excretory functions of mosquitoes. Moreover, recent transcriptomic studies have uncovered potential roles of mosquito Malpighian tubules in the metabolic processing and excretion of blood meal metabolites (eg, heme), which offer additional physiological functions and molecular targets in Malpighian tubules to consider for disruption (Esquivel et al., 2014, 2016; Overend et al., 2015).

In conclusion, the renal excretory system of mosquitoes promises to be an exciting area of future investigation not only to improve our understanding of the cellular and molecular mechanism of Malpighian tubule and hindgut function but also to enhance our capabilities for controlling mosquito vectors of emerging and reemerging diseases.

REFERENCES

Albers, M.A., Bradley, T.J., 2011. On the evolution of saline tolerance in the larvae of mosquitoes in the genus *Ochlerotatus*. Physiol. Biochem. Zool. 84, 258–267.

Benoit, J.B., Denlinger, D.L., 2010. Meeting the challenges of on-host and off-host water balance in blood-feeding arthropods. J. Insect Physiol. 56, 1366–1376.

Beyenbach, K.W., 2003. Transport mechanisms of diuresis in Malpighian tubules of insects. J. Exp. Biol. 206, 3845–3856.

Beyenbach, K.W., 2012. A dynamic paracellular pathway serves diuresis in mosquito Malpighian tubules. Ann. N. Y. Acad. Sci. 1258, 166–176.

Beyenbach, K.W., Hagedorn, H.H., 2005. Osmotic and ionic regulation by mosquitoes. In: Marquardt, W.C., Black, W.C., Freier, J.E., Hagedorn, H.H., Hemingway, J., Higgs, S., James, A.A., Kondratieff, B., Moore, C.G. (Eds.), Biology of Disease Vectors, second ed. Elsevier, Burlington, MA, pp. 347–362.

Beyenbach, K.W., Masia, R., 2002. Membrane conductances of principal cells in Malpighian tubules of *Aedes aegypti*. J. Insect Physiol. 48, 375–386.

Beyenbach, K.W., Petzel, D.H., 1987. Diuresis in mosquitoes role of a natriuretic factor. News Physiol. Sci. 2, 171–175.

Beyenbach, K.W., Piermarini, P.M., 2011. Transcellular and paracellular pathways of transepithelial fluid secretion in Malpighian (renal) tubules of the yellow fever mosquito *Aedes aegypti*. Acta Physiol. 202, 387–407.

Beyenbach, K.W., Wieczorek, H., 2006. The V-type H^+ ATPase: molecular structure and function, physiological roles and regulation. J. Exp. Biol. 209, 577–589.

Beyenbach, K.W., Oviedo, A., Aneshansley, D.J., 1993. Malpighian tubules of *Aedes aegypti*: five tubules, one function. J. Insect Physiol. 39, 639–648.

Beyenbach, K.W., Pannabecker, T.L., Nagel, W., 2000. Central role of the apical membrane H^+-ATPase in electrogenesis and epithelial transport in Malpighian tubules. J. Exp. Biol. 203, 1459–1468.

Beyenbach, K.W., Baumgart, S., Lau, K., Piermarini, P.M., Zhang, S., 2009. Signaling to the apical membrane and to the paracellular pathway: changes in the cytosolic proteome of *Aedes* Malpighian tubules. J. Exp. Biol. 212, 329–340.

Beyenbach, K.W., Yu, Y., Piermarini, P.M., Denton, J., 2015. Targeting renal epithelial channels for the control of insect vectors. Tissue Barriers 3, e1081861.

Bradley, T.J., 1987. Physiology of osmoregulation in mosquitoes. Annu. Rev. Entomol. 32, 439–462.

Brown, D., Breton, S., 1996. Mitochondria-rich, proton-secreting epithelial cells. J. Exp. Biol. 199, 2345–2358.

Cabrero, P., Pollock, V.P., Davies, S.A., Dow, J.A., 2004. A conserved domain of alkaline phosphatase expression in the Malpighian tubules of dipteran insects. J. Exp. Biol. 207, 3299–3305.

Cabrero, P., Terhzaz, S., Romero, M.F., Davies, S.A., Blumenthal, E.M., Dow, J.A., 2014. Chloride channels in stellate cells are essential for uniquely high secretion rates in neuropeptide-stimulated *Drosophila* diuresis. Proc. Natl. Acad. Sci. U.S.A. 111, 14301–14306.

Cady, C., Hagedorn, H.H., 1999a. The effect of putative diuretic factors on in vivo urine production in the mosquito, *Aedes aegypti*. J. Insect Physiol. 45, 317–325.

Cady, C., Hagedorn, H.H., 1999b. Effects of putative diuretic factors on intracellular second messenger levels in the Malpighian tubules of *Aedes aegypti*. J. Insect Physiol. 45, 327–337.

Chien, L.-T., Hartzell, H.C., 2007. *Drosophila* bestrophin-1 chloride current is dually regulated by calcium and cell volume. J. Gen. Physiol. 130, 513–524.

Chintapalli, V.R., Terhzaz, S., Wang, J., Al Bratty, M., Watson, D.G., Herzyk, P., Davies, S.A., Dow, J.A.T., 2012. Functional correlates of positional and gender-specific renal asymmetry in *Drosophila*. PLoS One 7, e32577.

Chintapalli, V.R., Kato, A., Henderson, L., Hirata, T., Woods, D.J., Overend, G., Davies, S.A., Romero, M.F., Dow, J.A., 2015. Transport proteins NHA1 and NHA2 are essential for survival, but have distinct transport modalities. Proc. Natl. Acad. Sci. U.S.A. 112, 11720–11725.

Clark, T.M., Bradley, T.J., 1996. Stimulation of Malpighian tubules from larval *Aedes aegypti* by secretagogues. J. Insect Physiol. 42, 593–602.

Clark, T.M., Hayes, T.K., Holman, G.M., Beyenbach, K.W., 1998. The concentration-dependence of CRF-like diuretic peptide: mechanisms of action. J. Exp. Biol. 201, 1753–1762.

Clements, A.N., 2000. The Biology of Mosquitoes. CABI Publishing, Oxfordshire.

Coast, G., 2007. The endocrine control of salt balance in insects. Gen. Comp. Endocrinol. 152, 332–338.

Coast, G.M., 2009. Neuroendocrine control of ionic homeostasis in blood-sucking insects. J. Exp. Biol. 212, 378–386.

Coast, G.M., Garside, C.S., Webster, S.G., Schegg, K.M., Schooley, D.A., 2005. Mosquito natriuretic peptide identified as a calcitonin-like diuretic hormone in *Anopheles gambiae* (Giles). J. Exp. Biol. 208, 3281–3291.

Cohen, E., 2013. Water homeostasis and osmoregulation as targets in the control of insect pests. In: Ephraim, C. (Ed.), Advances in Insect Physiology. Oxford, UK, pp. 1–61 (Chapter 1).

Day, J.P., Wan, S., Allan, A.K., Kean, L., Davies, S.A., Gray, J.V., Dow, J.A., 2008. Identification of two partners from the bacterial Kef exchanger family for the apical plasma membrane V-ATPase of Metazoa. J. Cell Sci. 121, 2612–2619.

Drake, L.L., Boudko, D.Y., Marinotti, O., Carpenter, V.K., Dawe, A.L., Hansen, I.A., 2010. The aquaporin gene family of the yellow fever mosquito, *Aedes aegypti*. PLoS One 5, e15578.

Drake, L.L., Rodriguez, S.D., Hansen, I.A., 2015. Functional characterization of aquaporins and aquaglyceroporins of the yellow fever mosquito, *Aedes aegypti*. Sci. Rep. 5, 7795.

Duchesne, L., Hubert, J.F., Verbavatz, J.M., Thomas, D., Pietrantonio, P.V., 2003. Mosquito (*Aedes aegypti*) aquaporin, present in tracheolar cells, transports water, not glycerol, and forms orthogonal arrays in *Xenopus* oocyte membranes. Eur. J. Biochem. 270, 422–429.

Esquivel, C.J., Cassone, B.J., Piermarini, P.M., 2014. Transcriptomic evidence for a dramatic functional transition of the Malpighian tubules after a blood meal in the Asian tiger mosquito *Aedes albopictus*. PLoS Negl. Trop. Dis. 8, e2929.

Esquivel, C.J., Cassone, B.J., Piermarini, P.M., 2016. A de novo transcriptome of the Malpighian tubules in non-blood-fed and blood-fed Asian tiger mosquitoes *Aedes albopictus*: insights into diuresis, detoxification, and blood meal processing. PeerJ 4, e1784.

Finn, R.N., Chauvigne, F., Stavang, J.A., Belles, X., Cerda, J., 2015. Insect glycerol transporters evolved by functional co-option and gene replacement. Nat. Commun. 6. http://dx.doi.org/10.1038/ncomms8814.

Giannakou, M.E., Dow, J.A., 2001. Characterization of the *Drosophila melanogaster* alkali-metal/proton exchanger (NHE) gene family. J. Exp. Biol. 204, 3703–3716.

Halberg, K.A., Terhzaz, S., Cabrero, P., Davies, S.A., Dow, J.A., 2015. Tracing the evolutionary origins of insect renal function. Nat. Commun. 6, 6800.

Hart, S.J., Knezetic, J.A., Petzel, D.H., 2002. Cloning and tissue distribution of two Na^+/H^+ exchangers from the Malpighian tubules of *Aedes aegypti*. Arch. Insect Biochem. Physiol. 51, 121–135.

Hayes, T.K., Pannabecker, T.L., Hinckley, D.J., Holman, G.M., Nachman, R.J., Petzel, D.H., Beyenbach, K.W., 1989. Leucokinins, a new family of ion transport stimulators and inhibitors in insect Malpighian tubules. Life Sci. 44, 1259–1266.

Hegarty, J.L., Zhang, B., Pannabecker, T.L., Petzel, D.H., Baustian, M.D., Beyenbach, K.W., 1991. Dibutyryl cAMP activates bumetanide-sensitive electrolyte transport in Malpighian tubules. Am. J. Physiol. Cell Physiol. 261, C521–C529.

Hegarty, J.L., Zhang, B., Carroll, M.C., Cragoe, E.J.J., Beyenbach, K.W., 1992. Effects of amiloride on isolated Malpighian tubules of the yellow fever mosquito (*Aedes aegypti*). J. Insect Physiol. 38, 329–337.

Hine, R.M., Rouhier, M.F., Park, S.T., Qi, Z., Piermarini, P.M., Beyenbach, K.W., 2014. The excretion of NaCl and KCl loads in mosquitoes: 1. Control data. Am. J. Physiol. Regul. Integr. Comp. Physiol. 307, R837–R849.

Izumi, Y., Furuse, M., 2014. Molecular organization and function of invertebrate occluding junctions. Semin. Cell Dev. Biol. 36, 186–193.

Jagge, C.L., Pietrantonio, P.V., 2008. Diuretic hormone 44 receptor in Malpighian tubules of the mosquito *Aedes aegypti:* evidence for transcriptional regulation paralleling urination. Insect Mol. Biol. 17, 413–426.

Jentsch, T.J., 2005. Chloride transport in the kidney: lessons from human disease and knockout mice. J. Am. Soc. Nephrol. 16, 1549–1561.

Jonusaite, S., Donini, A., Kelly, S.P., 2016. Occluding junctions of invertebrate epithelia. J. Comp. Physiol. B 186, 17–43.

Kang'ethe, W., Aimanova, K.G., Pullikuth, A.K., Gill, S.S., 2007. NHE8 mediates amiloride-sensitive Na^+/H^+ exchange across mosquito Malpighian tubules and catalyzes Na^+ and K^+ transport in reconstituted proteoliposomes. Am. J. Physiol. Renal Physiol. 292, F1501–F1512.

Kersch, C.N., Pietrantonio, P.V., 2011. Mosquito *Aedes aegypti* (L.) leucokinin receptor is critical for in vivo fluid excretion post blood feeding. FEBS Lett. 585, 3507–3512.

Kwon, H., Pietrantonio, P.V., 2013. Calcitonin receptor 1 (AedaeGPCRCAL1) hindgut expression and direct role in myotropic action in females of the mosquito *Aedes aegypti* (L.). Insect Biochem. Mol. Biol. 43, 588–593.

Kwon, H., Lu, H.L., Longnecker, M.T., Pietrantonio, P.V., 2012. Role in diuresis of a calcitonin receptor (GPRCAL1) expressed in a distal-proximal gradient in renal organs of the mosquito *Aedes aegypti* (L.). PLoS One 7, e50374.

Leiserson, W.M., Forbush, B., Keshishian, H., 2011. *Drosophila* glia use a conserved cotransporter mechanism to regulate extracellular volume. Glia 59, 320–332.

Linser, P.J., Dinglasan, R.R., 2014. Insect gut structure, function, development and target of biological toxins. In: Tarlochan, S.D., Sarjeet, S.G. (Eds.), Advances in Insect Physiology. Academic Press, San Diego, CA, pp. 1–37 (Chapter 1).

Linser, P.J., Neira Oviedo, M., Hirata, T., Seron, T.J., Smith, K.E., Piermarini, P.M., Romero, M.F., 2012. Slc4-like anion transporters of the larval mosquito alimentary canal. J. Insect Physiol. 58, 551–562.

Liu, K., Tsujimoto, H., Cha, S.J., Agre, P., Rasgon, J.L., 2011. Aquaporin water channel AgAQP1 in the malaria vector mosquito *Anopheles gambiae* during blood feeding and humidity adaptation. Proc. Natl. Acad. Sci. U.S.A. 108, 6062–6066.

Lu, H.L., Kersch, C., Pietrantonio, P.V., 2011. The kinin receptor is expressed in the Malpighian tubule stellate cells in the mosquito *Aedes aegypti* (L.): a new model needed to explain ion transport? Insect Biochem. Mol. Biol. 41, 135–140.

Marusalin, J., Matier, B.J., Rheault, M.R., Donini, A., 2012. Aquaporin homologs and water transport in the anal papillae of the larval mosquito, *Aedes aegypti*. J. Comp. Physiol. B 182, 1047–1056.

Masia, R., Aneshansley, D., Nagel, W., Nachman, R.J., Beyenbach, K.W., 2000. Voltage clamping single cells in intact Malpighian tubules of mosquitoes. Am. J. Physiol. Renal Physiol. 279, F747–F754.

Meyer, J.M., Ejendal, K.F., Avramova, L.V., Garland-Kuntz, E.E., Giraldo-Calderon, G.I., Brust, T.F., Watts, V.J., Hill, C.A., 2012. A "genome-to-lead" approach for insecticide discovery: pharmacological characterization and screening of *Aedes aegypti* D_1-like dopamine receptors. PLoS Negl. Trop. Dis. 6, e1478.

Morth, J.P., Pedersen, B.P., Buch-Pedersen, M.J., Andersen, J.P., Vilsen, B., Palmgren, M.G., Nissen, P., 2011. A structural overview of the plasma membrane Na+, K^+-ATPase and H^+-ATPase ion pumps. Nat. Rev. Mol. Cell Biol. 12, 60–70.

Nuss, A.B., Ejendal, K.F.K., Doyle, T.B., Meyer, J.M., Lang, E.G., Watts, V.J., Hill, C.A., 2015. Dopamine receptor antagonists as new mode-of-action insecticide leads for control of *Aedes* and *Culex* mosquito vectors. PLoS Negl. Trop. Dis. 9, e0003515.

O'connor, K.R., Beyenbach, K.W., 2001. Chloride channels in apical membrane patches of stellate cells of Malpighian tubules of *Aedes aegypti*. J. Exp. Biol. 204, 367–378.

Okech, B.A., Boudko, D.Y., Linser, P.J., Harvey, W.R., 2008. Cationic pathway of pH regulation in larvae of *Anopheles gambiae*. J. Exp. Biol. 211, 957–968.

Overend, G., Cabrero, P., Halberg, K.A., Ranford-Cartwright, L.C., Woods, D.J., Davies, S.A., Dow, J.A.T., 2015. A comprehensive transcriptomic view of renal function in the malaria vector, *Anopheles gambiae*. Insect Biochem. Mol. Biol. 67, 47–58.

Paluzzi, J.-P., Vanderveken, M., O'donnell, M.J., 2014. The heterodimeric glycoprotein hormone, GPA2/GPB5, regulates ion transport across the hindgut of the adult mosquito, *Aedes aegypti*. PLoS One 9, e86386.

Pannabecker, T.L., Hayes, T.K., Beyenbach, K.W., 1993. Regulation of epithelial shunt conductance by the peptide leucokinin. J. Membr. Biol. 132, 63–76.

Patrick, M.L., Aimanova, K., Sanders, H.R., Gill, S.S., 2006. P-type Na^+/K^+-ATPase and V-type H^+-ATPase expression patterns in the osmoregulatory organs of larval and adult mosquito *Aedes aegypti*. J. Exp. Biol. 209, 4638–4651.

Petzel, D.H., 2000. Na^+/H^+ exchange in mosquito Malpighian tubules. Am. J. Physiol. Regul. Integr. Comp. Physiol. 279, R1996–R2003.

Petzel, D.H., Hagedorn, H.H., Beyenbach, K.W., 1985. Preliminary isolation of mosquito natriuretic factor. Am. J. Physiol. 249, R379–R386.

Petzel, D.H., Berg, M.M., Beyenbach, K.W., 1987. Hormone-controlled cAMP-mediated fluid secretion in yellow-fever mosquito. Am. J. Physiol. 253, R701–R711.

Piermarini, P.M., Calkins, T.L., 2014. Evidence for intercellular communication in mosquito renal tubules: a putative role of gap junctions in coordinating and regulating the rapid diuretic effects of neuropeptides. Gen. Comp. Endocrinol. 203, 43–48.

Piermarini, P., Gillen, C., 2015. Non-traditional models: the molecular physiology of sodium and water transport in mosquito Malpighian tubules. In: Hyndman, K.A., Pannabecker, T.L. (Eds.), Sodium and Water Homeostasis. Springer, New York, pp. 255–278.

Piermarini, P.M., Weihrauch, D., Meyer, H., Huss, M., Beyenbach, K.W., 2009. NHE8 is an intracellular cation/H^+ exchanger in renal tubules of the yellow fever mosquito *Aedes aegypti*. Am. J. Physiol. Renal Physiol. 296, F730–F750.

Piermarini, P.M., Grogan, L.F., Lau, K., Wang, L., Beyenbach, K.W., 2010. A SLC4-like anion exchanger from renal tubules of the mosquito (*Aedes aegypti*): evidence for a novel role of stellate cells in diuretic fluid secretion. Am. J. Physiol. Regul. Integr. Comp. Physiol. 298, R642–R660.

Piermarini, P.M., Hine, R.M., Schepel, M., Miyauchi, J.T., Beyenbach, K.W., 2011. Role of an apical K, Cl cotransporter in urine formation by renal tubules of the yellow fever mosquito (*Aedes aegypti*). Am. J. Physiol. Regul. Integr. Comp. Physiol. 301, R1318–R1337.

Piermarini, P.M., Rouhier, M.F., Schepel, M., Kosse, C., Beyenbach, K.W., 2013. Cloning and functional characterization of inward-rectifying potassium (Kir) channels from Malpighian tubules of the mosquito *Aedes aegypti*. Insect Biochem. Mol. Biol. 43, 75–90.

Piermarini, P.M., Dunemann, S.M., Rouhier, M.F., Calkins, T.L., Raphemot, R., Denton, J.S., Hine, R.M., Beyenbach, K.W., 2015. Localization and role of inward rectifier K channels in Malpighian tubules of the yellow fever mosquito *Aedes aegypti*. Insect Biochem. Mol. Biol. 67, 59–73.

Pietrantonio, P.V., Jagge, C., Keeley, L.L., Ross, L.S., 2000. Cloning of an aquaporin-like cDNA and in situ hybridization in adults of the mosquito *Aedes aegypti* (Diptera: Culicidae). Insect Mol. Biol. 9, 407–418.

Pietrantonio, P.V., Jagge, C., Taneja-Bageshwar, S., Nachman, R.J., Barhoumi, R., 2005. The mosquito *Aedes aegypti* (L.) leucokinin receptor is a multiligand receptor for the three *Aedes* kinins. Insect Mol. Biol. 14, 55–67.

Plawner, L., Pannabecker, T.L., Laufer, S., Baustian, M.D., Beyenbach, K.W., 1991. Control of diuresis in the yellow fever mosquito *Aedes aegypti*: evidence for similar mechanisms in the male and female. J. Insect Physiol. 37, 119–128.

Pullikuth, A.K., Aimanova, K., Kang'ethe, W., Sanders, H.R., Gill, S.S., 2006. Molecular characterization of sodium/proton exchanger 3 (NHE3) from the yellow fever vector, *Aedes aegypti*. J. Exp. Biol. 209, 3529–3544.

Radford, J.C., Terhzaz, S., Cabrero, P., Davies, S.A., Dow, J.A., 2004. Functional characterisation of the *Anopheles* leucokinins and their cognate G-protein coupled receptor. J. Exp. Biol. 207, 4573–4586.

Raphemot, R., Rouhier, M.F., Hopkins, C.R., Gogliotti, R.D., Lovell, K.M., Hine, R.M., Ghosalkar, D., Longo, A., Beyenbach, K.W., Denton, J.S., Piermarini, P.M., 2013. Eliciting renal failure in mosquitoes with a small-molecule inhibitor of inward-rectifying potassium channels. PLoS One 8, e64905.

Raphemot, R., Estevez-Lao, T.Y., Rouhier, M.F., Piermarini, P.M., Denton, J.S., Hillyer, J.F., 2014a. Molecular and functional characterization of *Anopheles gambiae* inward rectifier potassium (Kir1) channels: a novel role in egg production. Insect Biochem. Mol. Biol. 51C, 10–19.

Raphemot, R., Rouhier, M.F., Swale, D.R., Days, E., Weaver, C.D., Lovell, K.M., Konkel, L.C., Engers, D.W., Bollinger, S.F., Hopkins, C., Piermarini, P.M., Denton, J.S., 2014b. Discovery and characterization of a potent and selective inhibitor of *Aedes aegypti* inward rectifier potassium channels. PLoS One 9, e110772.

Read, A.F., Lynch, P.A., Thomas, M.B., 2009. How to make evolution-proof insecticides for malaria control. PLoS Biol. 7, e1000058.

Rheault, M.R., Okech, B.A., Keen, S.B., Miller, M.M., Meleshkevitch, E.A., Linser, P.J., Boudko, D.Y., Harvey, W.R., 2007. Molecular cloning, phylogeny and localization of AgNHA1: the first Na^+/H^+ antiporter (NHA) from a metazoan, *Anopheles gambiae*. J. Exp. Biol. 210, 3848–3861.

Rodan, A.R., Baum, M., Huang, C.L., 2012. The *Drosophila* NKCC Ncc69 is required for normal renal tubule function. Am. J. Physiol. Cell Physiol. 303, C883–C894.

Romero, M.F., Chen, A.P., Parker, M.D., Boron, W.F., 2013. The SLC4 family of bicarbonate (HCO_3^-) transporters. Mol. Aspects Med. 34, 159–182.

Rouhier, M.F., Hine, R.M., Park, S.T., Raphemot, R., Denton, J.S., Piermarini, P.M., Beyenbach, K.W., 2014a. The excretion of NaCl and KCl loads in mosquitoes: 2. Effects of the small molecule Kir channel modulator VU573 and its inactive analog VU342. Am. J. Physiol. Regul. Integr. Comp. Physiol. 307, R850–R861.

Rouhier, M.F., Raphemot, R., Denton, J.S., Piermarini, P.M., 2014b. Pharmacological validation of an inward-rectifier potassium (Kir) channel as an insecticide target in the yellow fever mosquito *Aedes aegypti*. PLoS One 9, e100700.

Ruiz-Sanchez, E., O'donnell, M.J., 2015. The insect excretory system as a target for novel pest control strategies. Curr. Opin. Insect Sci. 11, 14–20.

Sawyer, D.B., Beyenbach, K.W., 1985. Dibutyryl-cAMP increases basolateral sodium conductance of mosquito Malpighian tubules. Am. J. Physiol. Regul. Integr. Comp. Physiol. 248, R339–R345.

Schepel, S.A., Fox, A.J., Miyauchi, J.T., Sou, T., Yang, J.D., Lau, K., Blum, A.W., Nicholson, L.K., Tiburcy, F., Nachman, R.J., Piermarini, P.M., Beyenbach, K.W., 2010. The single kinin receptor signals to separate and independent physiological pathways in Malpighian tubules of the yellow fever mosquito. Am. J. Physiol. Regul. Integr. Comp. Physiol. 299, R612–R622.

Scott, B.N., Yu, M.J., Lee, L.W., Beyenbach, K.W., 2004. Mechanisms of K^+ transport across basolateral membranes of principal cells in Malpighian tubules of the yellow fever mosquito, *Aedes aegypti*. J. Exp. Biol. 207, 1655–1663.
Stotz, S.C., Clapham, D.E., 2012. Anion-sensitive fluorophore identifies the *Drosophila* swell-activated chloride channel in a genome-wide RNA interference screen. PLoS One 7, e46865.
Terracciano, C.M., Hancox, J.C., 2013. ORM-10103: a significant advance in sodium-calcium exchanger pharmacology? Br. J. Pharmacol. 170, 765–767.
Tiburcy, F., Beyenbach, K.W., Wieczorek, H., 2013. Protein kinase A-dependent and -independent activation of the V-ATPase in Malpighian tubules of *Aedes aegypti*. J. Exp. Biol. 216, 881–891.
Tsujimoto, H., Liu, K., Linser, P.J., Agre, P., Rasgon, J.L., 2013. Organ-specific splice variants of aquaporin water channel AgAQP1 in the malaria vector *Anopheles gambiae*. PLoS One 8, e75888.
Veenstra, J.A., 1994. Isolation and identification of three leucokinins from the mosquito *Aedes aegypti*. Biochem. Biophys. Res. Commun. 202, 715–719.
Veenstra, J.A., Pattillo, J.M., Petzel, D.H., 1997. A single cDNA encodes all three *Aedes* leucokinins, which stimulate both fluid secretion by the Malpighian tubules and hindgut contractions. J. Biol. Chem. 272, 10402–10407.
Wang, S., Rubenfeld, A., Hayes, T., Beyenbach, K., 1996. Leucokinin increases paracellular permeability in insect Malpighian tubules. J. Exp. Biol. 199, 2537–2542.
Weng, X.H., Huss, M., Wieczorek, H., Beyenbach, K.W., 2003. The V-type H^+-ATPase in Malpighian tubules of *Aedes aegypti*: localization and activity. J. Exp. Biol. 206, 2211–2219.
Weng, X.H., Piermarini, P.M., Yamahiro, A., Yu, M.J., Aneshansley, D.J., Beyenbach, K.W., 2008. Gap junctions in Malpighian tubules of *Aedes aegypti*. J. Exp. Biol. 211, 409–422.
Wessing, A., Zierold, K., 1993. Heterogeneous distribution of elemental contents in the larval Malpighian tubules of *Drosophila hydei*: X-ray microanalysis of freeze-dried cryosections. Cell Tissue Res. 272, 491–497.
Wieczorek, H., 1992. The insect V-ATPase, a plasma membrane proton pump energizing secondary active transport: molecular analysis of electrogenic potassium transport in the tobacco hornworm midgut. J. Exp. Biol. 172, 335–343.
Wieczorek, H., Beyenbach, K.W., Huss, M., Vitavska, O., 2009. Vacuolar-type proton pumps in insect epithelia. J. Exp. Biol. 212, 1611–1619.
Williams, J.C., Hagedorn, H.H., Beyenbach, K.W., 1983. Dynamic changes in flow rate and composition of urine during the post blood meal diuresis in *Aedes aegypti*. J. Comp. Physiol. B 153, 257–266.
Wu, D.S., Beyenbach, K.W., 2003. The dependence of electrical transport pathways in Malpighian tubules on ATP. J. Exp. Biol. 206, 233–243.
Xiang, M.A., Linser, P.J., Price, D.A., Harvey, W.R., 2012. Localization of two Na^+- or K^+-H^+ antiporters, AgNHA1 and AgNHA2, in *Anopheles gambiae* larval Malpighian tubules and the functional expression of AgNHA2 in yeast. J. Insect Physiol. 58, 570–579.
Yamahiro, A., Piermarini, P.M., Beyenbach, K.W., 2008. Identification of Na-driven anion exchanger (NDAE) splice variants from Malpighian (renal) tubules of the adult yellow-fever mosquito. FASEB J. 22, 757.22.
Yu, M.J., Beyenbach, K.W., 2002. Leucokinin activates Ca^{2+}-dependent signal pathway in principal cells of *Aedes aegypti* Malpighian tubules. Am. J. Physiol. Renal Physiol. 283, F499–F508.
Yu, M.J., Beyenbach, K.W., 2004. Effects of leucokinin-VIII on *Aedes* Malpighian tubule segments lacking stellate cells. J. Exp. Biol. 207, 519–526.
Zelle, K.M., Lu, B., Pyfrom, S.C., Ben-Shahar, Y., 2013. The genetic architecture of degenerin/epithelial sodium channels in *Drosophila*. G3 (Bethesda) 3, 441–450.

INDEX

Note: Page numbers followed by "*f*" indicate figures, and "*t*" indicate tables.

B

C

D

E

J

K

L

M

N

O

P

Q

R

S

T

U

V

W

X

Y

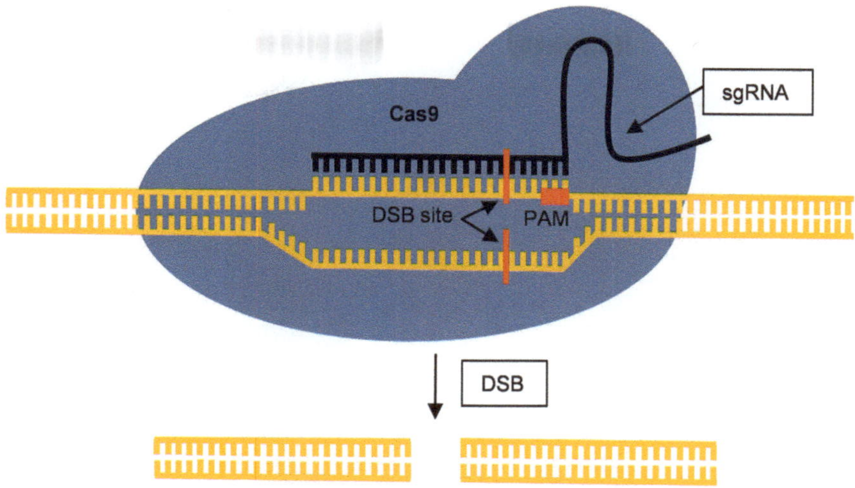

J. Overcash and Z.N. Adelman, Fig. 1 Model of sgRNA-directed Cas9 binding to DNA target. DSB (*red vertical lines*) are 3 bp upstream of the PAM site. The Cas9 protein cuts both strands of the targeted DNA generating a blunt ended double-stranded break, which will be repaired by one of several DSB repair pathways.

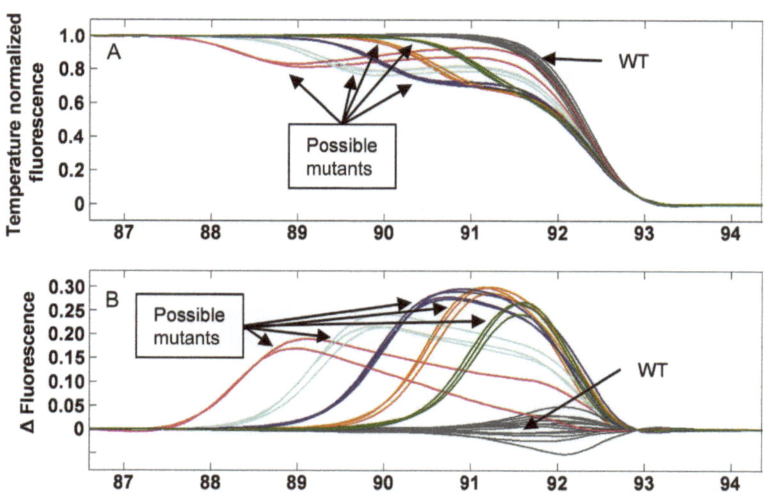

J. Overcash and Z.N. Adelman, Fig. 3 High-resolution melt curve analysis identification of mutations. PCR amplification of mosquito tissue from single individual (typically from a clipping of a leg) followed by melting in the presence of a fluorescent dye. As the dsDNA melts, the fluorescent dye dissociates from the dsDNA generating a melt curve. Melt curves can be viewed directly with different curves segregating from WT (A). Alternatively, WT curves can be subtracted from unknowns to generate difference curves (B). After putative mutant groups are identified, sequencing is required to confirm any insertions or deletions, as differences from WT may also be caused by preexisting polymorphisms.

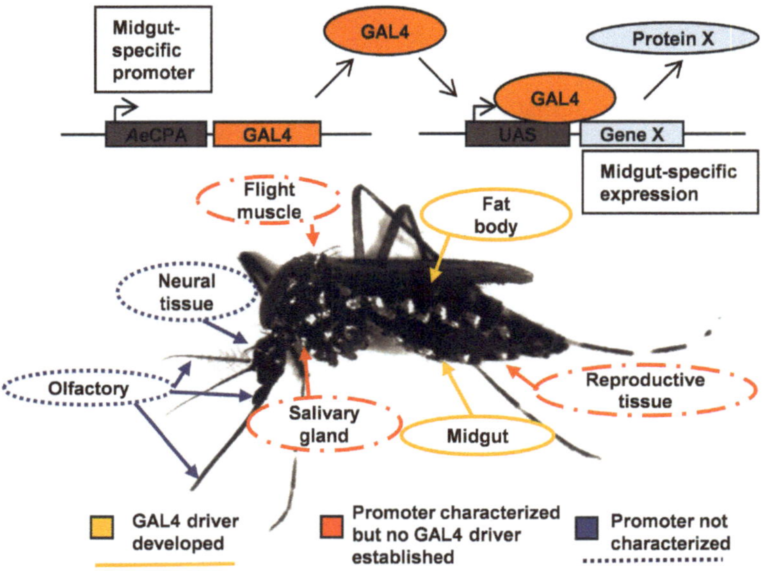

J. Overcash and Z.N. Adelman, Fig. 5 The GAL4–UAS system allows for tissue-specific expression of effector genes in mosquitoes. The upper portion of the figure shows a typical driver-responder pair, in this case a midgut-specific promoter (*Ae*CPA) driving the expression of GAL4; GAL4 in turn interacts with the upstream-associated sequence or UAS to promote the production of a second gene X (gene of interest). The bipartite nature of the GAL4–UAS system allows for the experimenter to interchange either the promoter of GAL4 expression or the gene being promoted by the UAS promoter. This allows for quick and easy crosses to be made to test the impact of a specific effector gene within a targeted tissue. The bottom portion of the figure shows specific tissue types in which the GAL4–UAS system has been successfully utilized in mosquitoes (*yellow*; *solid line*), tissue types in which possible promoters have been identified but the GAL4–UAS system has not been utilized yet (*red*; *dash-dot line*), and tissue types in which no promoters have been identified and the GAL4–UAS system has not been utilized yet (*blue*; *dotted line*).

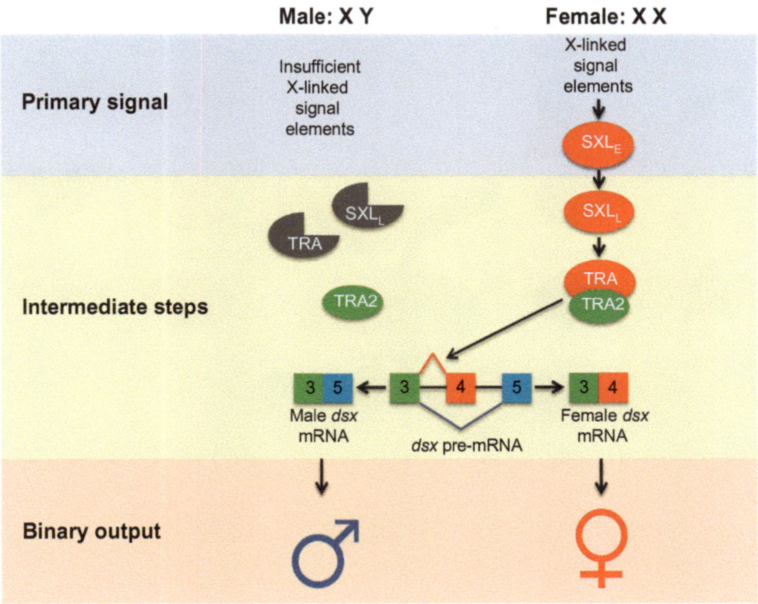

J.K. Biedler and Z. Tu, Fig. 1 The sex determination pathway in *Drosophila melanogaster*. *Ovals* indicate proteins in the pathway where colours *red* and *green* represent female-specific and nonsex-specific proteins, respectively. SXL_E and SXL_L refer to the early and late sex lethal proteins, respectively; TRA and TRA2 refer to the transformer and transformer-2 proteins, respectively. In males, insufficient X-linked signal elements result in the absence of SXL_E expression, which subsequently results in a truncated nonfunctional SXL_L protein (shown in *grey*). The lack of functional SXL_L results in a truncated and nonfunctional TRA (shown in *grey*). The lack of a functional TRA results in the male-specific splicing of *dsx* mRNA, excluding the female-specific exon 4. In females, a functional TRA and hence TRA/TRA2 complex directs female-specific splicing of *dsx* by activating a weak splice acceptor upstream of exon 4. Only select exons are shown that are alternatively spliced by TRA/TRA2 regulation. Exons are colour-coded *red* (female-specific), *blue* (male-specific), and *green* (nonsex-specific). TRA2 is expressed in both males and females. *fru* is also subjected to sex-specific alternative splicing by TRA but is not shown.

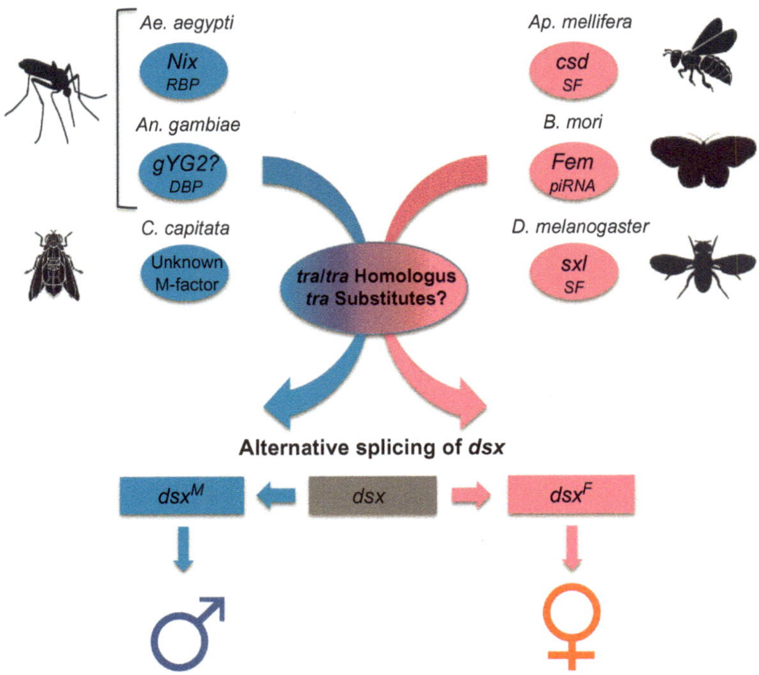

J.K. Biedler and Z. Tu, Fig. 2 Comparison of sex determination pathways of select holometabolous insects. *Blue-* and *pink-coloured circles* represent masculinizing (M-factors) and feminizing primary sex-determining signals, respectively. These primary signals determine the sex-specific spliceform of *transformer* or *transformer* substitutes produced. When these signals are not present (in the other sex), the 'default' splicing of *transformer* (or *transformer* substitutes, if present), and hence 'default' splicing of *dsx* take place, instructing development of the other sex. The type of molecule is indicated below the primary sex-determining signal gene name. In the case of *C. capitata* an M-factor is present but unidentified. In some species with an M-factor such as *Nix* in *Ae. aegypti*, the M-factor may directly modulate *dsx/fru* splicing. RNA-binding protein, *RBP*; DNA-binding protein, *DBP*; splicing factor, *SF*; Piwi-interacting RNA, *piRNA*.

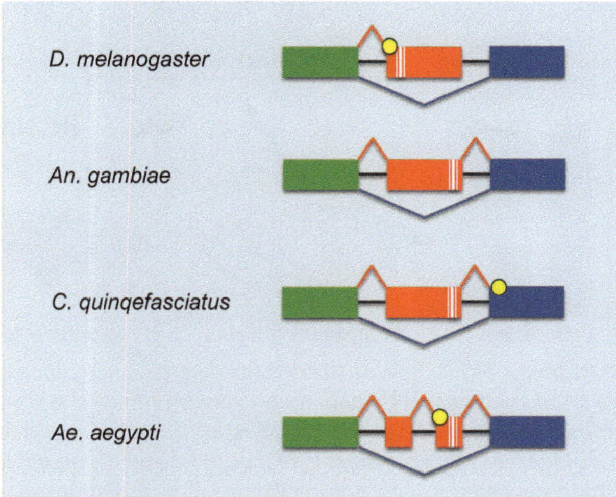

J.K. Biedler and Z. Tu, Fig. 4 Alternative splicing of *doublesex* pre-mRNA in *D. melanogaster* and mosquitoes. Simplified model of *doublesex* exon splicing is shown with *coloured rectangles* indicating homologous exons. *Red* exon in middle is the female-specific exon. In *Ae. aegypti*, the female-specific exon has gained an intron resulting in two exons. *Red* lines on top and *blue* lines on bottom connecting exons represent female-specific and male-specific splicing, respectively. *White* triplicate *vertical lines* indicate TRA/TRA2 binding sites and *yellow circles* indicate predicted weak splice sites (Price et al., 2015). Only select exons are shown that are spliced in a sex-specific manner.

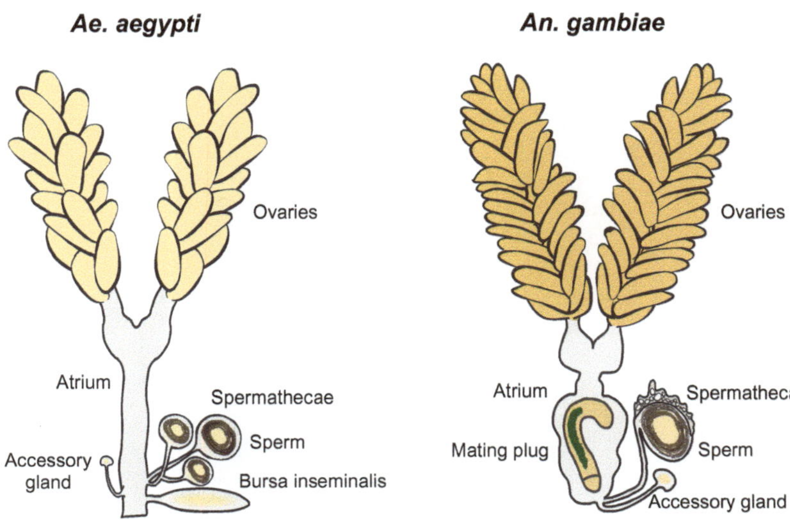

A. South and F. Catteruccia, Fig. 1 The reproductive tracts of female *Ae. aegypti* and *An. gambiae* mosquitoes. *Ae. aegypti* are characterized by the presence of multiple spermathecae, a bursa inseminalis and a more tubular atrium. By comparison, *An. gambiae* females have only a single spermatheca with a rounder atrium, shown here with a male transferred mating plug.

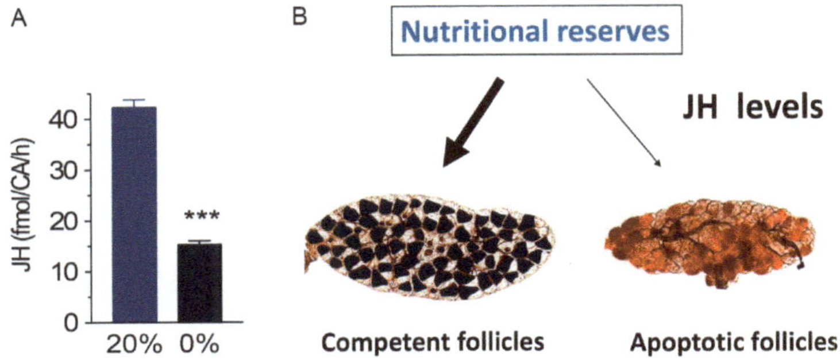

J. Zhu and F.G. Noriega, Fig. 4 Role of JH on controlling resource allocations in female mosquitoes. (A) JH biosynthesis levels depend on nutrient reserves (0% and 20% sugar meals). (B) Nutrients availability determine the levels of JH synthesis and reserve allocations into the ovaries. More nutrients result in higher JH (+) and decreased follicle resorption. Poor nutrition/starvation results in lower JH titres (−) and an increase of follicular resorption by apoptosis.

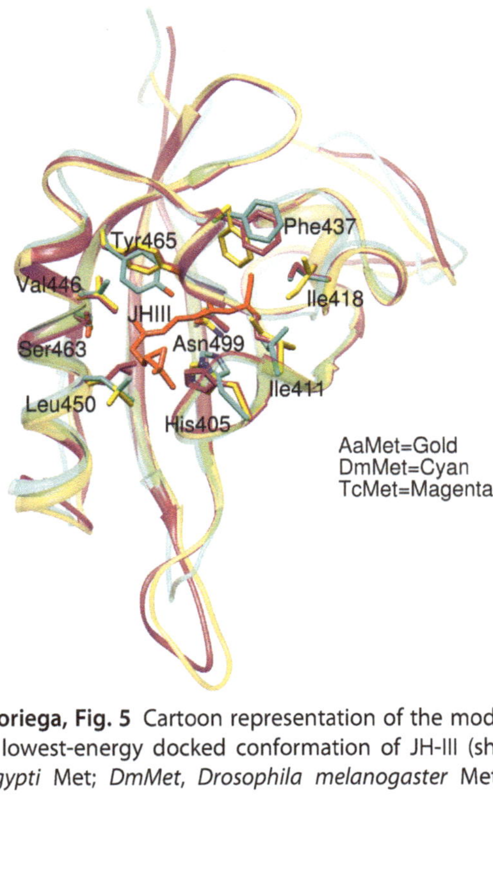

J. Zhu and F.G. Noriega, Fig. 5 Cartoon representation of the modelled AaMet PAS-B domain with the lowest-energy docked conformation of JH-III (shown as *red sticks*). *AaMet, Aedes aegypti* Met; *DmMet, Drosophila melanogaster* Met; *TcMet, Tribolium castaneum* Met.

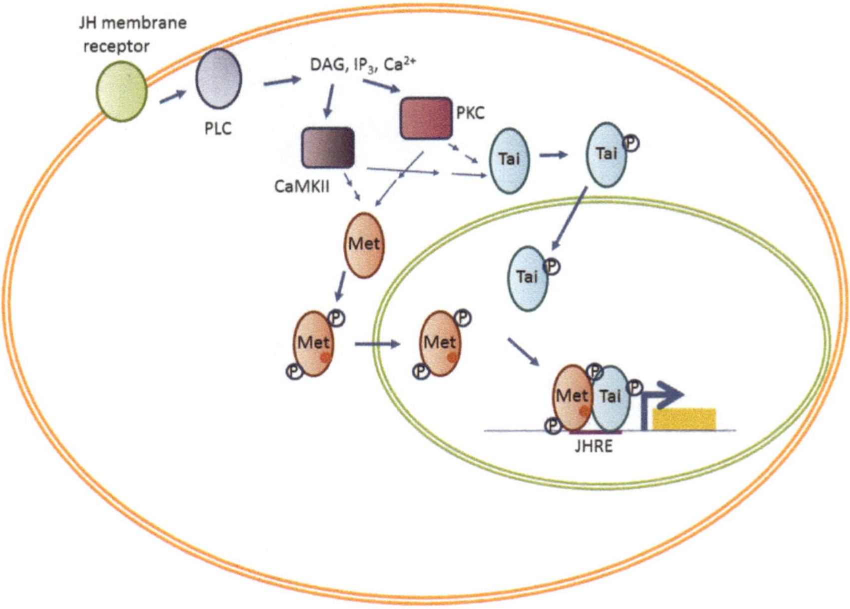

J. Zhu and F.G. Noriega, Fig. 6 Juvenile hormone signalling pathway. In the current model, JH binds to an unidentified JH receptor that is associated with the plasma membrane. As a result, PLC is activated, causing increases in DAG, IP_3 and Ca^{2+}. The second messengers activate CaMKII and PKC, resulting in phosphorylation of Met and Tai which in turn potentiates DNA binding of those two transcription factor to the JH-regulated genes.

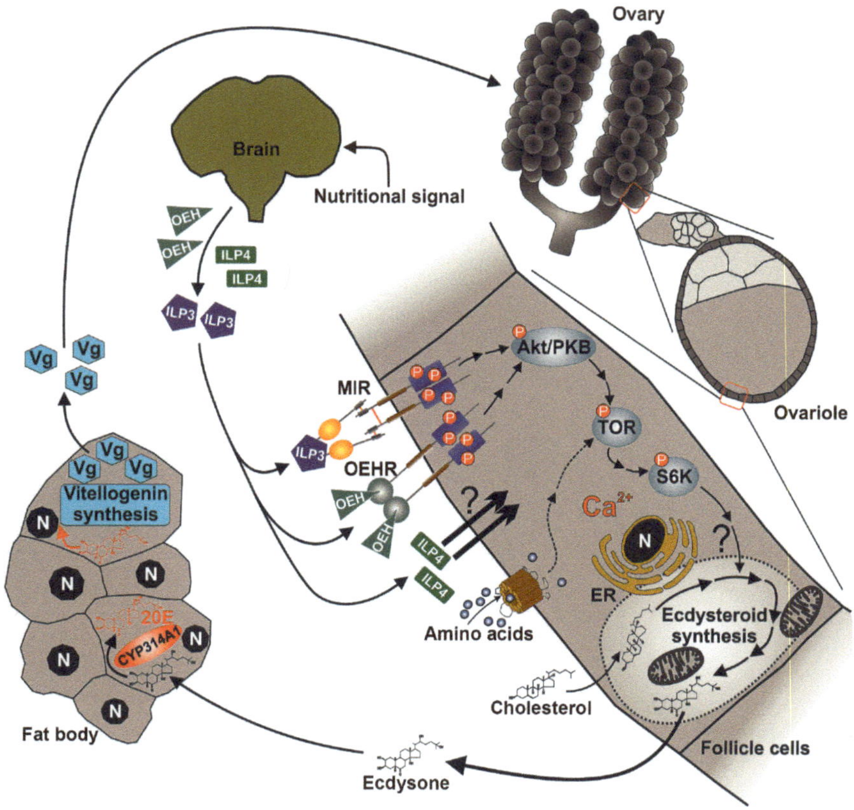

S. Roy *et al.*, Fig. 2 Ecdysteroidogenesis in the mosquito ovary. Nutritional signals from the blood meal triggers release of neuropeptides, ovary ecdysteroidogenic hormone (OEH) and insulin-like peptides 3 and 4 (ILP3 and ILP4, respectively) from the brain of an *Aedes aegypti* female. These neuropeptides bind their corresponding receptors in the membrane of ovarian follicle cells: OEH binds ovary ecdysteroidogenic hormone receptor (OEHR), ILP3 binds mosquito insulin receptor (MIR) and ILP4 binds to an unknown receptor. After binding, OEHR and MIR are phosphorylated and along with amino acid signalling activates insulin and AA/TOR signalling pathways, manifested by phosphorylation of Akt/PKB, TOR and S6 kinases. Activation of these pathways drives the multistep synthesis of ecdysteroids from cholesterol in the endoplasmic reticulum (ER) and mitochondria of the follicle cells. Ca^{2+} flux seems to be an important factor for the ecdysteroidogenesis although its precise role remains unclear. Ecdysone secreted into the haemolymph is converted by the cytochrome P450 enzyme CYP314A1 to the active form of 20-hydroxyecdysone (20E) in the peripheral tissues. 20E triggers *vitellogenin* (Vg) expression in the fat body and, once released into the haemolymph, Vg is taken by the oocytes and packed in yolk granules. Abbreviations: *20E*, 20-hydroxyecdysone; *Akt/PKB*, Akt/protein kinase B; *ILP3*, insulin-like peptide 3; *ILP4*, insulin-like peptide 4; *MIR*, mosquito insulin receptor; *N*, Nucleus; *P*, phosphorylation; *OEH*, ovary ecdysteroidogenic hormone; *S6K*, p70 ribosomal S6 kinase; *TOR*, target of rapamycin; *Vg*, Vitellogenin.

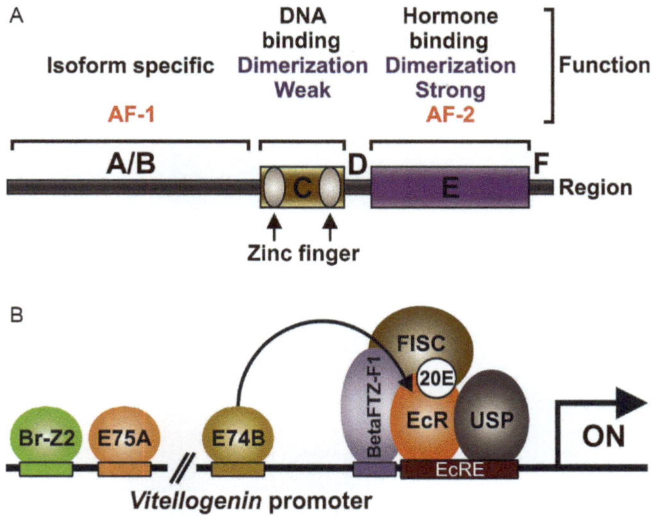

S. Roy *et al.*, Fig. 3 Molecular mechanism of 20E action. (A) Schematic representation of nuclear receptor protein structure. Nuclear receptors share six conserved domains, A–F. The variable, N-terminal isoform-specific (A/B) domain containing the activation function 1 (AF-1) is responsible for relatively weak transactivation. The conserved DNA-binding domain (DBD) (C) with two zinc fingers specifically binds DNA sequences, called hormone response elements, in the gene promoter region and partially participates in receptor dimerization. The hinge-region (D) is variable in sequence and connects the DBD and the following ligand-binding domain (LBD). It also contributes to intracellular trafficking and subcellular distribution of the receptors. The moderately conserved LBD (E) covers a wide range of functions from ligand-binding and ligand-dependent transactivation through its activation function 2 (AF-2), receptor dimerization, to coactivator and corepressor binding. The last, C-terminal domain (F) is variable in sequence and flexible in structure. (B) Transactivation complex on the promoter of the *vitellogenin* (*Vg*) gene. The nuclear receptor, ecdysone receptor (EcR), binds 20E with high affinity, heterodimerizes with another nuclear receptor, ultraspiracle (USP) and the heterodimer binds ecdysone response elements (EcREs) in the *Vg* gene promoter in the fat body of the *Ae. aegypti* female. Nuclear receptor BetaFTZ-F1 binds the histone acetyltransferase p160/SRC (FISC) in a ligand-independent manner and the complex is recruited to the EcR/USP heterodimer in the presence of 20E. BetaFTZ-F1/FISC/EcR/USP complex upregulates *Vg* transcription which is further enhanced by the binding of Br-Z2, E75A and E74B transcription factors to the *Vg* promoter.

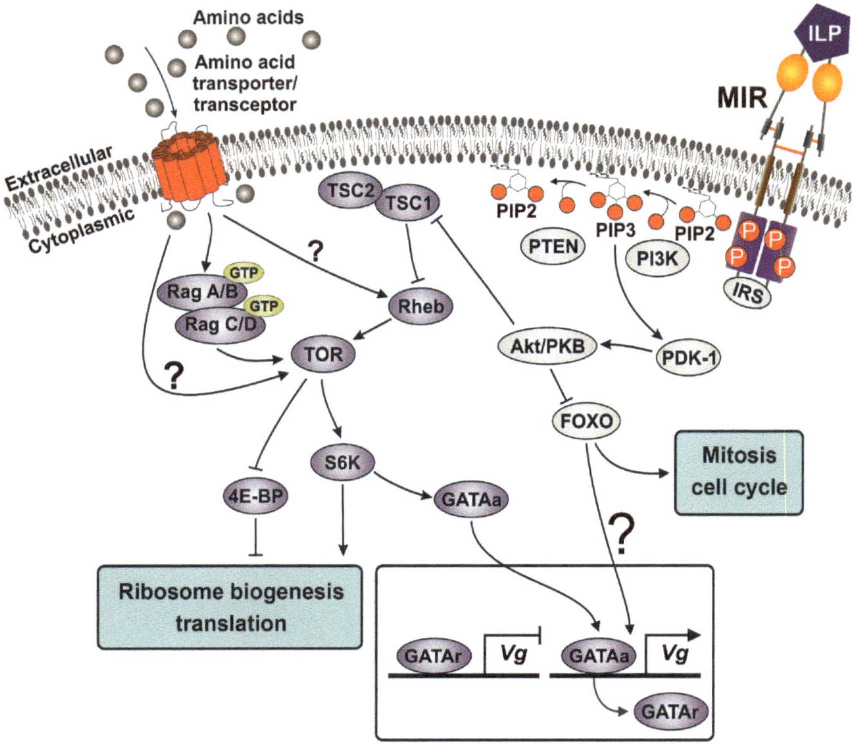

S. Roy *et al.*, Fig. 4 Insulin and AA/TOR signalling pathways. Amino acids released during digestion of a blood meal are sensed and transported into the fat body cells of *Ae. aegypti* where they participate in activation of the target of rapamycin (TOR) kinase via Ras homolog enriched in brain (Rheb). Activated TOR kinase phosphorylates downstream proteins and thereby derepresses ribosome biogenesis and protein translation by inactivation of the translational inhibitor eukaryotic translation initiation factor 4E-binding protein (4E-BP) and by activation of p70 S6 kinase (S6K). Activated S6K elicits massive translation of GATAa transcription factor that replaces repressive GATAr in the *Vg* gene promoter and participates in *Vg* transcription. Insulin-like peptides (ILPs) bound to the mosquito insulin receptor (MIR) localized in the membrane of follicle cells results in its phosphorylation and activation. Activated MIR transduces its signalling through a series of phosphorylation events to activate phosphatidylinositol 3-kinase (PI3K). PI3K phosphorylates phosphatidylinositol-4,5-bisphosphate (PIP2) to generate phosphatidylinositol-3,4,5-trisphosphate (PIP3), which activates Akt/protein kinase B (Akt/PKB). Active Akt/PKB can stimulate TOR signalling pathway through repression of the TOR-negative regulators, tuberous sclerosis complex 1 and 2 (TSC1/2). Forkhead box O (FOXO) transcription factor regulates cell cycle progression and mitosis, and Akt/PKB phosphorylation keeps FOXO from its translocation to the nucleus and likely repression of *Vg* transcription. However, the role of FOXO in Vg expression needs verification. Phosphatase and tensin homolog (PTEN) negatively regulates insulin signalling by dephosphorylation of the PIP3 to PIP2. Other abbreviations: *GTP*, guanosine-5′-triphosphate; *IRS*, insulin receptor substrate; *P*, phosphate/phosphorylation; *PDK*-1, 3-phosphoinositide-dependent protein kinase-1; *Rag A/B* and *Rag C/D*, Ras-related small GTP-binding protein GTPases; *Vg*, Vitellogenin.

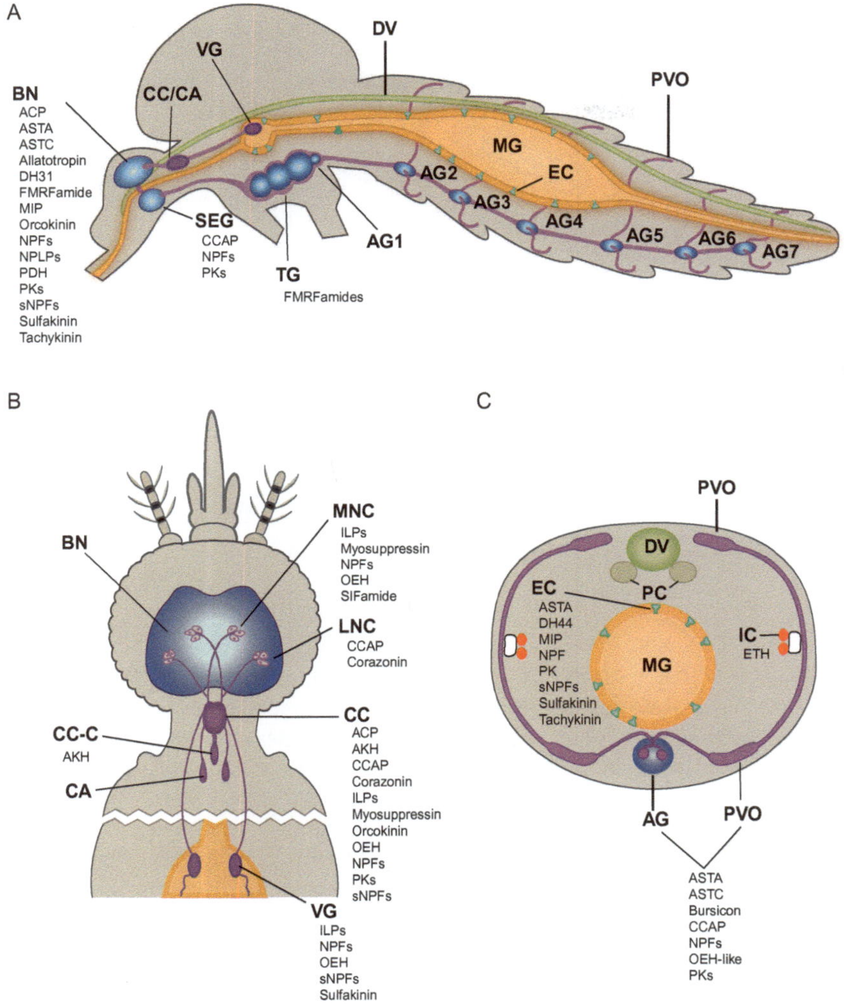

M.R. Strand *et al.*, Fig. 1 Schematic illustrating the sources of different peptide hormones in adult female mosquitoes as determined by immunocytochemistry or peptidomics. Sagittal view of the whole body (A), dorsal view of the head and thoracic regions (B), and a cross section through the third abdominal segment (C). Select regions of the nervous system, organs and cells are labelled in bold. Peptide hormones in Table 1 are listed alphabetically below detected sites. Abbreviations: *AG1–7*, abdominal ganglia 1 through 7; *BN*, brain; *CA*, corpora allata; *CC*, corpus cardiacum; *CC-C*, corpus cardiacum intrinsic cells; *DV*, dorsal vessel; *EC*, gut endocrine cells; *IC*, Inka cells; *LNC*, lateral neurosecretory cells; *MG*, midgut; *MNC*, medial neurosecretory cells; *PC*, pericardia; *PVO*, perivisceral organs; *SEG*, subesophegial ganglion; *VG*, ventricular ganglion. Peptide hormone abbreviations are listed in Table 1. Most of these data derive from studies of adults with the exception being for ETH, which is shown in (C) although data derive from studies of larvae (see text).

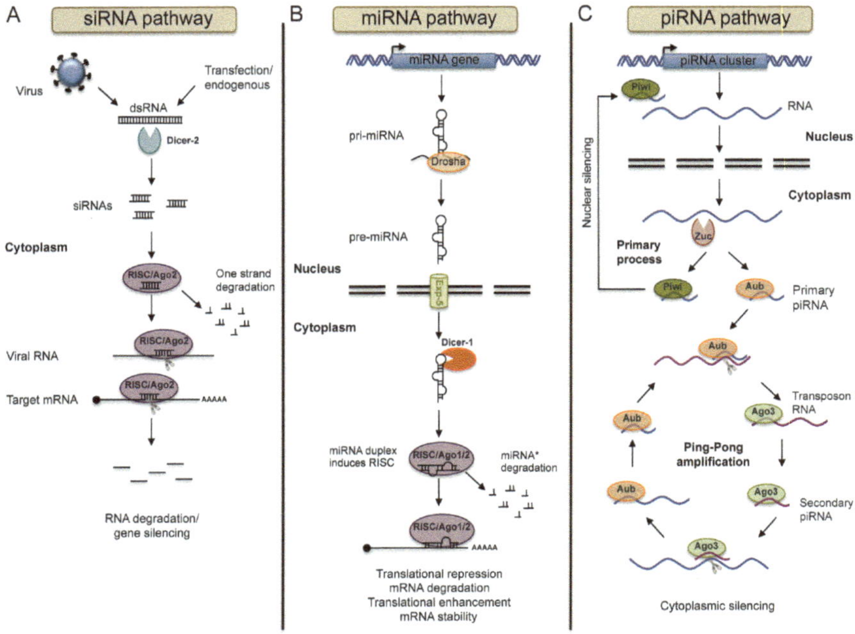

M. Hussain *et al.*, Fig. 1 RNAi pathways in insects; mostly based on *D. melanogaster* as a model. (A) In the *siRNA pathway*, the exogenous (with viral origin or by artificial transfection) or endogenous dsRNA is sensed by Dicer-2 and cleaved into 20–21 nt siRNAs that trigger the formation of the RISC complex in which Ago2 is involved. One of the strands is degraded and the RISC-guide strand binds to target RNA (viral or endogenous mRNA targeted for silencing) by complete complementarity leading to its cleavage. (B) In the *miRNA pathway*, a miRNA gene is transcribed into primary miRNA (pri-miRNA), which contains one or more stem-loop structures. The stem-loop is cleaved by the ribonuclease Drosha in the nucleus to form the precursor miRNA (pre-miRNA), which is transported into the cytoplasm by Exportin-5 (Exp-5). Subsequently, the hairpin head is removed by Dicer-1 and the resulting miRNA duplex is loaded into Ago1 or Ago2 as part of the RISC complex, in which one of the strands (miRNA*) is degraded. The mature miRNA ($\sim$22 nt) guides the RISC complex to mRNA targets by sequence complementarity (usually with mismatches). (C) In the *piRNA pathway*, piRNA clusters that mainly consist of repetitive and transposable DNA elements are transcribed and transported into the cytoplasm, where they appear to be cleaved by Zucchini (Zuc) into 24–27 nt piRNAs. These primary piRNAs are loaded onto Piwi and Aub proteins. Those associated with Piwi are translocated into the nucleus where they target the primary piRNA transcripts. On the other hand, the primary piRNAs associated with Aub target transposon transcripts in the cytoplasm. The cleavage products are transferred onto Ago3 and further processed into mature secondary piRNAs that target more transposon transcripts with the cleavage products loaded into Aub, trimmed into mature secondary piRNAs and enter in the cycle of Ping-Pong amplification. Therefore, the secondary piRNAs are involved in cytoplasmic silencing of transposon transcripts.

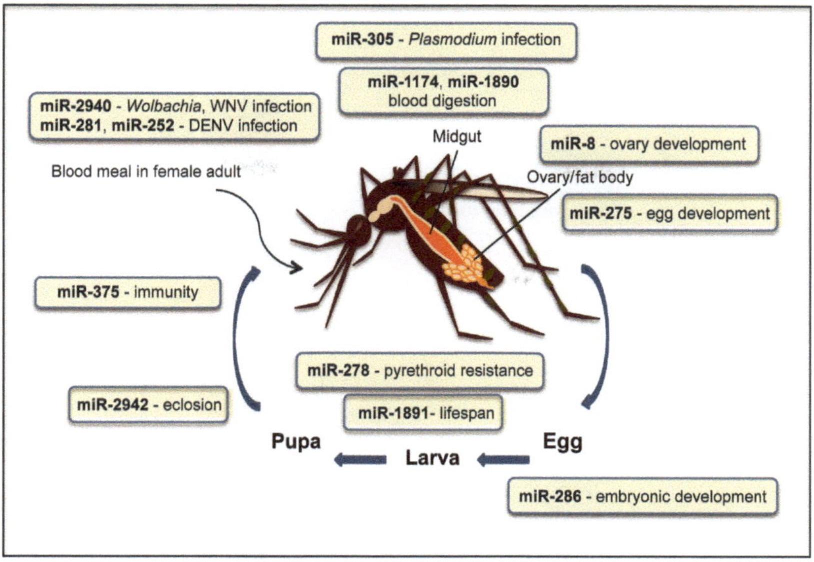

M. Hussain *et al.*, Fig. 2 A summary of miRNAs involved in various processes in mosquito biology. Refer to text for details.

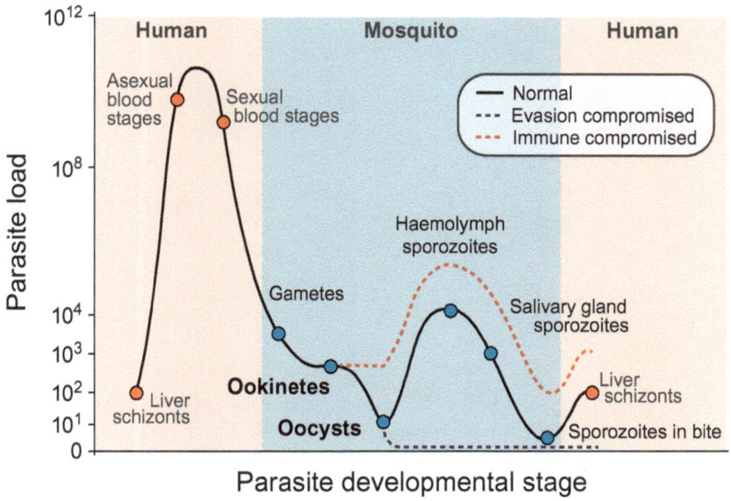

M. Povelones *et al.*, Fig. 1 *Plasmodium* population bottleneck in the *A. gambiae* midgut. Malaria parasites are confronted by robust reactions of the mosquito immune system in the mosquito midgut, especially during the ookinete-to-oocyst developmental transition. In laboratory conditions, of approximately 1000 ookinetes, the average number that survive through this stage and develop to oocysts is about 20 (*black solid line*). The killed ookinetes are eliminated by reactions of the mosquito complement pathway. Inactivation of this pathway increases the oocyst load between 5- and 50-fold with respective increases of the number of sporozoites (*red dashed line*). Conditions that disrupt the ability of parasites to evade the complement pathway completely abolishes transmission (*blue dashed line*). The numbers indicated are approximate values based on Vogel (2010).

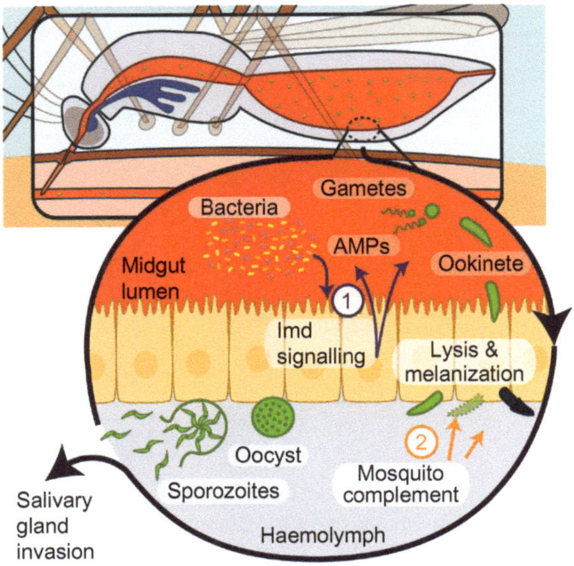

M. Povelones *et al.*, Fig. 2 *Plasmodium* life cycle in the *A. gambiae*. Schematic diagram of mosquito stages of the *Plasmodium* life cycle highlighting the site of parasite killing for the Imd pathway (1; *blue lines*) and the mosquito complement system as ookinetes enter the haemolymph (2; *orange lines*).

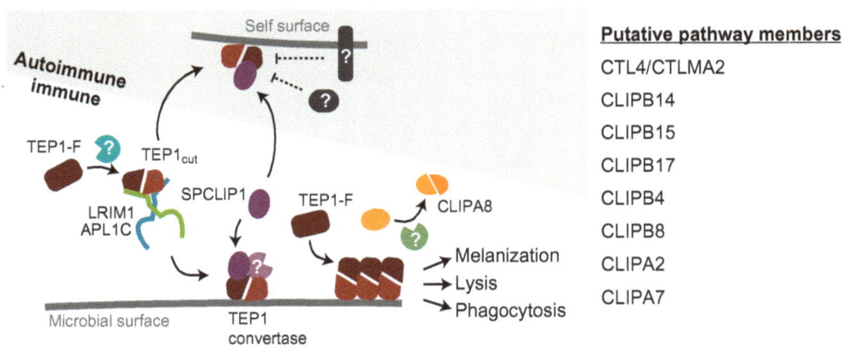

M. Povelones *et al.*, Fig. 3 Schematic model of *A. gambiae* complement pathway. *Black arrows* indicate known interactions or proteolytic reactions. Hypothesized components or interactions are indicated with a *question mark* or *dashed line*, respectively. Activating events are indicated with *arrows*. Inhibitory events are shown with a *flat line*. Mosquito complement components that are not yet placed within this model but whose function is known to depend on TEP1 (CTL4/CTLM2, CLIPA2), or highly expected to depend on TEP1 based on their RNAi phenotypes (CLIPB14, CLIPB15, CLIPB17, CLIPB4, CLIPB8 and CLIPA7) are listed in the space on the *right*.

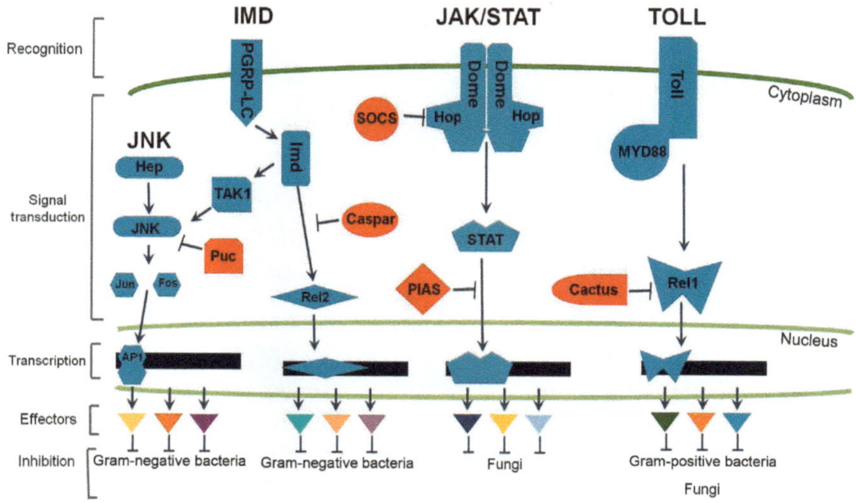

S. van Tol and G. Dimopoulos, Fig. 1 Role of endogenous microorganisms in immune stimulation. The IMD, JAK/STAT, and Toll immune signaling pathways are involved in controlling microorganisms naturally found in the mosquito midgut. Upon recognition of microbial-associated molecular patterns by each pathways' respective receptors, a signal cascade is induced resulting in the import of transcription factors into the nucleus. Once translocated into the nucleus, the transcription factors control the expression of a variety of genes including immune effectors. These immune effectors, and other genes, are then involved in regulating the microbial proliferation. Although a particular category of microbe can stimulate multiple immune responses, one pathway is stimulated predominantly in association with a particular category of microorganisms (Gram-negative bacteria stimulate both the JNK and Imd branches of the IMD pathway; fungi stimulate both the JAK/STAT and Toll pathways; and Gram-positive bacteria stimulate the Toll pathway). Receptors, signal transduction proteins, and transcription factors of immune pathways shown in *blue*; negative regulators of pathways shown in *red*; DNA represented by the *black rectangles*.

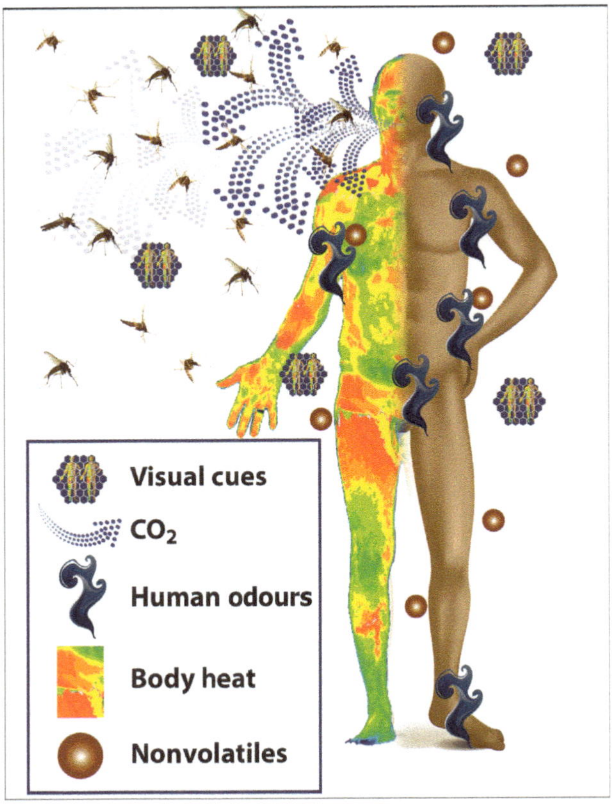

C. Montell and L.J. Zwiebel, Fig. 2 Graphic illustrating the sensory cues used by mosquito vectors to target human hosts depicting visual cues, CO_2, odours, body heat and nonvolatiles.

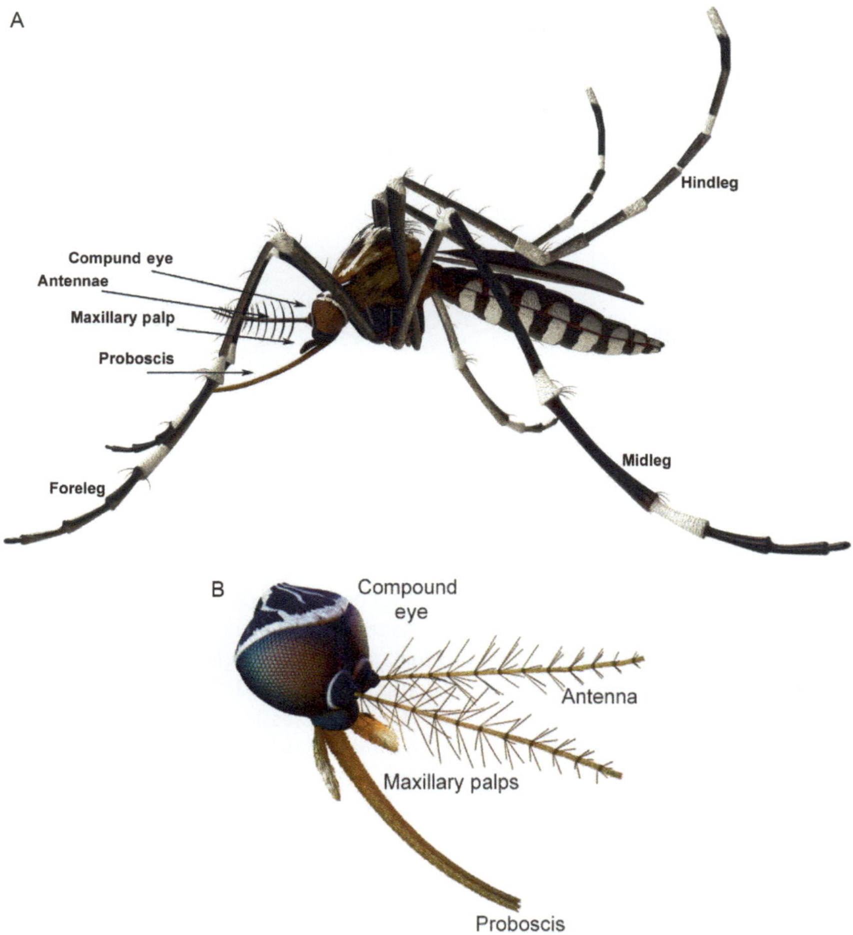

C. Montell and L.J. Zwiebel, Fig. 3 Main sensory organs in an adult *Aedes aegypti* that are used to detect cues emanating from human hosts.

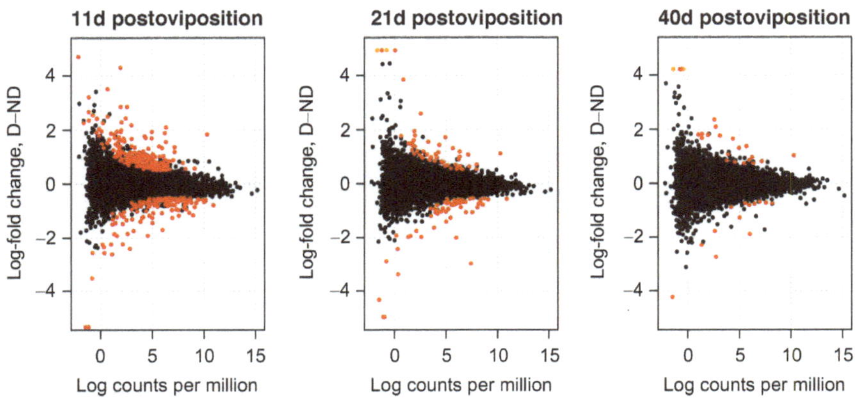

D.L. Denlinger and P.A. Armbruster, Fig. 2 Profiles of gene changes throughout diapause in embryos of *Ae. albopictus*, from Poelchau et al. (2013c). Differential gene expression at 11 (early diapause), 21 (middle diapause) and 40 days postoviposition (late diapause). Each point represents an individual gene plotted as log-fold change under diapause (D) relative to nondiapause (ND) conditions vs abundance after TMM normalization. Positive values represent genes with higher expression in diapause conditions and negative values indicate genes with higher expression in nondiapause conditions. *Red* indicates significant differential expression (FDR-corrected $p < 0.05$, absolute log twofold change >0.5), *orange* represents genes unique to either diapause or nondiapause.

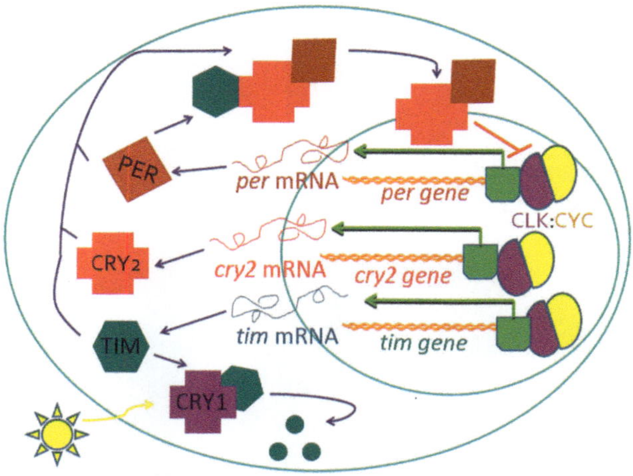

D.L. Denlinger and P.A. Armbruster, Fig. 3 Interactions of circadian clock genes in mosquitoes. Within the nucleus the proteins CLOCK (CLK) and CYCLE (CYC) form a heterodimer that acts as a transcriptional activator by binding to the E-box promoter region of the *period* (*per*), *timeless* (*tim*) and *cryptochrome2* (*cry2*) genes. The mRNAs encoded by these genes are translated in the cytoplasm where their protein products dimerize. The heterotrimer of PER, TIM and CRY2 then translocate into the nucleus where both PER and CRY2 likely inhibit the action of CLK:CYC, thereby suppressing the transcription of *per*, *tim* and *cry2*. The CRYPTOCHROME1 (CRY1) protein degrades TIM and itself in the presence of light. This results in increasing levels of *per*, *tim* and *cry2* mRNA throughout the day when CLK:CYC activity is uninhibited and decreasing levels of these transcripts during night. This scheme differs from the better-studied clock system in *Drosophila*, an insect that lacks involvement of CRY2. In mosquitoes, PER appears to assist CRY2 in nuclear translocation, whereas TIM helps to stabilize PER and CRY2. *Adapted from Meuti, M.E., Denlinger, D.L., 2013. Evolutionary links between circadian clocks and photoperiodic diapause in insects. Int. Comp. Biol. 53, 131–143.*

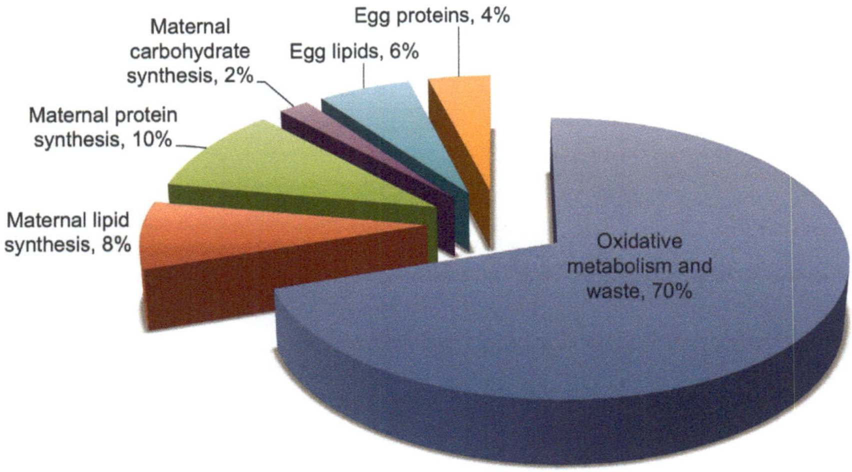

N. Petchampai and P.Y. Scaraffia, Fig. 1 Schematic representation of the fate of blood meal protein amino acids at the end of a gonotrophic cycle in *A. aegypti* mosquitoes. *Data obtained from Zhou, G., Flowers, M., Friedrich, K., Horton, J., Pennington, J., Wells, M.A., 2004. Metabolic fate of [^{14}C]-labeled meal protein amino acids in* Aedes aegypti *mosquitoes. J. Insect Physiol. 50, 337–349.*

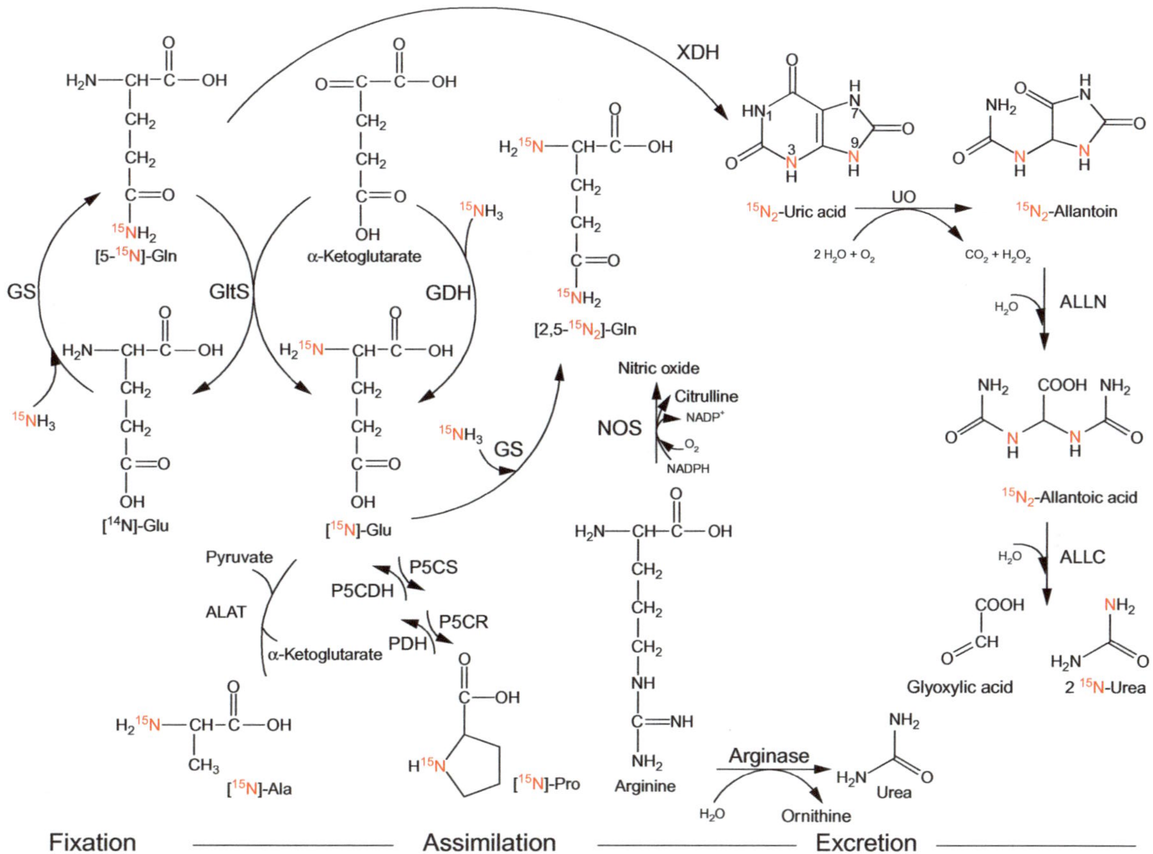

N. Petchampai and P.Y. Scaraffia, Fig. 5 See legend on next page.

N. Petchampai and P.Y. Scaraffia, Fig. 5 Multiple metabolic pathways involved in *A. aegypti* ammonia metabolism. Pathways were identified by using isotopically labelled compounds, mass spectrometry, specific inhibitors and RNAi. Abbreviations used in this figure: GS, glutamine synthetase; GltS, glutamate synthase; GDH, glutamate dehydrogenase; ALAT, alanine aminotransferase; P5CS, pyrroline-5-carboxylate synthase; P5CR, pyrroline-5-carboxylate reductase; P5CDH, pyrroline-5-carboxylate dehydrogenase; PDH, proline dehydrogenase; XDH, xanthine dehydrogenase; NOS, nitric oxide synthase; UO, urate oxidase; ALLN, allantoinase; ALLC, allantoicase. *Reproduced with permission from Scaraffia, P.Y., 2016. Disruption of mosquito blood meal protein metabolism. In: Adelman, Z.N. (Ed.), Genetic Control of Malaria and Dengue. Elsevier Academic Press, San Diego, pp. 253–275.*

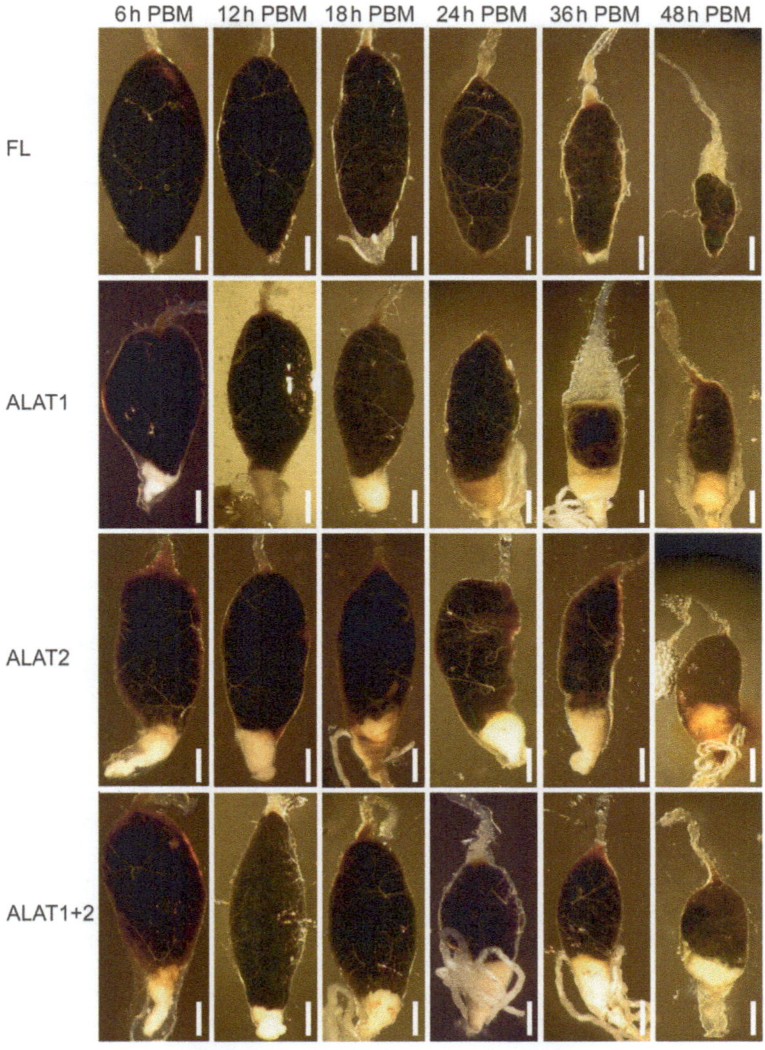

N. Petchampai and P.Y. Scaraffia, Fig. 7 Uric acid accumulation in the midgut of mosquitoes with alanine aminotransferase (ALAT) deficiency. Midguts were dissected from dsRNA firefly luciferase (FL) or ALAT injected mosquitoes at different time post blood meal (PBM). The posterior end of the midgut is down in this orientation. Scale bars, 500 μm. *Reproduced with permission from Mazzalupo, S., Isoe, J., Belloni, V., Scaraffia, P.Y., 2016. Effective disposal of nitrogen waste in blood-fed* Aedes aegypti *mosquitoes requires alanine aminotransferase. FASEB J. 30, 111–120.*

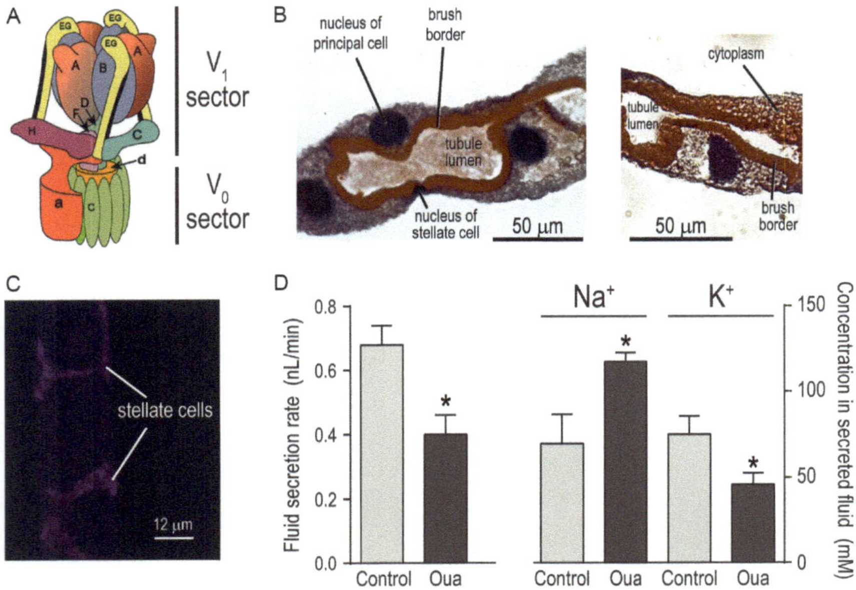

P.M. Piermarini, Fig. 5 ATPases in mosquito Malpighian tubules. (A) Schematic of the structure of the V-ATPase holoenzyme consisting of the membrane-bound V_0 sector and cytosolic V_1 sector. *Lowercase letters* indicate subunits of the V_0 sector; *uppercase letters* indicate subunits of the V_1 sector. The activity of the holoenzyme can be regulated via disassociation of the V_1 sector from V_0 sector. (B) Immunolocalization of subunit B of the V_1 sector in sections of Malpighian tubules isolated from adult female *Ae. aegypti*. Immunolabeling (*red stain*) primarily occurs in the brush border of principal cells but is also found in the cytoplasm of principal cells (see Section 2.2.1 for details). (C) Immunolocalization of the α-subunit of the NKA in Malpighian tubules isolated from larval *An. gambiae*. Immunolabeling (*blue*) occurs in stellate cells. (D) Effects of peritubular ouabain (Oua), an inhibitor of the NKA, on fluid secretion rates (*left*) and the concentrations of Na^+ and K^+ in the secreted fluid (*right*) in isolated Malpighian tubules of adult female *Ae. aegypti*. Values are means ± SEM ($N = 8$). *Asterisks* indicate significant difference ($P < 0.05$) from control tubules. *Panel (A) modified from Song, C.F., Papachristos, K., Rawson, S., Huss, M., Wieczorek, H., Paci, E., Trinick, J., Harrison, M.A., Muench, S.P., 2013. Flexibility within the rotor and stators of the vacuolar H^+-ATPase. PLoS One 8, e82207, with permission. Panel (B) modified from Piermarini, P.M., Weihrauch, D., Meyer, H., Huss, M., Beyenbach, K.W., 2009. NHE8 is an intracellular cation/H^+ exchanger in renal tubules of the yellow fever mosquito* Aedes aegypti. *Am. J. Physiol. Renal Physiol. 296, F730–F750, with permission. Panel (C) modified from Xiang, M.A., Linser, P.J., Price, D.A., Harvey, W.R., 2012. Localization of two Na^+- or K^+-H^+ antiporters, AgNHA1 and AgNHA2, in* Anopheles gambiae *larval Malpighian tubules and the functional expression of AgNHA2 in yeast. J. Insect Physiol. 58, 570–579, with permission. Panel (D) data from Hine, R.M., Rouhier, M.F., Park, S.T., Qi, Z., Piermarini, P.M., Beyenbach, K.W., 2014. The excretion of NaCl and KCl loads in mosquitoes: 1. Control data. Am. J. Physiol. Regul. Integr. Comp. Physiol. 307, R837–R849.*

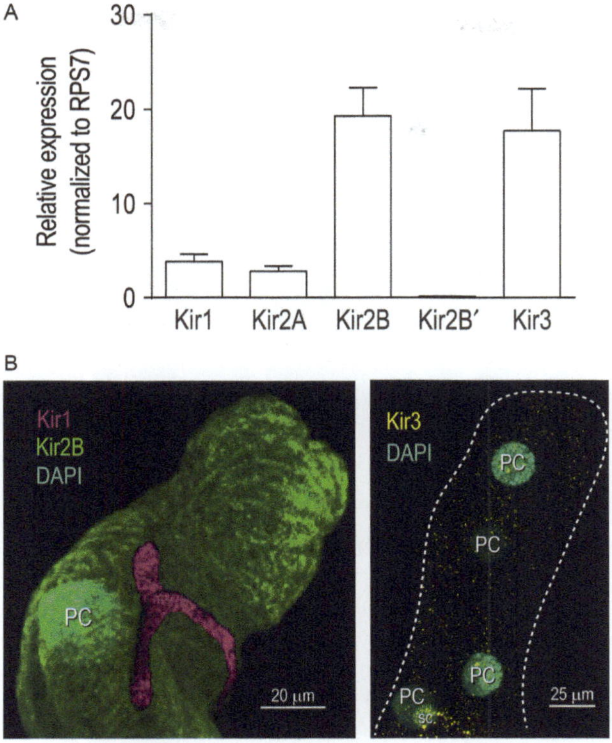

P.M. Piermarini, Fig. 6 Molecular and immunochemical expression of Kir subunits in mosquito Malpighian tubules. (A) Relative expression levels of Kir subunit mRNAs as determined by quantitative RT-PCR (qPCR) in isolated Malpighian tubules of adult female *Ae. aegypti*. Ribosomal protein S7 (RPS7) was used as a reference gene. The expression levels of *Kir2B* and *Kir3* are ~five times greater than those of *Kir1* and *Kir2A*. Values are means ± SEM ($N=3$). Data are previously unpublished. (B) Immunolocalization of Kir1 (*magenta*), Kir2B (*green*), and Kir3 (*yellow*) subunits in whole mounts of isolated Malpighian tubules from adult female *Ae. aegypti*. Nuclei are labeled with DAPI (*cyan*). Note the segregated immunolabeling of Kir1 and Kir2B to the basolateral surface of principal and stellate cells, respectively (*left*), and the punctate intracellular localization of Kir3 throughout the epithelium (*right*). 'PC' and 'sc' indicate nuclei of principal and stellate cells, respectively. The *dashed line* (*right*) outlines the boundary of the tubule epithelium. *Modified from Piermarini, P.M., Dunemann, S.M., Rouhier, M.F., Calkins, T.L., Raphemot, R., Denton, J.S., Hine, R.M., Beyenbach, K.W., 2015. Localization and role of inward rectifier K channels in Malpighian tubules of the yellow fever mosquito* Aedes aegypti. *Insect Biochem. Mol. Biol. 67, 59–73, with permission.*

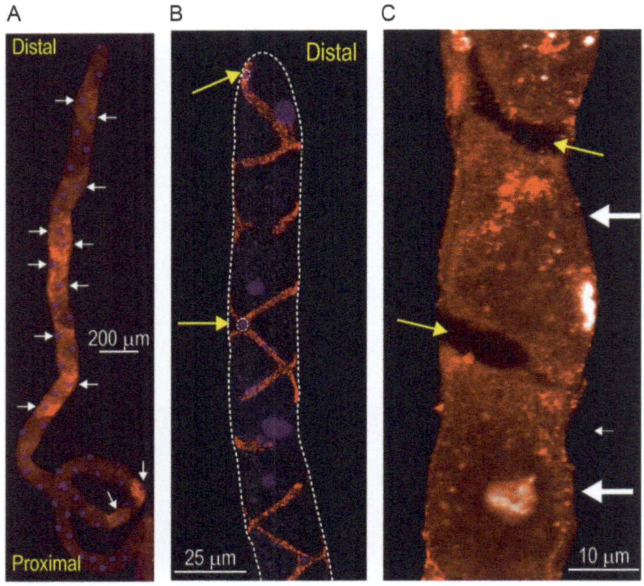

P.M. Piermarini, Fig. 9 Localization of neuropeptide receptors in mosquito Malpighian tubules. (A) DH_{31}-immunoreactivity (*red*) along the axial length of a Malpighian tubule isolated from an adult female *Ae. aegypti*. *Arrows* indicate representative principal cells expressing the receptor. Note that not all principal cells of the distal segment are labeled. Nuclei are labeled with DAPI (*blue*). (B) Binding of fluorescently labeled kinin peptides (*red*) to the Malpighian tubule of an adult female *An. gambiae*. *Yellow arrows* indicate representative stellate cells with bound peptide, indicating the presence of the kinin receptor. Principal cells do not bind detectable amounts of the peptide. Nuclei are labeled with DAPI (*blue*). The *dashed line* outlines the boundary of the tubule epithelium. (C) Binding of fluorescently labeled DH_{44} peptide (*red*) to the Malpighian tubule of an adult female *An. gambiae*. *White arrows* indicate representative principal cells with bound peptide, indicating the presence of the DH_{44} receptor. *Yellow arrows* indicate stellate cells that do not bind detectable amounts of peptide. Nuclei are labeled with DAPI (*white*). The author is grateful to Dr Kenneth Halberg (University of Glasgow) for providing the high resolution images used in panels (B) and (C). *Panel (A) modified from Kwon, H., Lu, H.L., Longnecker, M.T., Pietrantonio, P.V., 2012. Role in diuresis of a calcitonin receptor (GPRCAL1) expressed in a distal-proximal gradient in renal organs of the mosquito* Aedes aegypti *(L.). PLoS One 7, e50374, with permission. Panel (B) modified from Halberg, K.A., Terhzaz, S., Cabrero, P., Davies, S.A., Dow, J.A., 2015. Tracing the evolutionary origins of insect renal function. Nat. Commun. 6, 6800, with permission. Panel (C) modified from Overend, G., Cabrero, P., Halberg, K.A., Ranford-Cartwright, L.C., Woods, D.J., Davies, S.A., Dow, J.A.T., 2015. A comprehensive transcriptomic view of renal function in the malaria vector,* Anopheles gambiae. *Insect Biochem. Mol. Biol. 67, 47–58, with permission.*

Kolibri 360 GmbH
Ferdinand-Jühlke-Straße 7
99095 Erfurt, Deutschland
produktsicherheit@kolibri360.de